P9-EMF-720

solutions@syngress.com

With more than 1,500,000 copies of our MCSE, MCSD, CompTIA, and Cisco study guides in print, we continue to look for ways we can better serve the information needs of our readers. One way we do that is by listening.

Readers like yourself have been telling us they want an Internet-based service that would extend and enhance the value of our books. Based on reader feedback and our own strategic plan, we have created a Web site that we hope will exceed your expectations.

Solutions@syngress.com is an interactive treasure trove of useful information focusing on our book topics and related technologies. The site offers the following features:

- One-year warranty against content obsolescence due to vendor product upgrades. You can access online updates for any affected chapters.
- "Ask the Author" customer query forms that enable you to post questions to our authors and editors.
- Exclusive monthly mailings in which our experts provide answers to reader queries and clear explanations of complex material.
- Regularly updated links to sites specially selected by our editors for readers desiring additional reliable information on key topics.

Best of all, the book you're now holding is your key to this amazing site. Just go to **www.syngress.com/solutions**, and keep this book handy when you register to verify your purchase.

Thank you for giving us the opportunity to serve your needs. And be sure to let us know if there's anything else we can do to help you get the maximum value from your investment. We're listening.

www.syngress.com/solutions

SYNGRESS®

LEGO® Software Power Tools

Kevin Clague **Miguel Agullo** **Lars C. Hassing**

Technical Reviewer

Syngress Publishing, Inc., the author(s), and any person or firm involved in the writing, editing, or production (collectively "Makers") of this book ("the Work") do not guarantee or warrant the results to be obtained from the Work.

There is no guarantee of any kind, expressed or implied, regarding the Work or its contents. The Work is sold AS IS and WITHOUT WARRANTY. You may have other legal rights, which vary from state to state.

In no event will Makers be liable to you for damages, including any loss of profits, lost savings, or other incidental or consequential damages arising out from the Work or its contents. Because some states do not allow the exclusion or limitation of liability for consequential or incidental damages, the above limitation may not apply to you.

You should always use reasonable care, including backup and other appropriate precautions, when working with computers, networks, data, and files.

Syngress Media®, Syngress®, "Career Advancement Through Skill Enhancement®," and "Ask the Author UPDATE®," are registered trademarks of Syngress Publishing, Inc. "Mission Critical™," "Hack Proofing®," and "The Only Way to Stop a Hacker is to Think Like One™" are trademarks of Syngress Publishing, Inc. Brands and product names mentioned in this book are trademarks or service marks of their respective companies.

KEY	SERIAL NUMBER
001	HU2JMG8D4E
002	J6RDUM397T
003	QKV87BN5GS
004	XD3FYX6B7N
005	ZAQBVY4532
006	74KP7V76S6
007	PQ2AK89FE4
008	9BKMAS44FU
009	SC4ETMW6FH
010	P95BVX7F7Z

PUBLISHED BY
Syngress Publishing, Inc.
800 Hingham Street
Rockland, MA 02370

LEGO® Software Power Tools, Including LDraw, MLCad, and LPub

Copyright © 2002 by Syngress Publishing, Inc. All rights reserved. Printed in the United States of America. Except as permitted under the Copyright Act of 1976, no part of this publication may be reproduced or distributed in any form or by any means, or stored in a database or retrieval system, without the prior written permission of the publisher, with the exception that the program listings may be entered, stored, and executed in a computer system, but they may not be reproduced for publication.

Printed in the United States of America

1 2 3 4 5 6 7 8 9 0

ISBN: 1-931836-76-0

Technical Reviewer: Lars C. Hassing
Acquisitions Editor: Jonathan Babcock
Copy Editor: Darlene Bordwell
CD Production: Michael Donovan

Cover Designer: Michael Kavish
Page Layout and Art by: Shannon Tozier
Indexer: Rich Carlson

Distributed by Publishers Group West in the United States and Jaguar Book Group in Canada.

Acknowledgments

We would like to acknowledge the following people for their kindness and support in making this book possible.

A special thanks to Matt Gerber at Brickswest for his help and support for our books.

Karen Cross, Lance Tilford, Meaghan Cunningham, Kim Wylie, Harry Kirchner, Kevin Votel, Kent Anderson, Frida Yara, Jon Mayes, John Mesjak, Peg O'Donnell, Sandra Patterson, Betty Redmond, Roy Remer, Ron Shapiro, Patricia Kelly, Andrea Tetrick, Jennifer Pascal, Doug Reil, David Dahl, Janis Carpenter, and Susan Fryer of Publishers Group West for sharing their incredible marketing experience and expertise.

Duncan Enright, AnnHelen Lindeholm, David Burton, Febea Marinetti, and Rosie Moss of Elsevier Science for making certain that our vision remains worldwide in scope.

David Buckland, Wendi Wong, Daniel Loh, Marie Chieng, Lucy Chong, Leslie Lim, Audrey Gan, and Joseph Chan of Transquest Publishers for the enthusiasm with which they receive our books.

Kwon Sung June at Acorn Publishing for his support.

Jackie Gross, Gayle Voycey, Alexia Penny, Anik Robitaille, Craig Siddall, Darlene Morrow, Iolanda Miller, Jane Mackay, and Marie Skelly at Jackie Gross & Associates for all their help and enthusiasm representing our product in Canada.

Lois Fraser, Connie McMenemy, Shannon Russell, and the rest of the great folks at Jaguar Book Group for their help with distribution of Syngress books in Canada.

David Scott, Annette Scott, Geoff Ebbs, Hedley Partis, Bec Lowe, and Tricia Herbert of Woodslane for distributing our books throughout Australia, New Zealand, Papua New Guinea, Fiji Tonga, Solomon Islands, and the Cook Islands.

Winston Lim of Global Publishing for his help and support with distribution of Syngress books in the Philippines.

Acknowledgments

Contributors

Kevin Clague is a Senior Staff Engineer at Sun Microsystems, where he does verification work on their Ulta-SPARC V RISC processor. He also worked for Amdahl Corporation for 18 years as a Diagnostic Engineer. Kevin played with LEGO as a child, and got back into LEGO as an adult when his wife, Jan, got him the LEGO MINDSTORMS Dark Side Developer Kit for Christmas three years ago. Kevin soon got himself a LEGO MINDSTORMS Robotics Invention System 1.5 set, and has been having fun inventing LEGO creations ever since.

In 2001 Kevin got involved with authoring LEGO instruction books for Syngress Publishing, including *10 Cool LEGO Mindstorms Dark Side Robots, Transports, and Creatures: Amazing Projects You Can Build in Under an Hour* (ISBN: 1-931836-59-0) and *10 Cool LEGO Mindstorms Ultimate Builders Projects: Amazing Projects You Can Build in Under an Hour* (ISBN: 1-931836-60-4). In the process, Kevin developed the LPub program for creating professional quality building instructions using MLCad, L3P, and POV-Ray. More recently, Kevin has developed the LSynth program so that bendable LEGO parts can more easily be documented when creating building instructions.

Kevin would like to thank his wife, Jan, and children, Aaron, Tony, Allison, and Andrew for "ooohing" and "aaahing" over his LEGO creations.

Miguel Agullo was born in Spain and has lived abroad for long periods of time, from the Far East to South America, from central Europe to the U.S. Trained as a journalist and impressed with the candor and resourcefulness of the online LEGO community, he tries to give something back by regularly updating his Web site at http://www.geocities.com/technicpuppy with instructions for new models, new LDraw pieces, and anything he thinks is worth sharing with other LEGO aficionados. His building interests revolve around robotics, and specifically biomechanics: creating mechanisms that mimic the behavior of natural devices such as legs or arms. Miguel's creations include biped walkers, robots that jump, and a fully functional (including a brake!) LEGO motorcycle. His current hobbies include boating, biking, traveling, and learning Thai and Dutch (his wife is Thai and they live in Amsterdam—which also explains the boat and the bike). Miguel was a contributing author for *10 Cool LEGO Mindstorms: Dark Side Robots, Transports, and Creatures: Amazing Projects You Can Build in Under an Hour* (Syngress Publishing, ISBN: 1-931836-59-0).

Technical Reviewer

Lars C. Hassing lives in Århus, Denmark, only 90 km from Billund, where (naturally) he has a season pass to LEGOLAND. Lars works as a programmer at CCI Europe, where he builds large scale, multi-user desktop publishing programs for newspaper publishers in Europe and in the U.S. He is the author of L3P and L3Lab, which are programs he created for the LDraw community (www.ldraw.org). Lars is married with two young sons, which gives him a perfect excuse for playing with LEGO.

Contents

Foreword

Kevin Clague and I, Miguel Agullo—the authors of this book—consider ourselves typical adult LEGO aficionados, whose story is like that of many members of that fantastic community. We also consider ourselves extremely fortunate, because not everybody gets to write a book about an activity they enjoy. What Kevin and I share with the rest of the very talented international LEGO fan community is something very significant: LEGO has allowed us time and again to literally build our visions, brick by brick. The programs discussed in this book will allow you to do the same thing, but in a slightly different manner. Instead of building your LEGO visions with actual physical pieces, you will create virtual models, with an unlimited number of parts, using your computer.

A few years ago both Kevin and I, independent of each other, were thinking about how cool it would be to make a machine that walked on two legs, as humans do. This is not a terribly easy feat; it requires delicate balancing of weights in motion. Yet from the moment that we associated the LEGO product with the phrase *walking robot*, everything clicked. And I mean *everything*. We quickly discovered that those machines already existed, that we could build them, and that it wouldn't even be a big investment in any sense (materials, know-how, process, etc.).

However, this is not a book about robotics, and what happened next is very important. Not only did we quickly learn about LEGO robotics from the material available on the Web at the time, we were soon making a whole series of walking robots. What's more, we were also contributing to the LEGO community at large—which is essentially what this book is all about. This is a book about a specific set of cutting-edge computer programs made *by* fans, *for* fans. There is no commercial intent behind them, just a love of the hobby. In fact, all the programs covered in this book are available free of charge.

How good can they be if they are free? The answer is, they're truly awesome. Computer programs, like communities, have a life cycle. The online community built around recreating LEGO models using computers is now ripe for entering the mainstream. James Jessiman was one of the first to tackle this task. Just like Ole Kirk Christiansen, the original developer of LEGO, James's approach was to provide high-quality support for kids of all ages. Despite passing away at an early age, James's genius lives on at the heart of the system that has grown up around his original program that allowed LEGO fans to create virtual models using their computers.

His early efforts directly support an array of impressive programs that today make building virtual LEGO models almost as easy as building them in real life. In addition, these programs—from the overwhelming quality of MLCad to the ingenuity of LSynth and other model generators—offer many new possibilities. Other programs, such as L3P and LPub, link the LDraw virtual building system to programs like POV-Ray. POV-Ray is not a LEGO-centric application; it is instead the result of the work of another fan community, this one dedicated to creating top-end 3D computer renderings available to the mainstream user—again, for free.

It is important to credit the creators of these programs by stressing that indeed, these tools are completely free of charge to the user. The URLs for the Web sites where each program can be found and downloaded for free are as follows:

- **LDraw** http://www.ldraw.org
- **MLCad** http://www.lm-software.com/mlcad/
- **LSynth and LPub** http://www.users.qwest.net/~kclague/
- **L3P** http://home16.inet.tele.dk/hassing/index.html
- **POV-Ray** http://www.povray.org/

Kevin Clague has developed two of the programs covered in these pages and included on the accompanying CD-ROM: LSynth and LPub. These programs, like all the programs included on the CD-ROM, are available for free from Kevin's Web site to anybody with an Internet connection, not just to readers of this book. Making these programs freely accessible to everyone is something that Kevin feels very strongly about—a point on which our publishers at Syngress have always supported him.

The reader of this book will be immediately exposed to a system of building virtual LEGO models using a computer. As you will soon see, what at first might seem like a hobby or perhaps an intriguing toy is in fact a sophisticated *learning system.* When we were making our first tentative steps into the world of robotics, the LEGO system allowed us to actually build robots right away. The LEGO system as a whole, and the programs covered in this book, gave us exposure to a level of practical experience that is often difficult to get in many fields. This is the type of experience that everybody is exposed to as a child, whether they realize it or not. Children quickly assimilate the myriad experiences they encounter and channel them into a staggering variety of learned skills. The LEGO system offers a broad range of practical knowledge that is similar to this process in many ways and that can serve as a foundation for more specialized exploration and learning. In the case of this book and the programs it examines, we use LEGO to teach you about 3D computer graphics.

The LEGO-based 3D applications developed by fans have reached the degree of quality and usability that the modern computer user expects from commercial software. This book facilitates the critical stage in any learning system: the initial steps. From the very beginning, how to actually achieve this goal was a hot topic for discussion, since the software available is multifaceted and varied. After quite a bit of work, we have settled on a teaching and learning method that we hope will entice beginners, experts, LEGO fans, and people simply interested in 3D computer graphics.

We offer the reader a very specific path into creating virtual LEGO models as well as manipulating and rendering 3D images, by exploring several essential applications. At the same time, we illustrate the larger LEGO "world" as it relates to the various stages in the process. When we were first coming up with ideas for this book, we sincerely felt that the missing element in the LEGO community that has sprung up around these applications was a book that focused on providing the *overall* picture. The resources are there, but without a book like this, it will take the average reader quite a bit of time to learn how to use all these excellent applications and, more important, how to get them all to work with one another.

We started by centralizing some of the critical resources in one place: the CD-ROM that accompanies this book. Although being aware of the Web sites of this software's creators is absolutely necessary once we are users of the programs, it is arguably less so when we are trying to find out what the programs actually can *do*. The capabilities of each program are easy to find out with this book: Simply install the software and you will be up and running right away.

Beyond providing an easily accessible source for all the these applications under one roof, this book also offers a way to become very familiar with them in a relatively short time. The online resources for the hobby continue to grow and develop, but there is not yet a virtual LEGO academy of design. If there were, we think that this book could be its textbook.

This is a book for LEGO fans who want to build with LEGO using their computers. It shows them an easy way to do so and describes the robust system that ultimately supports it. This system will allow readers to share their creations, whether simple renderings, step-by-step instructions, or even animations, with other fans. Beyond that, the galaxy of possibilities is, as with real-life LEGO, infinite.

This is also an introductory textbook of sorts for people who would like to learn about 3D computer imaging, a field that is becoming more and more mainstream as technology advances. The problem for people interested in 3D computer imaging is taking the initial steps into the field, which can often be a daunting and difficult task. How do you get started? How do you make sense of it all precisely at the moment when nothing makes sense? Using existing and familiar-looking LEGO elements, you will acquire *instinctively* the basic concepts of computer 3D imaging—concepts used not only by the virtual LEGO modeling system described in this book but also by practically all computer 3D programs. This includes CAD systems used by architects and engineers as well as systems used to generate the graphics for your favorite video games and movies.

In short, this is a book about a system that offers very articulated insights into many different and fascinating fields: engineering, architecture, photography, animation, modeling, *Star Wars*, and the famed LEGO Pirates, to name just a few! It is up to you to choose your theme. The tools are ready and at your disposal—and with this book, so is the instruction manual.

—*Miguel Agullo*
Amsterdam, November 2002

Chapter 1

A LEGO CAD System

Solutions in this Chapter:

- The Software Power Tools Suite of Applications
- LEGO as a Learning Tool

☑ Summary

☑ Solutions Fast Track

☑ Frequently Asked Questions

Introduction

This book is a gateway into a world populated by virtual LEGO models such as the power drill pictured on the book's cover. As you might have noticed already, the power drill is not a *real* power drill; it is a LEGO model. As you might also have noticed, the image of the power drill on the cover is not a photograph; it is a computer-generated rendering. It has been created from scratch, literally pixel by pixel, using a home computer no different from the computers that sit in many homes throughout the world. *That* is what this book is about: creating LEGO 3D models inside virtual spaces that exist only inside computers. Why would you want to know how to do that? There are many reasons, but we will give you three basic ones:

- Because it's fun.
- Because it's easy.
- Because, like the best things in life, it's free.

The power-drill model was actually created by one of the best LEGO modelers of technical machinery in the world: Jennifer Clark, a top Scottish engineer if there ever was one. How this image came to be is a good example of what this book is all about. Jennifer created a computer file of the model with virtual LEGO bricks using two programs called LDraw and MLCad. She then e-mailed the file to Miguel Agullo, who lives in Amsterdam and is one of the co-authors of this book. Miguel ran her file through another program developed by Kevin Clague to create the finished image of the power drill that you see on the cover. Kevin, the second co-author of this book, had worked with Miguel in the past on other LEGO-related projects. So as you can see, the final power-drill image was the result of the efforts of several people (all LEGO fans), using a variety of LEGO-related programs.

The CD-ROM that accompanies this book contains all the software necessary to go through exactly the same process by yourself at home on your personal computer, as well as build a limitless variety of LEGO models made out of an endless supply of virtual LEGO parts. We will show you how to install all the programs from the CD-ROM onto your computer in Chapter 2.

Building models with computer programs in this manner is known as *computer-aided design*, or *CAD*. The programs that we discuss in this book are all LEGO CAD applications that we collectively refer to as the Software Power Tools. The role of a real-life power tool is to make the job it is intended to perform easier for the user. The Power Tools programs serve the same purpose:

They make your virtual LEGO modeling much more simple and enjoyable. All the programs included on the CD that accompanies this book are *freeware*, which means they are free to use. We have brought them all together in one place and show you how to use them all together.

Creating virtual LEGO models is a multifaceted hobby. There is no single "correct" way to go about it. This book offers some answers, but it mostly aims to stimulate the imagination with potential scenarios. Learning how to use the software contained on the CD is easy, and that is what this book focuses on: showing you how to use the actual programs themselves to create virtual LEGO models and images of them. Soon you will be creating very engaging virtual LEGO models and fabulous renderings like those on the cover of the book. At that point it is up to you to make them *memorable*. As you will see, it isn't difficult once you know your way around the programs.

NOTE

None of the programs covered in this book is an official LEGO product. All these programs have been developed by LEGO fans around the world and made available to the LEGO community free of charge.

LEGO is one of the most well-known toys in the world, so it won't come as a surprise to learn that a lot of work has gone into making it such a high-quality product. As you will see later in the book, LEGO's obsession with offering only the best possible product is actually a very important influence that has been carried on by LEGO fans. It's not that everybody who works and plays with LEGO is a perfectionist; rather, when using LEGO, you become a perfectionist by default.

The LEGO line of toys is at its core a *building system*. It allows us to build models by securely attaching together a great variety of parts that interlock with each other in several different ways. This makes the system as a whole robust and flexible—or, in other words, easy to use. At the same time, LEGOs are also toys, and as such they are targeted primarily (but not exclusively) to kids. Children are by nature a tough sell when it comes to toys; they expect to be entertained by their playthings, and not just once, but every time they pick them up to play.

How has LEGO survived for so long and retained its appeal for both children and adults? That is a very interesting question. Perhaps the best answer is that the LEGO system appeals directly to the user's imagination. LEGO parts come in a great range of shapes and sizes that can be attached, one way or another, to all the other LEGO parts. If we take a closer look at the variety of shapes and the multiple ways each of them can attach to the others, we realize that although the LEGO system is perfectly organized, the building possibilities it offers are limitless.

If we continue analyzing the way LEGO parts connect to one another, we find some truly amazing coincidences. LEGO, as a building system, works in a fashion that is very similar to many computer programs. What makes LEGO easy to use, apart from the fact that the pieces are constructed of high-quality plastic and the connection system is well designed, is the fact that it is a modular system that allows users to take many tiny pieces and create something that is much larger and more complex with them. Imagine for a moment that you had access to an unlimited supply of LEGO parts. When assembling your models, you would find that many of the processes involved would closely mimic processes you use every time you use your computer. You could *copy, cut,* and *paste* different parts, saving you the time involved with performing repetitive building tasks. Given this unlimited access to parts, you could easily *duplicate* and *save* your models, *deleting* those that you didn't like or were finished with. Needless to say, LEGO and computers mix particularly well. This book, and the Power Tools programs it describes, will give you access to that limitless supply of LEGO parts we just mentioned as well as showing you how to take your finished models to the next level.

NOTE

For more of Jennifer Clark's amazing LEGO creations, check out her Web site at www.genuinemodels.com.

The Software Power Tools Suite of Applications

Explaining the applications covered in this book in any detail beforehand is rather confusing because they become more specialized as the book progresses. Additionally, the book is necessarily linear, whereas the software system is not. This section briefly mentions *what* exactly each application does and leaves it to the following chapters to more fully flesh *how* they actually do it. The applications all gravitate around a common *file format*, defined by the late James Jessiman for his program LDraw. Of Jessiman's original ideas has sprung a whole community of users and applications that take advantage of a common file format to add further functionality to the building and rendering of virtual LEGO models. Let's take a look at each of the applications in the order they appear in the book as well as the topics of the last two chapters in the book, which are not dedicated to discussions of actual applications.

LDraw

All the applications discussed in this book revolve around the LDraw system, created by James Jessiman. Chapter 3 presents a brief introduction to the LDraw file format, which is at the root of the virtual LEGO parts that we will use to build our models as well as the way in which those parts are created and organized. The original LDraw executable (program) has become obsolete, but due to its author's ingenuity, the LDraw file format is still *the* standard used today in LEGO CAD. Various applications use both the LDraw file format and the LDraw parts library for a variety of purposes. The parts library is a collection of files that make up a catalog of over 2,000 types of virtual LEGO parts.

MLCad

MLCad is what is known as a *modeler*, which is a program that allows us to see a graphical representation of our virtual LEGO models and the parts they consist of. Created by Michael Lachmann, MLCad is a superb modeler that combines the LDraw parts library and file format with a standard Windows interface. MLCad looks and performs as well as or better than many commercial CAD applications used by architects and engineers. Not only can we use MLCad to create any LEGO model we want (whether as small as two parts or as large as 10,000 parts), it also provides many sophisticated options to our building process, such as adding step-by-step instructions to our models, providing ways to use third-party software to generate details for our models, or even allowing the use of non-LEGO elements in a model. Because MLCad is the tool you will use for the majority of time you're actually building your models, we have dedicated Chapters 4, 5, and 6 to discussing its use. Chapter 6 also offers a more detailed exploration of the LDraw file format.

LSynth

LSynth is a program developed by Kevin Clague, one of the co-authors of this book. LSynth allows users to create and incorporate very detailed flexile parts into their virtual LEGO models, such as hoses, rubber bands, chains, and treads, to name just a few. These parts can then be used in standard LDraw models. This functionality was lacking in LDraw and MLCad; LSynth is an excellent example of a program that was written to meet a specific modeling need. Chapter 7 covers LSynth in detail and was written by the creator of the program himself, Kevin Clague.

L3P and POV-Ray

L3P and POV-Ray are two applications that are good examples of how to take your LEGO modeling to the next level. Both are discussed in Chapter 8. POV-Ray (which stands for *Persistence of Vision*) is a very sophisticated program that simulates lightning environments in images; it can be used to create amazing effects with your finished LDraw images. It was not created with LEGO fans in mind, however, and the standard file format that LDraw and MLCad use is not compatible with POV-Ray. This is where L3P comes into the picture. L3P, a fantastic program written by Lars C. Hassing, converts standard LDraw files into a format that POV-Ray can use, allowing you to create finished images of your virtual LEGO models that are the equal of anything you might see on TV or in the movies. We feed POV-Ray the basic geometric data of our LEGO models (translated by L3P) and it renders photorealistic images like the one on the cover of this book. POV-Ray is a good example of a non-LEGO-specific application that we can use to further our modeling experience.

NOTE

We would like to say a very warm "thank you" to Lars Hassing for his help in reviewing Chapter 8 as well as for creating such an amazing program. For more details on L3P as well L3Lab, refer to his Web site at http://home16.inet.tele.dk/hassing/l3p.html.

LPub

LPub is another program written by Kevin Clague. Discussed in Chapter 9, LPub allows users to create professional-quality building instructions for their custom LEGO models, whether real-life models made with actual LEGO bricks or virtual models created in MLCad. LPub also will automatically create images showing all the parts that need to be added in each particular step. Created originally to help with the publication of Syngress' *10 Cool Mindstorms* series of project-based books, LPub allows users to manage many of the functions of several of the Software Power Tools programs, such as adding flexible elements to files via LSynth or automatically generating a series of photorealistic building steps generated through POV-Ray.

The LEGO Community

Chapter 10 takes a look at the LEGO community. A wealth of support for LEGO fans is available online, and this chapter introduces readers to the key players in the LEGO online community. Through these sites, the reader will be able to find resources catering to his or her tastes. Sharing models, construction tips, modeling ideas, and critiques are daily routines in the many forums, fan Web sites and other available resources. These discussions are not restricted to specific LEGO themes. There are also daily discussions about education, commerce, and many other topics made easily accessible, fun, and in general more effective through the common touchpoint of LEGO.

AT-ST Building Instructions

Chapter 11 contains a full set of instructions for building a LEGO AT-ST robot, created by Kevin Clague primarily out of LEGO Mindstorms parts. This set of instructions was created entirely with LDraw, MLCad, LSynth, L3P, POV-Ray, and LPub. Aside from being a neat bonus for readers of this book, Chapter 11 also serves as an excellent example of the possibilities offered by LPub and the Power Tools suite of applications in general.

LEGO as a Learning Tool

There is one big lie to which many first-time LEGO users fall prey. This is the misconception that LEGO is only a toy. LEGO is not just a toy, it is also a learning tool. In fact, LEGO is a learning tool ahead of its time, which is one reason kids have always liked it so much.

We have already discussed two ways in which LEGO is ahead of its time. First, kids like it and have for generations. If you think about the future, you'll realize that today's children represent what is *really* coming down the line. All clichés aside, the children *are* the future. Beyond that obvious metaphysical truth, the hard data points to the fact that LEGO, a toy created in 1949, works in a way similar to the computers we are using in 2002. Of course, LEGO is a building system, and *any* building system resembles a computer program to some extent. But the LEGO system not only uses vague overall building concepts—it replicates many of the functioning details of computers almost exactly.

LEGO bricks are not simple blocks; they carry functional binding mechanisms (often quite a few). Two LEGO parts can be connected to one another in several different ways and stay that way until we separate them. Just as computer bytes are kept together by the *computer system*, LEGO bricks are kept together by the *LEGO system*.

What does this mean? It means that LEGO is capable of teaching us the ways of the future. For several years now, executive training schools have been using LEGO bricks to teach corporate managers about strategy and organization. Recently, the LEGO company itself developed, along with several key partners, courses for training the enterprise managers of the future using the LEGO bricks of today—or, more precisely, the LEGO bricks of 1949.

This is by no means the only time the LEGO company has engaged in educational pursuits. The LEGO group of companies encompasses many different products often offered in different markets with different packaging. For instance, the LEGO bricks we are all familiar with are not only used in courses for executives but also in the Dacta line of educational products for primary and secondary schools. This official LEGO offering helps teachers design classrooms activities around specific themes with the support of Dacta material. This offering includes LEGO bricks in some lessons, but there are also other options, such as hot-air balloon kits, to name just one.

This is where the content of this book comes back into the picture. Learning how to create LEGO models inside your computer is not only easy and engaging, it actually provides a great way to learn about computer 3D graphics in general. The image of the power drill on the cover is not that far away from the special effects found in top-notch Hollywood films. It's not just how it *looks*. What is more important is that many of the same technical principles, methods, and concepts apply to both the image on the book's cover and a frame from the movie *Jurassic Park*. Halfway through the book, you will be ready to explore the world beyond LEGO-based 3D software. More important, you will have the knowledge to quickly gather and adapt more information relating to your interests.

The software included on the CD-ROM, along with this book, offers an excellent way to take a first dip into the 3D computer design pool of knowledge. LEGO-based 3D software offers many possibilities, not the least of which is an easy gateway into other specializations, such as animation or even real-time visualization. As you will see throughout this book, these LEGO-based programs not only perform well with one another, they can also be combined with other non-LEGO programs. Think of LEGO as a starting point. Once you have mastered the programs covered in this book, where you go is up to you. The sky is the limit.

Summary

The Software Power Tool suite of applications is in fact a group of freeware applications, created by LEGO fans, that allow users to build virtual LEGO models on their computers and perform a variety of special functions to the finished images of their creations. All these applications are available for free on the Web, but the goal of this book is to bring them all together in one place and give readers clear instructions for using them, both individually and in conjunction with one another. The applications covered in this book include the following:

- LDraw
- MLCad
- LSynth
- L3P
- POV-Ray
- LPub

We will do almost all the actual building of virtual LEGO models in MLCad, using the parts and file format created for the original LDraw program. These two programs offer users an unlimited supply of LEGO parts to model with and a virtual building space that includes a variety of tools for manipulating these pieces and putting them together. The building possibilities are quite literally infinite.

The remaining applications either meet modeling needs not offered by LDraw and MLCad (LSynth and LPub) or allow users to take their creations outside the realm of LEGO CAD and incorporate the functions of other programs into their model files (L3P and POV-Ray).

An important concept to grasp before moving on to the next chapters is the fact that both the LEGO building system and the LEGO CAD system that grew out of it are learning tools in and of themselves. They provide an excellent way to introduce novices to a variety of concepts that can be found running through such diverse fields as engineering, computer programming, architecture, animation, and 3D design. This book, along with the programs it covers, will give the reader a firm grounding in many key CAD concepts.

Solutions Fast Track

The Software Power Tools Suite of Applications

☑ The group of freeware applications that we refer to as the *Software Power Tools* consists of the following programs:

- **LDraw** Created by the late James Jessiman, LDraw is the foundation on which the majority of the LEGO-related applications in this book are built. Although the actual LDraw executable is no longer in use, the LDraw file format and parts library are the standard still in use today in most LEGO CAD applications.
- **MLCad** MLCad is a modeler that allows users to view graphical representations of their LDraw files on screen and build models using parts from the LDraw parts library. Its main benefit is its handy Windows graphical user interface.
- **LSynth** LSynth is a program that allows users to incorporate flexible parts into their LDraw files.
- **L3P and POV-Ray** L3P is a handy program that converts LDraw files to a format that POV-Ray can use. POV-Ray is a program that allows us to perform advanced lighting effects on our finished model images.
- **LPub** LPub allows users to create professional-quality sets of building instructions for their LDraw models, complete with images of the parts that will be included in each step.

☑ All the applications covered in this book are created and supported by amateur fans. They are the equal of, and in some cases superior to, many of today's commercially available CAD programs.

LEGO as a Learning Tool

☑ The LEGO system is more than just a toy. It is also an excellent way to teach the fundamentals of organization and construction, whether to executives in large companies or children in school classrooms.

☑ In many ways, the LEGO system is similar to computer systems. Just as computers organize and connect bytes of data, the LEGO system organizes and connects individual LEGO bricks to create something that is larger and more complex than the sum of its parts.

☑ The LEGO CAD system offers an excellent way to break into a variety of disciplines, including 3D design, animation, and architecture.

Frequently Asked Questions

The following Frequently Asked Questions, answered by the authors of this book, are designed to both measure your understanding of the concepts presented in this chapter and to assist you with real-life implementation of these concepts. To have your questions about this chapter answered by the author, browse to **www.syngress.com/solutions** and click on the **"Ask the Author"** form.

Q: Is this a book for LEGO fans?

A: Not necessarily. LEGO fans will find many tools described in it useful from the start. But readers simply interested in 3D computer graphics will find a LEGO-based way of learning while having fun.

Q: Will this book turn me into a LEGO fan?

A: We hope so! Seriously, all you need to start creating fantastic virtual LEGO models and illustrations is this book, the software included on the CD-ROM, and a Windows PC. Familiarity with real-world LEGO helps but is by no means mandatory.

Q: I've got a non-Windows PC. What am I to do?

A: The book describes a Windows PC-only suite of programs. However, most of the components and processes can be duplicated on other platforms. The best way to learn about available software and new releases for your system is to visit www.ldraw.org.

Q: You mentioned that all these programs are freeware. Does this mean they are open source software?

A: Many of the LEGO-centric software applications are freeware but not open source. This means that the programs themselves are free to use, but the code has not been made public for people to modify it. The file types are all non-proprietary so that new applications can be developed and the parts library augmented.

Q: Is this official LEGO software?

A: No. This is fan-created software. The LEGO company does not officially authorize or endorse these programs in any way. However, the intellectual property issues regarding the end user are pretty much sorted out.

Q: Is it really that easy to create pictures like the one on the cover of this book?

A: It's easy once you know how to use the programs. It's also pleasurable, entertaining, and highly rewarding, in a LEGO sort of way.

Chapter 2

Installation

Solutions in this Chapter:

- Installing the Software

☑ Summary

☑ Solutions Fast Track

☑ Frequently Asked Questions

Introduction

This chapter walks you through the installation of the LDraw-compatible CAD tools described in this book. These tools include:

- **LDraw Parts Library** A large database of LEGO parts in LDraw file format.
- **MLCad** A Windows CAD program based on LDraw parts.
- **LSynth** A tool to make LDraw models that represent real-world, bendable LEGO parts.
- **L3P** A tool that converts LDraw format files to POV-Ray format for use with the POV-Ray application.
- **POV-Ray** Draws photorealistic images using advanced computer graphics techniques.
- **MEGA-POV** Works in combination with POV-Ray to draw outlines around your LDraw parts.
- **LPub** Interfaces to L3P, POV-Ray, and MEGA-POV to automate the generation of step-by-step building instructions.
- **Examples** These are LDraw files that are used in several of the following chapters as examples.

To start the installation process, put the CD-ROM that came with this book into the CD drive on your computer. In a short while, the LEGO CAD Power Tools Installer program (see Figure 2.1) should automatically start running.

NOTE

If the installer program does not automatically run, double-click the **My Computer** icon on the Windows Desktop, then double-click your **CD-ROM** drive icon, then double-click the **Installer.exe** program icon.

The Power Tools Installer program lets you install any or all of the packages listed in the "Install Options" section of the Power Tools Installer window. For those of you who are new to the world of LDraw-compatible CAD tools, simply click the **Install** button to start the installation process.

Figure 2.1 The Power Tools Installer

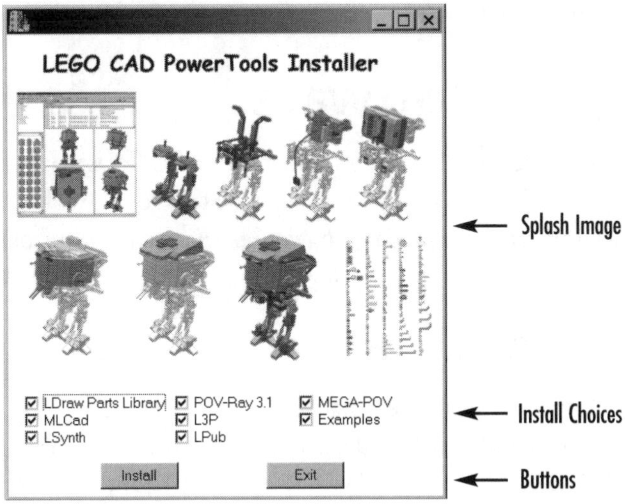

For readers who already have some of these tools installed, uncheck the boxes for the packages you already have installed. For example, if you already have LDraw and MLCad installed on your computer, uncheck the **LDraw Parts Library** and **MLCad** check boxes before clicking the **Install** button.

Once you click the **Install** button, the Power Tools Installer runs the installation package for each of the packages that are checked. Each installation package is its own self-contained installation program that performs the usual steps of program installation, including the following:

- Informational dialog window that tells you what you are installing
- Installation directory dialog window that lets you choose where to install the program and/or data
- Program shortcut dialog window that lets you decide where to place program shortcuts in the Start menu
- Confirmation dialog window that lets you see the choices you made
- Installation progress dialog window that shows you the status of your package installation
- Installation complete dialog window

After each installation process completes, the Power Tools Installer runs the next installation package until all the packages are installed. All but the POV-Ray installer can be uninstalled either via a menu item in the Start menu or by using

Start | Settings | Control Panel | Add/Remove Programs. The POV-Ray program can only be uninstalled using Add/Remove Programs.

Installing the Software

The Power Tools Installer lets you install a number of different installation packages. The first section that follows walks you the installation of the LDraw package. Each of the sections that follow describes the packages that will be installed and lists the chapters of this book that cover using each particular program.

The LDraw Parts Library

The LDraw Parts Library installation package installs LDraw 027, the complete official parts library and the unofficial parts library from the ldraw.org Web site, the centralized LDraw resources site. We look at the LDraw Parts Library in more detail in Chapter 3. In case you've never installed software before, we'll walk you through the installation of the LDraw package:

1. Figure 2.2 shows the informational window (the first step in the installation process) for the LDraw Parts Library installation package. This window should pop up when you click **Install** in the Power Tools Installer window. Click **Next** to continue with the installation process.

Figure 2.2 The LDraw Information Window

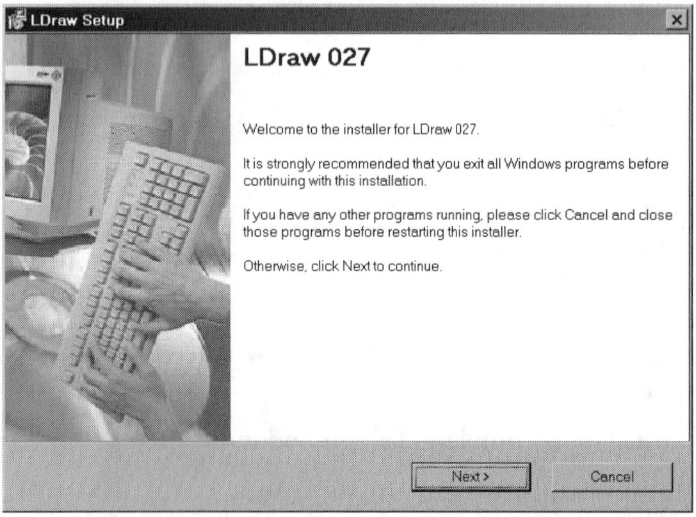

2. Figure 2.3 shows the Installation Folder dialog window that lets you decide where to install the LDraw Parts Library. This dialog window allows advanced users to control where the LDraw Parts Library is installed. If you are fine with the default location (C:\LDRAW) or you don't know of a better place to install it, click the **Next** button to continue.

Figure 2.3 The Installation Folder Dialog Window

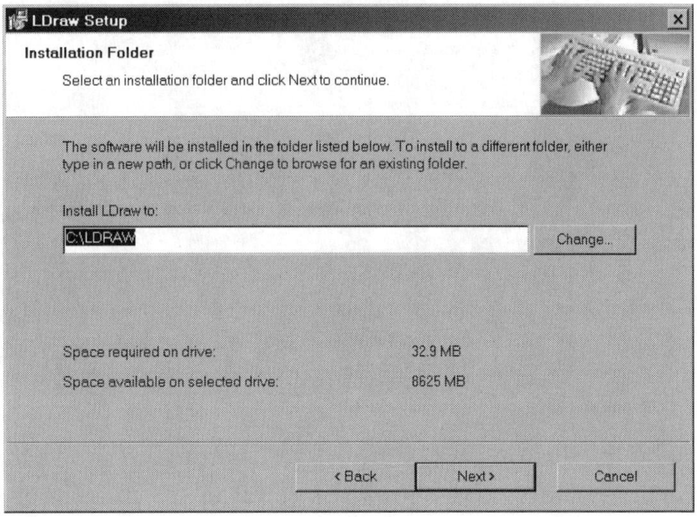

NOTE

You can override the default installation location either by typing an alternate directory into the **Install LDraw to:** field or clicking the **Change...** button to bring up a dialog window that lets you point and click your way to the desired installation location. When you are satisfied with the installation directory, click the **Next** button to continue.

3. The next dialog window that pops up is the Ready to Install window (see Figure 2.4) that shows you the installation information you have selected. If you are satisfied with the installation settings displayed, click the **Next** button to start the installation. If you want to change the installation folder, you can go back to the Installation Folder dialog

window by clicking the **Back** button. If you changed your mind and don't want to install at this time, click the **Cancel** button, which takes you back to the Power Tools Installer program.

Figure 2.4 The Ready to Install Window

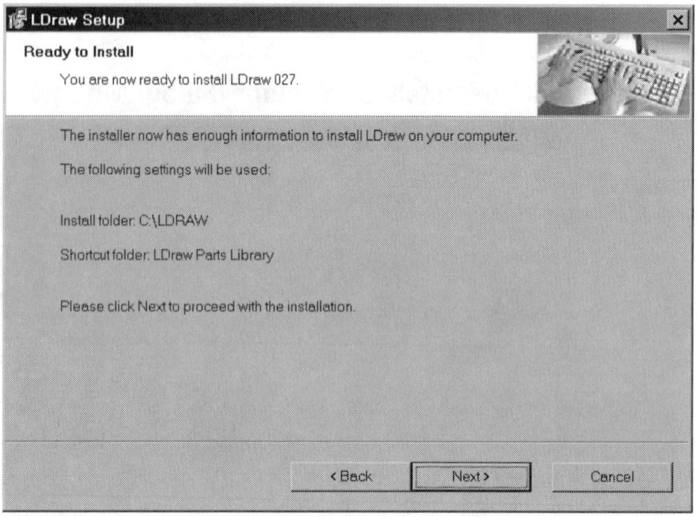

4. The installation process now starts. Figure 2.5 shows the Installing Files dialog window that shows you status of the installation progress.

Figure 2.5 The Installing Files Dialog Window

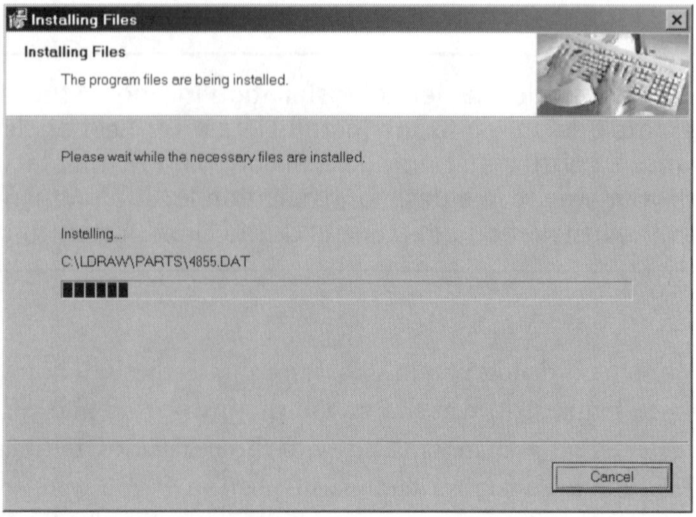

5. After the parts have been installed, the LDraw program MKLIST is run to make a list of the installed parts for LDraw CAD programs to use. On some Windows environments, the MKLIST program window closes automatically. On others, when MKLIST finishes, the window looks as shown in Figure 2.6. Close this window by clicking the **X** button in the upper-right corner of the window.

Figure 2.6 The MKLIST Finished Window

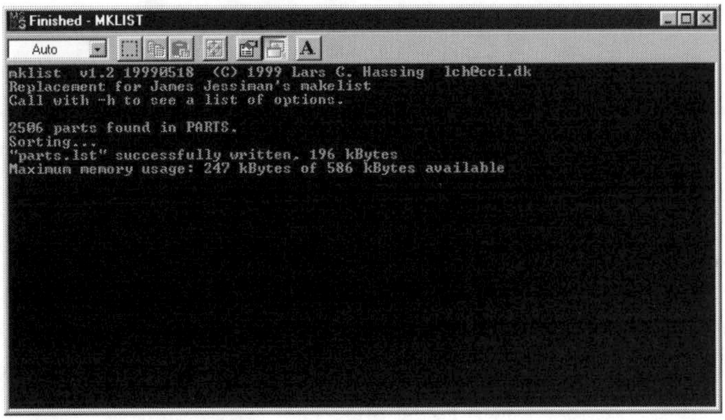

6. When installation completes, the Installed Successfully window pops up (see Figure 2.7).

Figure 2.7 The Installed Successfully Window

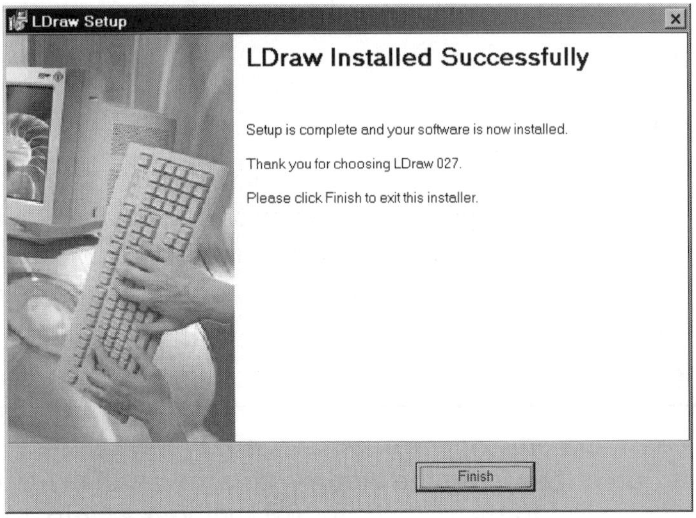

7. Click the **Finish** button to complete the LDraw Parts Library installation and move onto installing the next installation package. As each new installation package program is run, novice users should just click the **Next** buttons until they get to a **Finish** button. Do this until all the installations are complete. Advanced users might want to override installation directory defaults or Start menu shortcuts before starting installations.

LSynth

LSynth is a program that creates bent LDraw parts that are not directly available in the LDraw Parts Library. LSynth can be used to create hoses, electrical cables, flexible axles, rubber bands, treads, and chains. LSynth is covered in depth in Chapter 7.

L3P

L3P is a program that translates LDraw designs into POV-Ray scripts for rendering with the POV-Ray program, a photorealistic image-drawing package. L3P and POV-Ray are described in Chapter 8.

LPub

LPub is a program that automatically produces step-by-step building instructions directly from your LDraw design files. It provides a graphical user interface to L3P and POV-Ray for drawing building instruction pictures. LPub is covered in Chapter 9.

POV-Ray

POV-Ray is a world-class program that renders pictures with advanced ray-tracing computer graphics techniques. Its development is all done on a volunteer basis, and the program is available free of charge at the POV-Ray site: www.povray.org. POV-Ray is discussed in Chapter 8.

NOTE

The past and present members of the POV-Ray team have done a fantastic job of providing the world with a powerful ray-tracing program. We'd also like to thank them for packaging it all into a bulletproof installation package.

MEGA-POV

MEGA-POV is a program that works with POV-Ray to draw outlines around the parts in your LEGO design. This can make the individual parts in a large design easier to identify. MEGA POV is discussed in Chapter 8.

Examples

Some of the chapters in this book use example LDraw files to explain features of the programs being described. The Examples installation package allows you to install these examples onto your computer so you can use them. When installed, these files can be found in the LDraw\Models\PowerTools directory.

MLCad

MLCad is a Windows program that you use to enter and edit LDraw designs. MLCad is the cornerstone application of the LDraw-compatible LEGO CAD world.

Summary

This chapter describes how to install the various programs described in this book. The Power Tools Installer program lets you choose the packages you want installed and runs each of the installation packages you select. Once the packages are installed, you will be ready to proceed through the remaining chapters, learning the basic and advanced capabilities of these powerful programs. The trick to getting though the installation process is just clicking the Next buttons until you see a Finish button and waiting for the next installation package to start.

Solutions Fast Track

Installing the Software

☑ The Power Tools Installer lets you choose the packages and programs you want installed.

☑ Install all the packages (the default setting) unless you know you already have some packages or programs installed on your computer.

☑ Installation begins with the LDraw Parts Library. Once its installation is complete, the order of the install is as follows:

1. **LSynth** A program that makes LDraw models of bendable LEGO parts such as hoses, cables, and rubber bands.
2. **L3P** An LDraw-to-POV-Ray format converter.
3. **LPub** An automatic step-building instruction generator for showing others how to build your models.
4. **POV-Ray** An image renderer that makes photorealistic drawings of your LEGO designs.
5. **MEGA-POV** A program that works in conjunction with POV-Ray to draw outlines around your LDraw parts.
6. **Examples** The example LDraw files used in this book.
7. **MLCad** The powerful and easy-to-use Windows application that lets you enter and edit LDraw design files.

Frequently Asked Questions

The following Frequently Asked Questions, answered by the authors of this book, are designed to both measure your understanding of the concepts presented in this chapter and to assist you with real-life implementation of these concepts. To have your questions about this chapter answered by the author, browse to **www.syngress.com/solutions** and click on the **"Ask the Author"** form.

Q: Do I have to install all the packages provided on the CD-ROM?

A: If you already have some of these programs or packages installed, you do not need to reinstall them. To follow along with all the chapters in this book, you need to have all these programs and packages installed on your computer.

Q: Are there some packages that *must* be installed?

A: You will not make it very far in this book if you do not install the LDraw Parts Library and MLCad.

Q: Do I have to install the Examples package?

A: No, you do not. If you are satisfied with reading about these examples rather than working with them, you do not need to install the Examples package. You will be able to create many of these examples yourself as you follow along with the text.

Q: I've put the CD-ROM into my CD drive, but it doesn't appear to be starting. What's going on?

A: Try double-clicking the **My Computer** icon on your Windows Desktop, then double-click your **CD-ROM drive** icon, then double-click the **Installer.exe** program icon. That should get things moving for you, and the installation should begin.

Q: When would I *not* want to install to the default directories?

A: Advanced users might not want to install to the default directories due to organizational needs or limited disk space. People with small amounts of free space on their C: drives might want to install the packages on a second hard drive that has more free space.

Chapter 3

LDraw: A Virtual LEGO System

Solutions in this Chapter:

- Created by Fans, Supported by Fans
- A Lego CAD System
- A Multitude of Resources

☑ Summary

☑ Solutions Fast Track

☑ Frequently Asked Questions

Introduction

LDraw lies at the heart of the LEGO-based 3D illustration package described in this book. We chose LDraw precisely because it is the standard and most developed software in the LEGO CAD fan community. This software has quite a unique history and, consequently, some unique features. This chapter gives you a brief background and general overview of LDraw, the program that started it all.

This information is especially important for the first-time user. The rest of this book is a very practical guide dealing with how to use several applications to create 3D LEGO models using a computer. Building with virtual LEGO is very easy, but you will still have to concentrate on the finer points of the process to become a skillful modeler. In the following chapters we sometimes focus on *why* functions work the way they do, but we spend most of the time on *how* they work. To be focused from the start, you will be much better off knowing a few essential concepts about the LDraw software before you begin using the other programs.

First, you have to know that although LDraw is installed on your system, you will only really be using it *indirectly*. What does this mean? It means that you will not be directly using the original LDraw program but rather the *system* it has spawned. The software suite described in this book is the result of the work of many fans, both developers and users and testers. Many authors have contributed elements to these programs. In this book, we bring together some of these elements and show you how to use them to create 3D models using your computer. LDraw is the catalyst that binds them all together.

How can a program that we do not use directly have such an enormous influence on what we create? In certain ways, LDraw resembles the product that inspired it: LEGO. LEGO is fun to play with because a great deal of work has gone into its design. Even basic LEGO bricks are not as simple as they appear at first glance. For one thing, they contain several locking points that work flawlessly, almost seamlessly, every time you connect one brick to another. This and many other high-quality characteristics of LEGO bricks are the result of seriously intelligent design. If it is extremely easy to take this level of quality for granted, that is because the system is also designed to be *easy to use*.

The highly successful LEGO building system is the result of the ideas of Ole Kirk Christiansen, a Danish carpenter who lived in the first half of the twentieth century. Ole was bent on making the highest-quality toys possible; he simply would not settle for anything less. In the second half of the twentieth century, James Jessiman, an Australian programmer and avid LEGO fan, picked up on

Christiansen's original ideas and, with the creation of LDraw, set the rules for a high-quality LEGO-based building system using a computer. The important thing to understand is that James did not simply replicate the design of the LEGO parts using a computer. From the very beginning, he incorporated into the software the quality expectations found in the original LEGO toy. This means that the resulting product is both robust and easy to use. Read on to see how James Jessiman accomplished both objectives.

Created by Fans, Supported by Fans

Second only to its incredible quality is the fact that LDraw is an unofficial, fan-created software system. Tragically, James Jessiman, not only the creator of LDraw but also the key inspiration for all the LEGO CAD programs and materials that followed (including this book), died in 1997 at the young age of 26, barely two years after releasing his software to the world.

This was obviously a turning point in LDraw's history. The odds against its survival quickly piled up. Like most freeware, LDraw had no marketing budget and little initial support beyond a few of the creator's acquaintances. Not only that, at the time of Jessiman's passing, the software had become technically obsolete and was being overwhelmed by the newly emerging Windows operating system.

Yet the program and its specifications didn't disappear. There are probably a few factors that helped its survival, chief among them the emergence of the Internet and the ability to easily disseminate software all over the world online. However, the one key element was Jessiman's ingenious design guidelines for the LDraw file format and general program functions. They allowed the software's users to fight back against the odds and ultimately win, establishing the software in its justly deserved place at the top of the LEGO CAD pile—all, of course, still free of charge.

The Initial Release

James Jessiman released the first version of LDraw in 1995. At around the same time, Microsoft introduced Windows 95. Most developers, including James, were still creating applications for the DOS operating system (which was more reliable and familiar than Windows 3.1, the version of Windows that preceded the much more successful Windows 95). There cannot be a more fitting testament to Jessiman's immense and barely tapped talent than the fact that his DOS application has become the de facto LEGO CAD standard for the new millennium—long after DOS, Windows 3.1, and Windows 95 have become obsolete. As the international LEGO community achieves maturity, its CAD efforts are articulated around Jessiman's original vision for LDraw. This is a truly impressive accomplishment.

How does a program that a fan created come to have such an enormous influence on a community of users and developers? LDraw is a freeware program, like all applications described in this book. What this means is that the program is free to use. This factor might not seem to be all that important, but it was one of the key reasons for LDraw's growth and expansion. Since LDraw is a task-oriented piece of software offered for free, many people tested it just to see what it was all about. Once they actually used LDraw, they were hooked. Some well-known programs (especially videogames) started highly successful commercial lives this way. At this stage, the software passed its first critical test: Since it was good, many people clamored for more. Since programmers tend to know other programmers, word spread and some initial users came back asking how they could *modify* the program to fit their needs.

At this stage, Jessiman faced a very modern challenge: How exactly to manage his intellectual property. For an example of how other companies have answered this question, take a look at the LEGO company itself. LEGO puts out a product that lets users *create*. If a user invents a custom model and takes a picture of it, what part of that picture belongs to LEGO and what part belongs to the user? LEGO has taken (not surprisingly) a very intelligent approach to this dilemma: The company allows users to display any and all creations they have built with LEGO, as long as they make it perfectly clear that they are not *official* LEGO products. This suits all sides well, except maybe the pirates—tough luck for them!

Since James Jessiman did not have the backing of a global brand and was developing software, not plastic bricks, he used a different way to satisfy both creators and users. He retained the property of the LDraw software but allowed (and encouraged) full access to the file format. This way, he kept the intellectual property of his program but let people write their own programs and files using the same resources that he had used. This solution was quite generous, but also smart: It allowed James to retain the intellectual property of the program he had developed while at the same time sharing a common pool of users and testers to develop his software more fully.

LDraw Today

And develop they did. As we explain in Chapter 10, the international LEGO fan community was waiting for something like the Internet to happen—a way to share information and spread the word about the software. Although not all

LEGO fans use the LDraw software, its influx of users and developers keeps growing. With their input, the LDraw software system has grown stronger. As you will see in this book, LEGO-based 3D software is also part of a larger pool of 3D software developed for and by fans. Much of this software has reached a level of quality comparable to some of the best software available commercially. The relatively closed (and very experimental) world of computer-generated 3D graphics is being cracked open by programs such as these, which simplify and demystify the process, bringing it down to a level that is accessible to all. LDraw-based software is a key part of this scene (among other things, because it is one of the easiest ways to learn the discipline) and will increasingly continue to grow in influence as more people are exposed to it.

Additionally, the availability of the software and modern telecommunication tools has come close on the heels of a huge step forward in the development of computer hardware. Three-dimensional illustration relies *heavily* on hard-core computer equations; some software can literally require many millions of operations per second, putting even the fastest computer processor through its paces. In the past, the casual user required not only specialized software but also top-end (and thus expensive) hardware to create 3D images. Today, the situation is completely reversed. The software is easily available for free, and it will produce outstanding results in even medium-grade computers.

The LDraw resources are organized around two Web sites. The LDraw.org Web site is the beacon for all things related to LEGO-based construction software, including programs that do not follow the LDraw standard. The trustees of James Jessiman's memory have done a great job in providing all sorts of support to help LEGO fans use and further develop Jessiman's original work. Apart from the obvious benefit of having a central repository of information and resources on the LDraw.org site, users and developers find the CAD forums of the LEGO Users Group Network, or Lugnet, site a good place to meet people with a similar interest in all things LEGO, as well as peers willing to lend a hand or review additions to the LDraw system. You can find more information about these and other LEGO CAD resource sites in Chapter 10.

A LEGO CAD System

CAD stands for *computer-aided design*. The term refers to engineering tools such as the ones used in 3D industrial design to model objects using computers. CAD applications provide not only a visual representation of models (whether buildings, vehicles, or something else) but also technical information regarding materials, element count, and so on. Reading this book, you will first learn how to use the programs to create models using computers. Once you have mastered these techniques, we will take a look at some of the advanced capabilities, such as the creation of instruction steps and animation sequences.

In order to get to that level of knowledge, you must first understand a few basic concepts behind all LDraw-based software. You can think of this software as a cross between a graphics program and a database. We will first explore how this software works in practical terms, by taking a look at the parts library, which is a fundamental part of the LDraw system. We will then briefly look at the importance of having an "open" file format.

LDraw System Basics

Programs like LDraw and MLCad allow us to create 3D models using a computer—models that are made of existing elements based on LEGO parts. There are certainly several ways to accomplish this goal. James Jessiman took a three-pronged approach with his creation of LDraw. Instead of basing the system around a program, his approach actually gave equal or more relevance to two other elements of the system: the file format and the parts library. Thus, the LDraw system is composed of three essential elements:

- The file format
- The program(s)
- The parts library

Starting in Chapter 6, you will be introduced to the file format in detail and will learn how to tweak it to accomplish all sorts of special effects in your models. The hands-on use of the actual modeling programs (starting with MLCad) begins in Chapter 4. For now, what is important is that you become aware of the role of these other programs in the general LEGO CAD picture. To understand the overall LDraw scheme, the best place to start is the parts library. The role of the parts library is probably the element of the LDraw system that is easiest for beginning users to grasp. You can think of the parts library as a heap of virtual LEGO parts, much like the ones many of us had as children (and hopefully kept as adults!). In the parts library, you will find the basic material for your models. After all, you cannot build anything without parts.

The parts library has several advantages over a pile of real-life LEGO parts. It is actually a lot more like an organized catalog of part types linked to a warehouse with an infinite supply, available in any color we choose. We will look more closely at the parts library later in this book. For now, let's assume that we have the parts; what else do we need to build our computer models? We need a way to put the parts together and a way to store the models for later editing or perusing.

It would seem that the most obvious way to put these parts together would be to come up with a program that did just that. However, Jessiman took a slightly different approach in developing LDraw. Instead of creating the program side of LDraw first, he decided to define the file format first. As it turns out, this was a brilliant approach. Once he wrote the smart specifications for the file format, everything else came naturally—so much so that over seven years later we are still exploring the possibilities contained in those LDraw file format specifications. Additionally, it must be noted that when Jessiman died, he was still working on new ways to make the LDraw file format (and thus the system) even more efficient. Despite some advanced details of the system that are arguably "incomplete," his LDraw system still performs in a flawless manner. When we call it an impressive achievement, we do not exaggerate.

A Virtuous Cycle

Let's see how all this works in practice to give you a clearer picture of what we are talking about. How can a file format sustain a parts library *and* a suite of programs? The answer is that the LDraw file format allows us to store both *parts* for the library and *models* made out of parts from the library in the same format. Figure 3.1 shows how the three elements—the files themselves, the parts library, and the program to assemble the parts—work together.

Figure 3.1 The LDraw Scheme

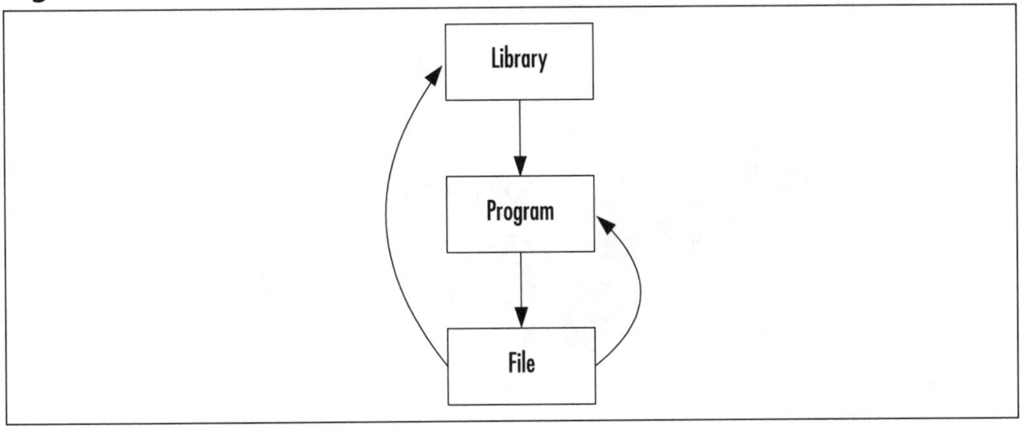

As we explained earlier, to create virtual LEGO models in a computer, you first need the library of virtual LEGO parts. You then need a program that allows you to put them together. Finally, you need a file format to store your finished models. By designing the LDraw file format so that it can *also* be used to define the LEGO parts in the library (allowing programs to use finished models as parts or submodels of larger models), James set up a *virtuous cycle*. Virtuous cycles are composed of elements organized in such a way that they reinforce each other automatically, strengthening the system as it is used.

Let's see the real-life consequences of this approach. By setting up LDraw as shown in Figure 3.1, Jessiman recruited the potential help of every user of his LDraw system. The LDraw executable was actually the program that read the files. Users of the initial release of LDraw had to type the model files into a text editor. When LDraw read those files, it produced an image of the model on the computer screen. This by itself was pretty revolutionary because it let LEGO fans share their creations among each other in a way that had not been possible since LEGO was first created. Even this limited software release was enough to plant the seed for the cycle to take place. Along with the initial LDraw executable, Jessiman included a parts library with three parts in it and the specifications for the file format.

NOTE

The original LDraw parts library contained only the three parts pictured in Figure 3.2. These parts have been rendered in LDraw (we added the name lettering with an outside program). Not surprisingly, these initial parts were virtual replicas of the first brick types that gave birth to the LEGO system as we know it today.

Figure 3.2 The Initial LDraw Parts Library

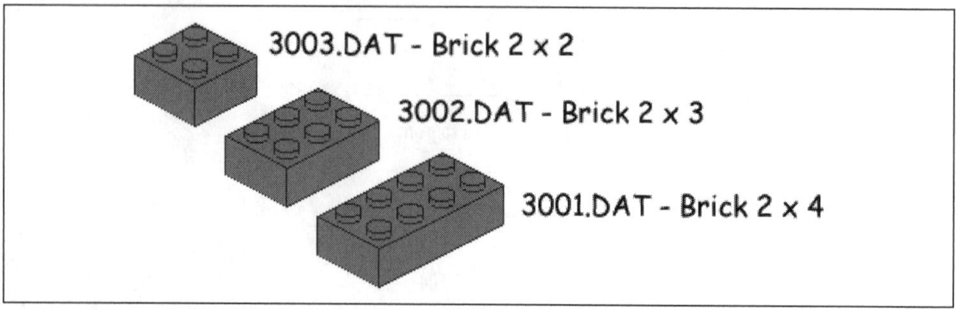

Where is the seed for the cycle in that package? The seed was in the file format: If the users liked the program (and there was no reason for them not to), they would inevitably start contributing parts to the library, since they would need the parts for their own creations. The more parts in the library, the more attractive the cycle became to new users and the more potential developers there were to add parts to the existing library. From the original release of three parts, the library now holds well over 2,000 *official* parts today, with many more constantly being developed in a peer-reviewed process.

A Multitude of Resources

Close to 10 years later, the LDraw system is not only healthy and growing, it has become much more sophisticated as well. The LDraw file format is what brings it all together, but it is no longer the only "seed" of the cycle. Several other programs have been developed to manipulate LDraw files. These include modelers like MLCad as well as format translators such as L3P. Format translators widen the possibilities available to us as modelers, because they allow us to interact with non-LDraw (and non-LEGO-based) applications.

Strictly within the LDraw realm, things have changed quite a bit as well. The LDraw system is still used, but the actual model building is generally done via MLCad (which we look at in the next three chapters) or other similar programs that use the LDraw library and file format. However, it is not only programs and new parts that have come out of the LDraw fan community. It is possible that in the near future, "parallel" libraries will coexist with the LDraw parts library. These new libraries will not store parts; instead, they will hold such diverse things as different effects that better simulate the conditions of real-life LEGO modeling or elements that allow users to better organize their models and create instruction steps for them. Development on some of these components has already begun, with many more unexplored possibilities still to be realized.

Consider this book your first "LDraw kit." When somebody (usually a child) gets his or her first LEGO kit, they are being exposed to a wide and gloriously intricate building system, even though they might not be aware of it right away. Similarly, in your first trips into the world of LDraw, you might not be completely aware of the complexity of the system you are using, either. However, once you get your feet wet and gain a basic understanding of these programs, you will quickly see that there is a wealth of opportunity for those looking to take the next step.

Other LEGO-Based Software

As we explain more fully in Chapter 10, deciding where to draw the line in terms of content was a major concern in writing this book. We finally decided to give readers a clear introductory path into the LDraw system and some of the major programs that have sprung up in its wake. This being the case, we decided to mention but not cover in detail everything not directly related to the LDraw system. As the experienced users of the software know well, once you are aware of the basics, there are simply too many options to try and explore them all in detail. For that reason, we encourage the reader to explore the other resources mentioned in Chapter 10 of this book.

In Chapter 10, the reader is exposed to information related to a fantastic fan community with a wealth of resources not easily matched in other hobbies. Many talented people have contributed all sorts of key elements to the LEGO and LDraw systems, from new part definitions to new programming languages for LEGO robots. Apart from being based in some way on products put out by the LEGO company, these elements all share another characteristic with LDraw: They are created by fans, for fans, and are offered free of charge.

We apologize to the developers of some of these other LEGO-based programs for not reviewing their often excellent, unique, and incredibly engaging software in the detail it deserves. Our main objective with this book is to introduce as many people as possible to the wonderful world of virtual LEGO-based modeling. It is our sincere belief that the LEGO community as a whole gets better as more people become involved. We think James Jessiman would agree.

Summary

This chapter provides some background on LDraw and its place in the LDraw system as a whole. This information is important for the first-time user because it is essential to understand that LDraw is the basic system that sustains the rest of the software described in this book. Although we do not directly use the LDraw program as we work through this book, we do use and rely on its file format and other key components.

The most important aspect of LDraw (aside from its quality) is the fact that it was created by a LEGO fan and thoroughly tested and further developed by other fans. James Jessiman, creator of the original LDraw software, passed away at a young age. In LDraw, he has left behind a technical legacy of impressive elegance and absolute rigor.

The system that Jessiman designed relies on three elements: a parts library, a program to create models, and a file format to store them. Jessiman started by designing a file format that, when completed, would allow the other two elements to happen naturally in the software development cycle. Since the file format accommodates parts from the library as well as models made of those parts, from the beginning there was the possibility to add new parts to the library, which made the program more attractive.

Thus, the more LDraw users, the better the LDraw system became. Today, the LDraw system has grown in both the number of parts in the library as well as the number of auxiliary programs, extensions to the file format, and a host of other goodies waiting for users to discover them.

Solutions Fast Track

Created by Fans, Supported by Fans

- ☑ LDraw is a *system* designed to create virtual LEGO models with computers.
- ☑ LDraw's great success lies in a three-pronged approach that ultimately has created a wide and sophisticated building system that resembles the real LEGO system in many ways.
- ☑ The key resemblance between the LDraw system and the LEGO system is the obsession with top-quality results and nothing else. This approach has ultimately proved successful in both cases.

A LEGO CAD System

☑ The LDraw system relies on a parts library that holds the basic building blocks, program(s) to put them together, and a file format that stores the completed models.

☑ By designing a file format that could store both models and parts for the library, James Jessiman set the successful specifications for a *whole* virtual LEGO CAD building system.

☑ By design, this system gets better as more people use it. It has already evolved quite a bit, and the future looks very bright.

A Multitude of Resources

☑ LDraw is not only a highly successful software system; it is also a great exponent of the pool of resources available to LEGO fans.

☑ These programs are generally created by fans, free of charge; those fans often accept (and even ask for) collaboration and user input.

Frequently Asked Questions

The following Frequently Asked Questions, answered by the authors of this book, are designed to both measure your understanding of the concepts presented in this chapter and to assist you with real-life implementation of these concepts. To have your questions about this chapter answered by the author, browse to **www.syngress.com/solutions** and click on the **"Ask the Author"** form.

Q: Is LDraw a program?

A: LDraw defines both a program and a file format to create and view LEGO models in virtual computer space. The file format and the general program structure have outlived the executable itself as new applications based on it have been developed. LDraw is most often thought of as a complete system of LEGO CAD.

Q: What is freeware?

A: Freeware are programs whose authors have allowed their free use and distribution. They do not carry any moral obligation to pay for their use. Not all freeware programs are open source. In fact, the LDraw program code is not public. But its file format is, which is a key element in its success.

Q: Who created LDraw?

A: Its author, James Jessiman, died in 1997, but the LDraw legacy has been embraced by the international LEGO community. The Web site dedicated to all things LDraw, LDraw.org, has become one of the pillars of this community.

Q: How can I get support for LDraw?

A: The international LEGO community, largely structured around the sites www.lugnet.com and www.LDraw.org, is full of incredibly talented and generous individuals who not only have created free tutorials and additional programs but will also answer any and all LEGO-related questions, no matter how complicated. In fact, the complicated questions posed on these sites are generally the ones that draw the most attention!

Q: Am I expected to fully understand the concept of "virtuous cycles" before using the programs in the next chapter?

A: No. The concepts covered in this chapter are general working principles. It is useful to be aware (even vaguely) of them. They will be of help when you're learning the finer details of creating virtual LEGO models. Overall, the important message contained in this chapter is that the top quality we often take for granted in the LEGO and LDraw building systems is in fact deeply rooted in a very intelligent and accomplished design. Take advantage of it!

Chapter 4

MLCad 101

Solutions in this Chapter:

- The MLCad Interface
- Building a Simple Model
- Working with MLCad
- A Closer Look at the Parts Library

☑ Summary

☑ Solutions Fast Track

☑ Frequently Asked Questions

Introduction

This is the first truly hands-on chapter of this book. By the end of the chapter, you will be building virtual LEGO models on your computer screen.

This chapter introduces you to the MLCad interface and basic modeling functions by walking you through the construction of a simple model. Since you will be using existing LEGO-like pieces, you will quickly find that building models in MLCad is similar to building models with real LEGO bricks. In a way, it might feel like you are playing a video game—a game that is both interactive and intuitive. Since MLCad features a very logical and simple-to-use interface, learning where all of the modeling tools are is a relatively easy chore. One thing is for sure: You will never run out of parts!

Building on the groundwork laid with James Jessiman's LDraw, Michael Lachmann created MLCad to provide the LEGO community with a user-friendly application with a straightforward approach to the job of computer 3D modeling. Like most LEGO-based software, MLCad is a program made freely available by its creator, but Michael Lachmann still retains the source code for the program. He regularly puts out new versions of the software and, more important, maintains a dialogue with MLCad users via the Lugnet forums.

NOTE

The LEGO Users Group Network's Web site (www.lugnet.com) is the most comprehensive Internet resource for the modern LEGO fan. You can find out more details in Chapter 10.

For most people, some apparent wizardry is involved in creating 3D images inside a computer. For practitioners, the real wizardry comes in the form of the actual programs created by enormously talented individuals such as James Jessiman and Michael Lachmann. They bring a tremendous amount of computer power to our desktops, harnessed in such an intuitive manner that all that is required of us is the desire to build something. All the puzzling and sophisticated mathematics involved with computer-aided design (CAD) are present in these applications, but we use them without being aware of them. Magic indeed!

The MLCad Interface

The MLCad interface is all contained in a single window, which we can somewhat customize to our particular needs. The key point, however, is that everything we will use to create MLCad models—the tools, the models themselves, and the parts library—is accessed from a single console. Many functions require navigating through one or two additional dialog windows, but these will be very specific to the function. Most of your modeling time, whether you're sketching model ideas or finishing up the intricate details of your masterpiece, will be spent working with the main window. So let's dive right in.

The Main Window

MLCad is a standard Windows application. Therefore, to start MLCad, use the mouse to double-click on the MLCad program icon (see Figure 4.1) on your desktop. The first time you run MLad you will be presented with an LDraw Base Path window. Click **Browse**, then click on the LDraw folder on your hard drive, then **OK** in the next two windows. MLCad should start right up. MLCad has a single-screen interface layout, pictured in Figure 4.2. It runs inside a window with the appropriate Window Close, Maximize, and Minimize buttons with which you are probably all too familiar. Its menu also displays some standard Windows commands—File, Edit, and Help; some of MLCad's keyboard shortcuts for these options will be familiar to long-time Windows users. Even the toolbars exhibit some of the same icons you might recognize from other Windows applications. Of course, MLCad also has its own specific commands. You will learn about each of these commands in this and the next chapters. First, let's see where exactly on this initial screen everything is.

Figure 4.1 MLCad's Desktop Icon

The interface might seem a bit cluttered to the rookie, but it is actually very practical and its use will quickly become obvious. The main window consists of two areas. The smallish gray area at the top holds the menu bar and the moveable toolbars. Think of this area as the place where you keep your tools. The bottom part is your virtual playground. The two windows on the left, the Available Parts window and the Parts Preview window, allow you to browse and manage the parts library. The large section on the right that appears here as four white empty windows (called *panes* in MLCad speak) is known as the *modeling area*. This is where you will visually build and manipulate your models. The window right above the modeling panes lists the parts for the active model that you are working on. At the very bottom of the screen is a single gray line known as the *status bar*. This bar provides varied and often very useful information, depending on what is underneath your mouse pointer—for example, part names from the parts library, XYZ coordinates in the modeling panes, or information about commands used when navigating the toolbars.

Figure 4.2 The MLCad Main Window

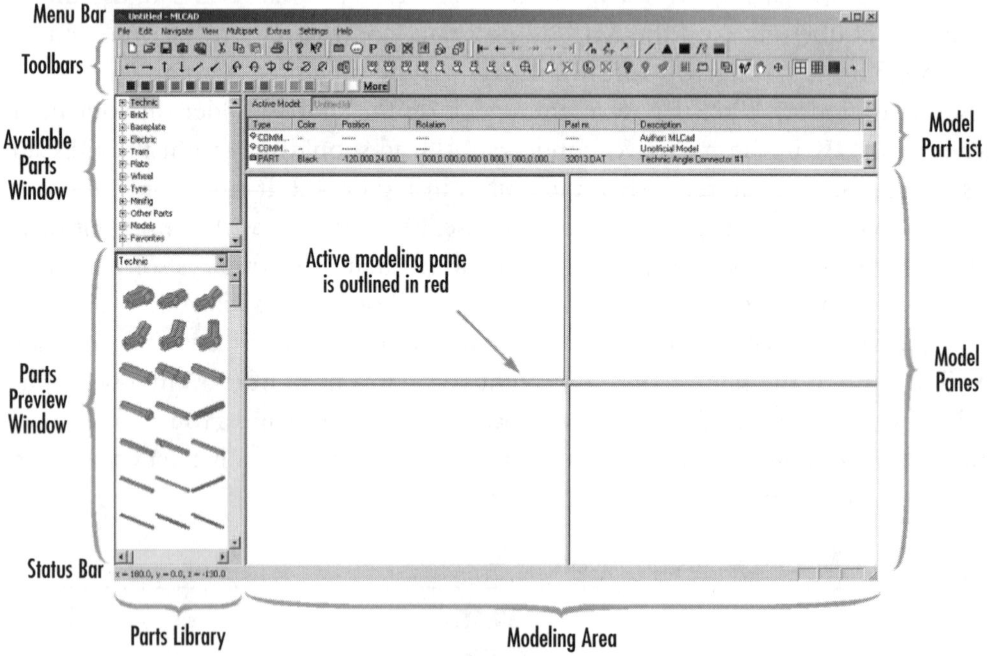

The interface itself is quite customizable. Any of the windows on the lower part of the screen can be resized to easily adapt the program to your needs. You resize a window by placing the mouse pointer over the appropriate window frame (the gray outline around each of the panes) and then clicking and dragging it to resize the panes. In Figure 4.3, the modeling panes are being resized. The mouse cursor changes to a four-pointed arrow icon and the new configuration is shown using gray lines. Once you let go of the mouse button, the frames from the previous configuration will disappear.

Figure 4.3 Resizing MLCad's Windows

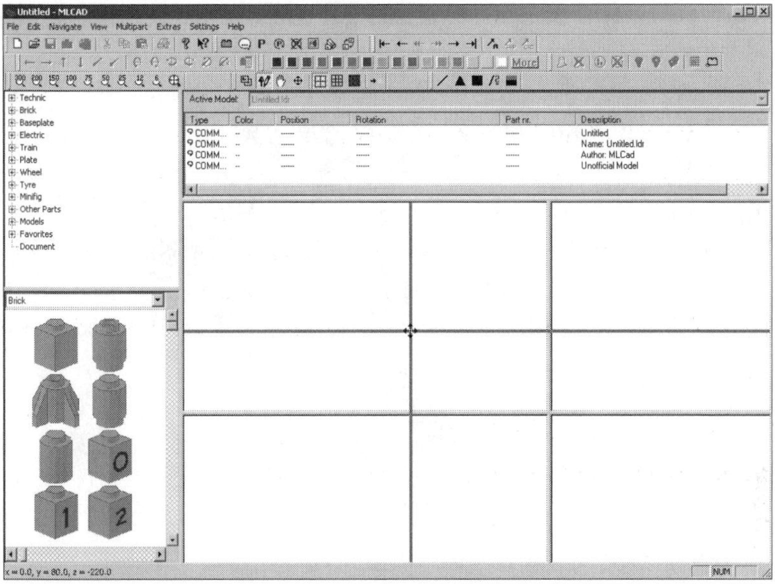

Figure 4.4 shows the model of a fairly intricate mechanism (a double differential drive). To give the user a closer look at the model, one of the modeling panes has been resized to occupy much of the whole lower screen. The two parts library windows have also been reduced. The toolbars (but not the menu bar) have been docked on all four sides of the screen, and one of them has been set floating in the modeling panes. These toolbars can also be turned on and off. This is a good example of how you can customize the main MLCad screen to suit your needs.

The Toolbars

Now that you know how the console is laid out, let's look at the available tools. In this section we take a look at the functions of each toolbar, to provide you with a reference to the buttons that activate the program's commands.

The Main Toolbar

The main toolbar (shown in Figure 4.5) contains buttons for file and edit commands and gives you access to the Help system.

Figure 4.4 A Customized Interface

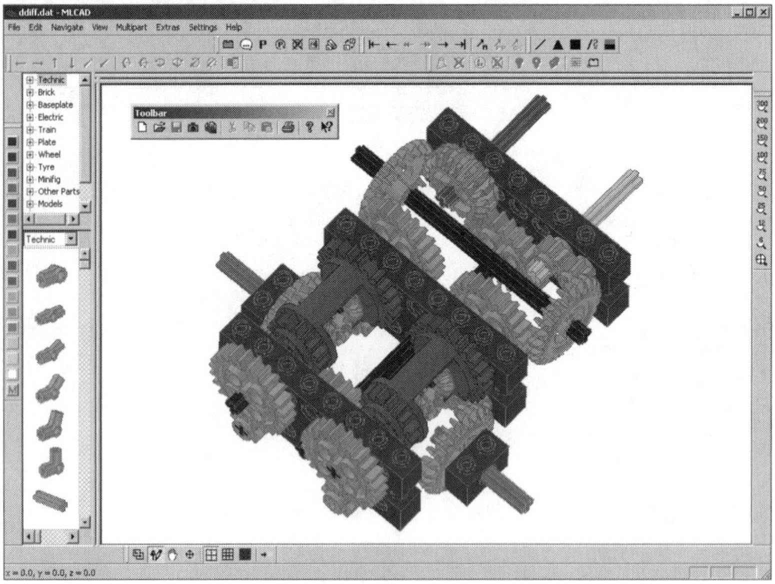

Figure 4.5 The Main Toolbar

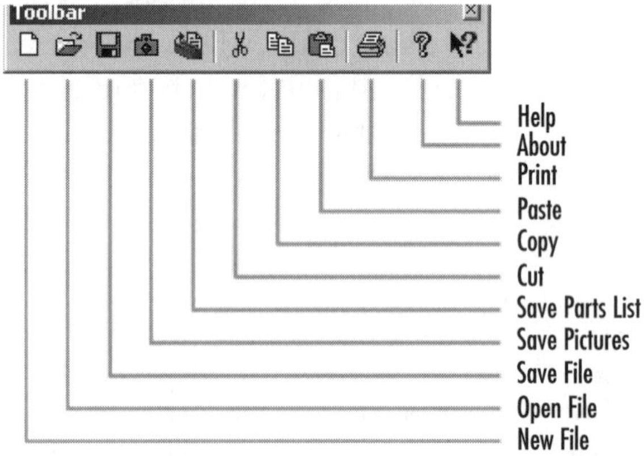

The Viewbar

The Viewbar (shown in Figure 4.6) has eight buttons. The first four buttons control the mode in which the program is running. The next buttons set the size of the grid, which is a modeling aid. The rightmost button activates a relatively new MLCad function that's helpful when you're creating step-by-step instructions for your models.

Figure 4.6 The Viewbar

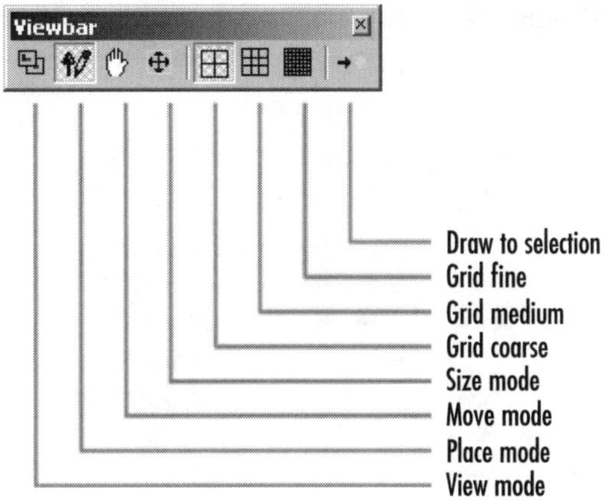

The Zoombar

When you are editing your models, the Zoombar's buttons (shown in Figure 4.7) change the zoom factor for all the modeling panes at the same time. The number on each button indicates the zoom percentage; the higher the number, the higher the magnification, and thus the larger the model will appear in our screen. The Zoom Fit button adjusts the zoom factor of each modeling pane so that a view of the complete model is shown in all of them.

Figure 4.7 The Zoombar

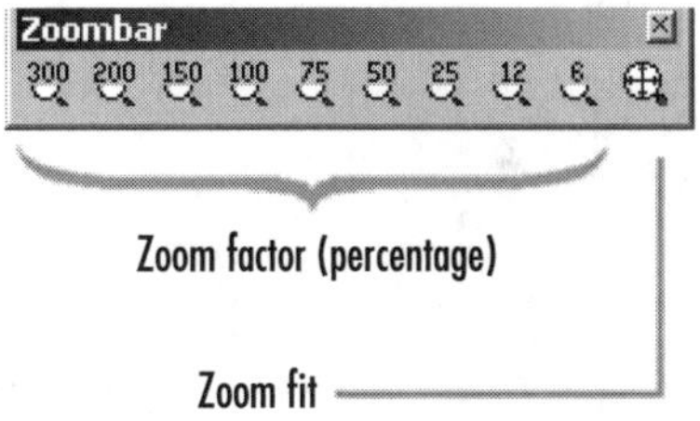

The Editbar

LDraw files were initially created with the idea of providing LEGO fans with a way to share their models. As you will learn later on, MLCad is not only a virtual LEGO modeler; it also provides support for viewing the step-by-step instructions of the models. The Editbar's buttons (shown in Figure 4.8) add such support capabilities to your model files.

Figure 4.8 The Editbar

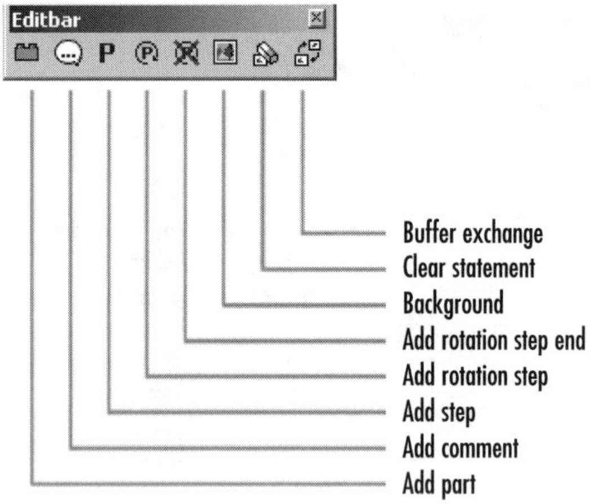

The Movementbar

The Movementbar (shown in Figure 4.9) contains buttons for two sorts of commands. When viewing step-by-step instructions of the models, you navigate them using the first six buttons. Official LEGO instructions are printed in booklets; using these buttons is the MLCad equivalent of flipping back and forth through the virtual pages of a set of instructions for a given model. The last three buttons are used to select parts of the model when you are editing it.

Figure 4.9 The Movementbar

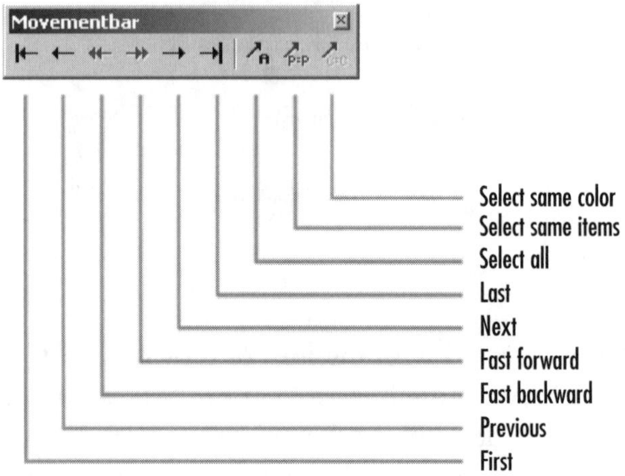

The Elementbar

The Elementbar's buttons (shown in Figure 4.10) are used to manipulate parts while building models. You move them with the straight-arrow icons, in the X, Y, or Z axis as indicated. The buttons with curved arrows are used to rotate the parts in the same axes. The Enter Position and Rotation button at the far right opens a dialog box that permits you to directly type the coordinates for your parts.

Figure 4.10 The Elementbar

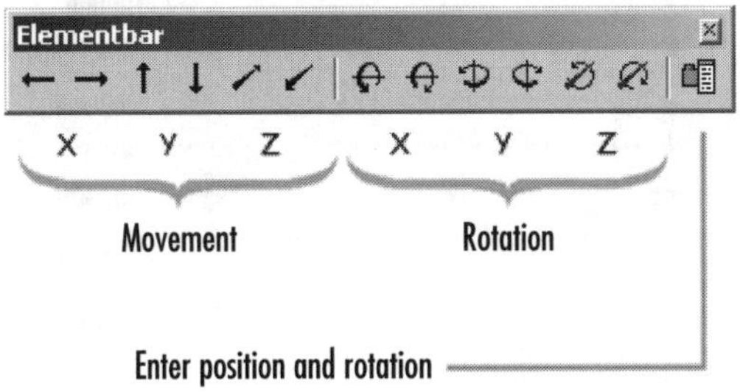

The Modificationbar

The Modificationbar's buttons (shown in Figure 4.11) come in very handy for editing models as well as for building step-by-step instructions. *Ghost techniques* are relatively esoteric commands used to build the step-by-step instructions of complex models made of smaller submodels. Group and Ungroup allow you to bundle together selected pieces of a model for many practical editing situations. The Hide and Unhide buttons let you make parts invisible—also useful in many editing situations when you need to see what lies behind some parts. Snap to Grid and Rotation Point are also editing tools.

The Colorbar

As you might have guessed already, the Colorbar (shown in Figure 4.12) lets you change the color of parts. This is a definite advantage over using real LEGO, which is always limited in the number of pieces available *and* in the color range of the available parts. Not only that—the More button lets you create your own custom colors for the parts.

Figure 4.11 The Modificationbar

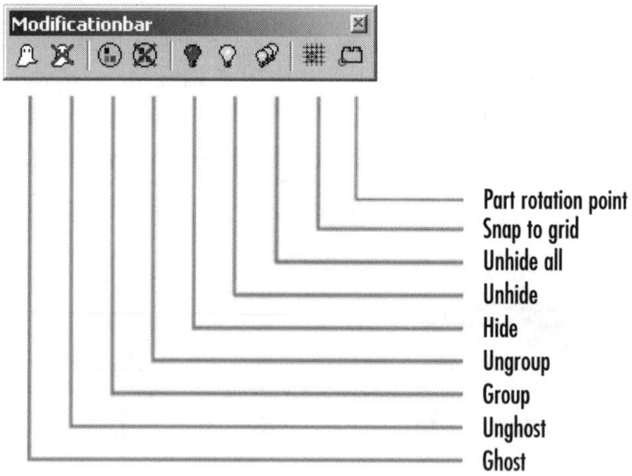

Figure 4.12 The Colorbar

The Expertbar

The Expertbar (shown in Figure 4.13) is truly for experts! Its commands generate basic geometric shapes that are used to create the parts for the library.

Figure 4.13 The Expertbar

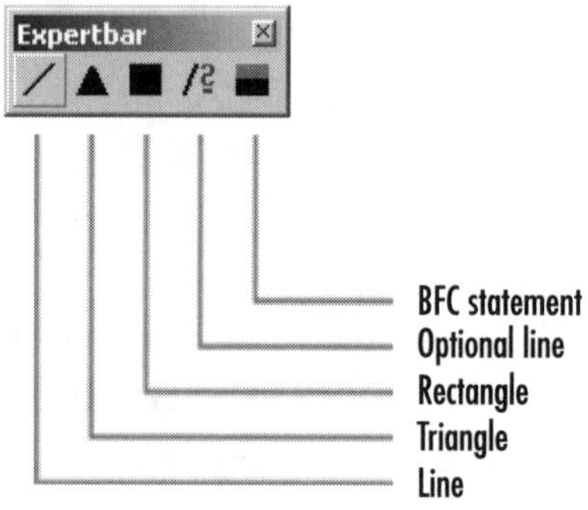

As mentioned earlier, the toolbars can be positioned anywhere on the screen—docked to any of the sides or left floating. They can also be turned on or off, either by clicking the appropriate X button at the top right of the floating toolbar or by right-clicking any of the toolbars. Right-clicking pops up a list of all the toolbars, as shown in Figure 4.14. The ones that have a tick mark next to them are visible. Clicking any name will change the status of the toolbar to on or off.

Figure 4.14 The Toolbar List

To move the toolbars, click anywhere on them except on the buttons—for instance, the "handle" on the left of each bar. Double-clicking it makes the toolbar a floating window. Double-clicking the floating window's top title area returns the toolbar to its former "docked" position.

The actual content of each toolbar is not customizable. Unlike other Windows applications, MLCad's buttons are tied to one specific toolbar and cannot be moved to another or turned off. This is one of those areas where time will undoubtedly bring improvement; we are sure that eventually Michael will get to this minor functional detail and allow even more customization of the interface.

NOTE

Even though it is much more reliable than many commercial applications, it is fair to say that MLCad has a few unpolished edges. The good news is that these edges are trivial, for the most part. We mention them throughout this and the following chapters and, if necessary, provide ways around them.

Building a Simple Model

Now that you know your way around the MLCad interface, let's create a simple model: a sports podium for our minifig champions, like the one shown in Figure 4.15. We will use this model to examine basic modeling concepts and how they are implemented in the program. Once you master these basic concepts, you can then use the podium for a variety of situations, so you can explore more sophisticated techniques *in context*. Later in the book we will populate the podium with minifigs and create a diorama around it.

Figure 4.15 The Sports Podium

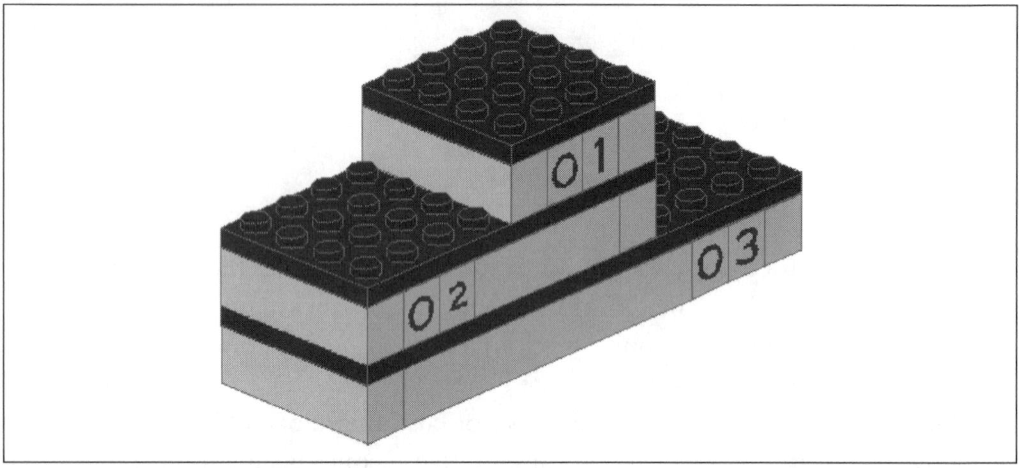

Before We Start to Model: The Rookie Checklist

The LDraw format and MLCad pack a lot more power than might be apparent at first sight. Working with them is often similar to playing a 3D videogame, such as a flight simulator. That makes the user a pilot of sorts. And like all pilots, you must follow a procedure to check to see that your plane is in working condition before take-off. To start creating a new model, you must prepare the program for it. The following checklist will help rookie pilots get the program ready for model-building take-off.

NOTE

If you are not creating a new file but rather are editing an existing model, skip Step 1.

1. Create a new file by clicking the **New File** button on the main Toolbar, as shown in Figure 4.16.

Figure 4.16 The New File Button

2. Set the mode to **Place** by clicking the **Place Program Mode** button of the Viewbar, as shown in Figure 4.17. In this mode, the program allows us to add parts to a model.

Figure 4.17 The Place Program Mode Button

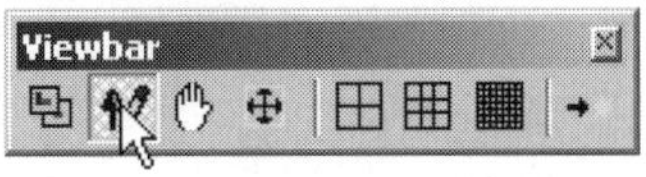

3. Set the grid to **Coarse** by clicking the **Grid Coarse** button in the Viewbar, as shown in Figure 4.18. As you will learn in the next chapter, the grid is an important modeling aid that helps you with the alignment of parts.

Figure 4.18 The Grid Coarse Button

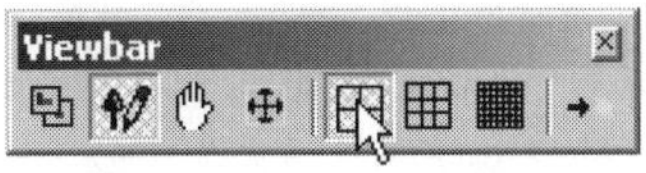

4. Adjust the screen area so that it looks approximately like the one in Figure 4.19. The important things to keep in mind are that you want four view panes of approximately the same size and you want the two parts library windows to occupy roughly a fourth of the horizontal space. We have also resized the view panes so that the Parts List window above them (shown in Figure 4.2) is not visible. The position of the toolbars is not important.

Placing the First Brick

The easiest way to add a part to a model is to drag it from the Parts Preview window into any of the modeling panes. To make a part appear in this window, we have to look for it in the library. There are over 2,000 parts in the library; to ease the task, MLCad organizes them in customizable categories known as *groups*. There are several ways to browse the library, but the easiest method in this case is to select the **Brick** group from the pull-down menu located between the Available Parts window and the Preview window, as shown in Figure 4.20.

Figure 4.19 The Console Set to Start Creating a Simple Model

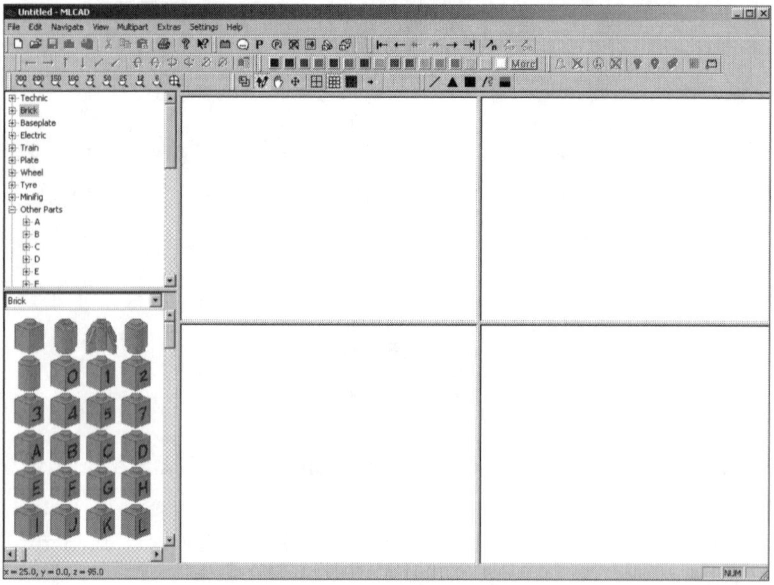

Figure 4.20 The Preview Window Group Pull-Down Selector

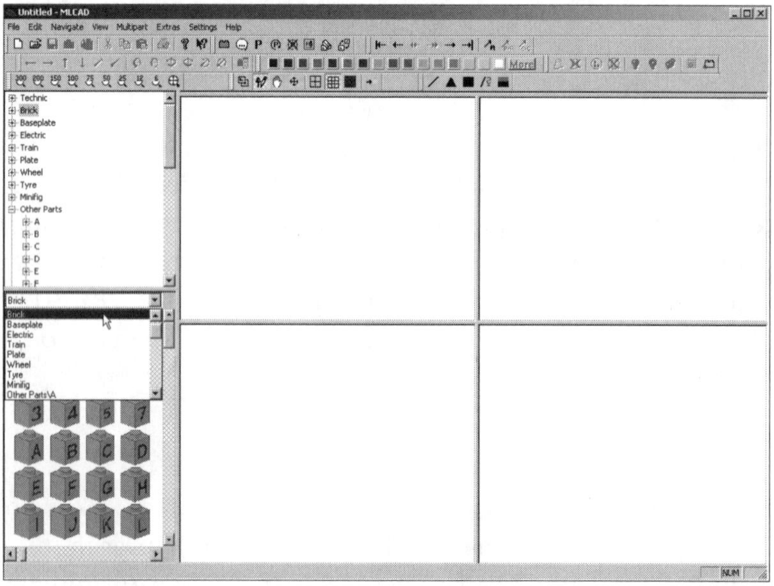

The Preview window will then show small images of the parts for that group. However, the Brick group is fairly large, so not all its parts will fit in the Preview window. The scrollbar on the right side of the Preview window allows you to

browse the rest of the group's contents. Even if you are a LEGO fan, you will probably soon find parts that you didn't know existed. Incidentally, this window has two zoom factors: one that shows many parts in less detail and one that shows fewer parts in more detail. You can switch between the two by right-clicking with your mouse anywhere in the window, as shown in Figure 4.21.

Figure 4.21 Two Different Zoom Factors in the Preview Window

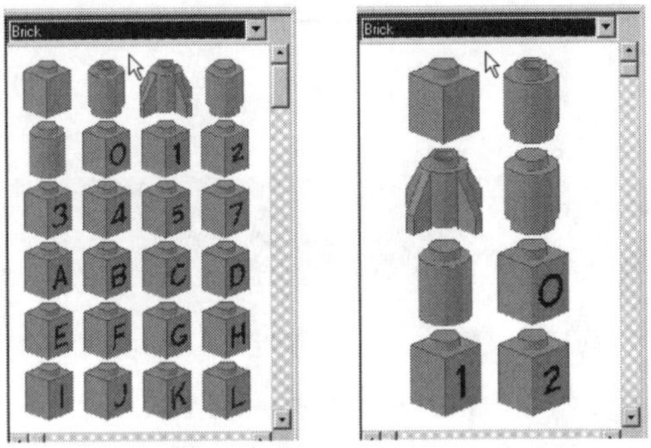

Locate a 1 × 8 brick, as shown in Figure 4.22. Although LEGO measurements are very easy to understand (see the sidebar that follows), there are often many variations for the same part. Notice that as you move the mouse pointer over the Preview window, the names of the parts appear in the status bar underneath.

Figure 4.22 A Preview Window with the Part Name in the Status Bar

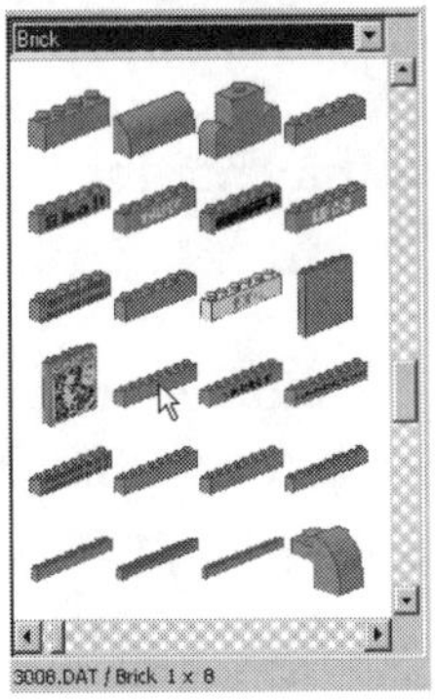

NOTE

If you include a large number of parts in your Preview window by making it larger, it might take MLCad some time to render all the parts on screen. Try to strike a balance between seeing many parts and the time it takes to display them; especially in the beginning, it is good to get acquainted with as many parts from the library as possible.

Designing & Planning...

LEGO Measurements

LEGO measurements are done using *studs*. Luckily for us, studs are generally visible and obvious. The parts names (which we get into later) always include a stud measurement. Generally, this measurement is a digit, followed by an × and another digit. Sometimes the measurements have an additional × and one more digit. The first digit is always the smallest. Thus, the 1 × 8 brick measures one stud wide and eight studs long. Close to the 1 × 8 brick in the Parts Preview window is the 1 × 6 brick family (refer back to Figure 4.22). The members of this family include some tall members of the 1 × 8 brick group, such as the 1 × 6 × 5 brick right next to our 1 × 8 specimen. The "× 5" part of the 1 × 6 × 5 designation, as you might have guessed, reflects the part's height. We'll return to this subject later in the chapter.

Once you have the desired part showing under your mouse pointer in the Parts Preview window, you are ready to go. Simply click on the part and drag it with your mouse into any of the modeling panes, and release the mouse button. Congratulations! You've just placed the first part of your first MLCad model. The process is captured in Figure 4.23.

Figure 4.23 Placing Your First Part

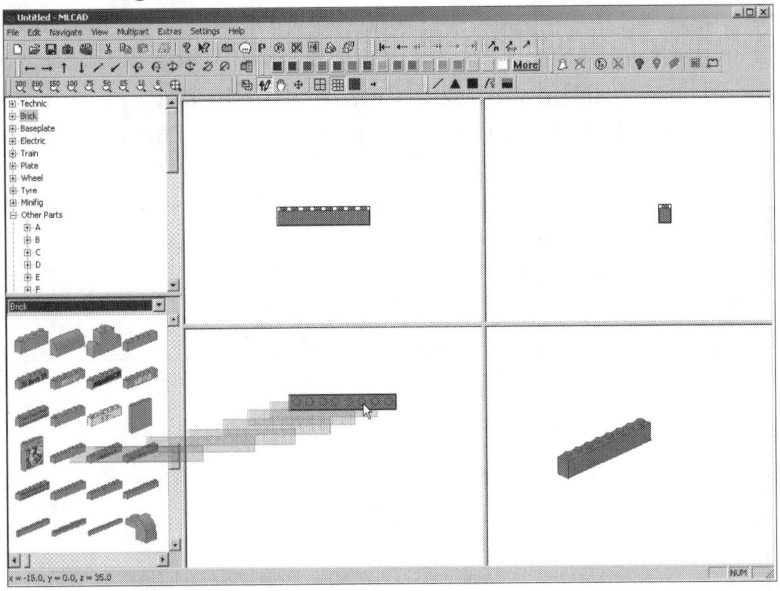

Designing & Planning…

Working with No Gravity

Modeling with virtual LEGO has some pretty amazing advantages over the real thing (and vice versa , of course!). Using MLCad is like being an astronaut in a virtual playground. Parts simply "float." This is generally to our benefit; it allows us to create models that would be impossible to create otherwise. But is that a good thing? This is a subject that LEGO fans will discuss forever.

Moving and Adding Parts

Try to get a feel for the virtual space by moving the part in the modeling panes. Take a look at Figure 4.24. Here you can see that the modeling panes show different views of the 1 × 8 brick we just pulled into them. Three of the views are flat: a front view (the top-left pane), a view from the left side (the top-right pane), and an overhead view (the bottom-left pane). The bottom-right pane shows a 3D perspective of the 1 × 8 brick. In the next chapter we describe the virtual space coordinates in more detail. For now, you can move the part by clicking and dragging it in any of the "flat view" panes. If you drag the part off the pane, it will simply return to the original position. Notice that as you move the mouse with its left button pressed, the part remains in its place, but a black outline (a "bounding box") of it moves with the mouse. This is a great way to use the current position of the part as a reference.

Figure 4.24 A Bounding Box

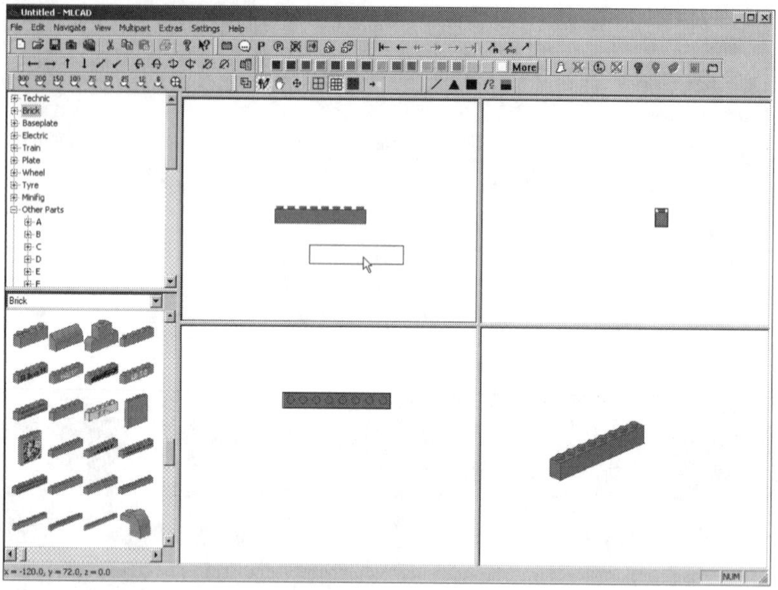

NOTE

You might notice that when you move the part, instead of sliding smoothly, it moves in short "jumps." This is the effect of the **Grid Coarse** setting, which allows the user to align basic parts with great ease. The next chapter explains this modeling aid in more detail.

Let's add a second part. We could go back to the Parts Preview window, but we can also make an instant copy of a part by clicking and dragging it (as though we were moving it) while holding down the **Ctrl** button on our keyboard. The bounding box we saw before is still there, but there is also a smaller box with a + sign inside (see Figure 4.25). If we let go of the mouse button, we will have two identical copies of the original part, as shown in Figure 4.26. Talk about magic!

Figure 4.25 The Move-Copy Function

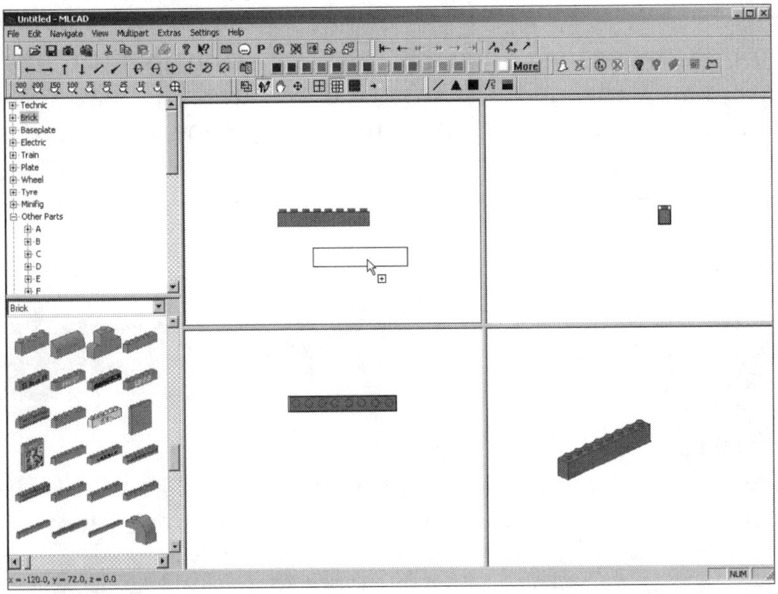

Figure 4.26 The Top Brick Is Selected

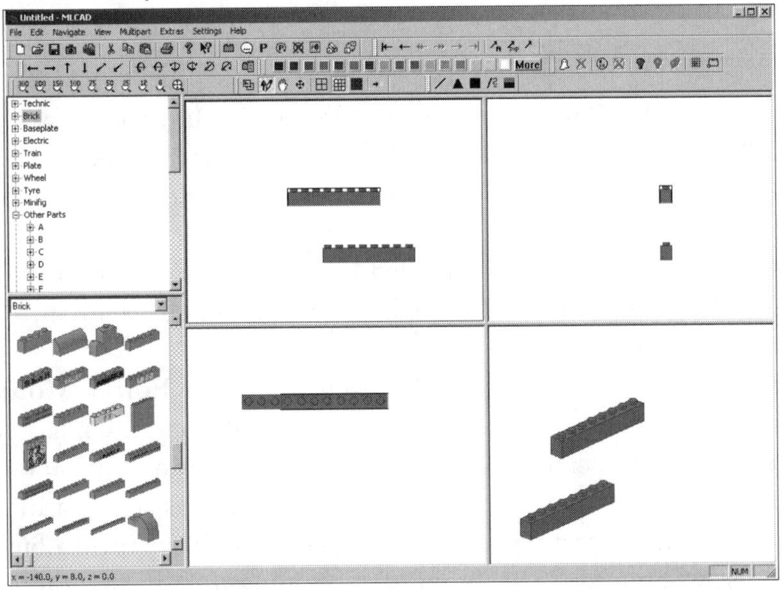

NOTE

If you try to move the part in a 3D view by dragging it with the mouse, you will simply change the point of view. To avoid dramatic changes, try not to click too far from the center of the pane, and move the mouse slowly.

Designing & Planning…

Working with No Density

Not having gravity is one thing; not having density is another. Unlike real-life LEGO plastic bricks, MLCad parts do not collide with each other—they simply overlap. As you will see shortly, there are easy ways around this characteristic, and a long-term solution is in the works. For now, we have to be alert to parts not positioned as they would be in real life.

Selecting and Modifying Parts

We now have two 1 × 8 bricks in our modeling panes. In order to work with one particular brick, you simply click it. The selected part will be surrounded by a thin-lined bounding box and will have a cross in its center. In Figure 4.26, the top brick is selected.

Unfortunately, the next brick we need for our podium is not of the 1 × 8 variety but rather the shorter 1 × 4 brick. Just as we can create a copy of a part, we can also modify its type.

Once you have selected which brick to change, select **Edit | Modify.** The Select Part dialog window pictured in Figure 4.27 will appear. Unlike the Preview parts window, when we modify a part here, we can only choose it by name; there are no previews. Luckily, we know we are looking for a Brick 1 × 4, so it is only a matter of sorting the list by clicking the **Description** button-header and browsing it for the appropriate part using the scrollbar on the right side of the window.

If we click **Brick 1 x 4** and then click **OK**, our model should look similar to the one pictured in Figure 4.28, with two bricks of different types. Note that they might appear in a different position on your screen.

Figure 4.27 Modifying the Part Type

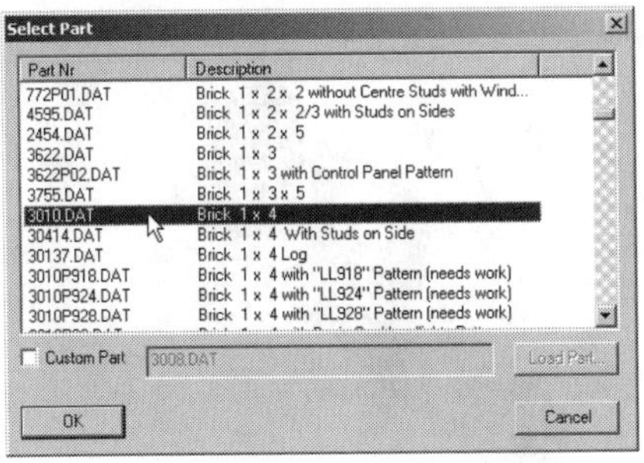

Figure 4.28 Bricks of Different Types

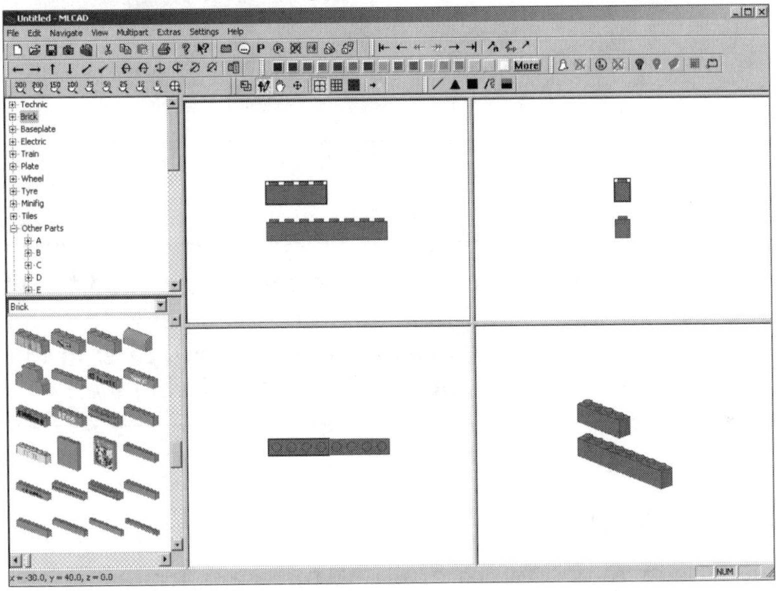

Now might be a good time to practice moving both the bricks in each of the flat view modeling panes. Try to place the bricks in all the positions shown in Figure 4.29. Remember that you can only move the bricks in the flat view modeling panes. Later we will use precise methods to position them, but when you're first learning to model with MLCad, it is much more important to quickly develop intuitive skills based on the parts' shapes.

Figure 4.29 Two Bricks in Different Positions

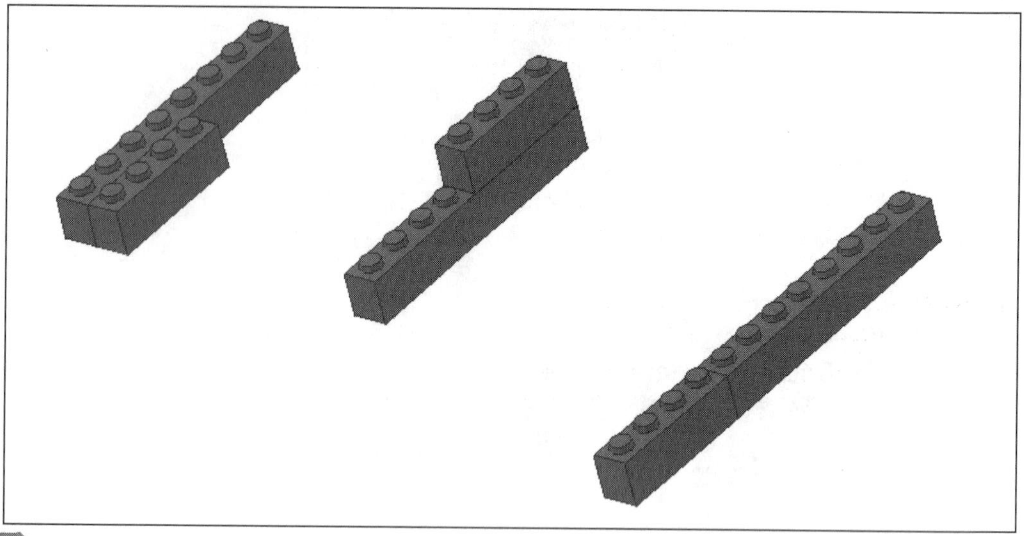

NOTE

Now that our model has more than one part, it is a good time to start saving it. Simply select **File | Save** and provide a name for the file (let's use *podium* for the sample model we are building). Later in the chapter we will go into more detail about opening and saving files.

Part Orientation

Once you learn how to position parts relative to each other, the next step is to learn how to orient them. The bricks we have used so far are parallel to each other. Let's rotate one of them so that they become perpendicular. Select the 1 × 4 brick. Click the **Y rotation axis** button of the Elementbar, as shown on Figure 4.30.

NOTE

To avoid time-consuming accidents, double-check that the grid is set to coarse *before* you perform any rotation—at least until you read the section on rotations and rotation points in the next chapter. Refer back to Figure 4.18 for a reminder on how to set the grid to coarse via the appropriate Viewbar button.

Figure 4.30 The Y Rotation Axis Button

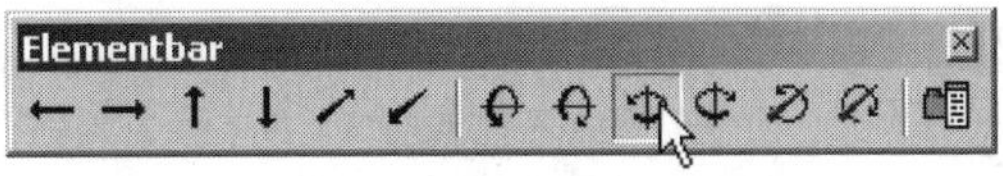

The selected brick will rotate 90 degrees; it will be perpendicular to its former orientation. Now position it next to the other brick, as it is shown in Figure 4.31. Make sure that the bricks are correctly aligned.

Figure 4.31 Perpendicular Bricks

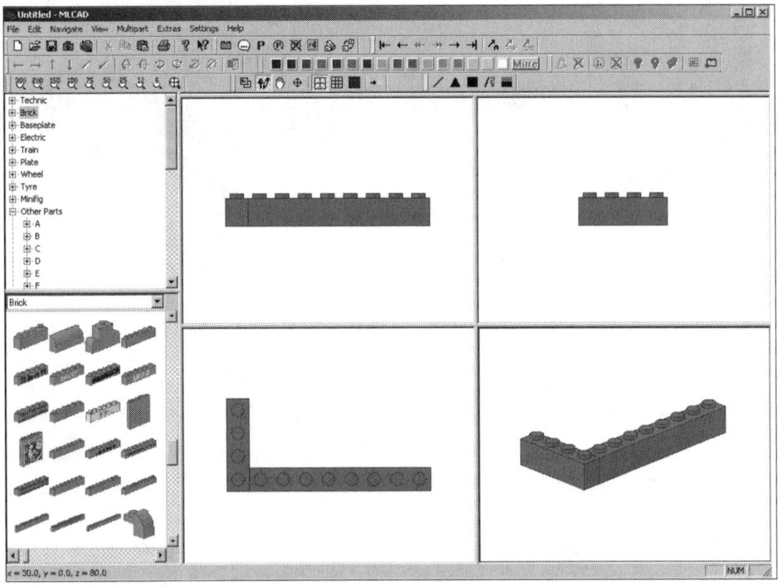

Let's add more bricks. Using the techniques we just used, place a longer 1 × 10 brick on the other side of the 1 × 4 brick. We suggest that you copy the 1 × 8 brick and modify it into a 1 × 10 brick using the technique we just discussed. You could also go into the Parts Preview window and select a 1 × 10 brick, then copy the 1 × 4 brick and position it on the other side of the new 1 × 4 brick. The model should look like the one in Figure 4.32.

NOTE

Copying and then modifying bricks is a time-saving technique. Once you get used to doing this, it will become much easier to copy and modify rather than going into the Library Parts Preview window every time you want to grab a new brick.

Figure 4.32 Four Bricks

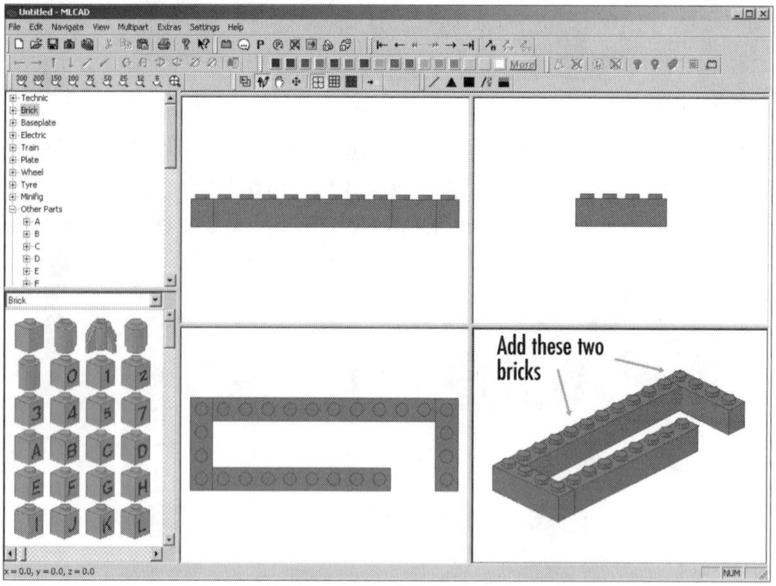

Next we want to add two "special" bricks. Go back to the Parts Preview window and return to the beginning of the Bricks group. Find the 1 × 1 bricks that have letters and numbers on them. Select the 1 × 1 brick with blue 0 on it and drag it into the modeling panes. Attach it to the podium, as shown in Figure 4.33.

Figure 4.33 Bricks with Numbers

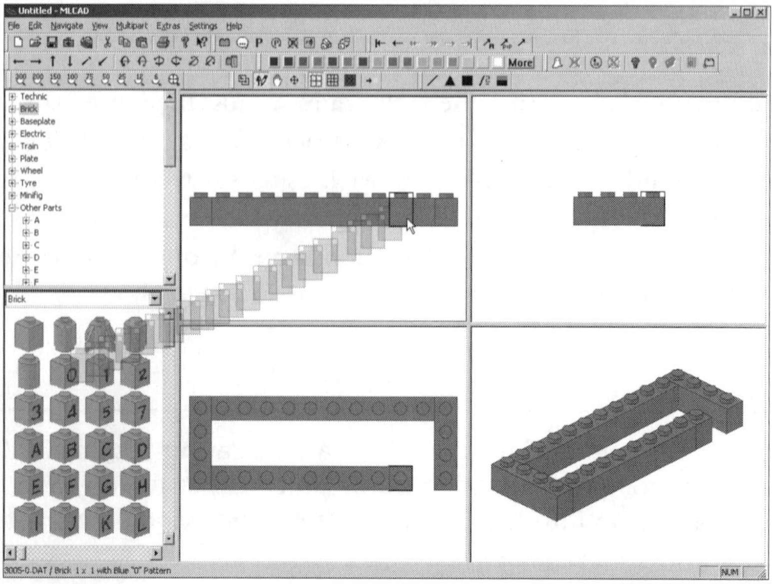

Something is wrong—the 0 is not showing! As you can see in the Preview window, the 0 appears on only one side of the brick. The brick is not orientated correctly; the *face* with the 0 on it is obscured by other parts and the perspective of our view of the model. We need to rotate the 1 × 1 brick as we did for the 1 × 4 brick when we placed it perpendicularly to the 1 × 8 brick. With the 1 × 1 "0" brick selected, rotate it on the Y axis as you did earlier with the 1 × 4 brick (see Figure 4.30). You might need to click the **Toolbar** button several times, but the face will eventually align correctly and the number will be shown. As shown in Figure 4.34, every time we rotate the 1 × 1 "0" brick, the face with the number will face a different direction, sometimes hidden by the other bricks beside it, sometimes not. If you keep clicking the **Y rotation axis** button, the brick will continue to turn. The 0 will disappear and reappear again every four clicks. Leave the 1 ×1 "0" brick with the 0 visible and facing outward, as shown in the example on the right in Figure 4.34.

Figure 4.34 Rotation of a Numbered Brick

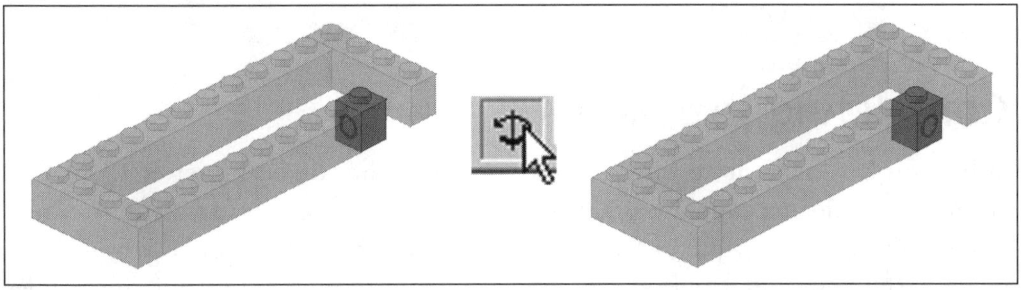

NOTE

Always be aware of the possibilities that each part holds. As we just saw, even the simplest shape might offer more possibilities than initially meet the eye.

Now place a copy of the 1 × 1 "0" brick next and to the right of the 1 × 1 "0" brick you just added, and change it to the type 1 × 1 Brick with Blue 3 pattern. The model should now look like the one in Figure 4.35.

Figure 4.35 The First Stage of the Podium

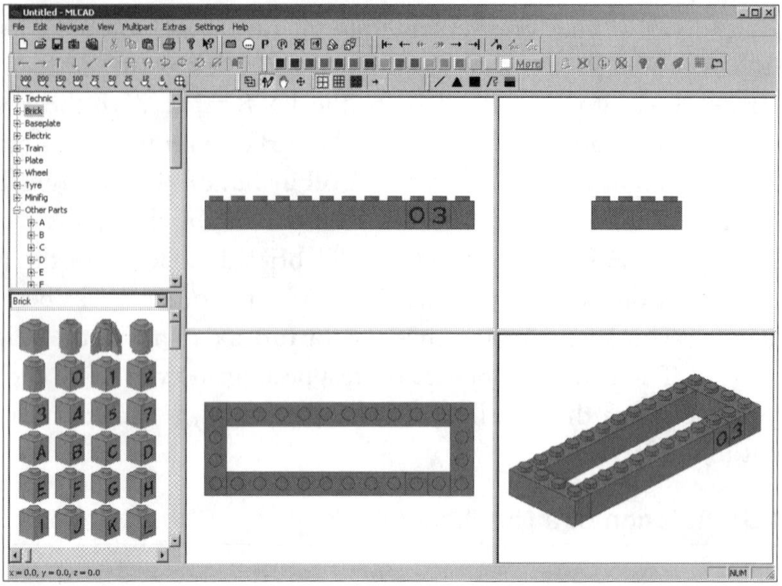

Completing the Model

What you have seen up to now is pretty much the bread and butter of LEGO computer modeling: adding parts, positioning them, and orienting them. As you will see in the next chapter, MLCad (supported by the LDraw format) provides us with a very sophisticated way to control each of these essential building fundamentals. Thus, it pays to become familiar with the basic concepts as quickly as possible: They really are all there is to modeling. There are certainly still more interesting tools, such as working with more than one part at the same time, but these are arguably secondary to the process. Without understanding how to add, position, and rotate parts, you simply can't build anything with MLCad.

Let's take a look at some of these secondary techniques as we finish the podium. Before adding more parts, let's change the color of the parts that we have already placed in the model. Changing the color of a single part is as simple as selecting it and clicking one of the colored buttons of the Colorbar. What makes the exercise more interesting is the fact that we are going to change the color of all the parts at the same time. As we mentioned before, to select a part, simply click it. To select more than one part, click each of the parts that you want to select while holding down the **Ctrl** key of your keyboard. When two or more parts are selected, they act as a block. For instance, if we select both numbered blocks by holding down the **Ctrl** button on the keyboard and selecting them both with the mouse, we can then move them both around as if they were a single part. Notice in Figure 4.36 how the parts both have individual bounding boxes around them. To deselect parts, simply click the empty space of any modeling part set to a flat view. Once we deselect the parts, the bond breaks.

Figure 4.36 Moving Both Numbered Bricks at the Same Time

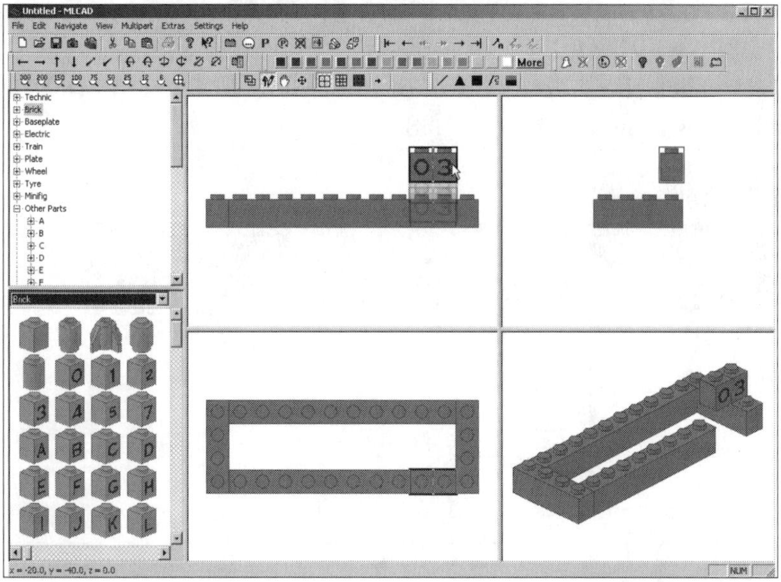

Grouping bricks into blocks in this manner is the simplest way to use two or more parts as one. As we will see in the next chapter, being able to group parts together in selections, groups, models, and submodels is a very important characteristic of the LDraw and MLCad modeling system. Let's take a look at another way to select multiple parts.

Since we want to change the color of all the red parts to yellow, we can use another handy multiple-part selection mechanism: the *Select Same Color* command. This is one of a series of new commands that Michael Lachmann has recently added to MLCad. These new commands can be triggered with the three buttons on the right part of the Movementbar (see Figure 4.37) or via the **Edit | Select** menu item. The other two commands are *Select all* (the button with the *A* underneath the arrow), which selects all the parts in our model, and *Select Same Type* (the button with *p=p* under the arrow), which selects parts of exactly the same type.

Figure 4.37 The Select Similar Color Button

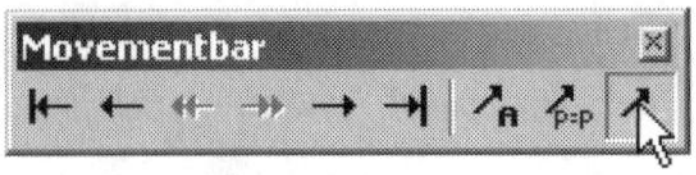

Select one brick from the podium (it doesn't matter which one, since they are all the same color) and click the **Select Same Color** button at the far right of the Movementbar, as was shown in Figure 4.37. All the red parts in the model will be selected, which in this case is to say that *all* parts in the model should be selected. Now click the yellow button of the Colorbar. The red bricks are now colored yellow, as shown in Figure 4.38.

Figure 4.38 Yellow Bricks

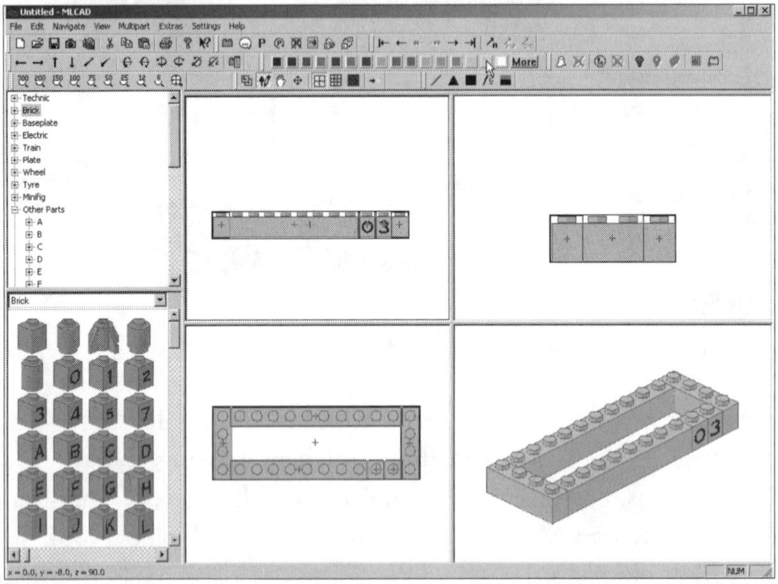

Now, let's add a different type of part. It is not a member of the brick family but of the plate family. Select **Plate** in the pull-down menu above the Preview window, and look for a 4 × 2 plate. Add it to the model, as shown in Figure 4.39, and color it black by clicking the **black** button on the Colorbar while the 4 × 2 plate is selected.

You can now build the remaining levels of the podium using the techniques we have just discussed to add, copy, position, color, and modify the parts. It's up to you to decide which is the best method to do so, but here we walk you through one way to go about it.

Figure 4.39 Adding a Black Plate

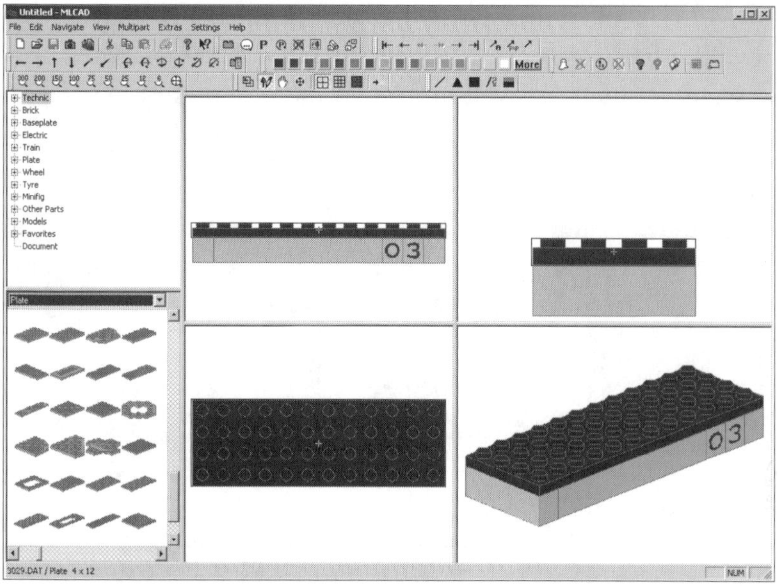

Designing & Planning…

Actual Colors Might Vary

The important part of this second phase of the exercise is not so much to replicate the steps exactly (colors might indeed vary due to the program's defaults) but to think in terms of selecting the right solutions to modeling problems from a set of different approaches. For instance, changing the color of a group of parts is a trivial task, but choosing the best way to select the parts is what makes a skilled modeler. Many MLCad functions, such as selecting several bricks to use as a group, can be accomplished in a number of ways. Use whichever method seems more natural to you. Incidentally, it is also worth noting that not every function of MLCad that's available when working with single parts is available when working with a selection of multiple parts. For instance, we can color more than one part at the same time, but the *Select Same Color* function is unavailable if we have more than one part selected—even if they are the same color. This will be yet another factor in choosing the method that is best for the task at hand.

To create the second level of the podium (shown in Figure 4.40), you can copy the 1 × 4 brick and add it to the left side of the 4 × 2 plate as shown, directly above the first 1 × 4 brick. Since two other 1 × 4 bricks are involved in this second stage, we might as well make two more copies of it and position them accordingly, as shown.

Figure 4.40 The Second Stage

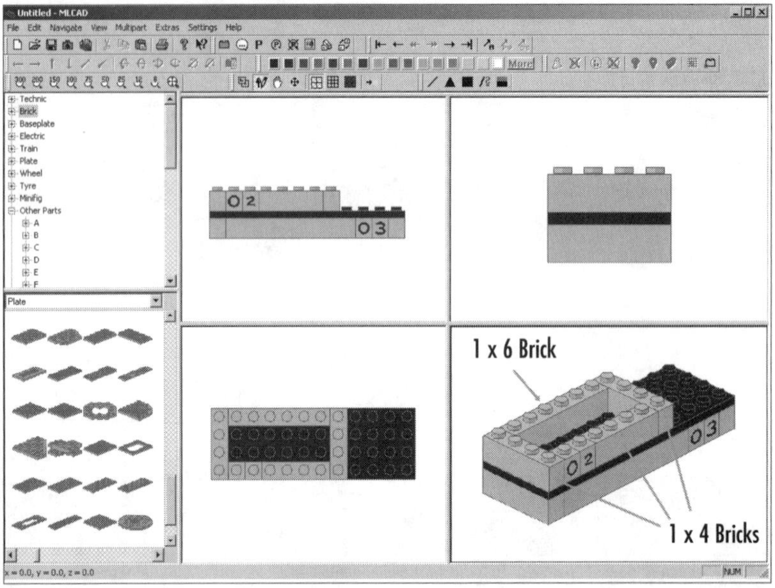

You then need to add a 1 × 6 brick, as shown. You can do this by modifying one of the 1 × 4 bricks into a 1 × 6 brick or by simply dragging a new 1 × 6 brick from the Parts Preview window into the modeling panes. Finally, the two numbered 1 × 1 bricks are copied and placed as shown. The brick with the 3 pattern is changed into one with a 2 pattern via the **Edit | Modify** command.

Building the third level involves repeating the same methods. For instance, make a copy of the 4 × 12 plate and modify it into a 4 × 8 plate. Place it on top of the second-level bricks, as shown in Figure 4.41. Copy one of the 1 × 4 bricks from the second stage and place it on top of this plate. Copy it again to make the other 1 × 4 brick. Add a 1 × 2 brick. Copy the numbered bricks and change the 2 into a brick with a 1.

Once you add the topmost 4 × 4 plate, as shown in Figure 4.42, the podium will be complete. You have now been exposed to a good number of the typical modeling functions and situations available in MLCad. Now that you have first-hand experience with the program, it is time to start widening your view to provide you with the larger picture. Before we explore the modeling functions in more detail in the next chapter, let's finalize the description of the program's interface and its basic administrative functions. Once you are familiar with them, they will never get in the way of your building—just as with real LEGO.

Figure 4.41 The Third Stage

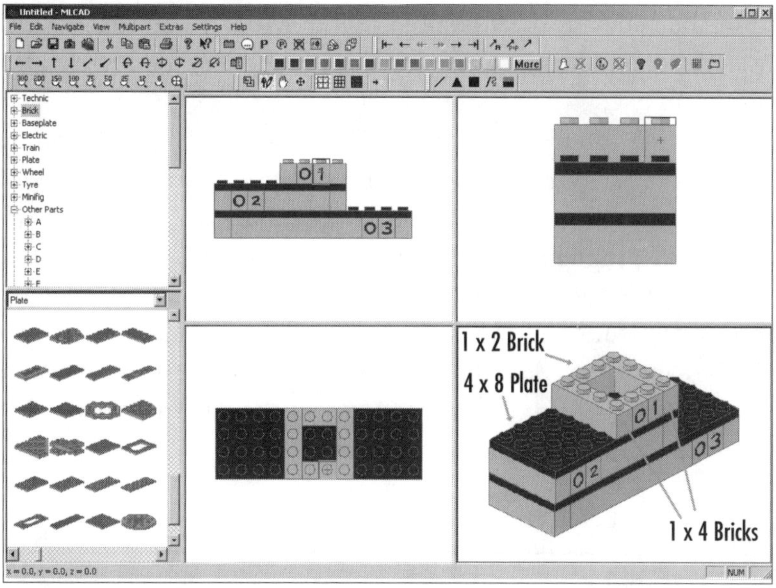

Figure 4.42 The Final Podium

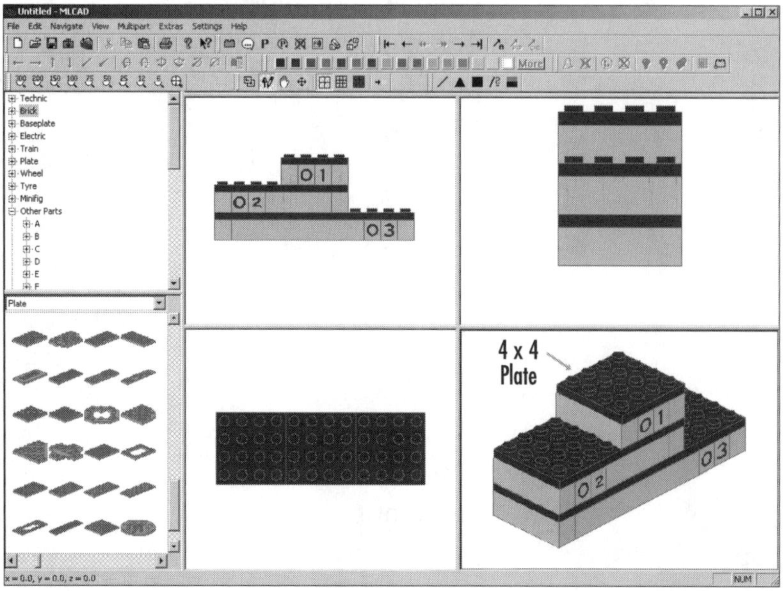

Working with MLCad

MLCad is a gigantic step forward in user-friendliness for the LEGO-CAD world. A lot of that improvement comes from its relatively simple interface. As you have seen, a single screen contains all the tools you will use to build your virtual LEGO models. The only other windows in MLCad are dialog boxes related to specific commands, such as the one used by the Modify command we used in the podium model's construction (refer back to Figure 4.27). Table 4.1 lists all the programs shortcuts.

Each of MLCad's functions and commands can be accessed in several different ways. As in other Windows applications, we can "drive" MLCad using only the computer's mouse. This is the easiest but generally not the fastest or most precise way to use any program, and MLCad is no exception. Its commands can be also accessed via other means. For more accuracy and/or speed, we will use a combination of keyboard and mouse actions or, in some cases, we might even type coordinates and other data directly into the program. In practice this means that we can throw models together pretty quickly and then later refine them as much as we need, without much more effort.

Table 4.1 MLCad Shortcuts

Key or Key Combination	Shortcut For
A	Rotate selection
C	Change part color
I	Add new part (or copy and paste previous part)
P	Modify selection
Cursor left	Move –X
Cursor right	Move +X
Cursor up	Move +Z
Cursor down	Move –Z
Home	Move +Y
End	Move –Y
PgUp	Previous
PgDn	Next
+	Zoom in
–	Zoom out
Ctrl + PgUp	First

Continued

Table 4.1 MLCad Shortcuts

Key or Key Combination	Shortcut For
Ctrl + PgDn	Last
Ctrl + C	Copy selection
Ctrl + D	Duplicate selection
Ctrl + X	Cut selection
Del	Delete selection
Ctrl + V	Paste copy buffer
Ctrl + M	Modify selection
F1	Help
F2	Activate View mode
F3	Activate Edit mode
F4	Activate Move mode
F5	Activate Size mode
Ctrl + cursor left (Keypad 4)	Rotate anti-clockwise along the Y axis
Ctrl + cursor right (Keypad 6)	Rotate clockwise along the Y axis
Ctrl + cursor up (Keypad 8)	Rotate anti-clockwise along the X axis
Ctrl + cursor down (Keypad 2)	Rotate clockwise along the X axis
Ctrl + Home (Keypad 7)	Rotate anti-clockwise along the Z axis
Ctrl + End (Keypad 1)	Rotate clockwise along the Z axis
Ctrl + G	Group selected items
Ctrl + Shift + G	Snap selections to grid

Saving and Opening Model Files

To save a file, we can press the appropriate keyboard shortcut (**Ctrl + S**), select the **File | Save** menu option, or use the **Save** button on the Toolbar. Figure 4.43 shows the three buttons used for creating a new file and saving and opening an existing file.

Files can only be saved if they have been altered since they were last saved. The program detects this status automatically and enables or disables the Save command appropriately. We can use the standard Windows command **File | Save as ...** to save new copies of an existing model. The **File | Save, File | Save as ...**, and **File | Open** menu paths all use a standard Windows file dialog box for file operations, as shown in Figure 4.44.

Figure 4.43 Administrative File Command Buttons of the Toolbar

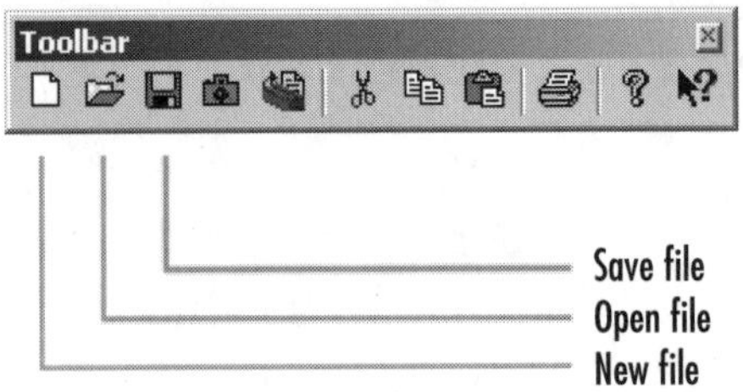

Figure 4.44 The File Operations Dialog Box

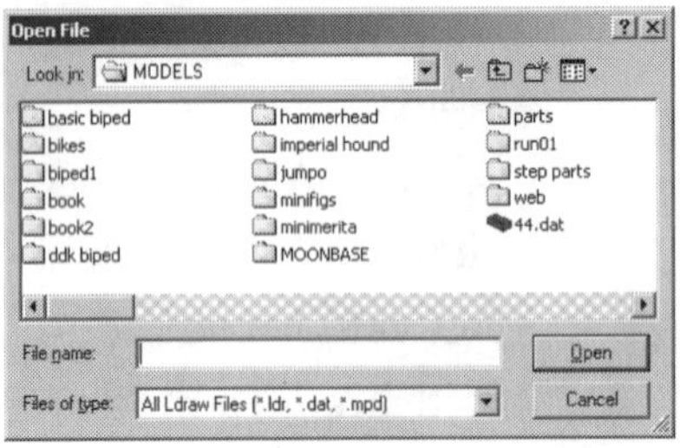

NOTE

LDraw's original file extension is .DAT. All the parts in the parts library carry this extension. However, this extension has become increasingly inconvenient over the years. As the Windows user base grew and Microsoft strengthened its file association features, the .DAT extension has proved to be too generic, in that many other non-LEGO programs now use it.

With the latest release of MLCad, the LDraw-based model files now carry the .LDR extension. Since the part library's extension is hardwired into the LDraw program, the parts will continue to carry the .DAT extension. Thus, all files compatible with the original LDraw format can now carry the extensions .DAT or .LDR. The internal file format is exactly the same.

We hope that this interim solution will eventually evolve into something more aligned with what we think was James's original intent: to make the parts library and model files *completely* interchangeable.

Additionally, MLCad has recently introduced an LDraw "flavor" (or format variation): Multi Model Files. These files carry the .MPD extension. We discuss .MPD files in the next chapter.

Using one of the save methods, save the podium model as *podium*, if you haven't done so yet. MLCad's default directory for disk functions in new installations is LDraw\models. When creating models, many users generate multiple files (for instance, alternatives to a particular model). It is always best to organize your models in folders. Thus, we suggest you first create a new folder called *powertools* (using the appropriate Windows methods) and then put your podium model file into this folder. *Podium* is a good name for the file.

Once you have saved the file, use the **File | New** command (**Ctrl + N**) or the appropriate Toolbar button (refer back to Figure 4.42) to create a new model. Doing so will empty the modeling space and let you start from scratch—always an emergency solution to keep in mind should you get into more trouble than you can handle with a particular model.

NOTE

MLCad does not yet have an Undo function. Save often and use **File | Revert** as a substitute. This command brings the model back to the state it was in when last saved.

Load the podium file again (via the **File | Open** menu option, for instance). Upon loading the file in a new installation of MLCad, something odd will happen. The model is there, but most of the buttons on MLCad's toolbars are grayed out, and the program will not let us add or edit the parts of the model in any way. Even the layout of the screen might look slightly different than when you last used it. Even more telling, the View program mode is activated in the Viewbar, as shown in Figure 4.45. To understand what is happening, we need to take a closer look at program modes.

Figure 4.45 View Mode Selected in the Viewbar

NOTE

If you want to edit the model right away, set MLCad to Place mode (refer back to Figure 4.17).

A Brief Look at Program Modes

LDraw was created primarily as a tool to let LEGO users share their creations. One of the major benefits of the LEGO system of building is its modularity. The LEGO user in Japan uses exactly the same elements as his or her Brazilian, Icelandic, and Australian counterparts do. If we simply send these users instructions on which parts to use and how to place them, they can easily replicate our creations to the very last detail. Before computers became commonplace, this process was very cumbersome. Sending a parts list only solves half the problem; we need to show *how* the parts are positioned as well.

The introduction of home computers radically changed this situation. In fact, the first efforts of James Jessiman with LDraw were directed strictly to create a program that would show models and their step-by-step instructions. The model building had to be done "by hand," creating a file by typing the part number and coordinates in a text editor. The LEGO CAD program would then read the file and draw the model on the screen. At this stage, this idea was already revolutionary; what was important was to have a tool that allowed the sharing of fan models, not ease of use. Nevertheless, James soon introduced a second program, LEdit, that helped users build the models.

MLCad incorporates both applications into one program. Not only can we create LEGO models with MLCad, we can also create and/or view the instruction steps for them—and this function encompasses a sizeable chunk of the program's commands. To set the program to show the instruction steps for a model, we activate View mode. If we want to edit a file, we activate Place mode. There are four program modes in MLCad:

- **View mode** This mode lets us view all the step-by-step instructions for a particular model file in MLCad.
- **Place mode** This mode is used to build or edit the models in MLCad, including *adding* the instructions steps.
- **Move mode** This mode is actually a simple visualization tool that is used in conjunction with Place mode.
- **Size mode** Like Move mode, this mode is actually a simple visualization tool that is used in combination with Place mode.

What just happened when we opened our saved Podium file into MLCad is that the program, set automatically to View mode, attempted to show the first instruction step for the podium. Since the model has no steps yet, the program shows all its parts in one freestanding "step."

NOTE

In a new installation, MLCad will switch to View mode every time we load a model file into the program. This is actually not such a good feature for users who are mostly interested in building models. Fortunately, this default can be changed. See the section "Customizing MLCad" in Chapter 6. In that same chapter, the sections on building step-by-step instructions explain how to add step commands to your models.

Only the relevant commands for instruction browsing are activated (i.e., the instruction step browsing button in the Movementbar). Part of Chapter 6 is dedicated to creating and viewing instruction steps for our models. There you will find a more through description of View mode and its functions.

Our actual model building will happen in Place mode, which deactivates the instruction browsing items in the Movementbar but allows access to all the modeling tools. This being the case, we need to click the **Place mode** button in the Viewbar so that we can continue working on the Podium file that we have just opened (refer back to Figure 4.17).

NOTE

When we create a new file via the **File | New** command, MLCad switches automatically to Place mode. This default cannot be changed—nor does it need to be.

It is important for the first-time user to realize that although View and Place modes cover two separate and key functions in MLCad—viewing the instruction steps and building the models, respectively—the roles of Move and Size modes are not nearly as important. In fact, they could very well be considered just a pair of tools used in Place mode.

In the next sections, we explain how to use the modeling panes to our advantage. Since Move and Size modes are closely related to the tasks involved, they will be covered in more detail later.

Working with the Modeling Panes

As you already know, the lower-right part of the MLCad screen interface represents the virtual space in which we build (and look at) our models. It includes a top window that shows the model data in text format and up to four lower windows that display the model from different points of view, as shown in Figure 4.46. We will see and work with the Model Part List windows in Chapter 6. For the time being, we want to concentrate on all the functions related to the modeling panes themselves.

Figure 4.46 The Modeling Panes and Model Parts List Window

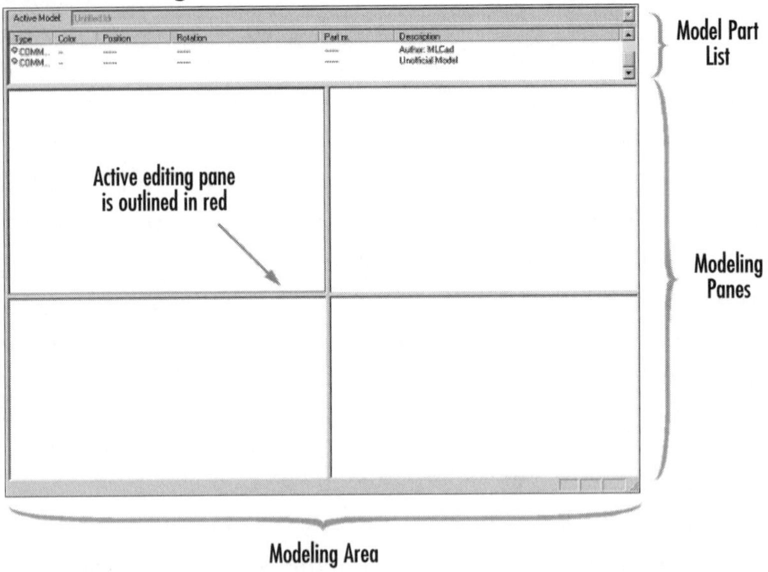

Why do we need four windows to peek into the "virtual" 3D space? When using real LEGO, we can easily move and rotate the LEGO parts and models. But when modeling inside a computer, it is much faster and easier to look at the model from different angles simultaneously as we build it. The basic elements of our models, the LEGO parts, help us figure out directions in this virtual space. However, as our models get more complex, we rely more and more on the standard coordinate system to find our way around. We will talk about this concept in detail in the next chapter. What is important for you to realize now is that the modeling panes are highly customizable tools—and that in fact they are adjusted often.

With the podium.ldr file loaded and the program in Place program mode, your screen should look similar to the one shown in Figure 4.47. The four view ports show the model from four different angles. The top-left view port shows it from the front, the top-right window from the left, the bottom-left window from the top, and finally the bottom-right view port shows it in a 3D-perspective view.

Figure 4.47 The podium.ldr File in Place Mode

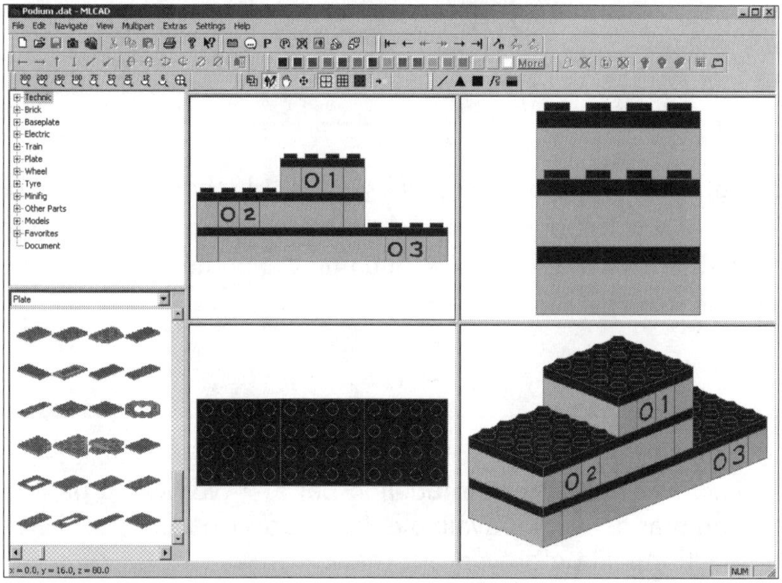

The next few sections cover some commands that allow us to set the modeling panes in ways that help our modeling. As we saw at the beginning of the chapter, modeling panes can be resized to occupy a larger section of the screen. Other helpful commands deal with *how* we look at the model: setting the view angle, zooming in and out, and repositioning the model within the panes.

The Modeling Panes Right-Click Menu

Each of MLCad's sections has a right-click menu. As we saw earlier, at the end of the toolbars section, when we right-click with our mouse in a toolbar, we are provided with a menu in which we can turn toolbars on or off (refer back to Figure 4.14). Later, we also saw how right-clicking with our mouse on the Parts Preview Window changes the zoom factor for the sample parts (refer back to Figure 4.21). When we right-click any of the modeling panes, we are presented with the menu pictured in Figure 4.48.

Figure 4.48 The Right-Click Menu for the Modeling Panes

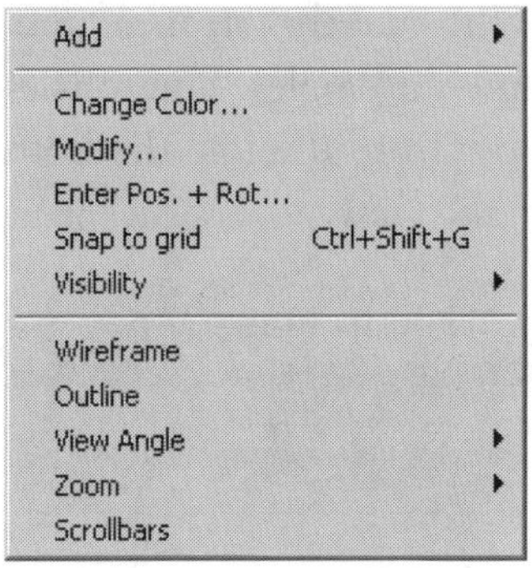

This menu is divided into three parts, separated by horizontal bars. The commands in the two top sections are related to modeling. You already know some of these commands, such as Modify, from building the podium. The rest are treated fully in the next two chapters.

NOTE

Some or all these modeling commands might not be available at all times when we right-click a modeling pane. If we have a part selected, all the commands will be available. If we don't have a part selected, the middle section will be grayed out (deactivated).

The section at the bottom of the menu allows us to set the content of each of the modeling panes according to our necessities. The following sections explain the uses of these five commands in the same order as they appear in the menu, starting at the top of the last part of the menu and proceeding downward.

Wireframe and Outline Rendering Modes

The first two pane functions, Wireframe and Outline, are not used often, since their utilization is somewhat specialized. They refer to how MLCad draws the virtual LEGO parts on the computer screen. By default, MLCad will show the parts "filled." That is, parts have a solid appearance, as in the upper-view ports of Figure 4.49. This is obviously the most comfortable way for most of us to recognize geometric shapes in a virtual space.

Figure 4.49 Shaded, Wireframe, and Outline Modes

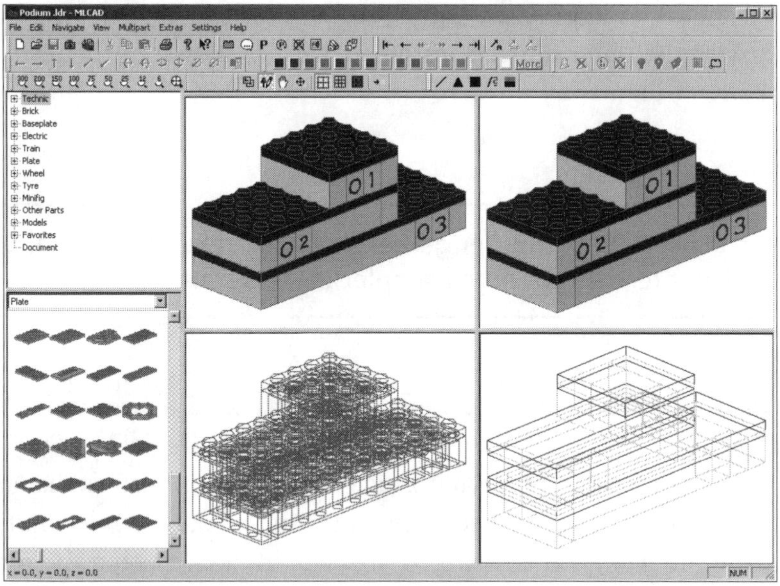

NOTE

To exactly replicate the screen shown in Figure 4.49, you first need to learn how to set the View Angle for each modeling pane. See the next section of this chapter for more information. However, you can activate Wireframe or Outline rendering in any modeling pane right away.

It takes a lot of computer power to calculate mathematically which parts of the shape are visible from a given point of view. When James Jessiman started working on MLCad about seven years ago, this was a real concern. Personal computers then were not very powerful, and even rendering to the screen a few complex shapes could take quite a bit of time. Things have changed since then, and personal computers now produce results almost instantly. However, very large models still strain even the fastest computer. When editing these models, we can find ourselves continuously waiting for the program to refresh the screen after a change, which can quickly become distracting.

In these situations, a practical way to speed up the refresh rates of the modeling panes is to render the model in Wireframe mode, shown in the lower-left modeling pane of Figure 4.49. The program simply draws the defining lines of the parts without filling the surfaces in between. Since the program does not have to continuously perform complex occultation operations, it runs much faster. For a simpler rendering, we can choose Outline, which renders the piece as simple box with the outer dimensions of the part, as shown in the lower-right modeling pane in Figure 4.49.

NOTE

Many so-called 3D computer games actually "fake" the virtual space and in fact run internally on much simpler 2D equations.

Since these two rendering modes make it pretty hard to work directly with models, they can be selected for each modeling pane individually. This flexibility allows us to set to nonsolid rendering modes only in those panels that are not critical to our modeling tasks. To set a panel to one of these rendering types, right-click the modeling pane to which you would like to assign the rendering mode and select the desired rendering mode in the menu shown in Figure 4.48. If neither Wireframe nor Outline are checked, the pane is rendered in default mode: solid.

NOTE

As you will see in Chapter 6, we can further customize MLCad's rendering options.

Apart from being useful with larger models, Wireframe or Outline can be handy in other situations. If we decide to create a part not yet in the library, looking at it in Wireframe rendering mode is like using X-rays. It can help us check the part's integrity as well as the integrity of models that include parts that connect with custom links. As we have mentioned before, the LDraw system allows us to use models as parts for other models. For instance, the Podium file could very well be used in a larger model of a stadium. If we had other submodels in that stadium (the athletes, for instance), it might be easier to position them using outline renderings. This way, we could focus on the general aspects of our overall model first (the composition of the scene) and not get distracted by unnecessary details. We can always go back to them later.

View Angle

The best way to work with 3D models inside computers is to view them from different angles at the same time. Our computer screens are two-dimensional: they have height (Dimension 1) and width (Dimension 2) but lack depth (Dimension 3). One way to overcome this limitation is to use a perspective view of the model. However, as the Dutch engraver M.C. Escher masterfully proved in his art, flat perspective views can easily trick our eyes in the most treacherous ways.

This fact becomes critical once we start building models. Using only flat perspective, we might find that our carefully built structures are floating in the air if we change the point of view or that the bricks we have assembled are not connected to one another at all! When using real LEGO in the three-dimensional world, the perspective is real, not faked: What we see is what we get every time around. In the two-dimensional world of the computer screen, MLCad helps quite a bit, but we have to stay alert and always be aware of how our model and its parts are positioned in the virtual space.

Think of this virtual space as a TV studio with our model in the center. As we have seen in previous figures, MLCad can show the model in up to four modeling panes at the same time, as monitors in a TV control room can. In each of these panes, we can set the angle from which we look at the model. It is as if we had four different cameras that can be placed anywhere around our model. To adjust the view in each view port, we use the View Angle command on the right-click menu of the modeling panes. The View Angle options are shown in Figure 4.50.

Figure 4.50 The View Angle Right-Click Menu

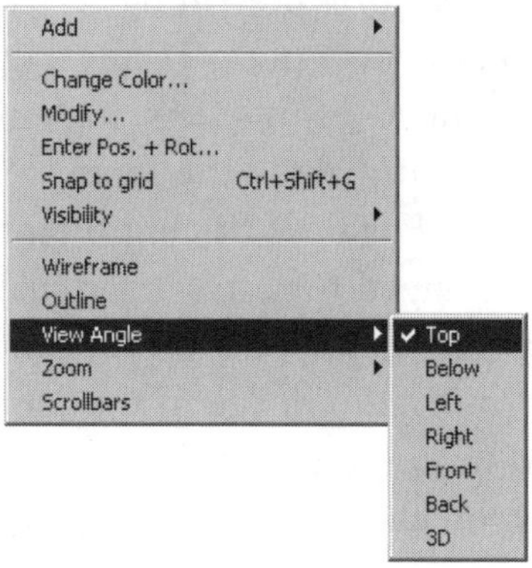

To make things easier for us, there are several preset positions. Choosing Top, Below, Left, Right, Front, or Back view angles will place the virtual camera in those positions in the virtual space. In Figure 4.51, the top-left view port is set to **Front** view angle; the top right pane to **Left** view angle; the bottom left to **Top** view angle, and the bottom right to **3D**. 3D is an isometric perspective that we have mentioned already and that we will examine further shortly. The "flat" preset views are perhaps less flashy than 3D but easier to work with: remember that you can't move parts by dragging them in modeling panes set to the 3D view angle.

Figure 4.51 Various View Angle Settings

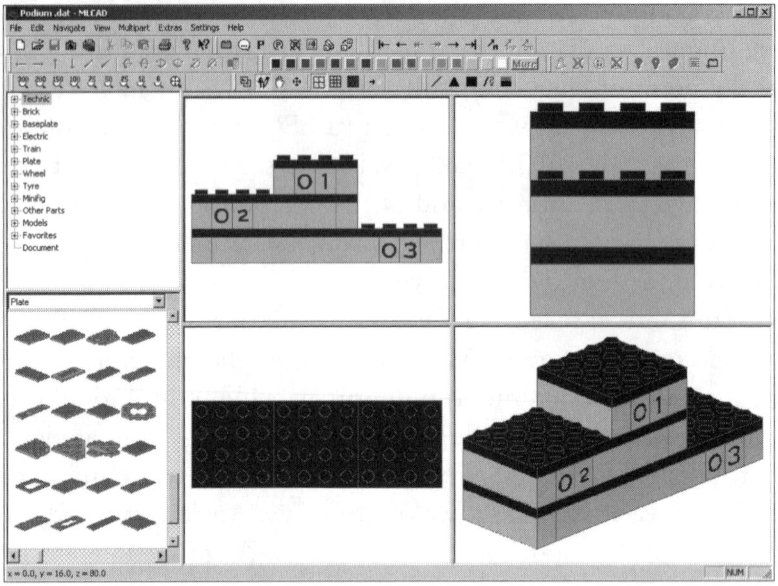

We encourage you to experiment with the Podium file in Place mode to become familiar with this system of representing a virtual 3D space. Learn how to change the view angle in any view port via the right-click menu. Modeling often requires frequent adjustments of the view angle to fit the pieces exactly the way you want them. Obviously, it makes little sense to try to place parts if we can't see where they are going! For now, the important things is to become aware of how view angles can be changed for each individual modeling pane and what advantages each of the view angles provide for your model building.

If the view angle is set to 3D, the image of the model in that particular modeling pane will be rendered in isometric perspective. This perspective lets you visualize the 3D model in a much more immediate way. When a pane is set to this view angle, you can change the point of view of the model by clicking and dragging the pointer anywhere in the view port. For better control, try not to click too far from its center. Figure 4.52 shows the four view ports set to the 3D view angle with four different points of view.

Figure 4.52 Four Points of View

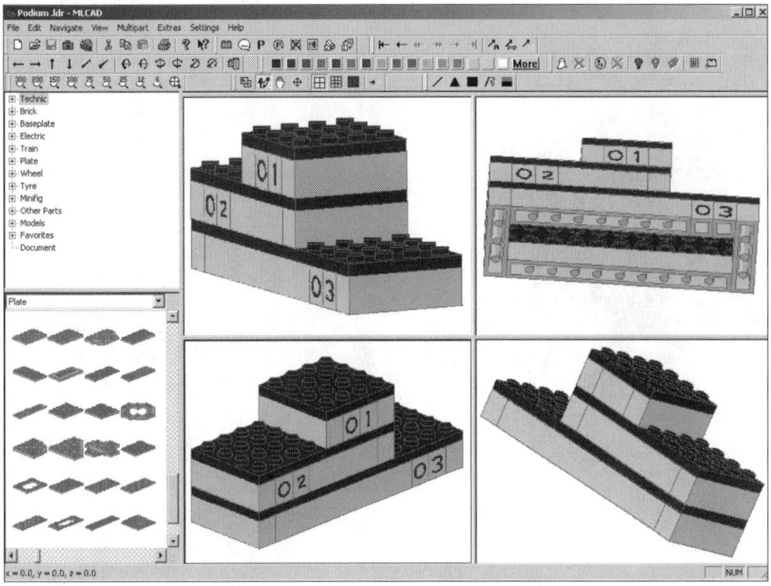

Try duplicating this on your own computer. First, set the angle for all the modeling panes to **3D** via the right-click menu (right-click in each pane and select **View Angle | 3D**). Then, click and drag anywhere in the panes to change the perspective. Finding the exact same points of view we want can get tricky, since we are using a 2D medium (mouse and computer screen) to navigate 3D space. If you lose your reference point (it happens), you can reset a 3D view port to the default point of view by switching it to another view angle setting—for example, **View Angle | Right** and then back to **View Angle | 3D**, as shown in Figure 4.53.

NOTE

When changing between view angles, you might need to adjust the zoom factor. The following section tells how to do this.

In any event, mastering this positioning technique is by no means as important as learning how to refer to our model via the flat-set view angles mentioned earlier (Top, Below, and so on). As you become more familiar with computer 3D modeling with LEGO and you master these advanced concepts, you will be better able to understand and navigate all the view angles, including the 3D view angle setting.

Figure 4.53 Resetting a Modeling Pane to the Default 3D View Angle

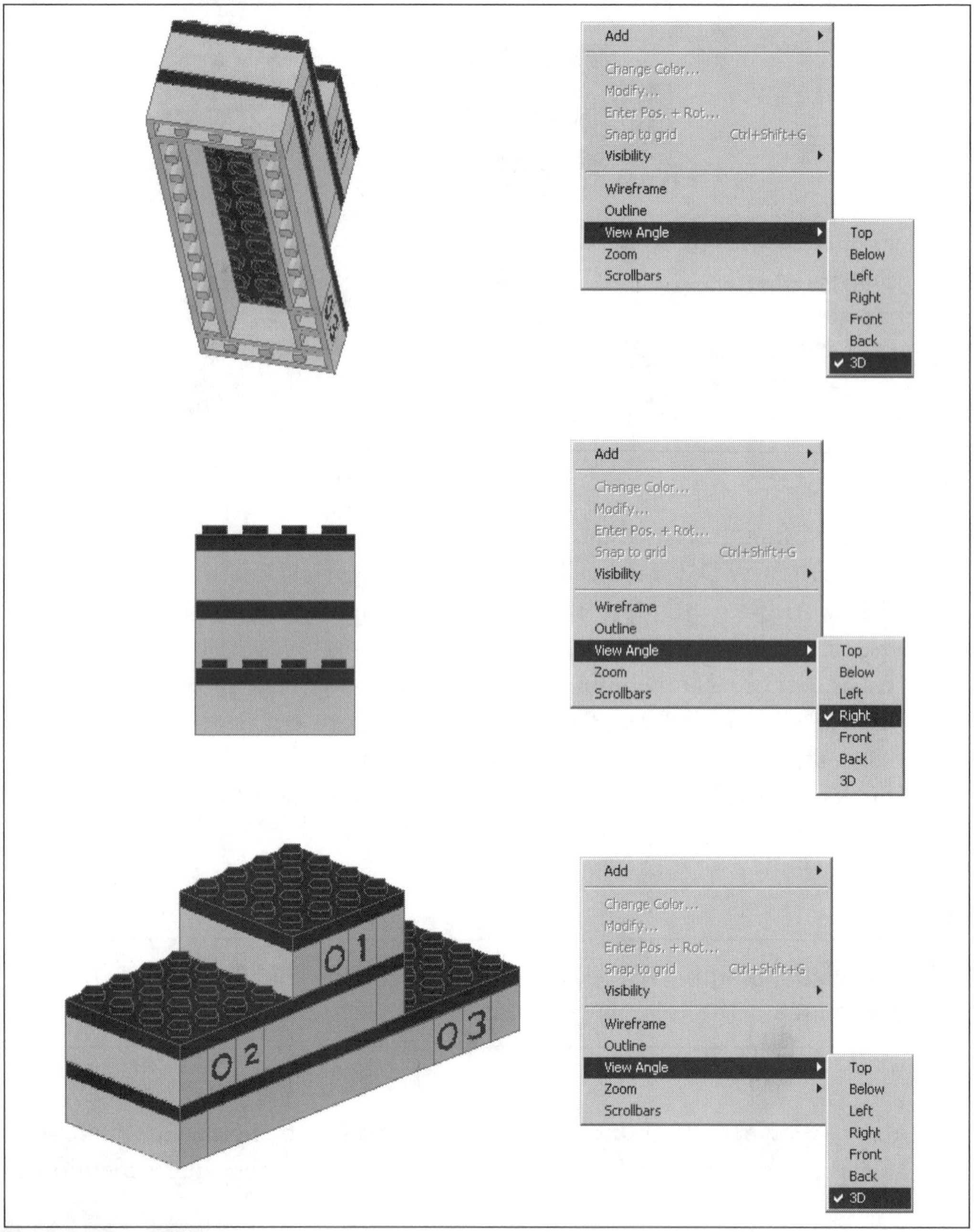

Zoom Functions, Including Size Program Mode

Some models have fewer than 10 parts, but others have over 1,000. The only way to fit models of all sizes in the modeling panes is to adjust the panes' zoom factors. There are several ways to zoom into or out of our model. One option is to use the buttons in the Zoombar (refer back to Figure 4.7). The higher the number on the icon, the higher the magnification and the larger the images will appear in the panes. The Zoom Fit button adjusts the zoom factor in every pane so that the model fits entirely in each of them. If we use the buttons in the Zoombar, we always set the zoom factor for *all* the modeling panes at the same time.

If we just want to adjust the zoom factor for a *single* pane, we can do so via the Size program mode, either by pressing the **F5** key on our keyboards or clicking the appropriate button in the Viewbar (refer back to Figure 4.6). Once in Size mode, the cursor changes to a double-sided arrow pointing up and down. Clicking and dragging your mouse pointer in any of the view ports will adjust the zoom factor for that view port, as shown on Figure 4.54. As you can see, the image in the top-left modeling pane has been "zoomed out" quite a bit so that the image of the podium model appears much smaller than the images in the other modeling panes.

Figure 4.54 Zooming Out in Size Mode

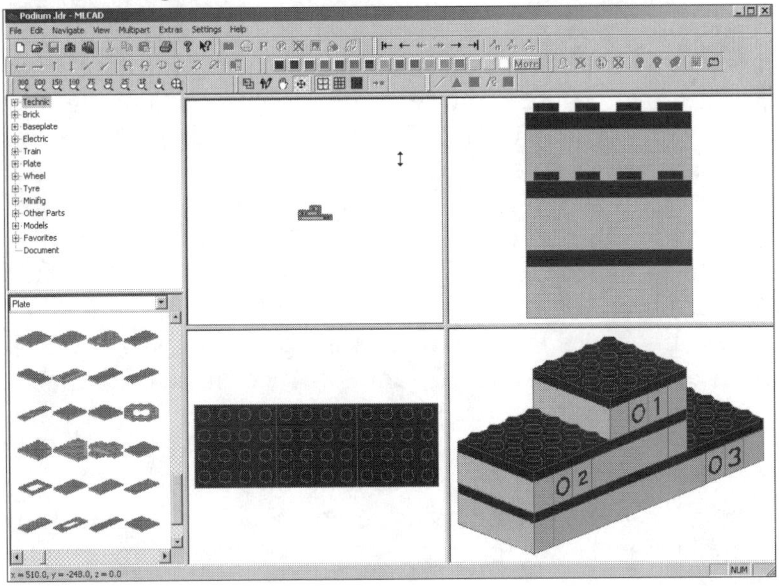

The one neat feature of Size mode is that it provides "zoom to part" support. If we switch from Place mode to View mode with a part or parts selected, the program not only adjusts the zoom factor as explained previously, but it also positions the model in the modeling pane so that the selected parts are centered on the pane. See the section on the Move program mode later in the chapter to get a better grasp on the implications of this concept.

Designing & Planning...

Scaling LEGO Models

One advantage to using existing elements is that we eliminate the need for scaling commands. Virtual LEGO parts do not change size relative to one another (neither do real LEGO parts, of course!). Scaling commands are basic tools for most computer 3D modeling applications. Using MLCad, we either stick to the part's scales (as in minifig scale) or we invent our own.

When someone asked the Japanese architect I. M. Pei how the large slanted-glass sides of his pyramid building at the Louvre in Paris were to be cleaned, he replied, "Simple—use water and soap." To make something bigger with LEGO, we take a similarly straightforward approach: Either we use more bricks or we use larger bricks. Or, as Pei himself would have proposed, we use a combination of the two.

Indeed, the art of *scaling* LEGO models—making bricks bigger or smaller but retaining the same features—is a key technique in the curriculum of many professional LEGO modelers, such as those who work in LEGO parks. The subject is also a favorite among many LEGO fans around the world. A good and common example is to build the same model of a car with two different widths. The minimum width is probably one stud (*microscale*), but most often the models are four or six studs wide or even wider.

If you are really interested in scaling techniques, here is some advice: Practice with real LEGO, and practice at scaling specific (and basic) shapes. A good exercise is to download the instructions for a LEGO sphere from the official LEGO site at www.LEGO.com. Create the sphere and then study ways of enlarging or shrinking it. The way the LEGO bricks are used for the ball's curved shape immediately suggests many possibilities. LEGO spheres, by the way, always make interesting office toys.

We can also set the zoom factor for an individual pane directly in Place program mode. Right-click anywhere in the pane and choose from the Zoom options, shown in Figure 4.55.

Figure 4.55 The Zoom Right-Click Menu

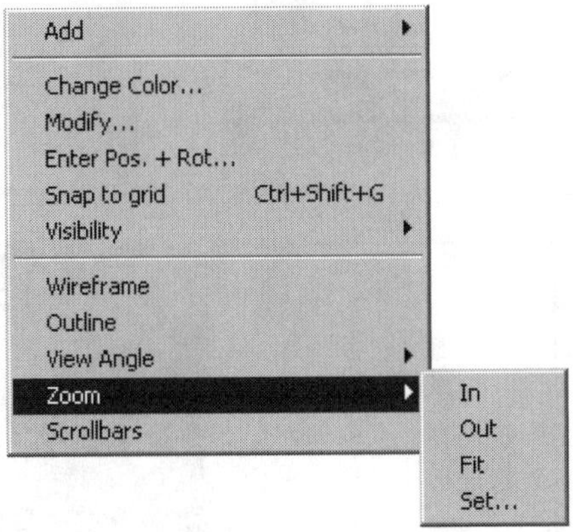

If we choose **Zoom | In** or **Zoom | Out**, the zoom factor for the pane will be adjusted up or down noticeably, but we might need to repeat it several times for large changes. **Zoom | Fit** adjusts the zoom factor to fit the whole model in the view port. **Zoom | Set** lets us directly type the zoom factor for the view port in a dialog window like the one shown in Figure 4.56.

Figure 4.56 The Set Zoom Factor Window

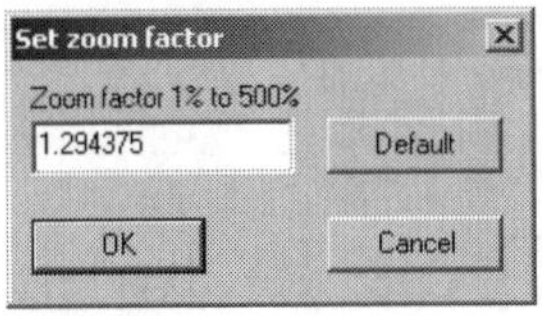

The Default zoom setting is 1, which is equal to the 100% setting in the Zoombar. In any event, there is generally no reason to need lots of precision for the zoom factor. We generally adjust it to see the model more clearly. Thus, once we have a "size" that is comfortable for us to work with, we know we have it right! Figure 4.57 shows MLCad with the podium.ldr file loaded and different zoom factors set on each view port. We recommend that you experiment with the different ways to set the zoom factor for the modeling panes. Unfortunately, our screens are often not big enough to fit our models, forcing us to use the zoom feature on a frequent basis.

Figure 4.57 Various Zoom Factors

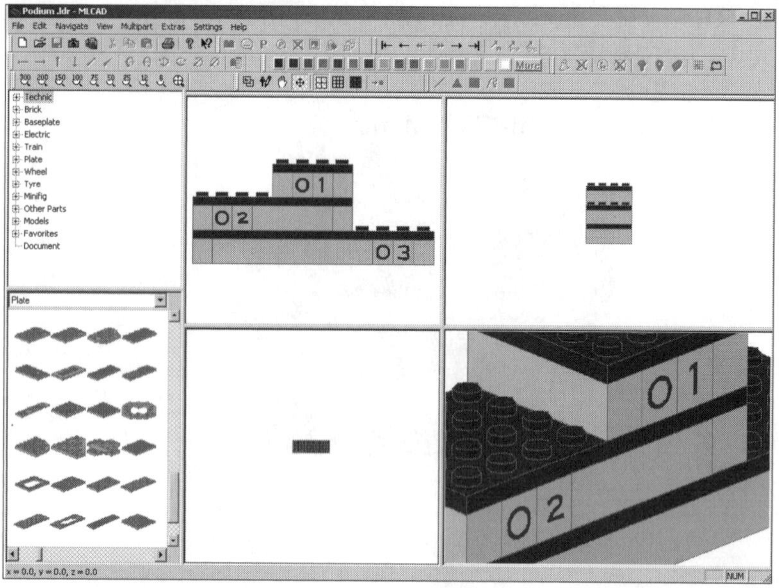

NOTE

The suggested range of zoom factors above the input box in Figure 4.56 is a bit confusing. Just remember that 1 equals 100 percent. Therefore, 2 equals 200 percent (in other words, is twice as large), and 0.5 equals 50 percent (half as large).

Model-Positioning Tools, Including Move Mode

As we mentioned, we use the zoom tools because our computer screens (not to mention MLCad's modeling panes) are generally smaller and flatter than we would like them to be. A problem often related to changing zoom factors is that we also often need to reposition the model within the pane. For instance, if we are using a large zoom factor and the model does not fit into the screen or modeling pane, we might need to move it in the view port. Figure 4.58 shows an MLCad screen zoomed in on the topmost 1 × 1 numbered bricks on the podium model, which at this magnification does not fit in the screen. But what if we wanted to look at another part of the model?

Figure 4.58 Large Zoom Factors Lead to Models That Don't Fit on the Screen

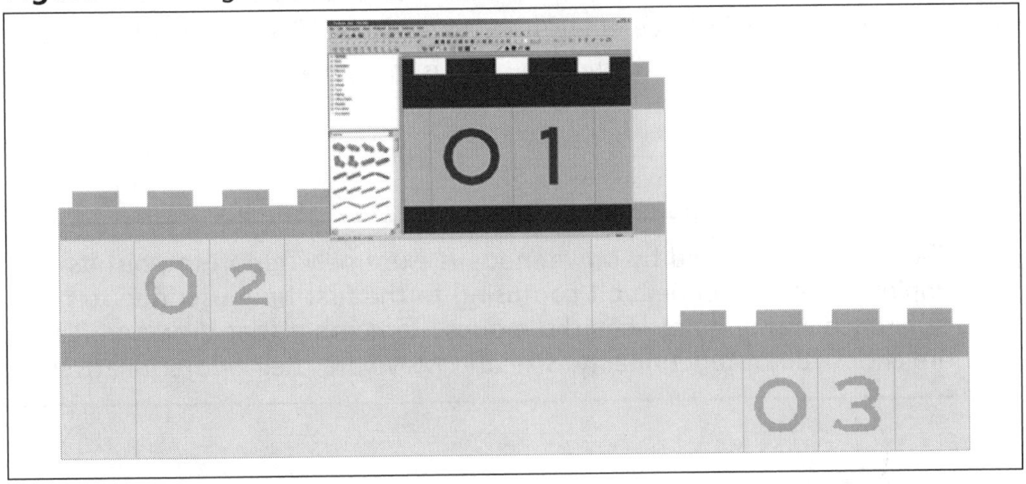

To move the model so that we can reposition it within the modeling pane, we have two options. One is to switch to Move program mode via the appropriate Viewbar button (refer back to Figure 4.6) or by pressing the **F4** key on your keyboard. The mouse cursor will change into a cross with arrows on each of its four ends. By clicking and dragging the mouse pointer on any of the view ports, we are able to reposition the model to fit our needs. See Figure 4.59.

Figure 4.59 Repositioning the Model with Move Mode

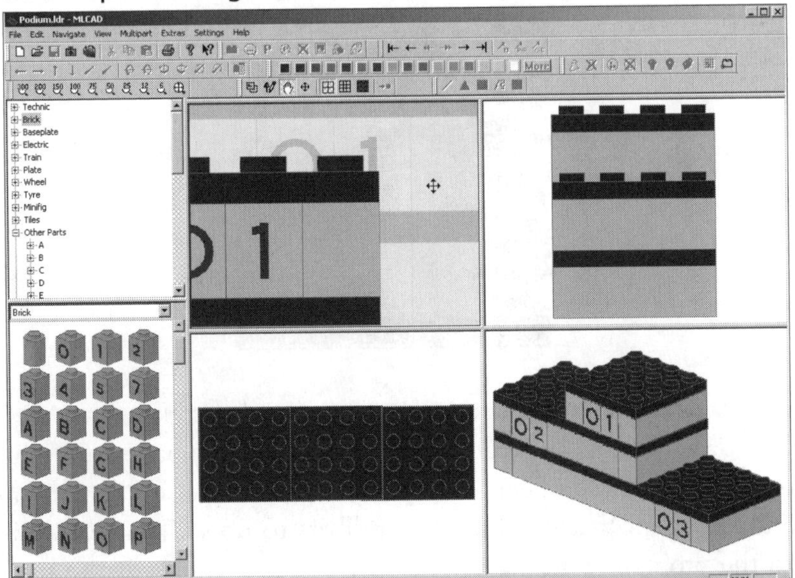

This is all that the Move program does—even less than what the Size mode does. In fact, we do not even have to switch to Move mode to reposition out models within a pane. In Place mode, if we press the **Shift** key of our keyboard, we can move our models inside the pane just as though we had activated Move mode.

NOTE

The function provided by Move mode is extremely important, but its implementation can be a bit confusing to the first-time user due to the difference in *scope* between the modes. Remember that View and Place are chock-full of functionality. Size and Move are their poorer relatives.

Scrollbars

Until quite recently, these were the only possible ways to reposition models in MLCad's modeling panes. However, Michael Lachmann has introduced a new and very neat modeling pane feature that gives us a constant control over the model's positioning within a pane. As you can see in Figure 4.60, Scrollbars is the last item on the right-click menu of the modeling pane area.

Figure 4.60 The Scrollbars Right-Click Menu Option

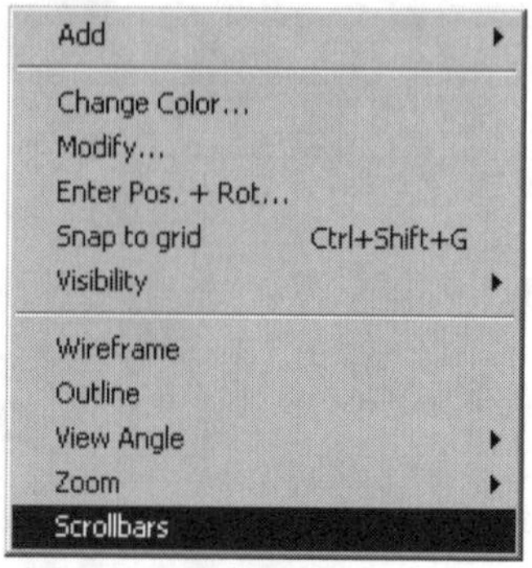

If you click **Scrollbars** in the modeling pane right-click menu, MLCad will add a set of horizontal and vertical scrollbars to the active pane—the one where you right-clicked. These scrollbars let you move the model inside the pane, just as in Move mode, except that you use the scrollbars to do so. This feature is a great addition to the program, offering yet another practical alternative to the Move program mode. Figure 4.61 shows a MLCad screen with the scrollbars activated in the lower-right view port.

Figure 4.61 A 3D Modeling Pane with Scrollbars Activated

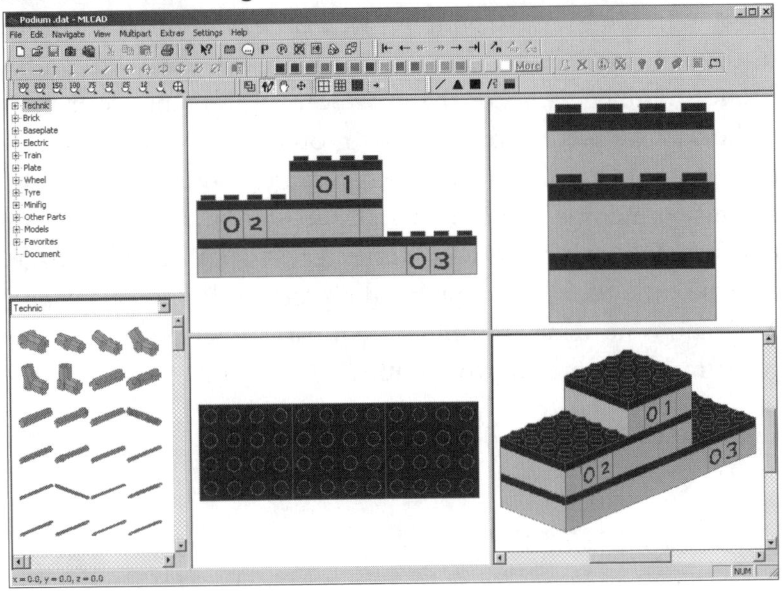

Counting this last and very practical function, you have now been exposed to all the available ways to display your model within the modeling panes. The content of this section has probably not been as exciting as building the podium model, yet these techniques provide you with a more secure approach to the (very neat) advanced modeling material coming up in the next chapter. Once you have full control over how to display your model in the modeling panes, you will be able to quickly relate to advanced modeling concepts. Repositioning the model to your needs becomes routine after you have spent a short while using the program.

A Closer Look at the Parts Library

We finish this introductory chapter to MLCad by visiting a very exciting place, the parts library. While building the podium, you saw some of the Preview window's basic functions. However, the library has over 2,000 parts and counting. We need better tools to make our way around this huge inventory.

Surfing the Parts Library

Browsing such a large number of parts is no simple matter, yet MLCad lets the user navigate the system with great ease. Navigation involves the use of both library windows. In the Preview window, we can search for parts by group (or category, as they are sometimes called). Using both windows, we can find parts by group, reference number, name, or shape. We can also organize our local copy of the library the way we prefer, creating new groups.

Let's take a look at the library windows in Figure 4.62. Like the rest of the MLCad interface windows, you can resize these windows as you like by simply clicking and dragging on any of their frames. Be aware that large image windows might take some time to refresh, because the program has to calculate in real time all the 3D equations necessary to render each part in perspective. This process does, however, make them easier to identify.

Figure 4.62 Parts Library Windows

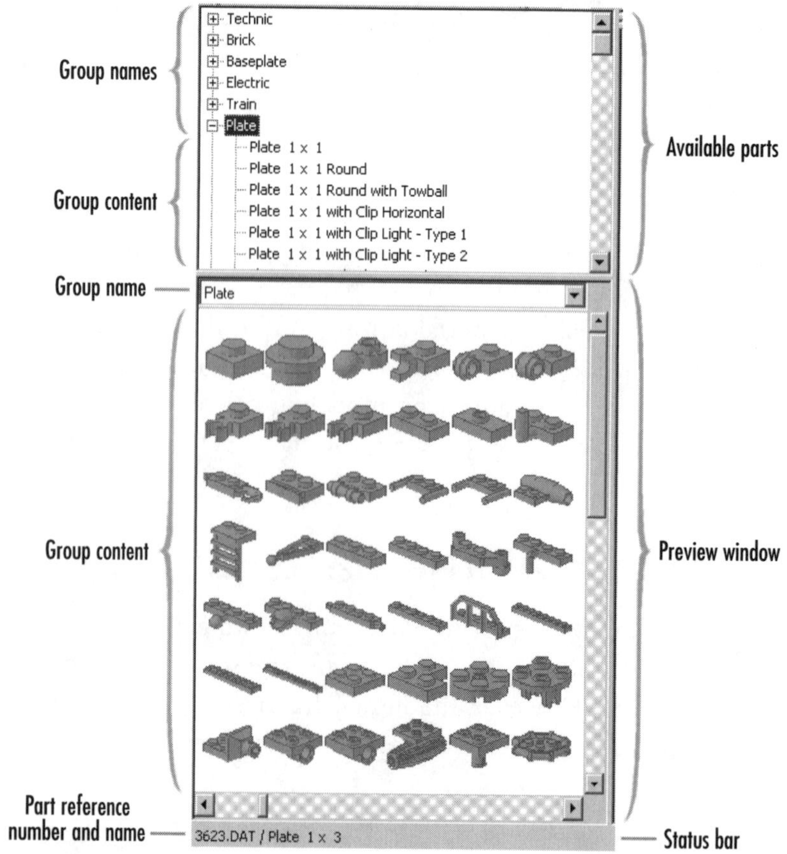

In your first forays into the library, it is advisable to view as many pieces at a time as possible. Even the most knowledgeable LEGO fan will discover some forgotten or new pieces while looking around. It is a fact that certain LEGO pieces quickly catch our imagination, sometimes inspiring a complete model with their shape or functionality alone. A large window increases the chances of this happening and gives us a broader perspective of a group's specific content. If we want to get acquainted with 2,000 or more pieces, we might as well see bunches of them at a time!

As mentioned earlier, we navigate the parts library through the two windows on the left side of the screen: the Available Parts window (top) and the Parts Preview window (bottom).

The Available Parts window lists the groups and the parts in each group by their names, in text form. It uses a tree-like structure to list the group names vertically. To the left of each group name is a small box. Clicking this box will list the group's contents in the upper window tree. In Figure 4.62, the content for the Plate group is listed below the group name on the upper (text) window. Clicking the icon again to the left of the group's name will make the contents list disappear. The box icons next to the groups' names have a plus (+) sign on them when their contents are not listed and a minus (–) sign on them when they are being shown, as demonstrated in Figure 4.62.

If, instead of using the icon, we click the actual name of a group or a part in a group, MLCad will display images of the part or parts in the lower Preview window. While we have the Plate group listed and shown in both the Available Parts and Preview windows in Figure 4.62, we can list two different groups in the two different windows. In Figure 4.63, the Available Parts window lists the contents for the Plate group, but the Preview window displays the contents of the Technic group. To do this, we use the pull-down window at the top of the Preview window to gain access to a different brick or group than the one displayed in the Available Parts window.

NOTE

In the Available Parts window, click the icons to the left of the group names for listings on the *top* Available Parts window, and click the actual names to list the groups on the *lower* Preview window. Additionally, the lower Preview window has a pull-down menu at the top that allows you to choose the group to preview.

Figure 4.63 Displaying Two Different Library Groups

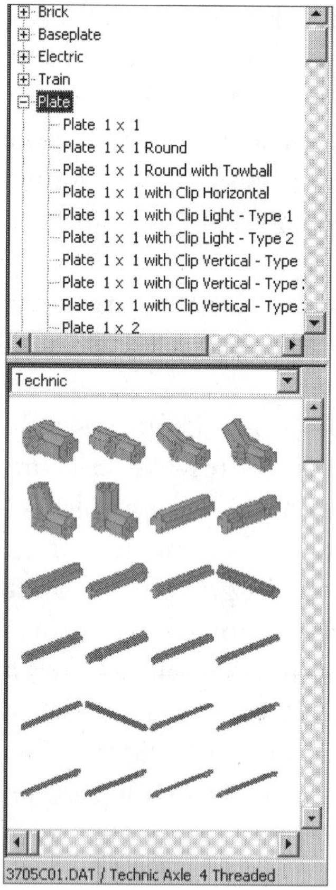

The Preview window only shows LEGO parts, without any further ID. But as you move the mouse pointer over the pieces, their part numbers and descriptive names are displayed in the status bar at the bottom of the Preview window. Right-clicking the lower windows will toggle between two zoom factors; a higher magnification will display fewer parts with more detail. Less magnification fits more pieces in the same space, but part details can be lost. Refer back to Figure 4.21 for an example of the two views.

How the Library Is Organized

Almost all real LEGO elements have a unique reference number engraved on each piece produced. This reference number is also used by all LDraw-based software to identify the part. Each LEGO part has its own .DAT file, stored in the LDraw\parts directory. The parts are named #LEGO*referencenumber*#.DAT. Unfortunately, this reference number does not itself bring much order to the library, since similar elements might have completely different reference numbers. This is because LEGO is a living product; in addition to providing a top-quality toy, the company tries to make a profit by, among other things, introducing new pieces regularly to incite fans to add new kits to their collections. New pieces are often variations of existing ones, increasing the complexity of the library.

LEGO officially categorizes pieces very broadly. Some lines, like the oversized Duplo kits for small children, apparently share few or no elements with the others, but all LEGO pieces are in fact designed to connect with at least several other LEGO pieces. This means that some large Duplo bricks do attach to some Mindstorms robotics elements. Not only that—many LEGO parts do not belong to a specific theme but instead to a scale. The minifig scale (not to be confused with larger figures also by LEGO) encompasses several lines of themes divided into overlapping categories. Castles lead to towns that lead to trains that lead to spaceships. The pirates share some accessories with Wild West folks, garbage collectors, and astronauts.

Thus, the best way to categorize pieces is not by any LEGO official themes or numbers (which are really meaningful only to the LEGO company) but rather by inventing a name for each part, just as many LEGO users have done as kids. Each official LDraw part file is named after the LEGO official part number, but it also has attached to it a descriptive name. This is the name that is displayed in the Available Parts window and in the Statusbar when we place the mouse pointer over the piece in the Preview window. Figure 4.64 shows some of the specialized LEGO parts and their corresponding .DAT filenames.

Most of the groups MLCad uses as a basis to navigate the library are actually search functions. For example, when we open the Plate group, all the program does is list the parts that carry the word *Plate* as the first word in their descriptive names.

Some groups are not search functions. The Models group lists as parts all the models in the LDraw\models directory. Remember that software based on the LDraw system allows models to be included as parts of other models, and this is one way of doing it. The Document group also reflects that possibility. We talk more about this topic in Chapter 6.

NOTE

Yet another reason to keep the LDraw\models directory clear is that rendering the Models group to the Library Parts Preview window can take a really long time if the folder is filled with models.

Figure 4.64 Specialized LEGO Parts

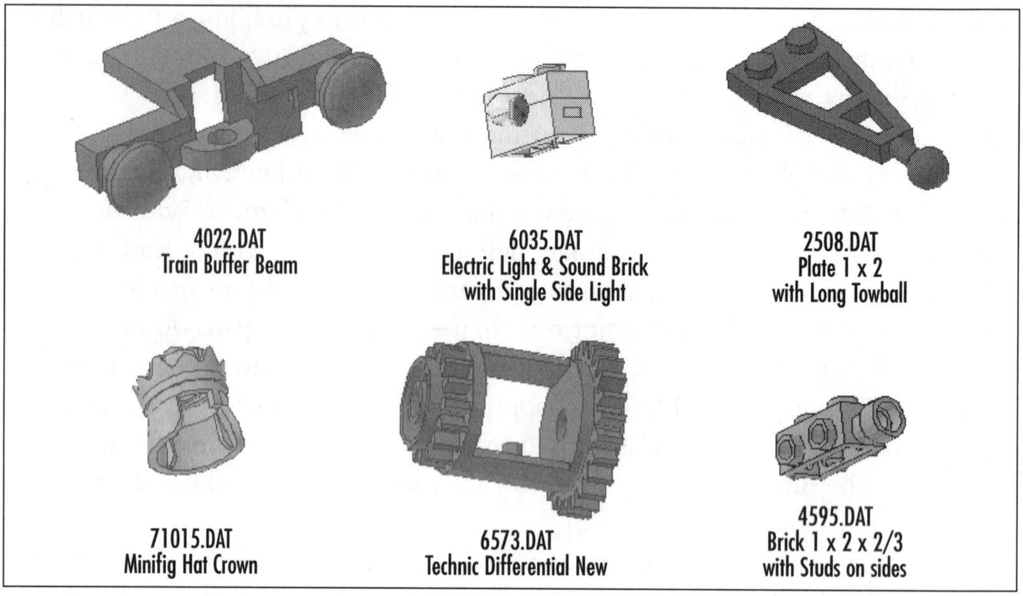

The Available Parts Window Functions

You can access several useful functions from the Available Parts window by right-clicking with your mouse anywhere inside it. Doing so will bring up the menu shown in Figure 4.65 Let's take a look at each of these functions.

Figure 4.65 The Available Parts Window Right-Click Menu

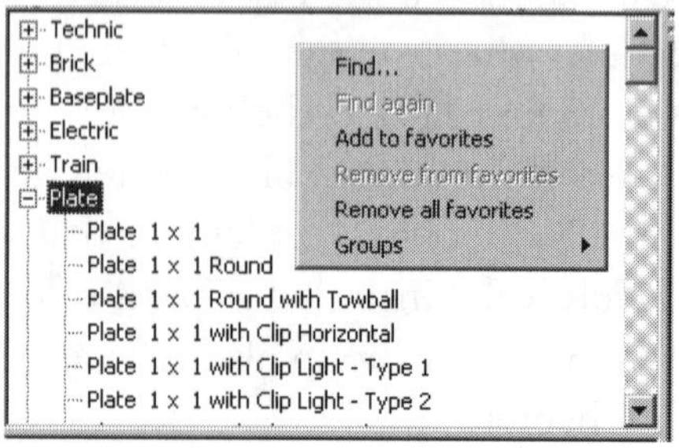

The Find Functions

The *Find...*and *Find Again* functions help us search for parts in the library. As we mentioned earlier, library groups are nothing more than search functions that display all available parts with the same initial word in their names. However, Find... and Find Again search all the words in the parts names, not just the initial one. If we select **Find**, MLCad will display the Search Part dialog window, shown in Figure 4.66.

Figure 4.66 The Search Part Window

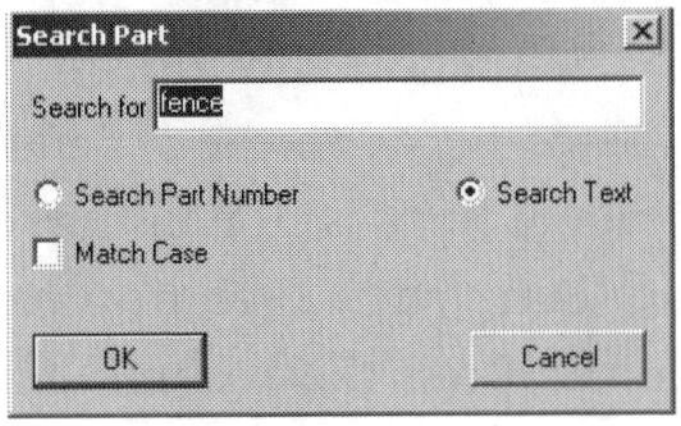

If we happen to know the part's number, we simply type it in and select the **Search Part Number** option before clicking **OK**. Otherwise, we can do a text search on the part names. Since this search will look at the entire name of the parts, not just the first word, it allows for a fairly vague search in cases where we do not have much information about the part we are looking for.

The Find Again function repeats the Find… search in case the first hit was not what we were looking for.

The Favorites Functions

The next three options of the Available Parts window's right-click menu, shown in Figure 4.65, all relate to the handy Favorites group. Select a part in the Available Parts window by clicking its name. Right-click the window and select the **Add to favorites** option in the menu. The part will now also be available through the Favorites group. The Remove from Favorites option reverses this process by removing a part that we have selected from the Favorites folder. Finally, we also have the option of clearing all the parts from the Favorites group by using the Remove All Favorites function..

The Group Function

The last option on the menu is the Groups function. Since this is really a customization feature, we will talk about it more in Chapter 6.

Group Names in the Parts Library

Becoming familiar with the majority of the parts available to you in the parts library as well as the modeling possibilities they offer is a task that is both enjoyable and time-consuming. Although some parts are quite unique, others span entire subfamilies that overlap with other part families. In most cases, this potential confusion is resolved by referring to a theme in the part names that is common to all of them. These themes do not necessarily follow the official LEGO themes. For instance, the Electric group holds all electric, electronic, and robotic elements except for those parts that are categorized as Train theme parts, which have their own group. The names adopted for the parts are good enough, but a proper ordering cannot be precise with only a single reference to each piece. For instance, the difference between the *bar* and *fence* names that are used for the two parts shown in Figure 4.67 is not all that obvious to the first-time user. Furthermore, the LEGO company's evolution has produced some truly puzzling situations: LEGO has phased out the Technic line, with its own group in the LDraw parts library, and its pieces are now scattered throughout different LEGO product lines.

Figure 4.67 A Bar and a Fence

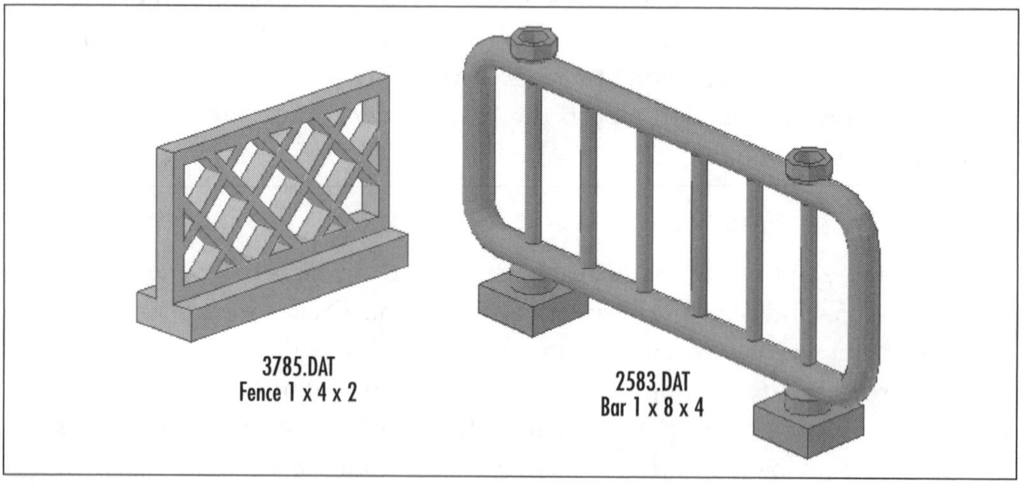

NOTE

Take it from a master. The artist Pablo Picasso often said, "I don't look for inspiration, I *find* inspiration."

The names used in the parts library not only refer to the theme or function of the part. They also carry additional information such as size and specific markings. As we saw earlier, substituting a part in a model for a different one is extremely easy. Thus, it is valuable to be aware not only of the overall library content but also of the different variations on some key parts. These variations are often a matter of size and/or texture.

LEGO Stud Measurements in the Parts Library

A benefit of using LEGO-based software to build 3D computer models is the fact that precise measurements are intuitive and easy to figure out. Anybody who has ever used LEGO is already familiar with its standard measurement unit: the *stud*. A stud unit is not the actual dimension of the LEGO circular stud by itself but rather its area of influence.

As you can see in Figure 4.68, the studs on top of the bricks lock into the female sockets on the bottom of the bricks. The stud unit takes into account the extra space needed by the socket end. Most (but by no means all) LEGO parts have studs and/or sockets, which you can easily count to find the dimensions of the part. The 1 × 3, 1 × 8, and 1 × 10 measurements for the bricks shown in Figure 4.68 correspond to their width in studs (1) and their length in studs (3, 8, and 10). Sometimes bricks also have a third measurement, as is the case for a 1 × 6 × 8 brick. In this case, the last measurement is for the height of the brick in studs.

Figure 4.68 LEGO Measurements, in Studs

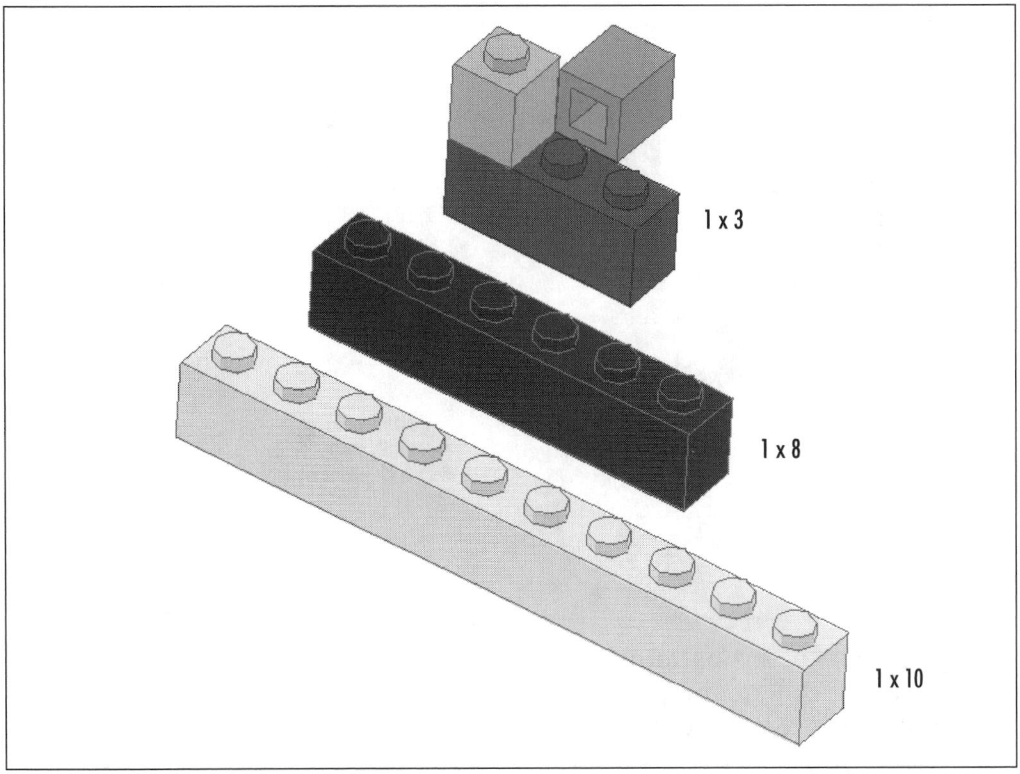

Part Patterns

In most 3D illustration programs, shape and texture are defined separately; but (as we saw with the numbered bricks in the podium) LDraw part files include their texture as well. These "textures" are really a pattern or logo of some kind on one or more faces of the part. This means that a plain LEGO brick has an LDraw part number different from the same brick with a decoration of some sort printed on it. In cases like this, the number of the "decorated" piece will be the number of the basic "plain" piece followed by a "P" and another two-digit number (and sometimes a letter) that refers to the specific decoration. For instance, 3010.DAT is the standard 1 × 4 brick. 3010P15.DAT and 3010P20W.DAT are variations on 3010.DAT that have different decorations on them, as you can see in Figure 4.69.

Figure 4.69 Variations on Part 3010.DAT

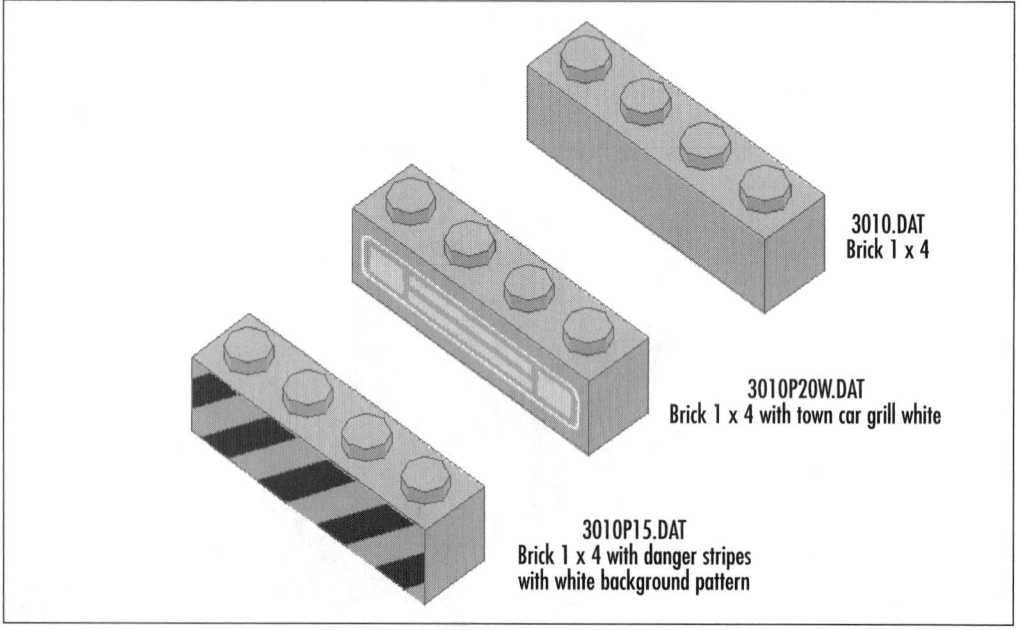

NOTE

The colors of the patterns cannot be changed in MLCad. In Figure 4.69, we could change the yellow bricks to different colors, but the town car grill would remain white and the danger stripes would stay red.

Summary

This chapter introduced you to MLCad's interface and general functionality. We started by providing a general view of MLCad's single console, which contains the tools you used to build and view models, the modeling area itself, and the parts library. The MLCad interface is both customizable and consistent. It takes only a short time to become familiar with it.

We then proceeded to build a simple podium model, taking a brief look at some standard modeling techniques to provide the reader with a feel for using MLCad. Next we described the administrative file functions available in MLCad in order to allow the user to learn how to save and load models, which are all standard Windows commands.

Editing models and viewing their instruction steps are two different MLCad functions, called *program modes*. Apart from View and Place program modes, two other program modes, Move and Size, are actually little more than specialized tools to adjust the models in the modeling panes that "look into" the virtual space where we create our models. Learning how to zoom in or out of our models and how to position them in the modeling panes is essential for 3D modeling.

Once you became aware of these two-dimensional pane functions, it was time to take a dip into the virtual 3D space. Learning how to look into it is key. Instead of relying on false perspectives, we see the models from different view angles simultaneously. We can have up to four modeling panes showing different views of the same model at any given time.

We saw some preset view angles that render "flat" views of our model from familiar positions: Front, Back, Left, Right, Top, and Below. As modelers, we need to be extremely comfortable with these view angles because we often work on our models using the mouse directly on the modeling pane.

We can also set the view port to a 3D view angle, which renders the models in perspective. These perspectives are more attractive but often less useful than the preset flat views. However, they also render the model in a more immediate and recognizable way.

The last part of the MLCad interface that we covered in the chapter is the parts library. In MLCad, we browse the parts library via two windows on the lower left side of the screen. The top one, the Available Parts window, provides a tree-like structure and search functions for the 2,000+ pieces. The parts are rendered in the lower windows, the Parts Preview window.

The naming convention for the parts integrates several standards: the official LEGO part number, a descriptive name, and the dimensions of the part. Despite this fact, the sheer number of elements contained in the library and their complexity make it extremely hard to create a completely logical organization. In other words, no matter how well we think we know the parts library, we will always be up for surprises, whether finding pieces we didn't know existed (often variations of other pieces) or bumping into pieces we had forgotten about.

However, MLCad allows a precise navigation of the library via search-based tools. Knowing how to use these tools and having a general knowledge of part classification basics, we can find what we are looking for—or something better.

Solutions Fast Track

The MLCad Interface

☑ MLCad uses a single-window interface divided into roughly two parts: the toolbar section on top and the modeling section on the bottom.

☑ The modeling section includes four modeling panes that display your model, a parts list window, and two library parts windows.

☑ Most of MLCad's commands can be accessed several ways. The toolbars provide a handy and easily identifiable way to access the commands.

☑ The MLCad interface is very customizable. The screen real state can be organized to suit your needs by resizing the internal windows and turning the toolbars on or off.

Building a Simple Model

☑ Remember the rookie checklist: Create a new file, set the program mode to Place, set the grid to coarse, and adjust the modeling panes and library.

☑ The easiest way to add a part to your model is to drag it from the Preview Parts window. Learn how to locate a specific part in that window first.

☑ Move parts by clicking and dragging them. Holding down the Ctrl key will produce a copy of the part.

☑ To change an existing part in your model for one of a different type, use the **Edit | Modify** command.

☑ To place parts in your models, you need to be able to move them and rotate them. Even simple 90-degree rotations widen the building possibilities tremendously.

Working with MLCad

☑ Program modes enable you to use MLCad for different functions such as editing models or viewing step-by-step instructions.

☑ View mode displays the instructions steps for models, providing controls to browse through the virtual booklets.

☑ Place mode allows you to edit and create models, whereas Size and Move modes are really simple modeling pane tools.

☑ The modeling panes are highly customizable tools that you will need to adjust often. Apart from resizing the pane, the actual view of the model in it can be changed from its apparent size to the view angle.

☑ Like all MLCad sections, the modeling panes have a handy menu that is activated by right-clicking in any of them. The menu allows you to change several settings for each individual pane.

☑ The modeling panes are like monitors in a TV control room: They provide different views of the model.

☑ There are several preset positions in which the model is rendered in a "flat" view. Additionally, the 3D view angle generates perspective views of the model that are flashier but less handy.

☑ To vary the zoom factor of an individual pane, you can use Size mode, or you can right-click any modeling pane and use the Zoom tools.

☑ You will often find your modeling panes and even computer screens too small to fit the entire model at the desired zoom factor. MLCad overcomes this issue by providing tools to reposition your models inside the modeling panes.

☑ The Move mode allows you to move directly with your mouse; so will pressing the Shift key while clicking and dragging the mouse in the pane or activating the scrollbars for a modeling pane.

A Closer Look at the Parts Library

- ☑ MLCad's parts library consists of two windows: the Parts Preview window and the Available Parts window. The Parts Preview window is visual and immediate; the Available Parts window lists the parts in text format only but packs more searching power.
- ☑ The more than 2,000 parts in the library are loosely organized along several criteria. They can be identified by the official LEGO number, a LDraw part name, and often, part dimensions.
- ☑ The right-click menu of the Available Parts window provides several handy find and favorites functions.
- ☑ The more parts you know, the better modeler you will be. However, learning the parts is a task that requires both imagination and knowledge of the LEGO universe. The best way to learn the parts in the library is to use them.

Frequently Asked Questions

The following Frequently Asked Questions, answered by the authors of this book, are designed to both measure your understanding of the concepts presented in this chapter and to assist you with real-life implementation of these concepts. To have your questions about this chapter answered by the author, browse to **www.syngress.com/solutions** and click on the **"Ask the Author"** form.

Q: What is MLCad?

A: MLCad is a standard Windows application that allows the user to easily create 3D models inside a computer, using existing elements based on LEGO bricks and parts.

Q: What can I use MLCad for?

A: For a LEGO fan, MLCad offers many possibilities. You can create any design that you want without worrying about running out of parts, or you can replicate real LEGO models. The program also allows you to build instruction steps so that other users can reproduce your models. For non-LEGO users, the program offers a very easy (and completely free) way to learn how to use

3D software. For the software developer, it offers an excellent base to create all sorts of cool add-ons.

Q: How difficult is MLCad to use?

A: MLCad is quite possibly one of the easiest ways to learn how to use 3D software from scratch. This is a field in which, in most cases, the initial learning curve is very steep, since the user is required to deal early on with sophisticated applied mathematics. Using LEGO parts in MLCad, the process becomes completely intuitive and allows the user to acquire a basic knowledge common to all 3D software.

Q: Point-and-click building? Isn't it boring after awhile?

A: On the contrary. The tool and the system are sophisticated enough to allow each user plenty of personal growth space. Exploring new building techniques, writing code, creating parts, and using the model files in other programs are only a few of the options available. Each of them explores and adds to the LEGO universe while allowing you to use LEGO to learn new things.

Q: MLCad is a freeware program, right? How good can it really be?

A: It is excellent. The program and format are very robust, even compared to commercial applications. MLCad provides very elaborate support for a few basic modeling functions. Since you use existing elements to create your models, the number of core tasks (moving parts, rotating them, etc.) is small—but you can perform each task to a very precise degree.

Q: Where can I get support for MLCad?

A: The best way to get support is to post messages in the www.lugnet.com CAD forums. Michael Lachmann, the creator of MLCad, often posts there. The program official's Web page is located at www.lm-software.com/mlcad/.

Chapter 5

Modeling with MLCad

Solutions in this Chapter:

- Our Virtual Playground
- MLCad's Precision Modeling Tools
- Advanced Modeling Techniques
- Working with Submodels
- An Introduction to Automatic Modeling

☑ Summary

☑ Solutions Fast Track

☑ Frequently Asked Questions

Introduction

This chapter is devoted entirely to the finer points of creating computer 3D LEGO models with MLCad. In the last chapter, you were introduced to the MLCad interface and general use. In this chapter, we provide a comprehensive view of all the modeling tools and concepts that are a part of MLCad, so that you will be able to use the program as a tool to achieve your modeling objectives, whatever they might be. The objective of this chapter is to get you to concentrate on your models, not on how MLCad works. MLCad provides extremely accurate control of each of the basic functions of virtual LEGO modeling: adding parts, coloring them, modifying their shape, and moving and orienting them. In the first part of this chapter, you will create a virtual minifig for your podium while you learn how to use these controls.

Once you are familiar with all the basic modeling tools, the next step is to learn how to combine two or more models into a larger single model file. LDraw and MLCad provide several ways of integrating models as parts of other models. These techniques might at first glance seem a bit complicated for the casual user, but in fact they are not. With a little organization, these techniques enable you to create whole worlds in a matter of seconds. If you are not *that* organized, they will still be very helpful on many occasions. To illustrate these techniques, we will show you how to place the minifig you will create in the first half of this chapter onto the podium you built in the last chapter.

Lastly, you will learn how to *automatically* build other elements to include with the podium model to create a virtual LEGO diorama.

Our Virtual Playground

In this part of the chapter, we explore all the advanced building tools and techniques MLCad offers while putting together a minifig like the one shown in Figure 5.1. Despite its small size, this model is not only much more complex than it seems at first glance, it is also very engaging; all of us can relate immediately to representations of the human body.

NOTE

The miniature figure, known among LEGO fans as a *minifig*, was introduced in 1978 and is considered LEGO's second most important design after the classic brick. LEGO has manufactured over 2 billion minifigs to date.

Figure 5.1 An Athlete Minifig

Before we learn how to tell the program to turn the minifig's head or move its arm, we must *ourselves* learn how to work with directions in the virtual space that all MLCad models inhabit. In the previous chapter, we discussed very simple modeling operations that allowed us to position parts using other parts as spatial references. As our modeling gets more precise and complex, we will trade intuitive building techniques for more efficient and precise ones. Since with these advanced techniques there is no direct relation between our mouse movements and the modeling process, we must refer to a standard coordinate system in order to describe our models and our modeling actions to the program.

Understanding View Angles

In the last chapter, we explained how the virtual space was like a TV studio with our model in the center and how the modeling panes act as monitors connected to cameras with free or preset positions inside this imaginary TV studio. As modelers, we need to become very familiar with this setup. If the mouse pointer is our hand in MLCad, the modeling panes are our eyes. We have to know where to look.

Take a look at Figure 5.2, which shows the minifig in MLCad, seen from four different view angles.

With a finished model like this, it is easy to identify which view angle the modeling panes are set to. In Figure 5.2, going clockwise from the upper-right modeling pane, they are set to Left, 3D, Top, and Front. Relating to the virtual space should quickly become a game, with plenty of opportunities to explore but a set of standard moves to fall back on. You will soon be thinking in model terms ("Let's move the left arm up") and looking for the view angle that best suits your needs.

Figure 5.2 View Angles

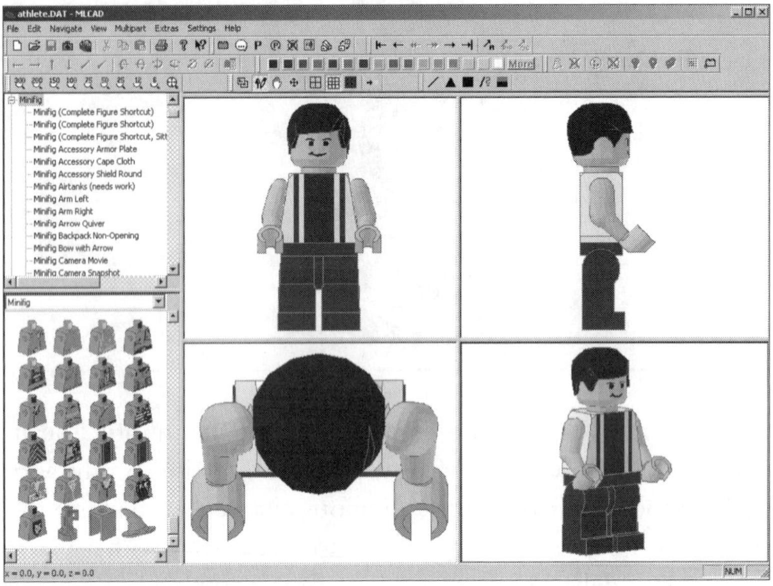

NOTE

Figure 5.2 is an example; we will tell you how to build the athlete shortly. You don't really need to use your computer until the section "Reading XYZ Coordinates in MLCad" a bit later in the chapter.

However, we cannot expect 3D modeling scenarios to stay this simple. For instance, the athlete could be aligned a different way; the view in the "front" view angle could actually correspond to the left side of the minifig. Preset camera positions cannot be changed, and the program has no way of knowing which is the orientation of the model in a file. Why, then, would anybody align their parts and models in such a confusing way? There are several reasons, but here is a good example that shows that there really is no "correct" way to align the models.

Imagine that we are building a virtual LEGO city. Once we had built all the buildings, it might still be obvious to us what the standard left, 3D, top, and front views were. What if we then added models of LEGO cars to the city, each facing in a different direction, based on which street they were placed on? Suddenly, it isn't so obvious what is front or left anymore, depending on which street we are using as the base for our perspective. Clearly, we need another way of dealing with orientation—a way that is independent of the alignment of the model we are basing it on.

To deal with these kinds of situations, which are in fact fairly common, we need to be aware of a second level of reference coordinates beyond the view angle setting. If the modeling panes are like screens connected to cameras that look *into* the modeling space, the XYZ reference system is a precise map of the *inside* of that space.

Designing & Planning...

Professional XYZ Spatial Perception

It is a fact that some people are better than others at visualizing shapes in space (think of architects and engineers, for instance). MLCad's approach to 3D modeling, combining LEGO parts with CAD techniques, provides a very powerful but very accessible tool for visualization. The basic techniques we have seen until now are limited but commonsensical. Becoming as familiar with the XYZ system we are about to discover might take a bit more time. Still, the approach is just as straightforward, just a bit more complex.

Once you master an understanding of the XYZ coordinate system, you will be suddenly find yourself very fluent in the field of general 3D computer graphics. This coordinate system is *the* 3D standard. For instance, most of the coolest 3D effects you see in the latest sci-fi movies relate *directly* to this standard. Knowing the XYZ coordinate system will allow you to visits industry Web sites and pick up professional magazines and actually understand pretty much everything that is being said! Such is the power of the XYZ coordinate system—it automatically makes us pros.

The XYZ Coordinate System: A 3D Compass

The modeling panes let us look at and navigate the 3D virtual space. But to create models in that space, we need the equivalent of the cardinal points of maps. With those points, we can plot specific positions in space and tell the program where to place the LEGO parts to form our model, just as any place on a map corresponds to a set of North/South and East/West coordinates.

You are already familiar with the view angles, the preset "camera" positions in our virtual TV studio. These preset positions also act as the cardinal points of MLCad's virtual space. Regular cartographic maps have two dimensions and thus employ four cardinal points. North-South defines one dimension (up and down on a map), and East-West defines the second (left and right on a map). Where 3D spaces differ from maps is that they add the third dimension of depth (think of it as "into" and "out of" a map), and therefore they use six cardinal points. Figure 5.3 serves as a compass for MLCad's 3D space.

Figure 5.3 A 3D Compass

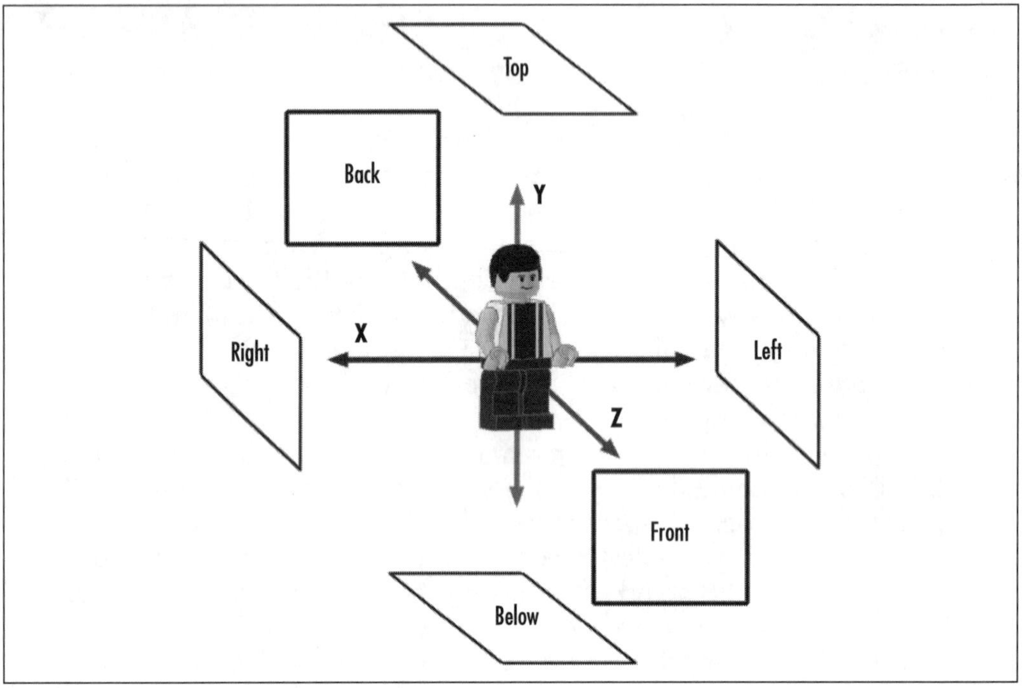

In the two-dimensional world of maps, we use two coordinates to plot a position. Movement is often described by referring to any of the four cardinal points (for example, "move South and East"). In virtual 3D spaces, instead of referring to any of the six view angles (our cardinal points), we commonly refer to the *axes* that connect opposite view angles.

As you can see in Figure 5.3, the imaginary horizontal line that connects the Left and Right view angles is the X axis, the vertical line connecting the Top and

Below view angles is the Y axis, and the line that connects the Front and Back view angles is the Z axis. When we're looking at the virtual space from the Front view angle, left-right movement is along the X axis. So, instead of saying we move the piece left or right, we say we move the piece "along the X axis." In the Front view angle, top-bottom movement is along the Y axis, and front-back movement is along the Z axis.

Reading XYZ Coordinates in MLCad

By now some readers might be more confused than *oriented*. Don't panic—in a moment we will begin positioning parts for the minifig using techniques that relate to this coordinate system, and everything will begin to make more sense. Let's start with a simple exercise that will show you how to use the Statusbar to find your bearings inside the modeling panes. First, create a new file. As you can see in Figure 5.4, when you move your mouse pointer in the modeling panes, the status bar at the bottom of the screen gives you the coordinates for the position of the mouse in X, Y, and Z values.

Figure 5.4 XYZ Coordinates in the Statusbar

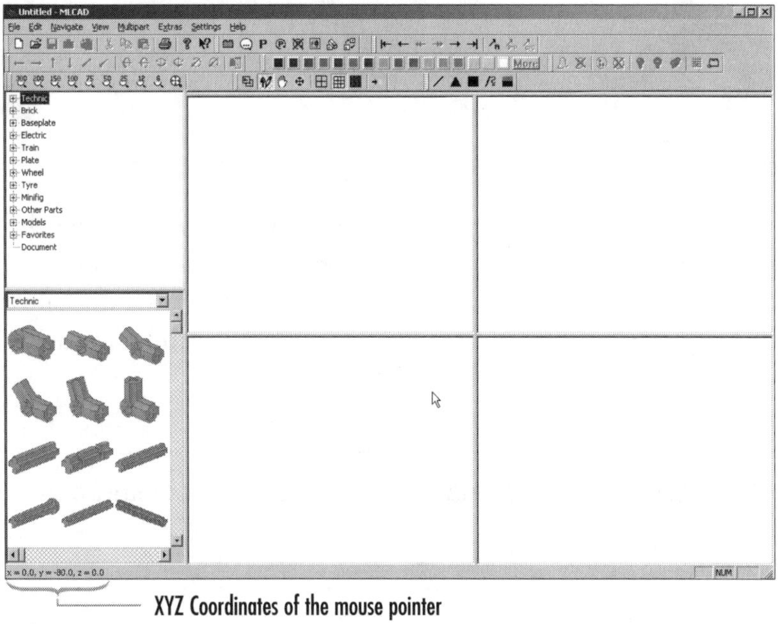

Move the mouse pointer while looking at these coordinates displayed in the Statusbar. Notice that if you move the pointer vertically or horizontally, only one of the coordinate values changes; you are moving it along one axis only. Now try moving along the same axis in other view angles. Try to match the different view angles and axes to the 3D compass in Figure 5.3, using Figure 5.5 to help you locate the view angles best suited for moving along the different axes.

Figure 5.5 Preset View Angles with Axis Positions

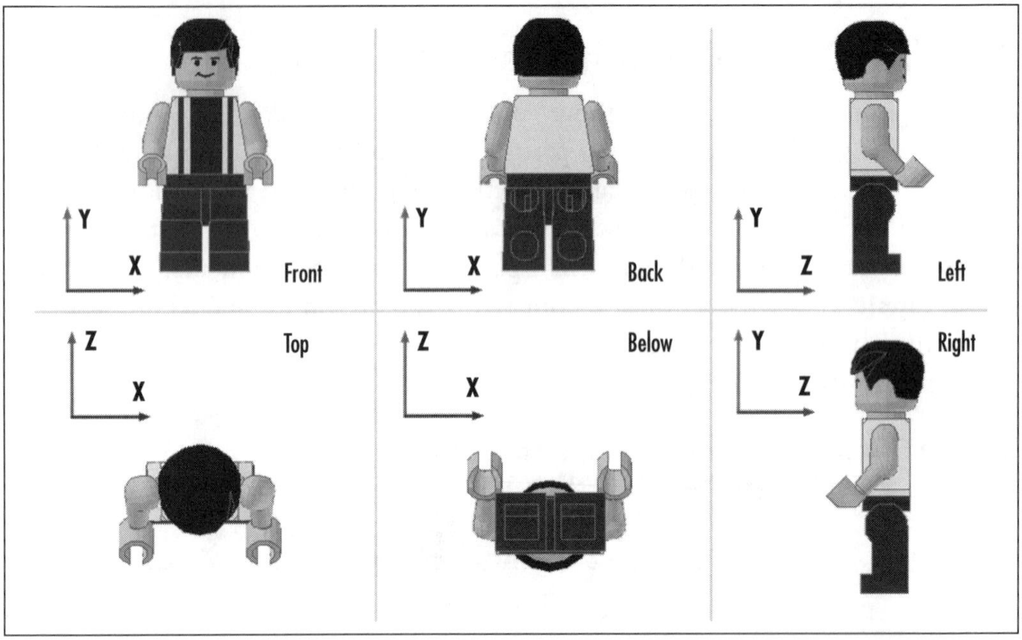

For instance, moving your mouse pointer up and down changes the values for the Y axis in Front, Back, Left, and Right views but not in Top and Below views. You don't need to spend much time on this task, but it is useful to always be aware of the locations of the different axes in relation to the view angles.

NOTE

The Statusbar does not display coordinates for modeling panes set to the 3D view angle.

Since the screen has no depth, we can only move our mouse in two dimensions in each of the preset view angles, whether X and Y, X and Z, or Y and Z. Figure 5.5 shows clearly which axes are available in each of the preset view angles.

The point where the three axes cross is the center of the virtual space, as you can see in Figure 5.3. But what does that mean in practical terms? It means that the center (or *origin*) of MLCad's virtual space is where the position along all the axes is 0. Try locating this position with your mouse pointer in one of the modeling panes. The Statusbar will tell your when you are there; the values for all the coordinates will equal 0, as shown in Figure 5.6.

Figure 5.6 The Center of MLCad's Virtual Space

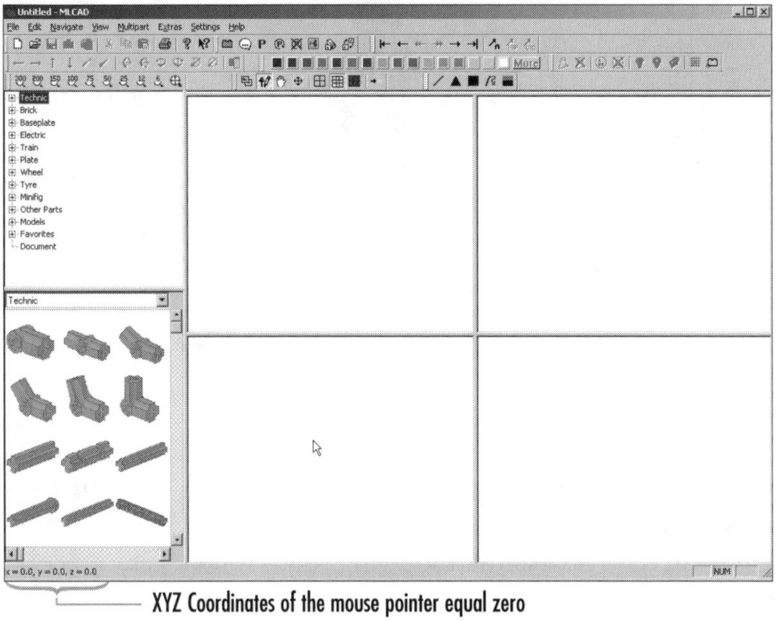

From the center, your mouse pointer position is defined in terms of positive or negative units. From the center to the top viewpoint, the Y axis has negative values; from the center to the bottom, the values are positive. Along the X axis, the positive values go from the center to the right, and the negative values go from the center to the left. Along the Z axis, movement from the center toward the back is positive, and it is negative from the center toward the front. This concept is illustrated in Figure 5.7.

NOTE

The orientation of the negative and positive axes in the LDraw system does *not* follow the standard used by most 3D software. Instead, the positive and negative ends of the Y and Z axis are switched.

Figure 5.7 Negative and Positive Values Along the Axes

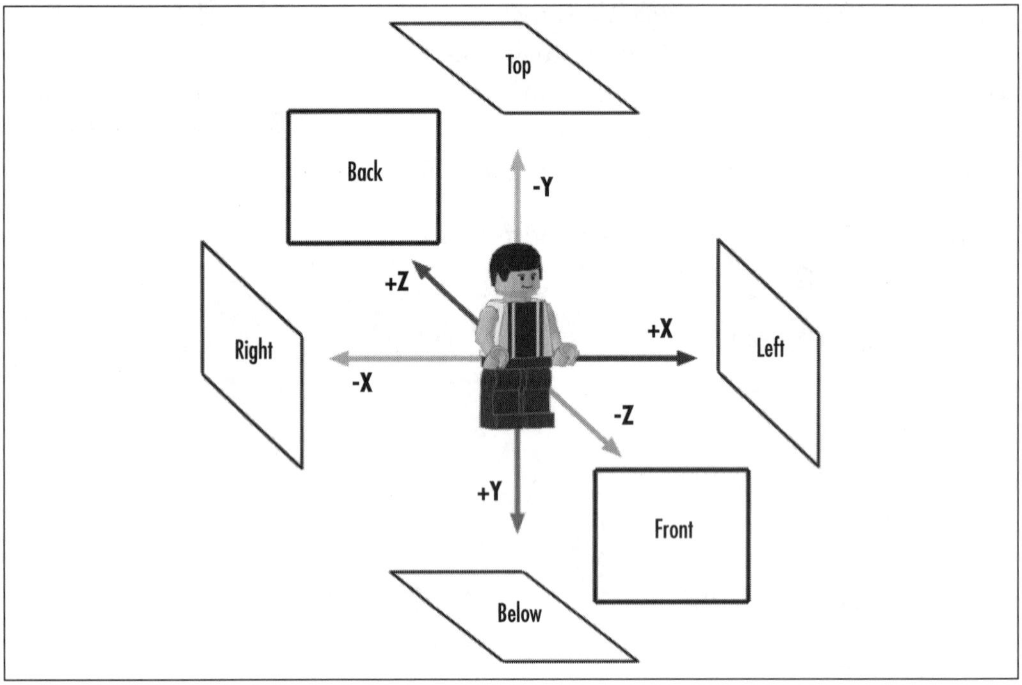

If Figure 5.7 seems a tad complex, it is because it *is* rather complex. The good news is that you do not need to understand it right away. Push on and everything will become clearer as you use the modeling tools. You can always come back and use this diagram as a reference.

Nevertheless, a good modeler always keeps the XYZ coordinate present in his or her mind. While modeling, you might be thinking in terms of "I am going to move this yellow brick right next to the blue one," but the tools you'll be using will have X, Y, and Z labels on them. This concept is very similar to playing a flight simulator game, where you receive information from the screen and incorporate it into a larger mental picture of the location of your plane.

As you build more models, you will acquire a very good knowledge of the XYZ compass. Using LEGO CAD facilitates the task of building inside a virtual space, because the design of the building blocks is so intuitive that directions and dimensions are extremely clear. However, complex models and interaction with other programs will end up forcing you to pay more and more attention to these standards.

MLCad's Precision Modeling Tools

Enough theory—let's start building. Since LEGO parts and models also provide a direct reference system, the XYZ coordinate system will become clear as our athlete minifig takes form while we explore MLCad's fine (in every sense of the word) modeling tools. If you haven't done so yet, create a new file, set the grid to Coarse mode, and resize the modeling panes to your taste.

Adding Parts to a Model

In the last chapter, we explained how to add a part to a model by dragging it from the library windows of the console. We also showed you how to display the part in those windows, via different library browsing and searching tools. It is fair to assume that most modelers start a large part of their MLCad creations that way, browsing the library for a part and dragging it into the modeling panes—but from there on the building tools can be used and combined in infinite ways.

For starters, you can add a part to a model via several methods. Other than dragging it from the Parts Preview window, you can add a file via the Editbar (see Figure 5.8) or via the **Edit | Add | Part ...** menu option (also available via the right-click menu of the modeling panes).

Figure 5.8 Adding a File Via the Editbar

However, for pure speed, nothing beats the shortcut. Press the **I** key on your keyboard. As with the options described above, you will be presented with the Select Part dialog window that you are familiar with from the last chapter; you used it to modify parts. Select the part **3626BPS5.DAT Minifig Head with SW Smirk and Brown Eyebrows pattern**, as shown in Figure 5.9.

Remember: There is no "best" method for adding a part. Pressing the **I** key is without a doubt the fastest way. The Select Part dialog window offers important options such as the Custom Part method, which allows us to load entire models as single parts of other models (we talk about this concept later in the chapter). On the other hand, the Library windows have better part-searching tools and the visual previews are more immediate. Thus, the "best" way to add a new part to your model is to use both methods, depending on the occasion.

Figure 5.9 Selecting a New Part

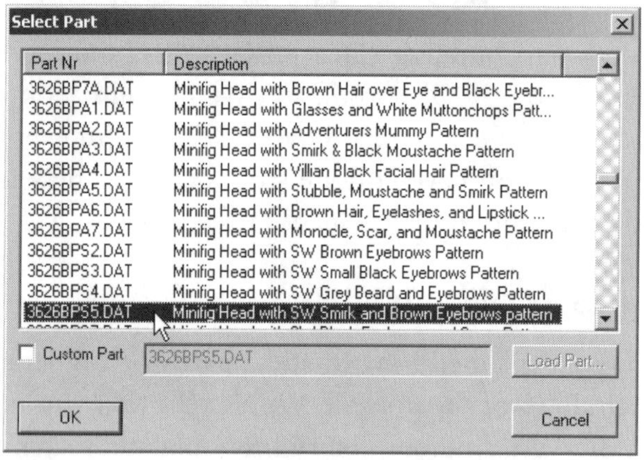

Designing & Planning…

Using Specialized Parts

The minifig in Figure 5.1 is made of 10 parts. Unlike the bricks we used in the last chapter, most of these parts are not very recognizable until they are put together. This discussion gives you very specific instructions, but using specialized parts like these still requires some extra attention. On the other hand, it also makes us better modelers; not only do these parts require more precision from us, they also force us to question some of the most basic LEGO principles, such as the stud-based attachment system.

Each subfamily of specialized parts has its own unique usage characteristics. For instance, the minifig family of parts includes many accessories for the minifigs, but the bulk of these are variations of the nine basic parts: heads, torsos, arms, hands, hips, and legs. The tenth part is the hair, which can also be a hat. Legs, heads, torsos, and legs present quite a few alternate designs, but these options are all based on different textures for those parts, as illustrated in Figure 5.9. Other parts families typically include shape modifications as well. For minifig legs, the only available shape modification is a wooden leg!

After placing the "3626BPS5.DAT Minifig Head with SW Smirk and Brown Eyebrows pattern" part into the modeling panes, your screen should look similar to the one pictured in Figure 5.10.

Figure 5.10 A Minifig Head

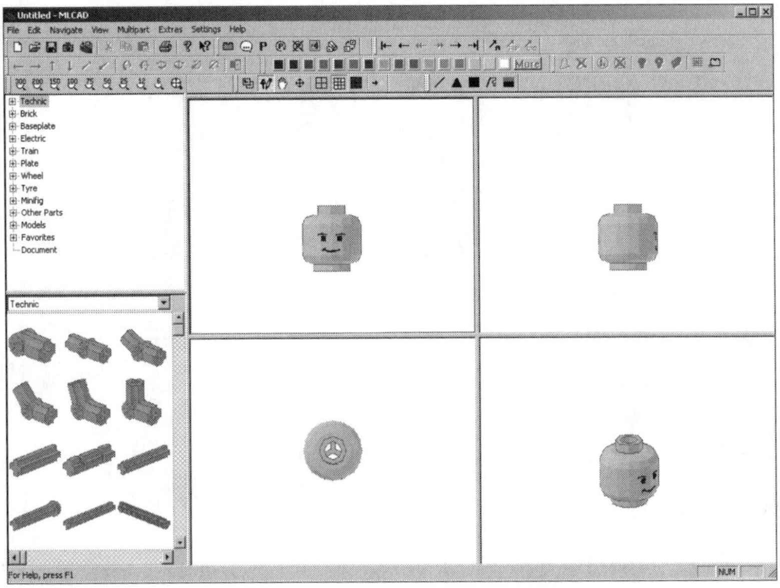

NOTE

The head we are using for the athlete is based on Harrison Ford's facial expression in the film *Star Wars.* LEGO put out this face pattern as part of the *Star Wars* theme line of LEGO kits; it was used in the Han Solo character minifigs.

Duplicating Parts

In the previous chapter, we saw how creating a copy of a part is as simple as dragging it with the mouse while holding down the Ctrl key. This method is very direct, but some circumstances call for another equally fast method. Adding hair to the minifig is one of those circumstances.

If instead of copying a part via the Ctrl key/move part method, we use the **Edit | Duplicate** menu option, or simply **Ctrl + D**. This way we create a copy *without* moving the part. This is one of those moments where the lack of density in MLCad can be a bit puzzling. Since the two parts are identical and occupy the same position, there is no way we can tell them apart. Yet they are both there; move the uppermost part slightly with your mouse and you will see something similar to Figure 5.11.

Figure 5.11 A Duplicate of the Head

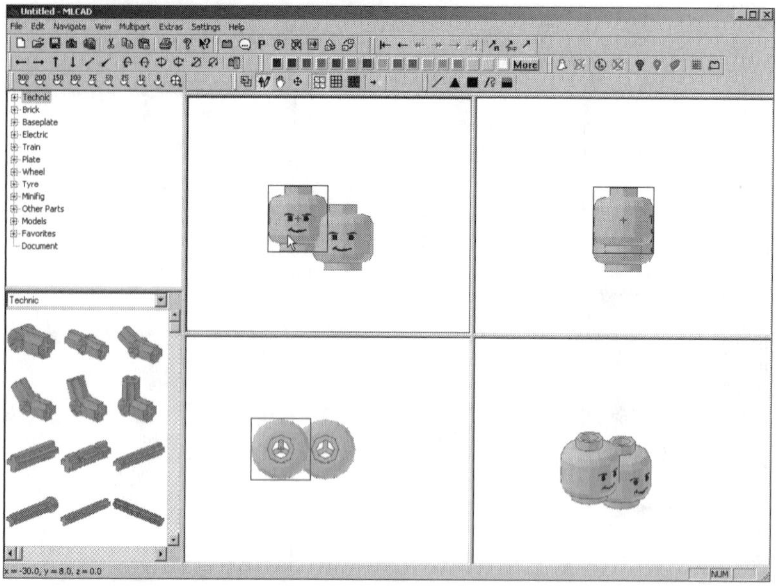

Delete the copy using the **Edit | Erase** menu option, or simply by pressing the **Del** key. Select the head and duplicate it again. This time, we do not want to move the copy (we already know it is there). Instead, we will change its part type.

As we saw in the last chapter, this task is done via the **Edit | Modify** command or the **Ctrl + M** shortcut. The command activates the Select Part dialog window with which we are quickly becoming familiar (refer back to Figure 5.9). Choose part **3901.DAT Minifig Hair male** and click **OK**. Your model should look similar to the one in Figure 5.12.

Understanding Part Origins

Each part has a specific *part origin,* which is the point in which the part will "lock" to the XYZ coordinate system. In Figure 5.13, we can see the part origins for the head and hair parts as tiny black boxes. When both part origins occupy the same coordinates, the parts combine gracefully. In this case, the hair attaches to the top of the head.

As you just saw, if a minifig *head* part and a minifig *hair* part occupy the same coordinates, they will combine correctly without further adjustments (the hair will appear on top of the head). This is a handy feature found in many specialized parts that share a common part origin, and it allows us to quickly build subassemblies such as the head/hair combination of our model.

Figure 5.12 A Minifig Head with Hair

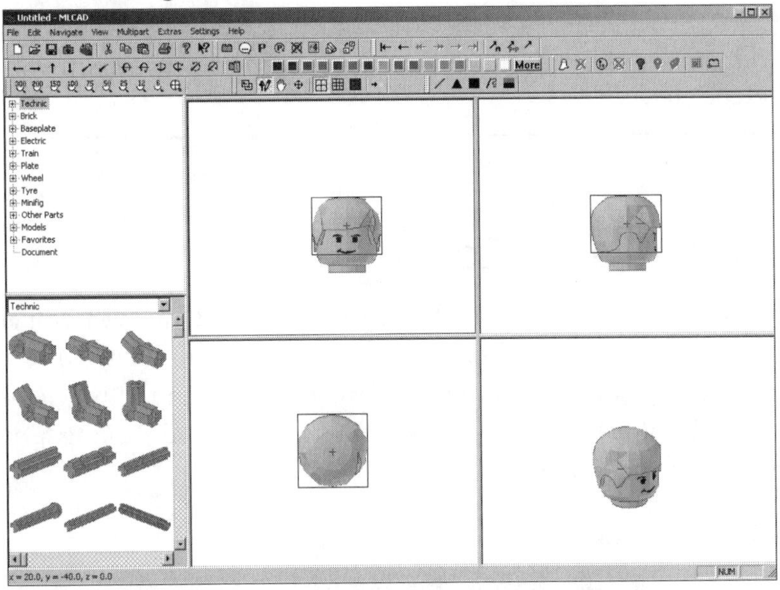

To understand part origins, we have to understand how our part is positioned in the virtual space. Say the head occupies the center of the virtual space, at coordinates 0,0,0; what does that mean? Those coordinates represent a single point in space, and the head part has volume. Think of it as similar to being given the geographical coordinates for a city. The latitude and longitude data corresponds to a single point—usually somewhere near the center of the city, which covers a larger area. Similarly, the head covers a larger area than its XYZ position coordinates.

Figure 5.13 Common Part Origins

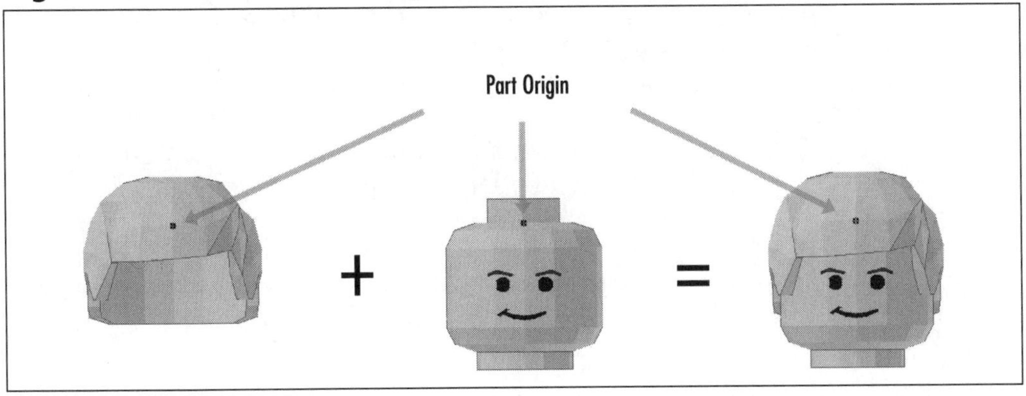

NOTE

Part origins are the default rotation points for parts, as you will see later in the chapter.

Changing a Part's Color

When you add a new part from the library into MLCad, the part will always be set to a default color. As we will see in the section "Customizing MLCad" in the Chapter 6, we can customize this default. But as we saw in the previous chapter, swapping a part's color is as simple as selecting the part and picking a new color from the Colorbar (see Figure 5.14).

Figure 5.14 The Colorbar

Designing & Planning…

Official LEGO Colors

Not that long ago, LEGO bricks came in eight colors: black, blue, dark gray, green, light gray, red, white, and yellow. This array seemed more than enough to several generations of kids and adults. Arguably, the introduction of transparent elements led the way for the introduction of new hues, and today there is an exciting variety of colors, including metallic ones.

The MLCad palette, part of which is visible in the Colorbar, follows the LDraw color standard created by James Jessiman. As you can see, the Colorbar's boxes are only vaguely related to the original LEGO colors. This is a question that LEGO fans will never agree on: Is it *that* important to use only official LEGO colors in MLCad?

Think about it for a minute. For anyone with any experience in computer graphics, this concept does not deserve a second thought. But for a LEGO user, this is a dream come true—not only do we have an infinite number of parts, we also have them in every color, including custom-made ones!

Clicking the **More** button on the Colorbar opens the Select Color window (see Figure 5.15), which can also be accessed via the **Edit | Change Color** menu option or by simply pressing the **C** key on the keyboard. Remember that you need to have a part selected first.

Figure 5.15 The Select Color Window

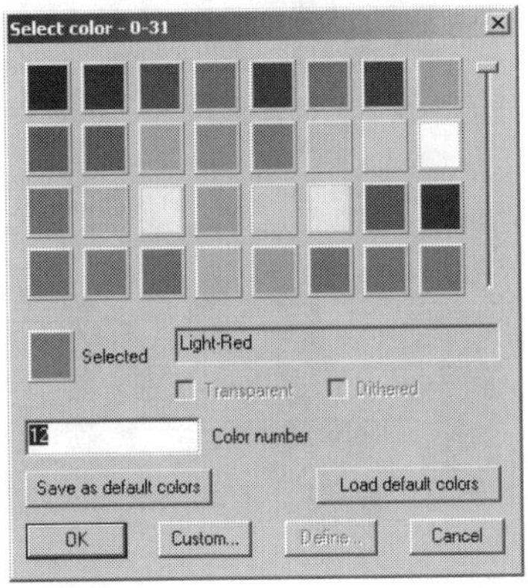

This window allows us to choose any of the available colors in the boxes or type the color number directly into the Color Number box. The slider shows color positions for up to 511 colors from which you can choose, which is far more than 99.9 percent of MLCad users will ever need. It is also possible to create custom colors by clicking the **Custom...** button. We will show you how to create new colors in a minute. Before we do that, let's see the other functions of the Select Color window.

The **Save as default colors** and **Load default colors** buttons let us manage the color palette to a limited degree. As we mentioned, 511 possible colors are available at any given time—more than most MLCad users ever need for their models. What is more interesting to know is that each "box" represents a fixed position in the palette of 511 colors, and each position is identified by a number. In the MLCad palette, position 0 is black, position 1 is blue, position 2 is green, and so on. The first 16 palette positions are always directly available via the Colorbar.

We can change the color of some of the boxes. Unfortunately, we can only change the color of positions between 64 and 256. We can customize our palette this way by clicking the **Define** button, which is only activated if we choose a color from that range of positions.

Creating Custom Colors

If we want to create a custom color, we click the **Custom** button. Clicking either the **Custom** button or the **Define** button opens a Custom Color Definition window, pictured in Figure 5.16.

Figure 5.16 The Custom Color Definition Window

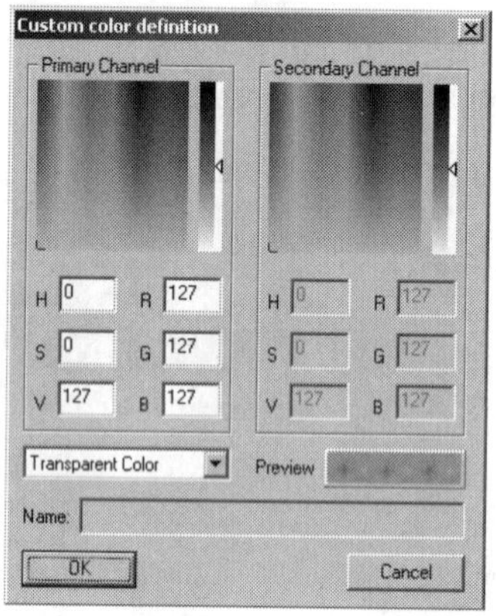

NOTE

The sophistication of the color definition options present in MLCad is due to the fact that MLCad takes full advantage of the coloring functions provided by Windows.

At the bottom of the Custom Color Definition window is the Name field, which only becomes active when the user is defining colors in the 64 to 256 positions. These apparently random figures are actually standard numbers we use when dealing with computer color.

Above the Name box is a Preview window that shows the new color being created. To the left of the Preview window is a pull-down menu that lets us select the type of color. The types available are *Solid Color, Transparent Color,* and *Dithered Color.* Transparent and dithered colors are offered in 24-bit mode as well. Most of the time we will stick to solid or transparent colors. Only when we start to refine our photorealistic rendering techniques will we start exploring other types of colors.

Above the color type menu and the Preview window are two color definition fields, Primary Channel and Secondary Channel. Except when dealing with dithered colors (a rare occurrence), we will only use the Primary Channel, on the left. The Primary Channel area offers three ways to define the custom color. We can type in H/S/V (Hue/Saturation/Brightness) values at left or R/G/B (Red/Green/Blue) values at right. But the easiest way to create a custom color is to pick it directly from the window at the very top of the color, which shows color gradations.

Designing & Planning…

Defining Color on Computer Systems

There are several ways to define colors in computer software. By assigning values between 0 and 255 for three basic components (generally red, green, and blue but also hue, saturation, and brightness), a computer screen can render over 16 million colors. This number is, in fact, roughly the same number of colors that the human eye can differentiate.

That is the theory. The practice is that the same color renders differently to different monitors and different media, which might use a different color scheme with less colors. So, unless you have a situation in which color hues are extremely important, it is hardly worth the effort to dig too deeply into the color definition boxes for MLCad's modeling purposes.

Once we define or pick a color, we simply click **OK**, and the part that we had selected will change to that color. For our athlete minifig, select his hair and pick black (see Figure 5.17).

NOTE

Colors can be our allies in the often deceiving 3D world. While building models, you can use bright, contrasting colors to clearly differentiate the parts to help you visualize complex shapes. You will also find that temporarily making a part transparent can assist you in seeing how well two parts are aligned with each other.

Figure 5.17 A Minifig Head with Black Hair

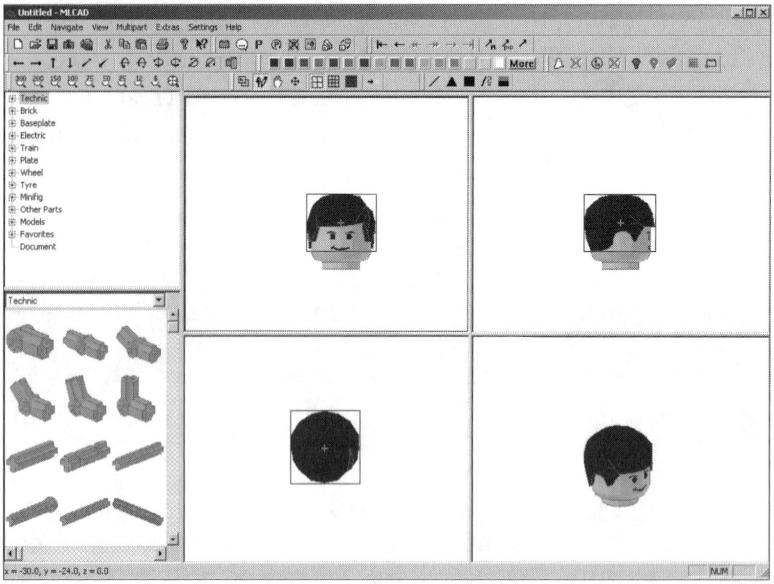

It is important to realize that most of the time you will be working with the colors shown in the Colorbar. Only when you decide to use transparent colors will you use the More button, and then only really to pick one of the existing transparent colors. Much later in the process, when we want to capture a very specific color for our renderings, we might start tinkering with the RGB values of custom colors.

Moving Parts

At the beginning of this chapter, we took a look at the XYZ coordinate system as a reference system that "maps" virtual space in MLCad for us. That reference system is very useful to us when we're moving parts in our model. In the previous chapter, we moved parts by simply clicking and dragging them with the mouse. There are several other ways of moving parts, and we access them using the XYZ coordinate system.

To illustrate this point, let's create a part for the torso of the minifig. We will use part **973P02.DAT Minifig Torso with Vertical Striped Blue / Red Pattern**, a classic that goes well with today's 1970s revivalist fashions. You *could* add the torso directly by dragging and dropping, as explained earlier. Instead, let's use the same technique that we used to add the hair. Select and duplicate the hair or the head and use the **Edit | modify** command to change it to **973P02.DAT**

Minifig Torso with Vertical Striped Blue / Red Pattern. Your model should look similar to Figure 5.18. Color the torso part white.

Figure 5.18 A Torso Part Added to a Model

Unlike the head and the hair piece, the torso does not fit automatically; it uses a different *part origin*. Thus, we need to move the part down—or in XYZ coordinate parlance, we need to move it positively along the Y axis (refer back to Figure 5.7). We can do so by simply clicking and dragging it with the mouse. Before we do this, we need to take look at a new concept: the grid.

The Grid

Even at this basic level, MLCad offers a very useful yet very discrete tool: the grid. It acts as 3D graph paper to which the parts "stick." When you move the torso with the mouse, you will notice that the part does not slide smoothly; rather, it jumps from a fixed position to another fixed position nearby—like jumping a piece from one square of graph paper to the next. If you change the settings of the grid by clicking the **Grid medium** or **Grid fine** buttons on the Viewbar (see Figure 5.19), the distances "jumped" will be smaller and the movement smoother because the graph paper now has smaller squares.

Figure 5.19 Grid Settings

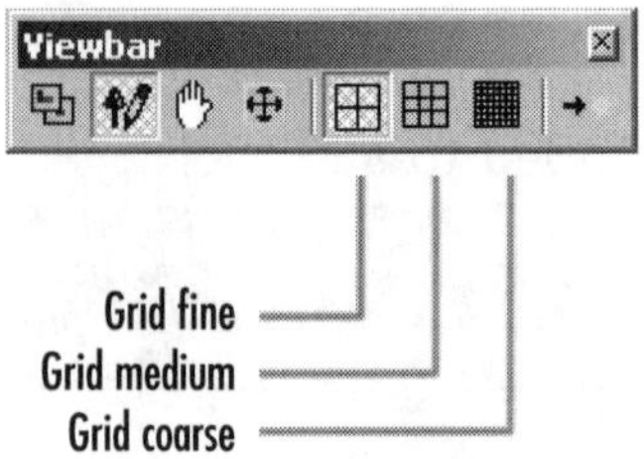

NOTE

Most of the time, smoother movement is actually not preferable, because the coarser grid modes facilitate the task of placing parts so that they "lock" together.

The grid is always invisible. Unlike other computer graphics programs, MLCad does not let us view the grid, just "feel" it. This apparent limitation is probably a deliberate decision by Michael Lachmann. Although there are times when being able to see the grid would be handy, most of the time we use the part shapes and connection points as references to position new parts. The grid helps us position those parts so that they interlock. The parts themselves provide the actual reference for connection points—just like real LEGO.

Using your mouse, move the torso to the position pictured in Figure 5.20. Position the parts so that the neck part of the head rests atop the shoulder platform of the torso part. Be aware that the zoom factor affects the relative size of the grid on the modeling panes, just as it affects the relative size of the models. A general view (small zoom factor) of a large model will blur the details. Likewise, it will make the grid spacing feel smaller. Large zoom factors (higher magnification) will bring the details into focus and increment the apparent size of the grid spacing.

Grid Alignment

As we saw in the last chapter, MLCad's virtual LEGO parts do not have density. They do not "bump" into each other; rather, they simply overlap. This is a significant difference from real-life LEGOs because it means that parts do not really "lock" with each other. In the models, the parts simply "lie" in positions where they would connect in the real world. This difference can lead to some problems when you're modeling in MLCad.

Figure 5.20 The Torso in the Correct Position

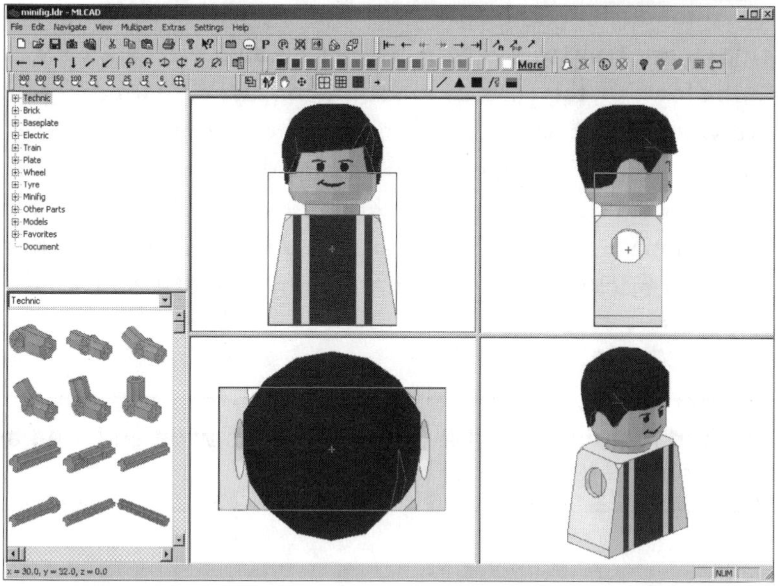

Most of the time, alignment is not a big issue. First, many misplacements are easily caught by simply looking at the model in the modeling panes. Misaligned parts generally stick out like sore thumbs. This is specially true for the simpler parts. For instance, misplaced bricks are very easy to identify, as shown in Figure 5.21. Notice how the two bricks appear to be attached "through" the two studs on the lower brick. Clearly, this would not be possible in the real world.

Figure 5.21 Misaligned Bricks

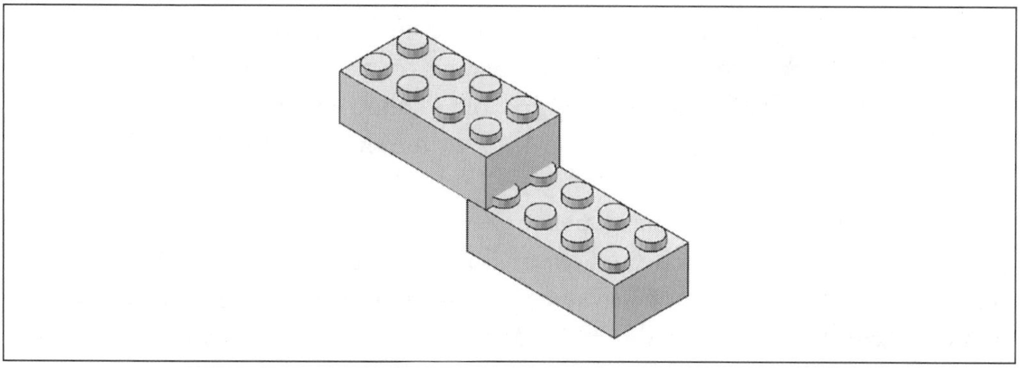

Furthermore, as mentioned before, the coarse grid steps are set by default to place many parts in locking positions. However, specialized parts and part rotations, as we will see shortly, often call for very small adjustments. In the next sections we explain how parts can be positioned in extremely precise ways, using fine or even custom grid settings.

The grid is a great ally to the MLCad user because parts always snap to the grid positions. However, the parts can become misaligned with the grid. This can happen if we switch often between different grid modes when moving parts. The settings for each of the grid modes can be changed, as explained in the section "Customizing MLCad" in the next chapter; but the default spacing along the X and Z axes is 10 units for coarse, 5 for medium, and 1 for fine. Due to the proportions of LEGO bricks, along the Y axis the units are 8, 4, and 1, respectively.

NOTE

A LEGO stud measuring unit is 20 MLCad units wide and long and 24 units high.

Thus, if we move a part in Coarse grid mode along the X or Z axis, we move it in 10-unit increments: 10, 20, 30, and so on. If we switch to Medium mode, we move it in 5-unit increments: 5, 10, 15, 20, 25, and so on. If we move the part in an odd number of increments in the Medium grid mode and then switch back to Coarse grid mode, the part will then move from coordinate 5 to 15 to 25 to 35 and will never properly align or "lock" with parts aligned to coordinates 10, 20, 30, and the like.

Let's look at a practical example of this concept. Using the minifig model in Figure 5.20, switch to Medium grid mode via the Viewbar, and then move the torso exactly one grid space to the right, as shown in Figure 5.22.

Now switch back to Coarse grid mode and try to reposition the torso correctly, as in Figure 5.20. You will only be able to align the parts in the positions shown in Figures 5.22 and 5.23, never back to the original we started with in Figure 5.20.

As explained earlier, the head and torso are both set to coordinate 0 along the X axis. With Coarse grid activated, we will move it to coordinates 10, 20, and so on (or –10, –20, and so on). If we switch to Medium grid and move the torso once, we will place it in coordinate 5. Once we switch back to Coarse grid, the torso will be able to jump from position 5 to position –5 or position 15, but never positions 0 or 10.

Figure 5.22 A Misaligned Torso

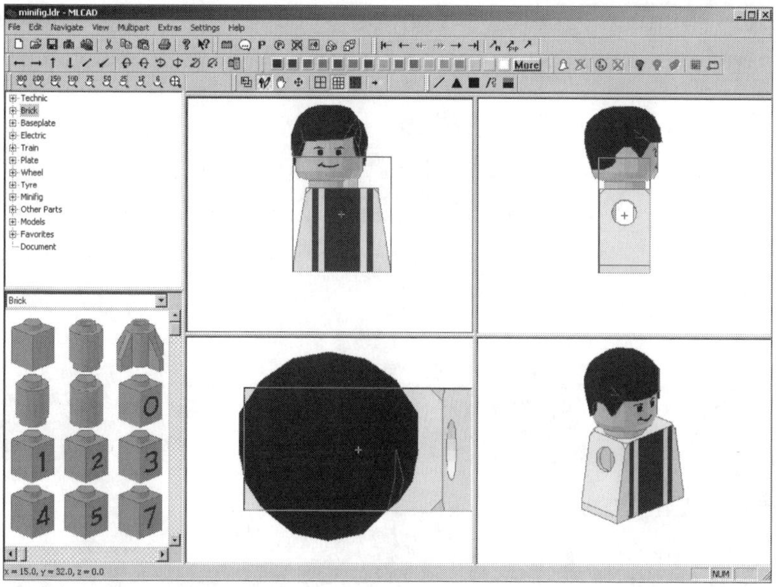

Figure 5.23 The Torso Is Still Misaligned

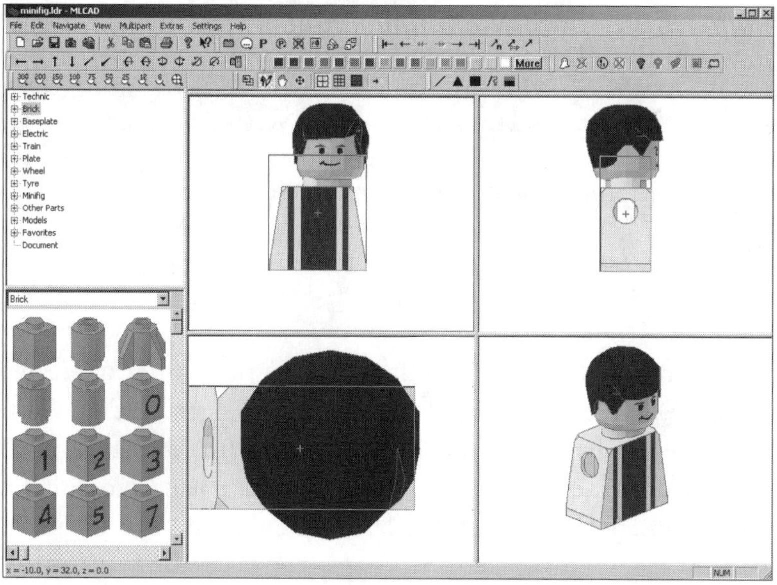

Snap to Grid

To properly align the torso again, we could switch back to Medium grid and realign the part by hand to the Coarse grid positions (or simply to the minifig head). A quicker way of achieving all this but leaving the Coarse grid activated is via the **Edit | Snap to grid** menu option (see Figure 5.24), also accessible through the Modificationbar or via the **Control + Shift + g** shortcut.

Figure 5.24 The Snap to Grid Button

Designing & Planning...

The LDraw Connection Database

There currently is an ongoing effort to create virtual snap-together characteristics for LDraw parts, collectively called the LDraw Connection Database. The proposed format would create a duplicate of each part file in the library and store in it all the possible connections for that particular piece. Once the "parallel" library is created, software developers could integrate support for it into their programs.

This is an official project of the LDraw.org organization. (Details can be found on the organization's Web site, www.ldraw.org.) It is a very interesting idea that offers lots of possibilities. If and when a LEGO-based CAD program incorporates such functions, the casual user, including many LEGO fans, will greatly benefit from an even easier-to-use building system.

This command realigns the part to the current grid settings, and it will actually move it! Since the head part is aligned with the Coarse grid setting, the torso part will be correctly aligned with it again, as in Figure 5.20. Although it might not seem like much to you yet, Snap to Grid will become a handy tool that makes up for part of the annoyances caused by the lack of density in virtual LEGO!

Other Ways to Move Parts

As we saw in the last chapter, one of the disadvantages of moving bricks directly with the mouse is that we cannot do so in modeling panes set to the 3D view angle, since the mouse does not perform part-related tasks in those panes. This does not mean we cannot move parts while looking at the model in a modeling pane set to the 3D angle, however. There are two other ways to move the parts without dragging them with the mouse. As we said earlier, we relate to these parts via the XYZ coordinate system directly, and they also use the grid settings.

We can move a selected part or parts in any modeling pane, including those set to the 3D view angle, by clicking any of the first six buttons of the Elementbar or by using the **Edit | Move** menu option. With each mouse click, we trigger movement along one axis only. Parts can be moved in two directions in each axis: negative or positive. To move a part from X coordinate 40 to coordinate 30, we would use the negative movement button for the appropriate axis (see Figure 5.25). If we were to move the part to X position 60 from X position 40, we would use the positive movement button.

Figure 5.25 The Move Part Buttons

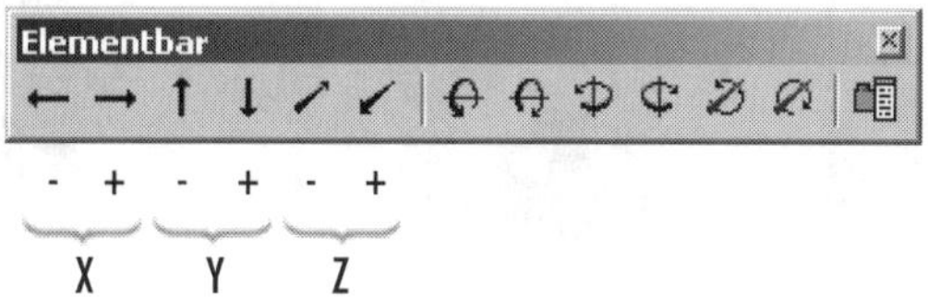

Each mouse click moves the part the number of units equivalent to one grid space. For example, the default setting in Coarse grid mode is 10 units for the X and Z axis. For a part placed at 0,0,0, moving it once on each of the positive X and Z axes will move the part to 10,0,10 XYZ coordinates. Additionally, we can use the keys listed in Table 5.1 to move parts in the same fashion.

Table 5.1 Movement Keys

Keyboard Key	Action
Cursor left	Move –X
Cursor right	Move +X
Cursor up	Move +Z
Cursor down	Move –Z
Home	Move +Y
End	Move –Y

Let's build the legs of the minifig using these techniques. Duplicate the torso part and convert it into a part of the type **970.DAT Minifig Hips**. Use the Elementbar buttons to position the part as shown in Figure 5.26.

Figure 5.26 Minifig Hips

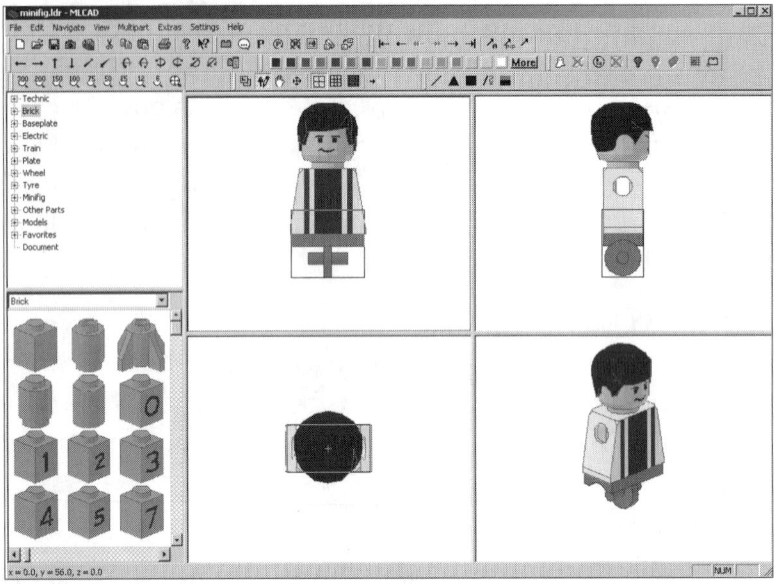

Let's now add the right leg (part **971.DAT Minifig Leg Right**). Use the cursor keys to move it this time. You should place the part as shown in Figure 5.27. This time, you need to switch to Medium grid mode to properly align the leg with the hips. Look closely at the upper-right modeling pane, set to the Left view angle, to see how the leg adjusts to the hip part.

The right and the left leg parts share the same part origin. Thus, to create a left leg, all you need to do is duplicate the right leg and change its type to **972.DAT Minifig Leg Left** and the new leg will appear aligned automatically, as in Figure 5.28.

Next, we will add the arms. Arms do not have a very clear connection position, nor do they share a part origin with the torso. However, you can duplicate the torso, turn it into a right arm (**975.DAT Minifig Arm Right**), adjust it 8 grid units along the Y axis, and it will be properly positioned. Remember, the default Coarse grid setting for the Y axis *is* 8 units. See Figure 5.29 for the correct position of the arm.

Figure 5.27 The Right Leg of the Minifig

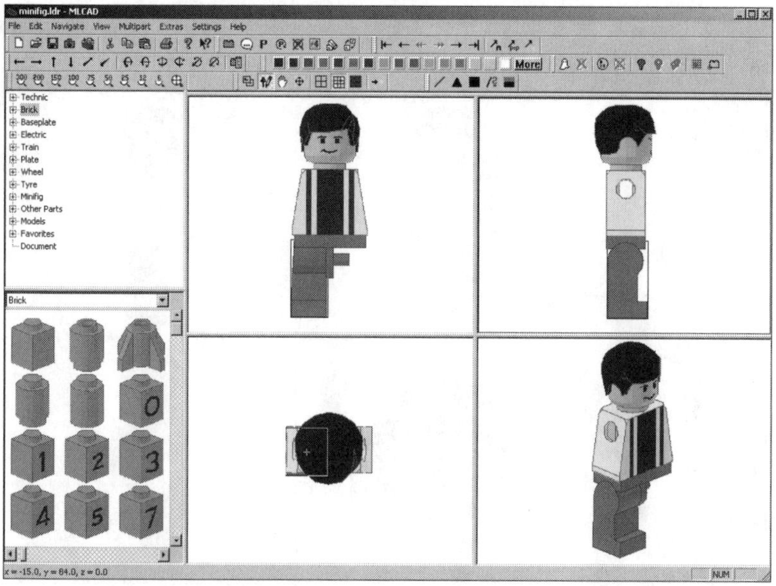

Figure 5.28 The Minifig with Both Legs

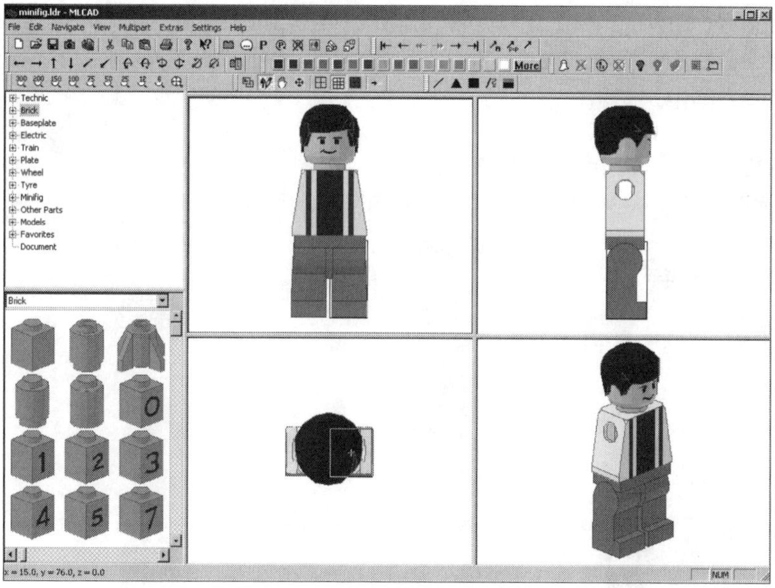

Figure 5.29 The Minifig's Arm Position

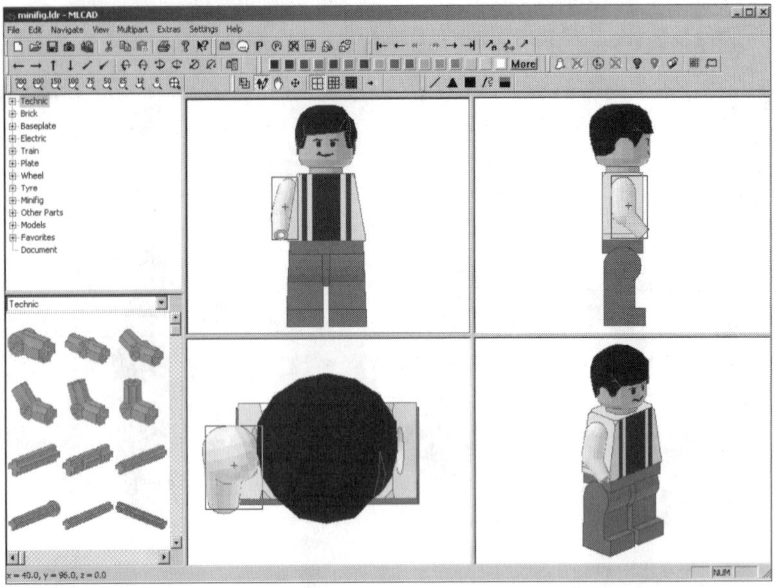

The left and right arm parts do share a common part origin between them. Once you have the right arm in place, simply duplicate it and convert it into part **976.DAT Minifig Arm Left**. Figure 5.30 shows the end result. The minifig now has two arms.

Figure 5.30 The Minifig with Arms

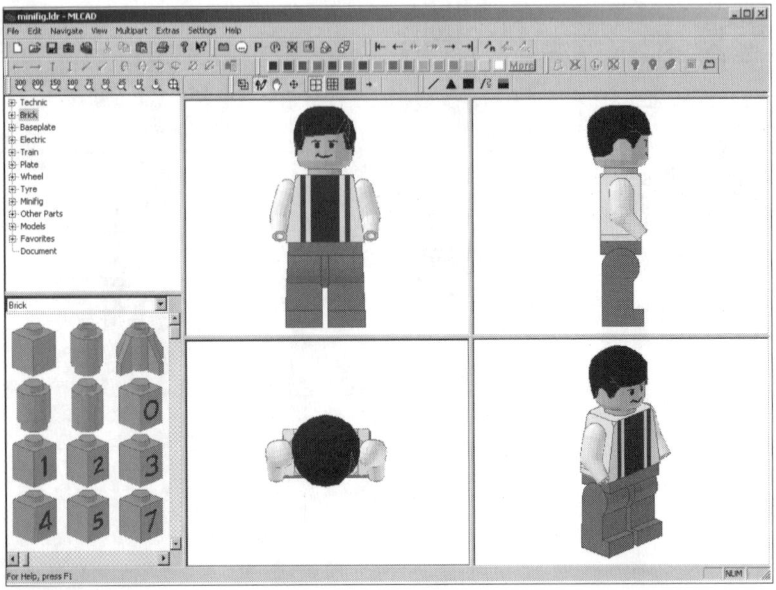

Advanced Modeling Techniques

Until now, we have covered basic modeling skills in detail: adding parts, changing colors, switching part types, and moving parts. Part movement itself requires awareness of MLCad's XYZ coordinate system, which describes the virtual playground where we perform movements. At this point, the reader has a very good grasp of some essential concepts related not only to MLCad but to 3D computer software in general.

In this section of the chapter, we cover part rotation and orientation. Rotations are essential (and common) part transformations, but they require a more sophisticated perception of the virtual space—a further twist (no pun intended) of our mental representation of the model inside the virtual playground. Movement is easier to understand than rotation. Almost anybody can easily describe or relate to movement, but explaining rotation involves abstract concepts such as axes, degrees, and rotation points.

Not to worry. You have already performed part rotations in the last chapter. What we will do now is dissect these rotations into their basic elements. Rotations simply require more input than movement. As with movement, there are default settings to assist us, but as modelers, we want to be especially alert to part rotations and orientations.

Mastering these techniques further opens our horizons as modelers. Rotation vectors and part orientation, covered at the end of the chapter, allow us to perform *very* neat tricks, seemingly without effort, which will impress our MLCad acquaintances. More important, however, rotation vectors and part orientation will cause us to start seeing our models as software-independent creations. Remember, MLCad is a great tool, but it is not the only one covered in this book nor the only one you should use as a modeler.

Understanding Rotations

Since most of the parts in the library are not symmetrical, rotating them is actually a common transformation when we use them in our models. As we saw in the last chapter, bricks need to be rotated to align them perpendicularly. By far the most common rotation is the 90-degree rotation, which we saw in the last chapter. In Figure 5.31, the red brick has been rotated from its initial position at right to the position at left. This is a 90-degree rotation.

To move a part, we must provide two things: direction and distance. To rotate a part, we provide direction and distance as well, but we call them *rotation axis* and *degree of rotation*. We also need to provide a third component: *rotation point*. Let's take a look at each of these concepts.

Figure 5.31 A Ninety-Degree Rotation

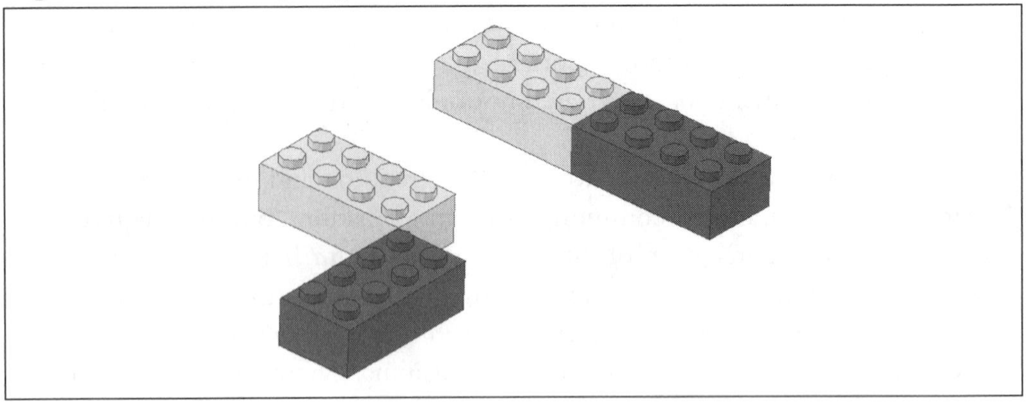

The Rotation Axis

We start with the direction of the rotation, known as the *rotation axis*. This concept relates directly to the 3D compass we looked at in Figure 5.3, but it is not as immediate as movement direction. For instance, when we see the minifig, it is immediately obvious which directions are length, width, and height. But defining axis of rotation is a bit trickier. In which direction do the arms rotate when we raise them? The arm rotates around an axle at the shoulder, and that axle is parallel to the "width" dimension. From Figure 5.3, we know that this dimension corresponds to the X axis. Thus, the rotation axis of the minifig arm is the X axis.

Now that we know which axis we want to rotate the part on, let's give it a try. With the right arm selected and the grid set to Coarse, rotate the minifig's right arm clockwise on the X axis. To do so, you can use the rotation buttons for the appropriate axis in the Elementbar (see Figure 5.32).

Figure 5.32 The Rotation Buttons of the Elementbar

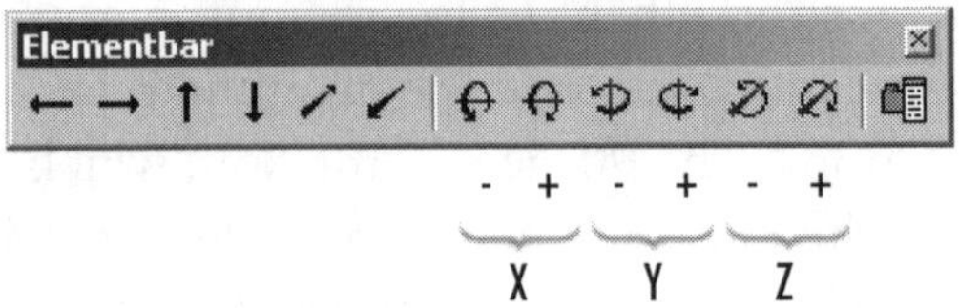

You can also use the **Edit | Rotate** menu options or the keyboard shortcuts found in Table 5.2 to accomplish this task.

Table 5.2 Rotation Keyboard Shortcuts

Key Combinations	Action
Ctrl + Cursor left (Keypad 4)	Rotates counter-clockwise along Y axis.
Ctrl + Cursor right (Keypad 6)	Rotates clockwise along Y axis.
Ctrl + Cursor up (Keypad 8)	Rotates counter-clockwise along X axis.
Ctrl + Cursor down (Keypad 2)	Rotates clockwise along X axis.
Ctrl + Home (Keypad 7)	Rotates counter-clockwise along Z axis.
Ctrl + End (Keypad 1)	Rotates clockwise along Z axis.

Once the arm is rotated, the minifig should look something like the one in Figure 5.33.

Figure 5.33 The Minifig with Raised Arm

If instead of the X axis, we had rotated the arm on the Y or Z axis, we would get the strange results shown in Figure 5.34.

NOTE

Defining the appropriate rotation axis is not always an easy task, but generally, errors are glaringly obvious! You will have a better grasp of what is happening once we have covered rotation points.

Figure 5.34 Rotating the Arm Along the Y and Z Axes

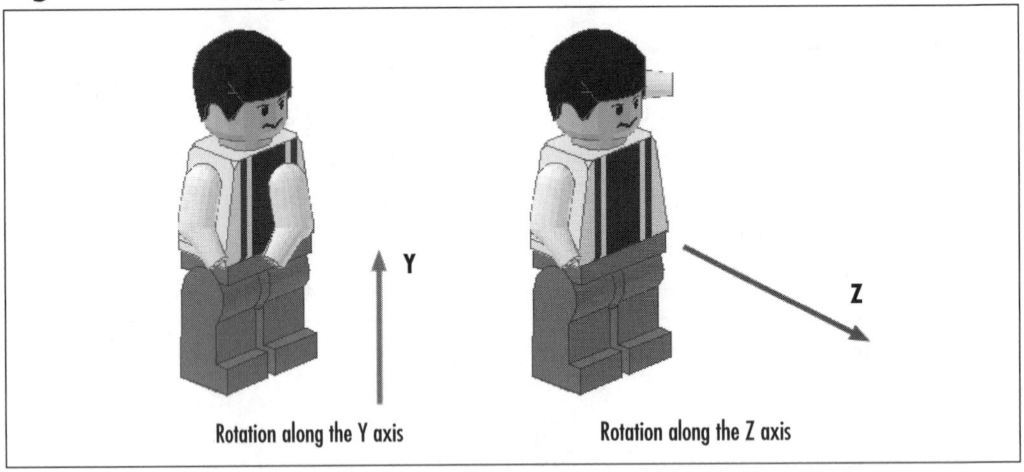

Rotation Angle

Once we know which rotation axis we want to use, the next step is deciding the distance the part will travel from its starting position. This distance is known as the *rotation angle* and is measured in degrees. A complete circle has 360 evenly spaced degrees. Thus, 90 degrees represent a quarter of a circle, and 45 degrees represent an eighth. Figure 5.35 shows other common angles.

Figure 5.35 Some Common Rotation Angles

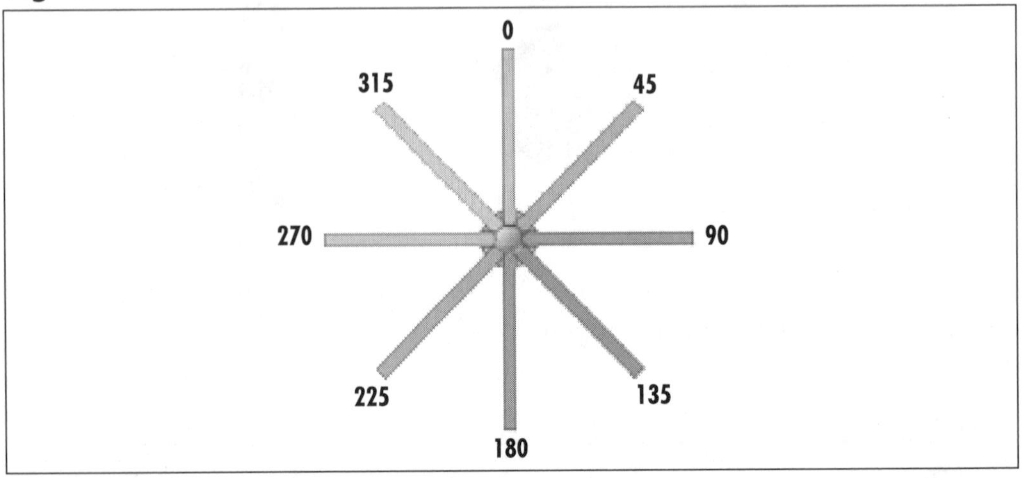

In order to tell the program how many degrees we want the part to rotate, we use the grid. Just as grid settings aid movement *directly* when we're moving parts in the modeling panes with the mouse and *indirectly* when we use the keyboard shortcuts or Elementbar buttons, the grid settings tell the program how many degrees of rotation to apply to the part. The default degree settings are:

- Ninety degrees when Coarse grid is activated
- Forty-five degrees when Medium grid is activated
- Fifteen degrees when Fine grid is activated

These settings can be changed, as you will see in the section "Customizing MLCad" at the end of the Chapter 6. For now, these default settings will give us plenty to work with.

Let's see how this works. With the minifig oriented as shown in Figure 5.33, click **Grid Medium** and rotate the part counterclockwise *once* along the X axis using any of the methods described earlier. The model should look like Figure 5.36.

Figure 5.36 The Minifig's Arm Is Semi-Raised

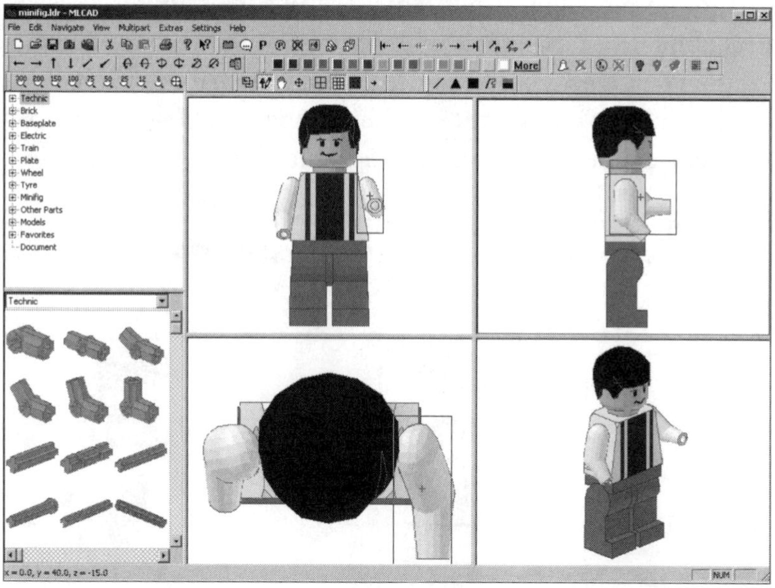

This time, the arm has traveled 45 degrees instead of 90 (as in Figure 5.33). Figure 5.37 shows the difference between these two rotation angles for the left arm.

Figure 5.37 Rotation Angles for the Minifig's Arm

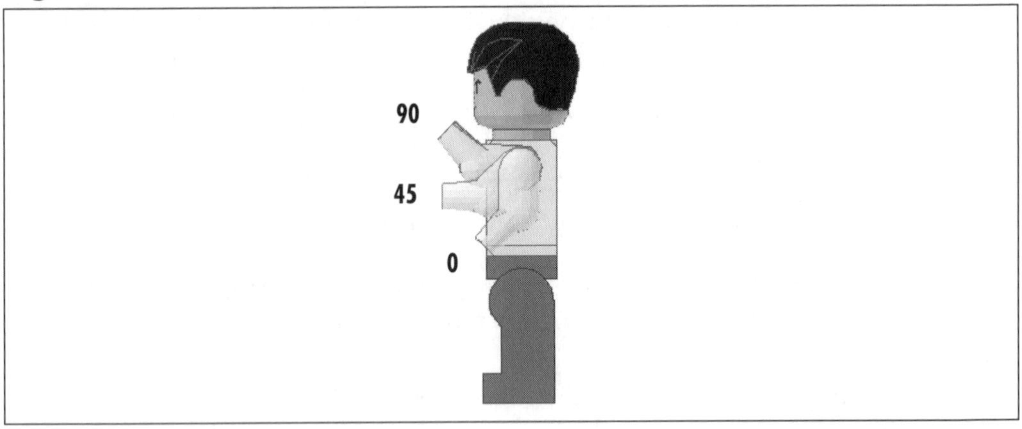

The arm's position in Figure 5.36 is actually very useful for completing the figure, because it offers a more practical way to add the minifig's hand. The minifig hand part is **977.DAT Minifig Hand** (there are no left and right hands). Add this part to the modeling panes whichever way you prefer. This part has an origin point completely misaligned with the rest of the minifig's parts. Even with the arm positioned at a 45-degree angle, aligning the hand with it will require the use of Grid Fine for movement and perhaps several 90-degree rotations. The minifig hand's correct position is shown in Figure 5.38.

Figure 5.38 The Hand Correctly Attached

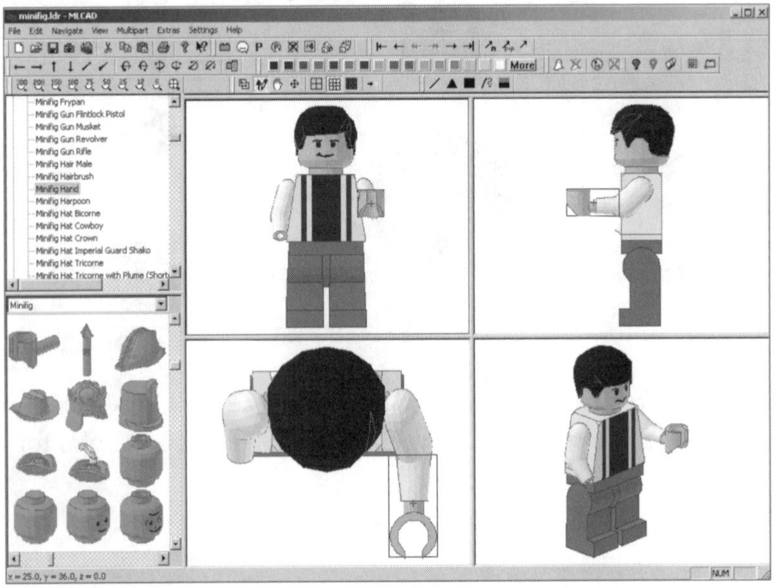

Now rotate the right arm into the same position and copy the hand to its end. You will still need to be in the Fine Grid setting to adjust the hand so that it aligns precisely with the socket at the end of the arm (see Figure 5.39).

Figure 5.39 Both Hands Attached

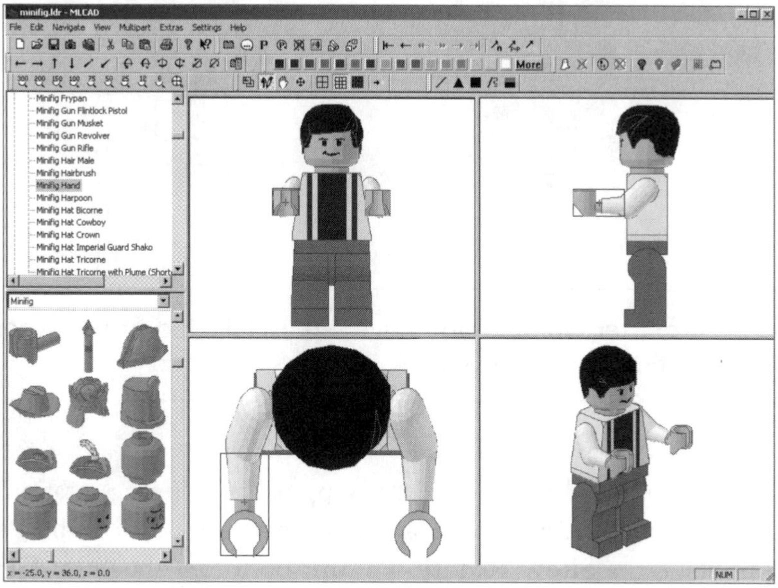

Rotation Points

The third component of a rotation is the *rotation point*, which you can think of as the axle around which we rotate the part. This is not the same as the rotation *axis*, which indicates the *direction* of the rotation. Like the part origin we talked about earlier, the rotation point is a single point to which the rest of the volume of the part relates. To use (and learn about) rotation points, click the **Rotation Point** button of the Modificationbar (see Figure 5.40).

Figure 5.40 The Part Rotation Point Button

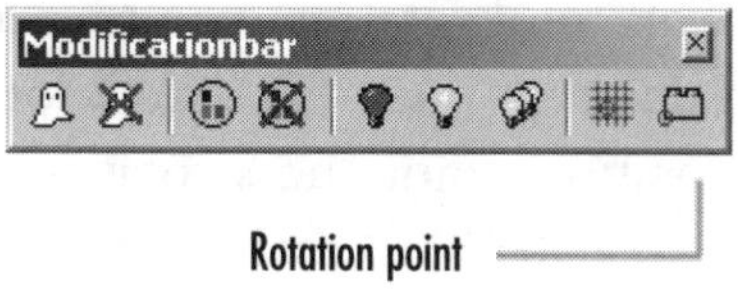

Clicking the Rotation Point button brings up the Rotation Point Definition dialog window, pictured in Figure 5.41.

Figure 5.41 The Rotation Point Definition Dialog Window

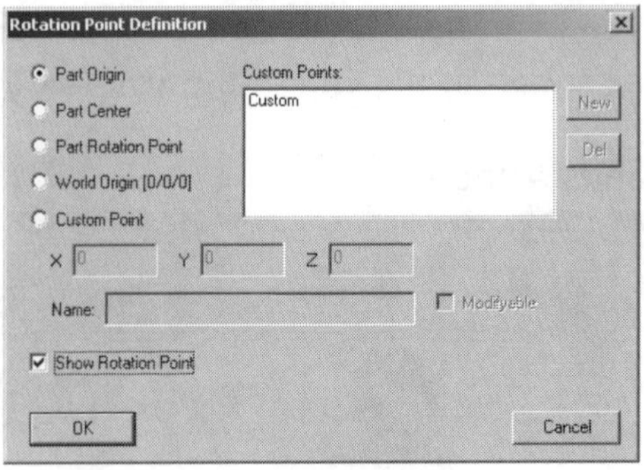

NOTE

It does not matter whether you have a part selected or not when you use the Rotation Point button.

As we can see, MLCad offers excellent support for defining rotation points. We will go through all the options shortly. For now, simply select the **Part Origin** option and the **Show Rotation Point** option, as shown in Figure 5.41, and then click **OK**. Now, when we select a part from the model, not only will get a black bounding box around it and a small cross indicating its center—we will also get an inverse box that shows the part's current rotation point. If you select the right leg from the model, you will see the rotation point represented by a small inverse square in the upper thigh, as shown in Figure 5.42.

If we rotate the leg 45 degrees along the X axis, the end result will look like Figure 5.43.

If we now select the minifig's right arm and click **Rotation Point**, we will see that it has a different rotation point than the leg, as shown in Figure 5.44.

As you can see, the rotation point and rotation axis set a local rotation axle for the part being rotated. This local rotation axle is called the *rotation vector*. Figure 5.45 shows the two different rotation vectors for the arm and the leg, both of which are defined by a rotation point and a rotation axis. In this case, the rotation axis is parallel to the X axis for both rotation vectors.

Figure 5.42 The Right Leg's Rotation Point

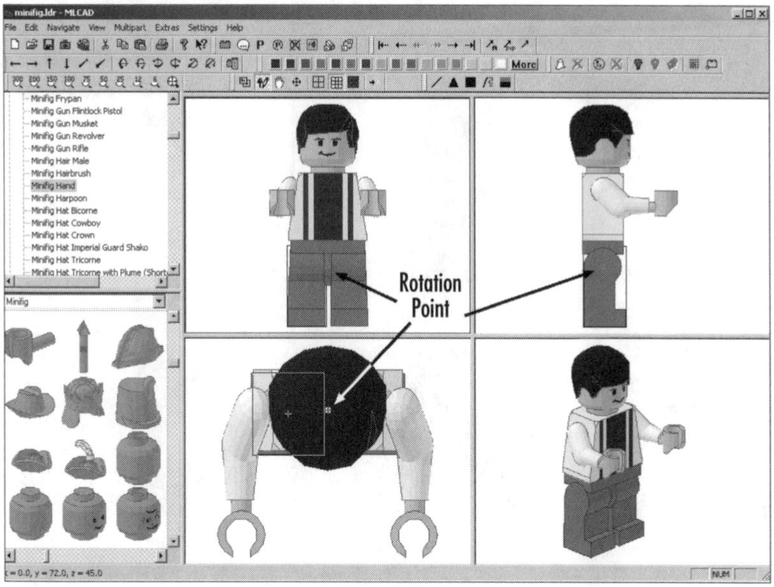

Figure 5.43 Leg Rotated

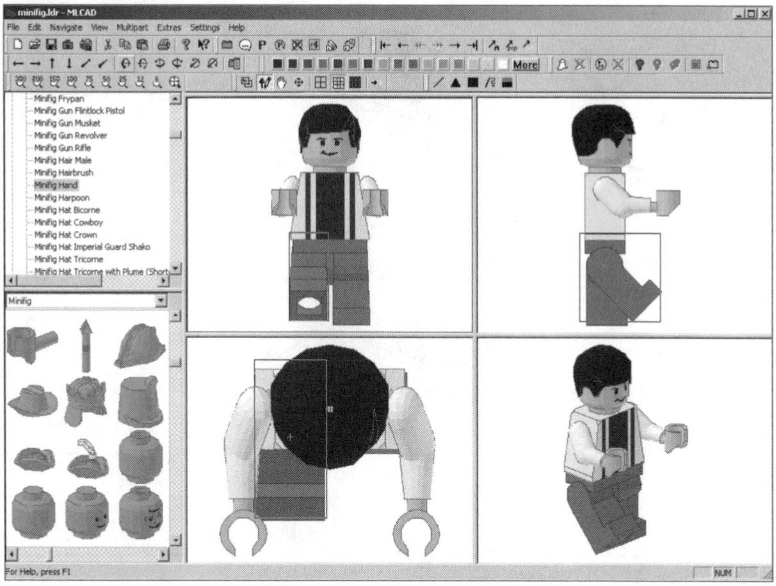

Figure 5.44 The Arms' Rotation Points

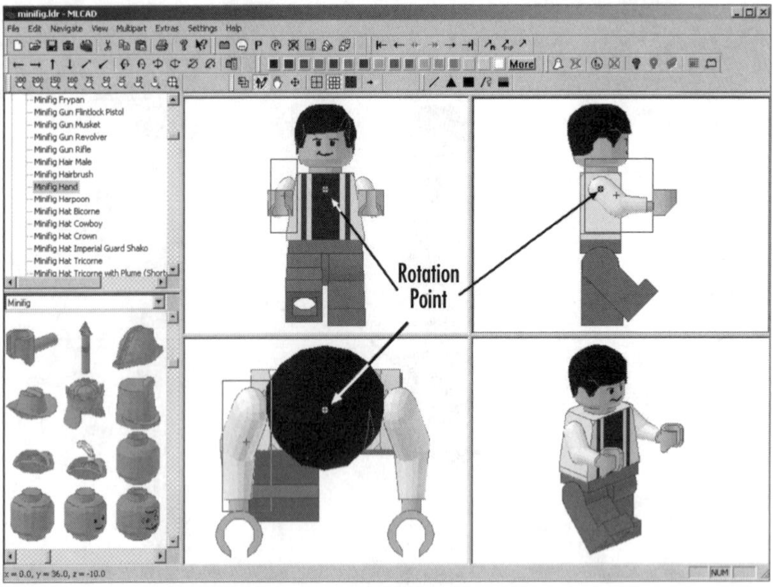

Figure 5.45 Rotation Vectors for Arms and Legs

NOTE

We will see rotation vectors in more detail in the section "Precise Part Positions and Orientations" later in this chapter.

Changing the rotation point is as easy as clicking and dragging it in a modeling pane as though it were a part. Figure 5.46 captures the action in the upper-right modeling pane.

Figure 5.46 Changing the Rotation Point Directly with the Mouse

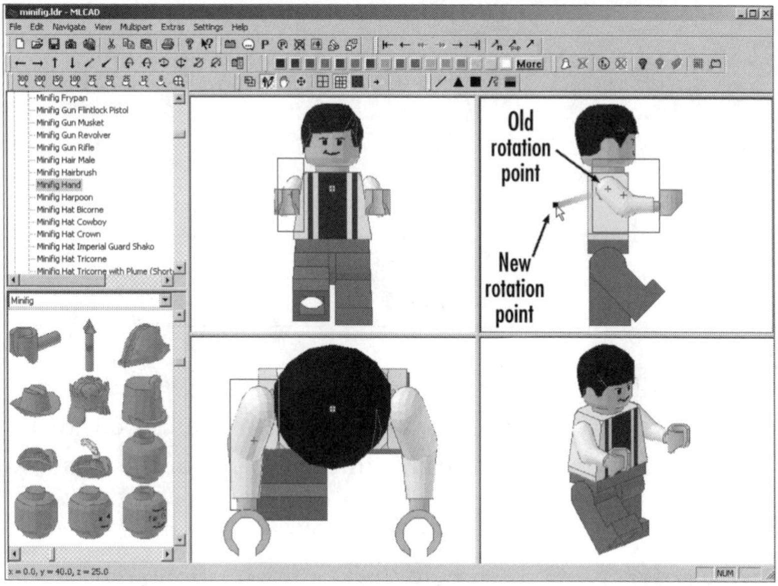

NOTE

Rotation points are not subject to the grid spacing.

As we saw earlier in Figure 5.41, the Define Rotation Point dialog window offers much more control over the rotation point. Before we get into those specific controls, you might very well be asking yourself why is this so important, if parts already have built-in support such as the common part origins. As we saw earlier with the minifig's hand, not all parts have that kind of support built in. Furthermore, some parts can use more than one rotation point. For instance, in Figure 5.47 we can see some obvious rotation vectors aligned with the studs and holes of a TECHNIC brick. Similarly, other parts and modeling situations require changes in the rotation point of parts and models.

Figure 5.47 A Technic Brick

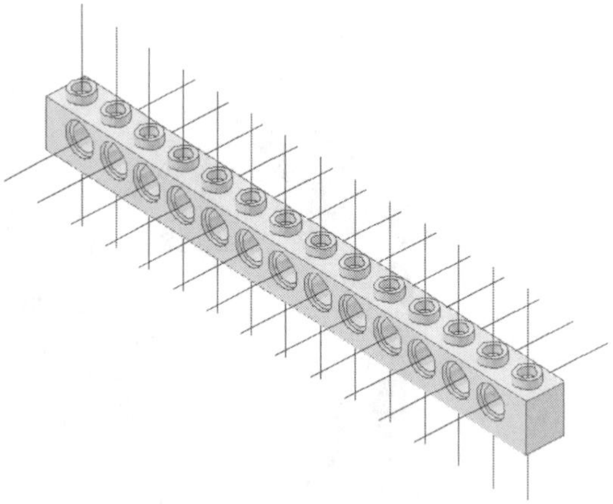

In the Rotation Point Definition dialog window shown in Figure 5.41, we can choose from a variety of preset rotation points, or we can create a set of our own. The preset rotation points are listed on the top left of the window:

- **Part Origin** This is the default setting.
- **Part Center** Not surprisingly, this sets the rotation point as the center of the part (which is always visible as a tiny cross when the part is selected).
- **Part Rotation Point** This setting is more a possibility than a feature. It allows part creators to define a rotation point other than the part origin and part center for a part, yet none of the available parts in the library use this feature at the moment. This situation could change in the future.
- **World Origin** This setting uses the center of the virtual space of the current model as the rotation point. As we saw earlier, the center is at XYZ coordinates 0,0,0.
- **Custom Point** This setting allows the user to define custom rotation points—not just one but as many as needed. This means that we can indeed define rotation points for each of the holes of the Technic brick in Figure 5.47. As we saw earlier, defining a custom rotation point can be as simple as moving it with the mouse in a modeling pane as though it were a part. By doing so, we also automatically switch the rotation point to custom. Additionally, in the Define Rotation Point dialog window, we can type the XYZ coordinates for the custom view port.

If we define and use several custom rotation points for a model, naming them is a good way to keep them organized. By default, MLCad names custom rotation points "Custom," but we can assign a custom rotation point any name we like by typing our chosen name into the **Name** box. The name will then appear in the Custom Points list in the top right of the window. The New and Del buttons manage the list, adding or deleting new entries to the custom rotation points list. There is no limit on the number of rotation points a model can have.

If we uncheck the Modifiable option of the Define Rotation Point dialog window, we are telling the program that the rotation point cannot be changed by dragging it with the mouse in the modeling panes. This is a great feature that prevents accidents when we are moving parts with the mouse in the modeling ports and we have the Show Rotation Point option activated.

Once you are done setting the rotation point in the Rotation Point Definition dialog window, click the OK button. The new setting will apply to all parts that you select from then on.

Multiple-Part Rotation

Let's see a practical example in which rotation points make our lives easier. This example involves selecting more that one part and is a good preparation for the next section of the chapter, which deals precisely with groups of parts. In our minifig model, we raised the arms to make it easier for us to fit the hand parts in them. However, once the hands are in, the minifig's stance is more like that of a criminal about to be handcuffed than that of a sports champion (see Figure 5.48).

What we want to do is rotate the right arm with the hand attached. Activate the **Part Origin** and **Show Rotation Point** options in the **Rotation Point Definition** window (refer back to Figure 5.41). Select the right arm. It should display a rotation point like the one you saw in Figure 5.44. Now, as you learned in the previous chapter, select the **right hand** of the minifig as well by clicking it while holding down the **Ctrl** key. The two parts are now selected, yet the Part Rotation point for the whole selection (the arm and the hand) is the same as that for only the arm, as shown on Figure 5.48.

If we now rotate the selected parts 4 degrees clockwise along the X axis using one of the methods we looked at earlier, the minifig's arm will move to come to rest by its side, with the hand part still attached to it, as shown in Figure 5.49.

Figure 5.48 Rotation Points for the Arm and Hand

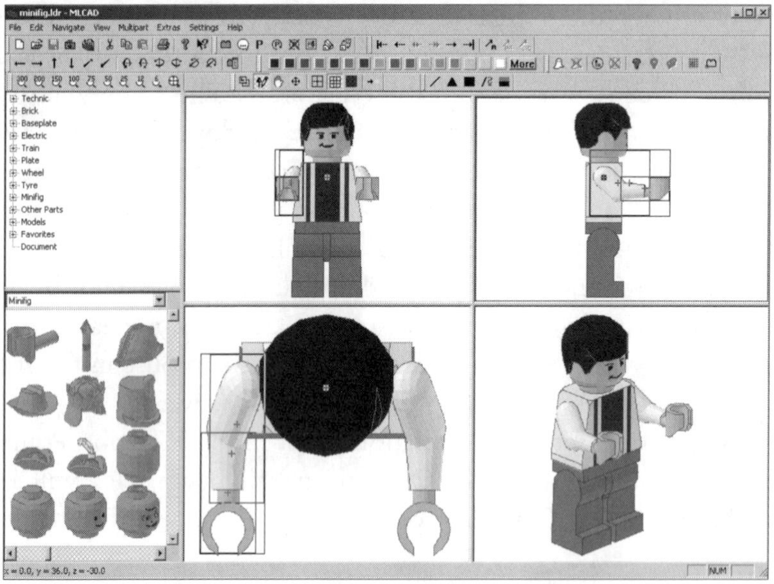

Figure 5.49 The Right Arm and Hand Resting at the Side

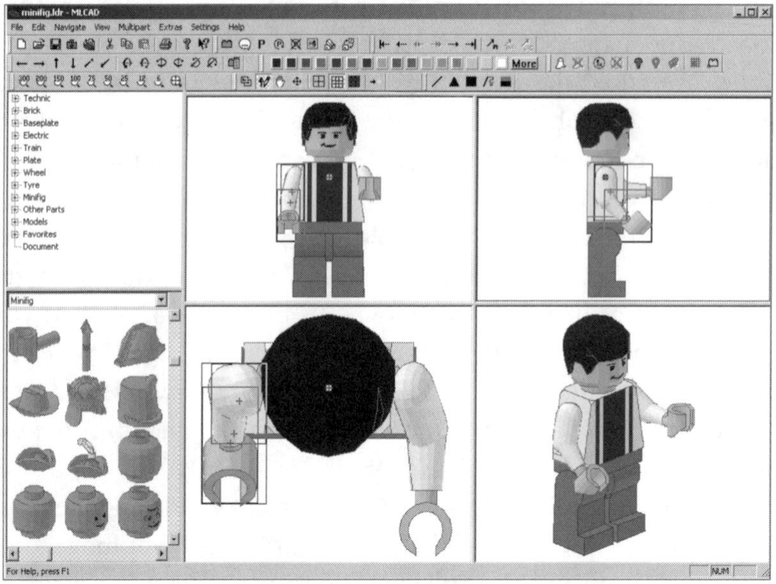

What could go wrong? Well, let's try the other arm. Select the left hand only. The rotation point is somewhere in the middle of the "palm" of the hand, as shown in Figure 5.50.

Figure 5.50 The Left Hand's Rotation Point

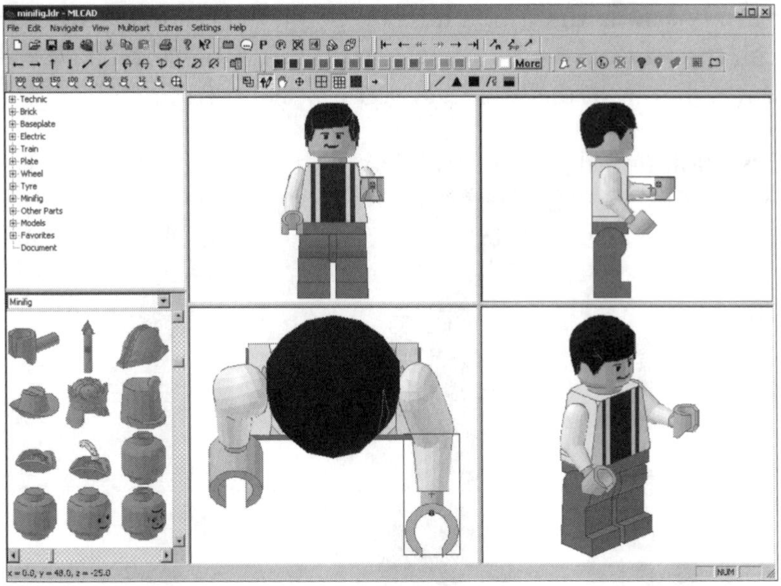

Now select the left arm as well. As shown in Figure 5.51, the arm is selected, but the rotation point of the hand (the first part selected) remains active.

Figure 5.51 Left Arm Selected After Left Hand

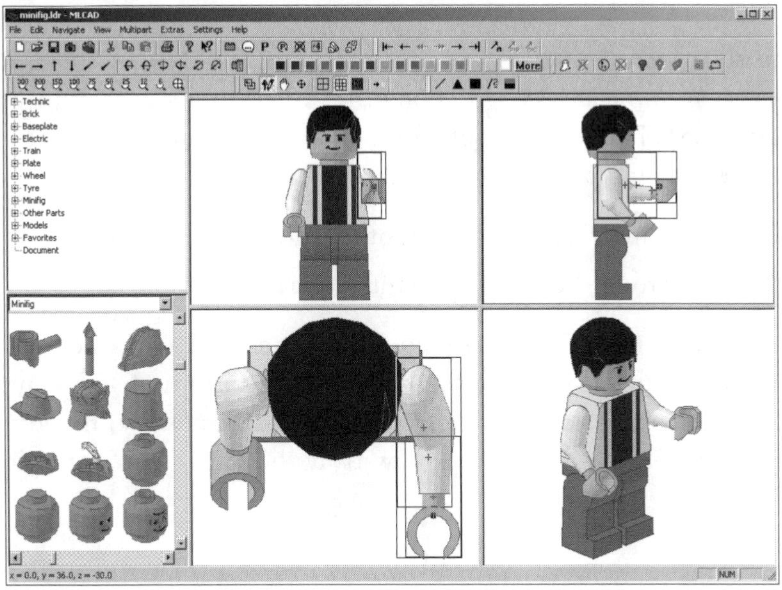

If we perform a 45-degree clockwise rotation in the X axis as we did with the other arm but using the hand's rotation point, the results are not exactly the same as with the previous arm. In fact, in performing this maneuver, we would pull the arm right out of the socket, as you can see in Figure 5.52.

Figure 5.52 Rotating the Hand and Arm on the Hand's Rotation Point

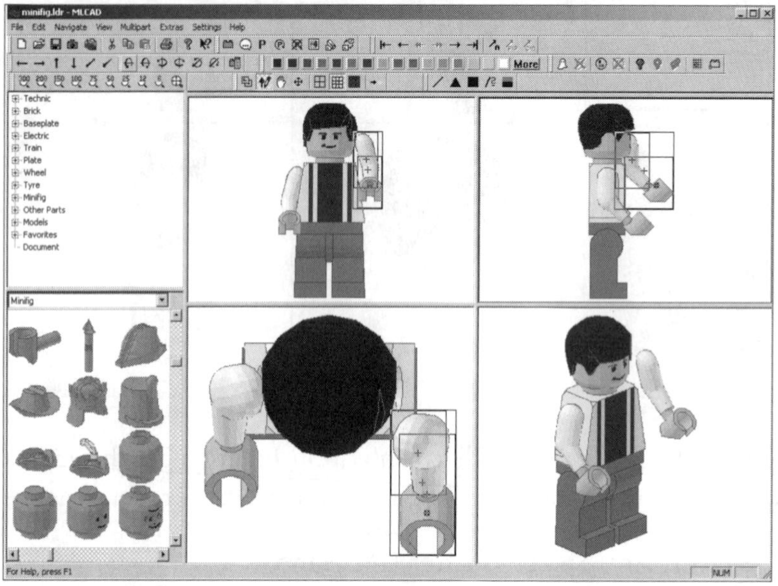

Of course, there are several ways to fix this problem. For example, we can rotate the arm again *counterclockwise* along the X axis to get it back to where it was in Figure 5.51. The point is not so much to avoid making mistakes, which are part of the learning process, but to realize that:

- Rotations are trickier than any other modeling tool in MLCad.
- Being aware of the rotation point is extremely important.

Pulling the arm out of a virtual minifig is not critical. Choosing the best rotation point for our purposes certainly is. With this in mind, let's complete the minifig. Using your new knowledge of rotation points, position the left arm and hand as shown in Figure 5.53. This is now truly the stance of a champion!

Figure 5.53 Left-Arm Stance

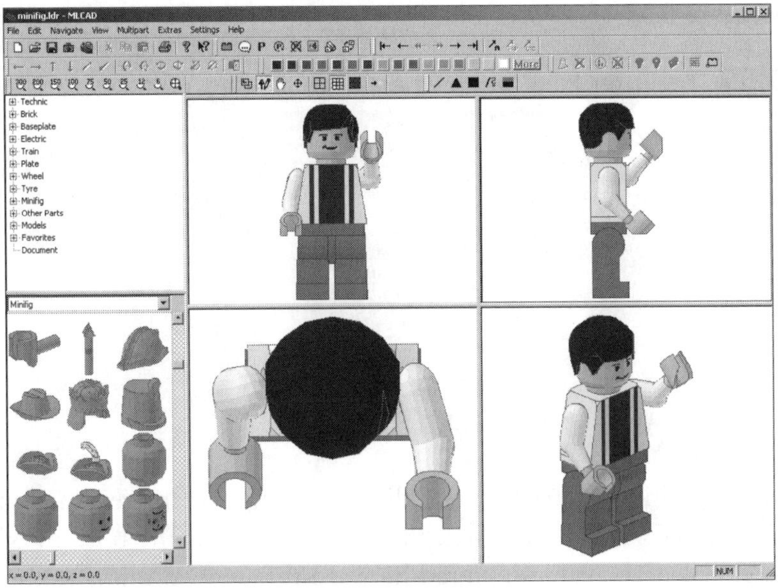

Precise Part Positions and Rotations

There is one last way to position and/or rotate our models and parts *exactly* as we want them. As we have mentioned before, it is possible to create LDraw files by simply typing the parts and their coordinates in a regular text editor, following certain format rules. MLCad also lets us type the coordinates and rotations of parts directly into the model—with the added benefit of showing us the results instantly!

To enter the coordinates of a part directly into MLCad, we select the part first. We can then use the **Edit | Move | Enter...** or **Edit | Rotate | Enter...** menu options or press the **Enter Pos. + Rot...** button on the Elementbar, shown in Figure 5.54.

Figure 5.54 Enter Position and Rotation Button of the Elementbar

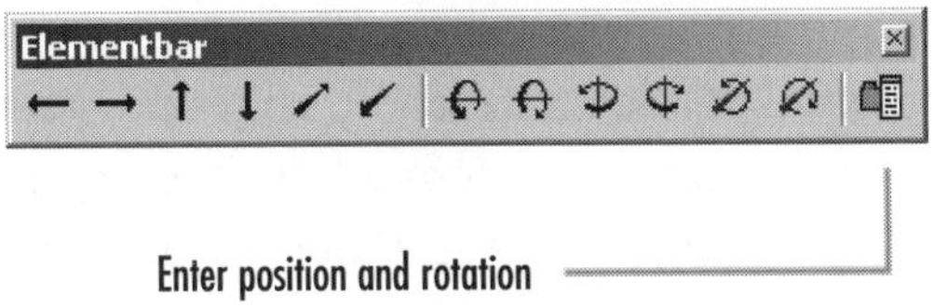

These actions open the **Enter Position & Orientation** dialog window, shown in Figure 5.55.

Figure 5.55 The Enter Position & Orientation Dialog Window

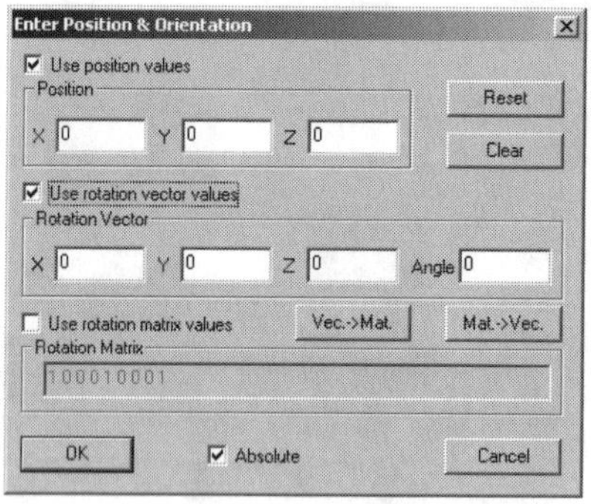

The XYZ position values boxes at the top of the screen are exactly that: the XYZ coordinates for our part. When the Absolute box at the bottom of the window is checked, the values entered here are the new coordinates for the part. When the Absolute box is deactivated (unchecked), the part or parts are offset (moved) by the values entered in the boxes. This allows us to move (or position) a part in all three dimensions at once. For instance, if we have a part located in XYZ coordinates 10,10,10 and we type a value of 10 for the X position with the Absolute box unchecked, the part will move to XYZ position 20,10,10.

Likewise, we can also type in values for part rotations. As mentioned, when the Absolute box is checked, we set the orientation of the part in space. When the Absolute option is inactive, the values describe the rotation of the part from its current position.

We can describe rotations via two different methods. Both are somewhat complex, but every modeler owes it to himself or herself to give them a shot because they are incredibly powerful tools.

Using rotation vector values allows us to describe any rotation with only four values. How is that possible? In Figure 5.45, we saw how rotation vectors were defined by the part rotation point and the rotation axis. In this dialog window, however, the rotation axis does not have to be aligned with the X, Y, or Z axis. The rotation axis vector will be a line defined by 0,0,0 coordinates and the coordinates we enter in the XYZ boxes of the Rotation Vector boxes of the Enter Position & Orientation dialog window. The degrees of rotation go in the Angle box to the right of these values.

This might sound like too much hassle, especially since you already know how to perform rotations in a different way. It's worth the effort, though. Let's do a practical exercise that will show the power of the tool. We want to rotate the hand itself from its current position to the one shown in Figure 5.1. As you can see in Figure 5.56, the hand is not aligned with any of the XYZ axes.

Figure 5.56 Rotation of a Part Not Aligned with the XYZ Axes

To rotate the hand with the tools we have examined until now, we first have to rotate the hand (or the arm and the hand) until it is a aligned with the X,Y, or Z axis, perform the hand's rotation, and reorient the hand as it is shown in Figure 5.56.

Another way of doing this is to directly define the rotation vector for the desired rotation of the hand. From the rotation steps we performed earlier in the chapter, we know that the lower arm is aligned at exactly 45 degrees from the Z and Y axes. Thus, we can easily define a rotation vector for the rotation of the hand. In Figure 5.57, we see that the rotation vector is defined by equal negative values in the Y and Z axes. These values define the orientation of a rotation vector that will be positioned where the rotation point is.

Once we have calculated them, we input the corresponding values in the X, Y, Z, and angle boxes of the Enter Position & Orientation dialog window (refer back to Figure 5.55). The rotation vector (shown in Figure 5.57) will be a line that goes from XYZ coordinates 0,0,0 to XYZ coordinates 0,-1,-1. The angle will be 45 degrees. Be sure you deactivate the **Absolute** option before you click the **OK** button in the **Define Position & Orientation** dialog window. The result is what you see in Figure 5.58. The hand has rotated 45 degrees along the designated rotation vector.

Figure 5.57 The Minifig Hand Custom Rotation Vector

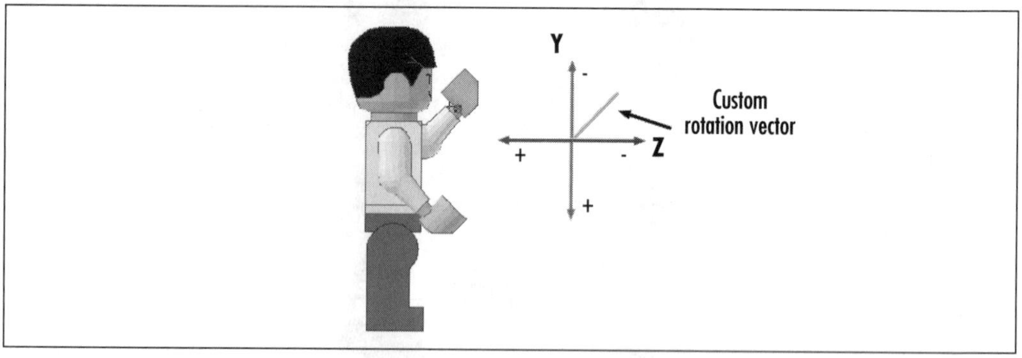

Figure 5.58 The Rotated Hand

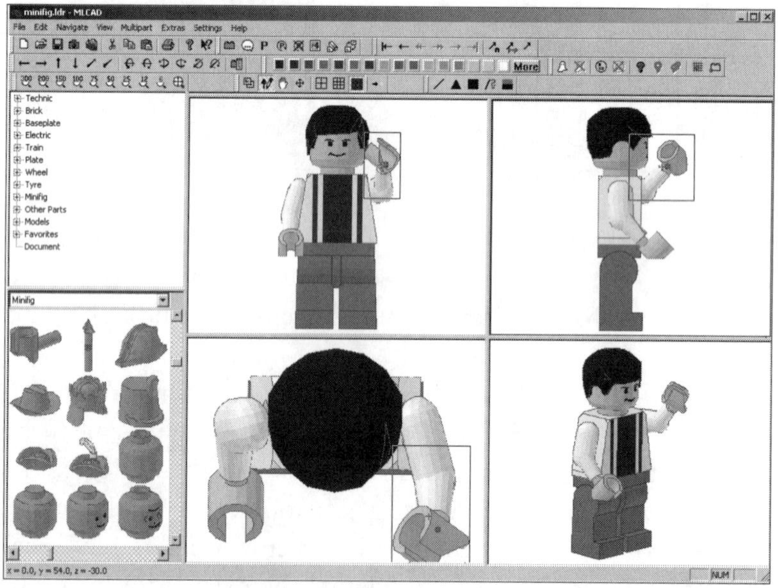

Before you rotate the hand, you might want to adjust its rotation point to the one shown in Figure 5.57, which is lower than the default part origin rotation point. After the hand is rotated, make sure that it is aligned to grid; you might need to make small movement adjustments to it.

NOTE

Even if you do not plan to use rotation vectors, it is good to know that an "absolute" rotation vector with X, Y, Z, and angle values of 0 will realign any part to the standard XYZ rotation axes.

The other way to define part rotations and orientations involves *rotation matrices*. Since they play an essential role in the LDraw file format, we describe these tools in Chapter 6.

Once you've understood rotation vectors, you have pretty much conquered the summit as far as modeling with virtual LEGO in MLCad is concerned (and already have an important grasp on general 3D modeling concepts). The next step up is to learn how to use completed models as parts of larger models. Save the minifig out into the c:\ldraw\models\powertools folder as a file named champ.ldr. We will be using it shortly, along with the podium we built in the previous chapter. Before we move on, we need to make you aware of one last command related to MLCad modeling: the Hide command.

Hiding Parts

When you work on complex models, it is not unusual for parts to become obscured and even hidden by other parts. For instance, if we are building a house with a detailed interior and exterior, we will reach a point were the outer walls will not let us see the inside of the house. Since the distance between the model and the modeling pane is set automatically by the program, we simply cannot "place a camera" inside the house.

Just as we can modify a part's color, position, orientation, and even type, we can also make it invisible. The technical term for doing this is known as *hiding* a part. The program will register the part internally, but it will not render it to the modeling panes. Let's look at an example. Select the torso of the minifig model and use the **Edit | Visibility | Hide** menu option or click the **Hide** button of the Modificationbar, shown in Figure 5.59.

Our minifig should now look like the one shown in Figure 5.60. Notice that we can now see the top of the minifig's hips, which had previously been hidden by the lower portion of the torso.

Figure 5.59 The Visibility Buttons of the Modificationbar

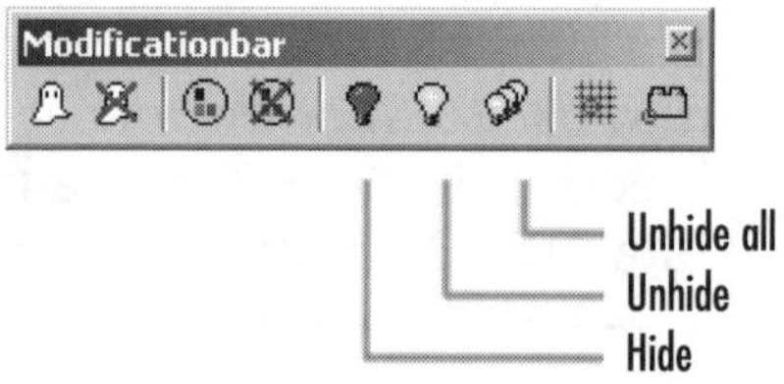

Figure 5.60 Minifig with a Hidden Torso

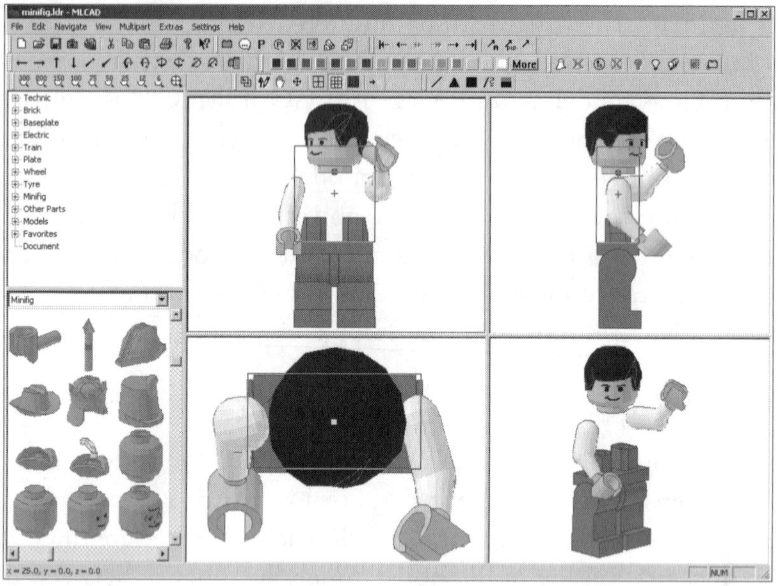

Notice also how the torso is hidden but remains selected. To unhide a part, use the **Edit | Visibility | Unhide** menu option. **Edit | Visibility | Unhide all** makes visible all hidden parts in a given model. You can also use the appropriate buttons of the Modificationbar (refer back to Figure 5.59).

NOTE

To select hidden parts, use the Part List window, as explained in Chapter 6.

Hiding is a *local* function, which means that a part's hidden/visible properties are not saved to a file. When we load a file that has parts that were hidden in it when we were last working on it, all its parts will be visible.

Working with Submodels

As we said earlier, becoming familiar with the Part Rotation & Orientation dialog window puts you at the top level as far as MLCad modeling knowledge goes. There are still plenty of features to discover in the program, but as far as putting together LEGO models, only a few techniques remain … and you'll find it's mostly downhill from here.

Until now, we have mostly used commands to work with individual parts. In this part of the chapter, we talk about the commands and techniques that are useful and relevant *only* when working with multiple parts. For the most part, MLCad's commands work the same way if we have one part or several parts selected. However, some commands exist for the sole purpose of facilitating the process of working with several parts at the same time. In any event, as you saw in the last chapter, working with multiple parts is a useful technique.

As usual, there are several ways to work with multiple parts in MLCad. In fact, a key characteristic of the LDraw system is that model and part files are interchangeable. Models can be used as individual parts of another model, and part files can be edited as though they were models. What Michael Lachmann added to this principle is further support on the editing level and on the multimodel level. In practical terms, this means that we can use groups of parts casually or with a high degree of precision and efficiency, depending on our purpose.

In this section you will first see how to select and group parts while editing a model, which will let you easily create copies of the athlete minifig we just built. We will then show you how to incorporate model files into other models by combining the podium and the minifig into a single model file. Finally, we will show you how to use a new method of creating efficient multimodel files invented by Michael Lachmann for MLCad. We'll end the chapter by building a complete LEGO diorama made of several different models integrated into one.

Selecting Various Parts

In order to work with more than one part, we first need to know how to select several parts. As we saw in the previous chapter, the easiest technique to select a part from any modeling pane (except those set to the 3D view angle) is to simply click it. To continue adding parts to the selection, press the **Ctrl** key and continue clicking other parts to add them to your selected parts. As we saw earlier, this is the preferred method if we want to use the rotation point of a specific part in the selection—start by selecting that part first.

While building the podium, we also took a look at the selection buttons of the Movementbar that can be used to select multiple parts. Using these buttons, we can select all the parts in the model or all the parts in the model of the same color or exact type. To select parts based on color or type, you need to have one part already selected. MLCad will use that part's color or type for the selection criterion. Figure 5.61 shows the Movementbar selection buttons. (Go back to Chapter 4 to see them in use.)

Figure 5.61 The Movementbar Selection Buttons

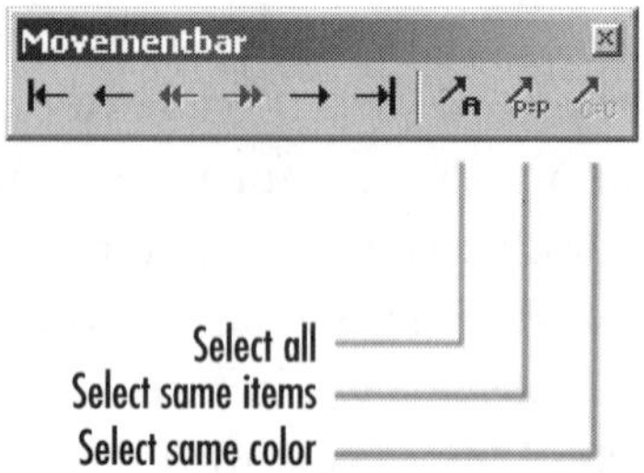

There is yet another way to select several items in the modeling panes. Click anywhere on empty space inside a modeling port set to a "flat" view angle and drag the mouse diagonally. While you are holding the mouse button down, MLCad will draw an inverse selection box, as shown on Figure 5.62.

Figure 5.62 A Selection Box

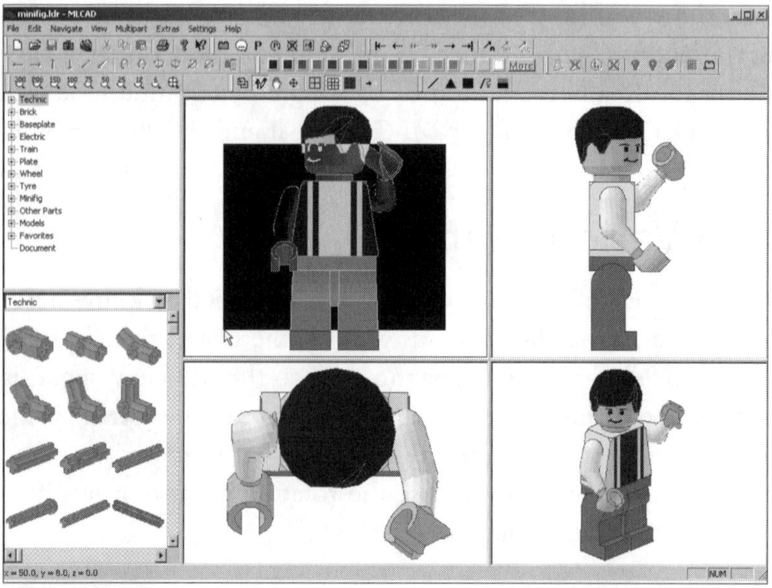

NOTE

Make sure you start by clicking empty space in the modeling pane or you will move a part instead of selecting a group of them!

Once you let go of the mouse button, any part that is at least partly inside the inverse selection box will be selected, as shown in Figure 5.63.

Figure 5.63 Selected Parts

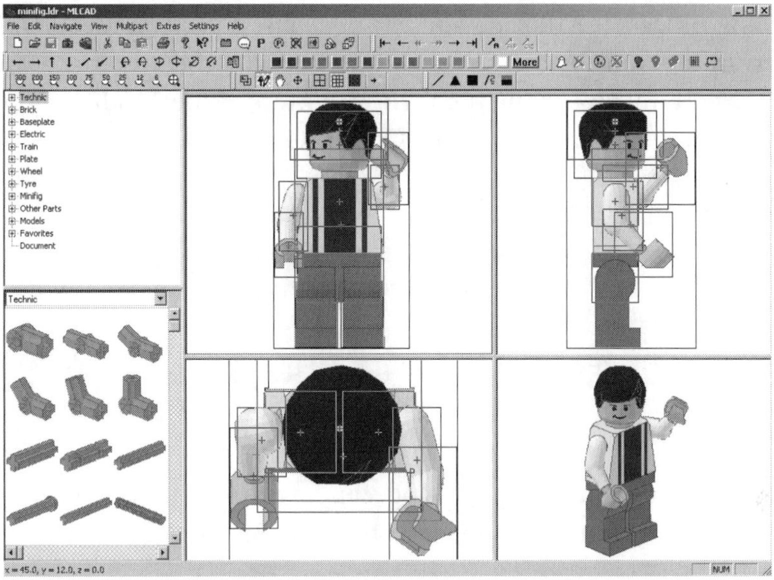

NOTE

To deactivate the selection, click empty space again in the modeling pane.

In models with many parts, selecting the right parts that you want to work with from among all the parts in a model can be a complicated task. In Chapter 6, we will look at some new methods that will simplify this task and that involve using the Parts List window.

As we said earlier, the use of many MLCad commands is the same whether we have one or several parts selected. For instance, if we click a color on the Colorbar, all selected parts will turn that color. Some commands will not work (for example, we cannot change the part type of multiple parts at once). Other commands might make some options unavailable or functional in a limited way.

Grouping Parts

Selecting multiple parts makes it easier to work with all the parts at the same time. In building models, it is sometimes useful to keep these selections together. There are several ways to use multiple parts, but the most useful method during the modelling process is MLCad's ability to *group* sets of parts together. To group all the parts in the minifig, first select all the parts and then use the **Edit | Group | Create** menu option, the **Group** button of the Modificationbar (see Figure 5.64), or the keyboard shortcut **Ctrl + G**.

Figure 5.64 Group Buttons

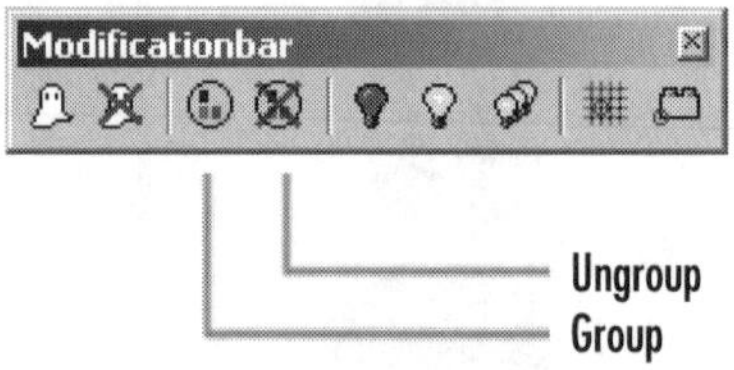

You are then presented with a dialog window that requests a name for the group. Let's call our Champ, as shown in Figure 5.65.

Figure 5.65 A Group Name

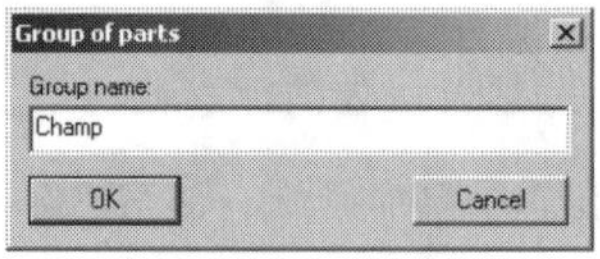

Once grouped, the parts still feature separate selection boxes, as shown in Figure 5.63. The difference is that the parts will remain grouped even if we deselect them. Anytime we want to work with the group again, all we need to do is select a part belonging to the group; by doing so, we select the whole group. To select a single part that belongs to a group, you have to ungroup the whole group first. To do this, you can use the **Edit | Group | Ungroup** menu option or the **Ungroup** button on the Modificationbar.

To see how using groups can help us, let's take a look at an example. Select the Champ group we just created by clicking any part of the minifig. Now, copy the whole group by moving it while pressing the **Ctrl** key, as illustrated in Figure 5.66.

Figure 5.66 Duplicating the Champ

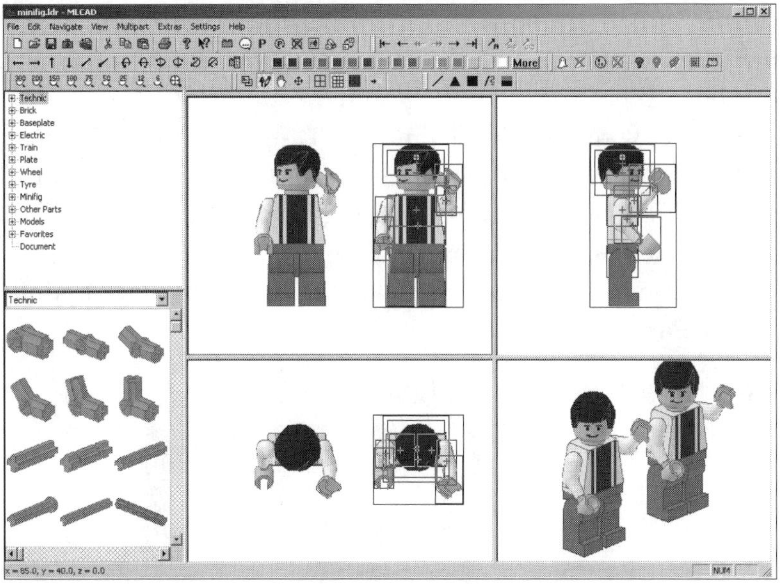

We now have two champs ready to go. Notice how the copy is not a group (although you can make it into one easily right after copying it, because all its parts are selected). Obviously, we do not need to group the parts in the minifig to copy them, but grouping allows us to treat as a single element all the parts that make up a group. For example, to place several minifigs in a scene, it is easier if we can move and select each minifig as a single element (a minifig group, like our Champ group) rather than assembling all the parts that the minifig consists of every time we want to add one to our model.

Originally, groups only worked when the file was loaded in MLCad; once the file was closed, any groups that had been created were lost (much like the "hidden" part feature). This made grouping a valuable but somewhat limited modeling tool. More recent versions of MLCad have expanded the function, and groups are now saved as groups in the model file. This functionality makes grouping a highly attractive option to modelers who want to work with sub-models in a file.

Using Models as Parts

The group feature allows us to turn part selections into groups that momentarily behave as parts, which is a handy feature. However, often the best way to use models as parts is to combine two or more separate model files into a new model. This is one of the original advantages of the LDraw system invented by James Jessiman: Models can be used as parts. Imagine you wanted to place the two champs we just created onto the podium that you built in the last chapter. Let's look at how to do that.

Save the model with the two champs as champs.ldr in the C:\ldraw\models\powertools folder. As we saw at the beginning of the chapter, we can add parts in a model via the **Edit | Add | Part** option or the **I** key on our keyboards, which opens the **Select Part** dialog window shown in Figure 5.67. Instead of choosing a part from the list, however, select the **Custom Part** option and then click the **Load Part...** button.

Figure 5.67 The Select Part Dialog Window

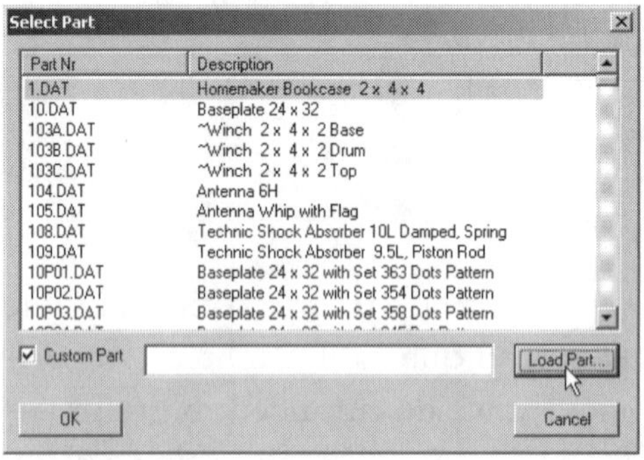

You will be presented with a standard file dialog window. In it, select the file **podium.ldr** that you saved earlier, in the C:\ldraw\models\powertools folder. Click **OK** and the podium will appear in your model, as shown in Figure 5.68.

NOTE

The podium might not be aligned correctly when it's inserted.

Notice how the podium, unlike the minifigs, has one bounding box around it, as though it were a single part. In fact, some commands—modify, certain preset rotation points—that are not available to grouped minifigs are available to models inserted as parts, as we have just done with the podium. The next step is to position our champ minifigs properly on the podium. Now, unless we want triplet members of the same team to take over the podium, we would probably want to achieve a model like the one shown in Figure 5.69.

Figure 5.68 The Podium Inserted in the Champs Model

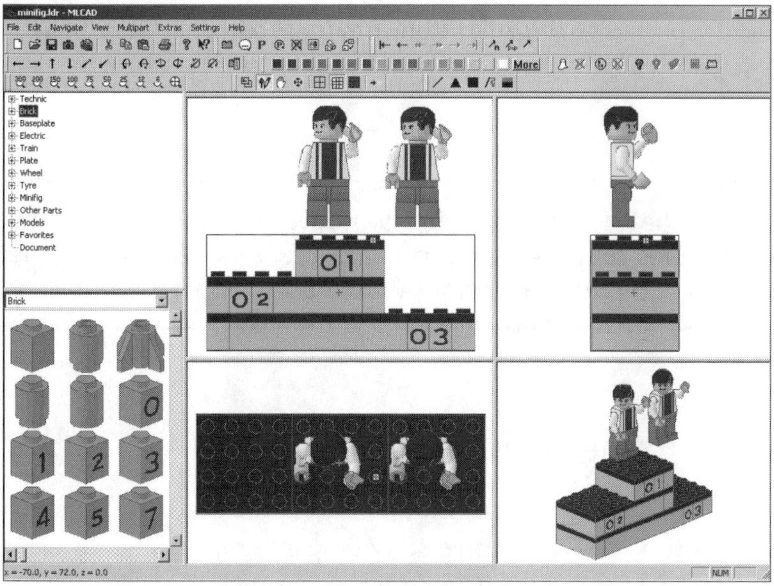

Figure 5.69 Three Champs on the Podium

The advantage of using groups instead of models as parts lies in the editing flexibility it gives us. Using a group as a part allows us to quickly switch between the group and the individual parts that make up the group. This makes it easy to change the appearance or stance of the minifigs and then position them on the podium as complete figures. The method of using models as parts is less flexible. It is impossible to modify the individual parts of the podium if it were inserted as a model into another model file. However, using the podium as a part means that it can be used repeatedly in many future models. Extracting individual groups from a file is not that easy.

NOTE

Remember that the Parts Preview window of the library section also lists the model files that are saved in the LDraw\models directory. They can be dragged directly from the window and used as parts.

Multipart Project Files

Multipart Projects are, like groups, another addition by Michael Lachmann (creator of MLCad) to the LDraw universe. Multipart Project files are slightly different from regular LDraw files. While they are loaded in MLCad, they behave very similarly to regular models. With the introduction of this file type (with the extension .MPD), Lachmann solved several of the issues involved in the creation of models with multiple submodels.

For one thing, Multipart Project files allow the best of both worlds: submodels are used as single parts, like the podium in our example, but they can also be edited as models and their individual parts can be modified, as we did for the minifigs we created. Additionally, Multipart Project files can be shared with other users as single files. In a regular LDraw file, if we simply insert a model as a part into another model, the file will only carry a reference to the model being used as a part. If we open the file in a different computer that does not have the file for the submodel stored on it, we will not be able to load it, and the main model will lack the part for which the model served.

With Multipart Projects, everything (including the main model and complete submodels) is contained in a single file. All in all, this is a much more efficient way of working with multimodel files.

To create a Multipart Project, we start by creating a new file in MLCad (a *regular* new file). We manage the Multipart Project's content via the Multipart menu. For instance, if you want to import a model into the file, select **Multipart | Import Model**, and you will be presented with a regular file dialog window. Select **podium.ldr** in that window, and click **OK**.

The podium is now loaded into the file and behaves like a model. We can edit any of its parts separately. See Figure 5.70.

Figure 5.70 The Podium Loaded as an Editable Model

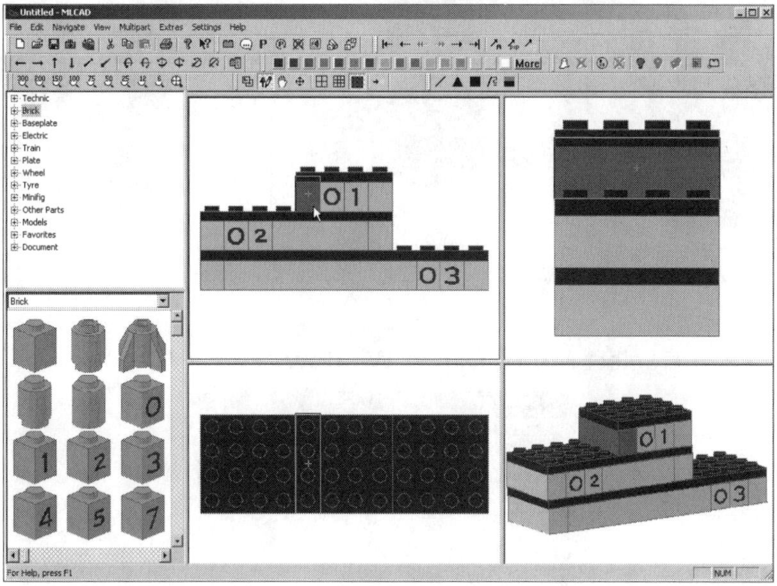

Leave the podium as it was and, using **Multipart | Import Model** again, import the minifig file as you did with the podium. As you can see from Figure 5.71, the minifig is now loaded as an editable model.

Multipart Project files have the capacity to store several models *separately*; one of these models is the main model that uses the other submodels as parts. This gives us the advantages of direct edit and the advantages of using the models as parts and not groups, which lack certain commands. Use the **Multipart | Select Model** menu option to navigate the content of the multipart project. You will be presented with a Select Model dialog window like the one shown in Figure 5.72.

In Figure 5.72, we see the contents of the .MPD file. It contains the podium.ldr model we imported earlier, the minifig.ldr we just imported, and a third file called Untitled.ldr. Select the **Untitled.ldr** file and click the **OK** button. You will find yourself in an empty MLCad screen with empty modeling panes.

Figure 5.71 The Minifig Submodel

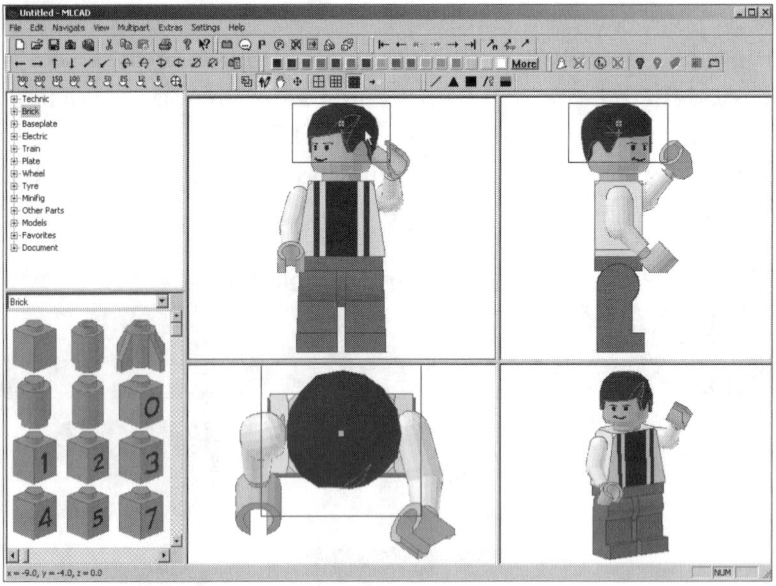

Figure 5.72 The Select Model Dialog Window

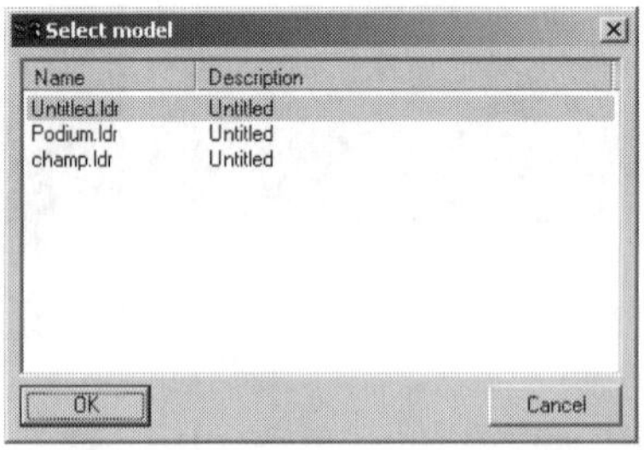

You are now actually working on the main model, which is empty—but you can use the other submodels as parts. Select the **Document** group (category) in the Parts Library window. This group contains all the models in the MPD file; the Preview Window will even display the models!

Drag the two models into the modeling panes and position them as shown in Figure 5.73.

You can also add regular parts to the model. To add more submodels, simply import them as you just did or create them from scratch via **Multipart | New Model**. This will present you with a Model Information dialog window like the one shown in Figure 5.74.

Figure 5.73 The Podium and a Champ

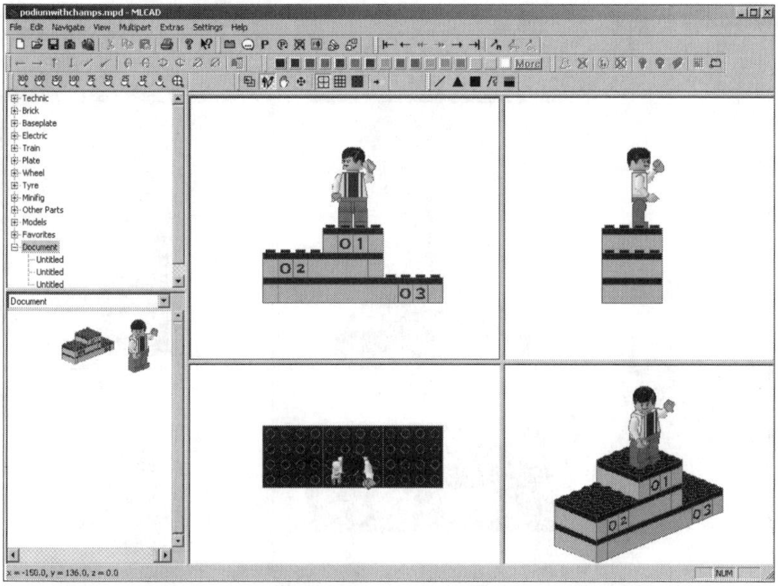

Figure 5.74 The Model Information Dialog Window

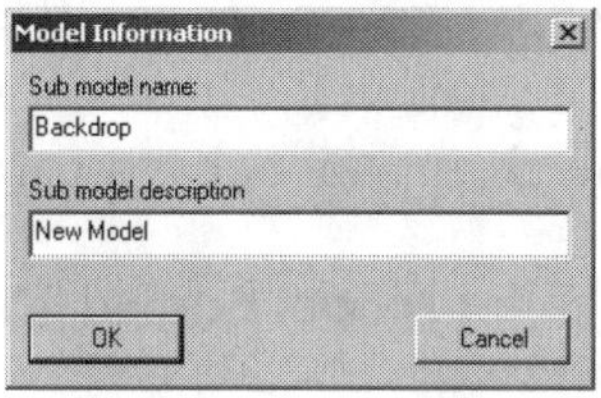

You can then proceed to create a model that will then be available as a part or submodel through the document library group. In Figure 5.75, a backdrop for the podium has been created as an independent model and is ready for inclusion as a part of the main model. You can find the file in the CD that accompanies this book.

The **Multipart | Change Model Info** menu option allows us to rename the models and/or add information about them via the Model Information dialog window shown in Figure 5.74. To change a model's information, you must first select it using **Multipart | Select Model**. Once the model is loaded on the screen, use the **Multipart | Change Model Info** menu option.

The **Multipart | Remove Model** menu option allows us to delete a model from the multipart project file, and the **Multipart | Sequence** menu option allows us to reorder the list shown in the Select Model dialog window shown in Figure 5.72.

Figure 5.75 The Podium with a Backdrop

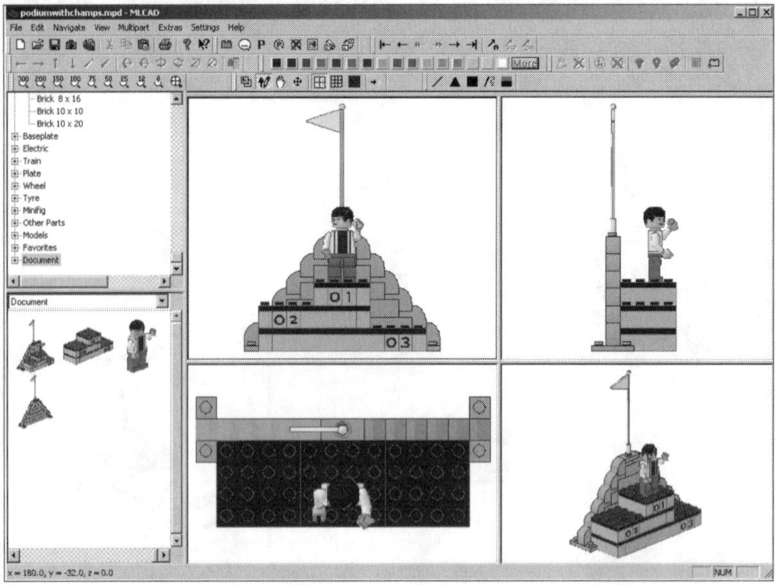

These file management tools are very helpful because one of the advantages of using Multipart Project files is that we can use the same model several times, and we can do it two different ways: as identical copies or as different models. In Figure 5.76, we have used the champ model as a part three times.

Figure 5.76 The Champ Minifig Model Copied Three Times

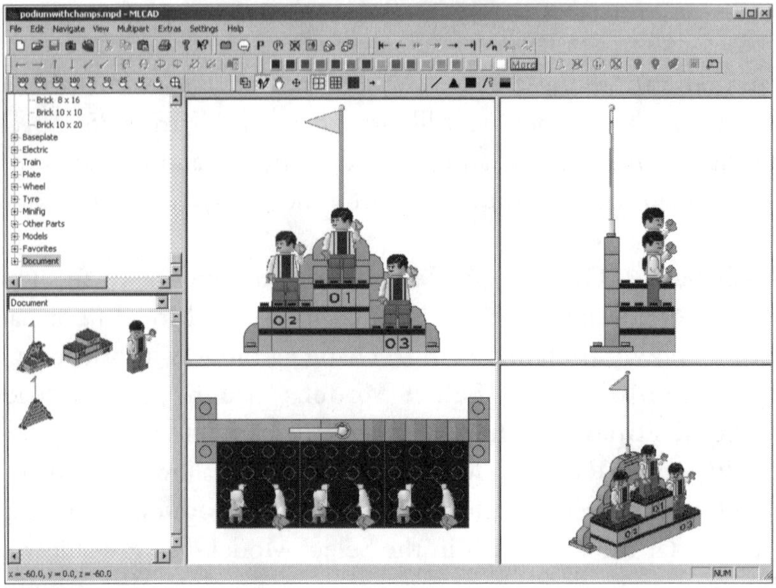

The disadvantage of this method for this particular model is that it looks like three triplets just won a sporting event! In other models, identical copies might be desirable. Think of multiple freight cars in a train, for instance. According to the official documentation, this is the example that Michael Lachmann had in mind when he came up with the MPD format.

If we want a more varied cast of champions on our podium, as shown in Figure 5.70, we can also import the champ minifig three times and then modify each submodel separately. Figure 5.77 shows all the champs from Figure 5.70, with each of the minifigs modified enough to make them unique.

Figure 5.77 Three Champs on the Podium, MPD Style

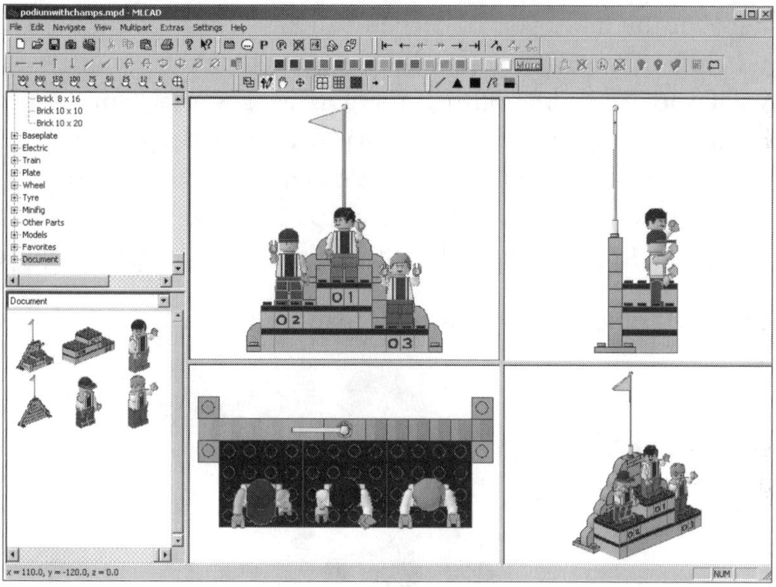

As we mentioned earlier, Multipart Projects are not standard LDraw files (their extension is .MPD). Just as the **Multipart | Import Model** menu option allows us to import models into an MPD file, **Multipart | Export Model** creates a separate standard LDraw file for each model contained in the multipart project file. It will not, however, create a regular LDraw file of the main model that contains all the submodels.

Save the Multipart Project file for our example under the name **podium with champs**. In the next section, we will create a "natural" setting for our creation.

An Introduction to Automatic Modeling

In this chapter, we have covered all MLCad's modeling techniques used for creating models with LEGO-based parts. To make yourself a *complete* modeler, however, you must now learn about the material from which you create your models: the virtual LEGO bricks. These might look like LEGO ABS plastic on the screen, but in fact they are made of computer bits and bytes.

This is important to realize because no matter how fantastic your MLCad models are, you will eventually have to take them elsewhere, using other programs to distill from them exactly what you need—whether a set of instructions, a photorealistic rendering, or an animation sequence. Many of these processes can be automated, but the more you understand about what is going on with your MLCad files, the more control you will have over them.

As you will soon see, some of the programs we cover later in the book simply add to our model files parts or functions that MLCad lacks. For instance, LSynth is an application that allows us to use flexible elements in our models. MLCad does not include this support, which involves placing many small subelements one after another following a curved path, like the hose shown in Figure 5.78.

Figure 5.78 A Flexible Element with a Custom Curve

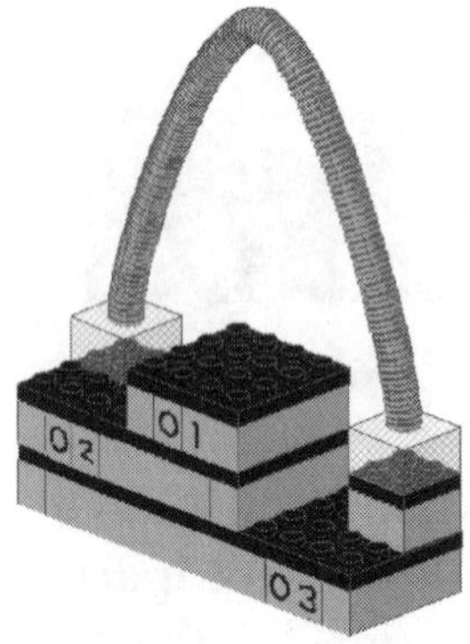

Generally, flexible elements are achieved through specialized programs that address very specific tasks (such as creating curved hoses). These programs are called *model generators* because we tell them what we want and they provide a custom part (or model) according to the specs. For example, Chapter 7 describes LSynth, a bendable part generator for tubes and rubber bands, written by Kevin Clague.

MLCad offers three model generators that allow us a variety of options for modeling that are not available in the standard LDraw library. These generators are:

- Fractal Landscape generator
- Picture Model option
- Rotation Model option

These tools are accessed through the **Extras | Generators** menu. Let's take a look at each of them.

Fractal Generator

The **Extras | Generators | Fractal Generator** creates virtual "landscapes" for us. Through the dialog window shown in Figure 5.79, we can select the size and height of the model landscape, which in this case is a series of hills. We can also specify whether we need a base plate included and what should be the size for the largest parts used in creating the hills. The landscapes can be colored according to height, as on some maps: lower lands are blue, hills brown to green, and high mountains white. Instead of bricks, we can use plates, making the slopes "gentler." The **Compute new landscape** button changes the *fractal seed* used to generate the basic shape of the landscape. Clicking it generates new basic shapes.

Figure 5.79 The Fractal Landscape Generator

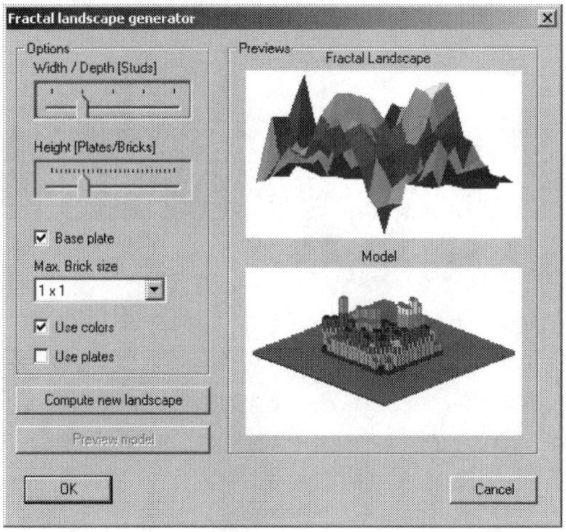

NOTE

Using high width/depth studs settings creates large, very cumbersome model landscapes.

Once we are satisfied with the landscape, we click **OK** and MLCad will generate it for us automatically. In Figure 5.80, we have created a hilly terrain with a flat valley.

Figure 5.80 An Automatically Generated Landscape

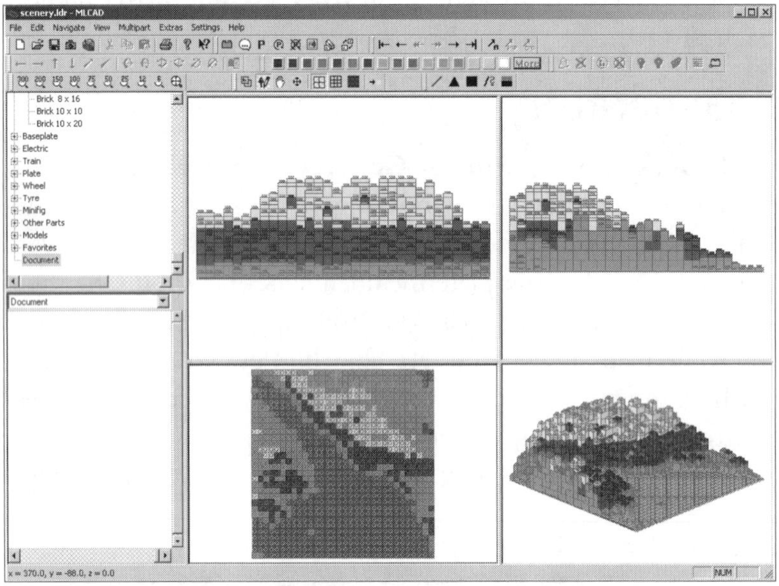

Designing & Planning…

Fractals and 3D Modeling

Fractal mathematics were developed to model and measure natural scenes. Many 2D and 3D illustration programs use them to create all sorts of natural *structures*, from entire landscapes to the individual leaves of a plant. The program automatically generates each of these structures using a single number as a "seed." The program uses the seed in recurrent calculations that *always* produce different but very natural-looking results. To vary the seed in MLCad's Fractal Landscape generator, simply click the button **Compute new landscape**, and MLCad will automatically generate a new landscape out of a new seed.

Picture Model

The **Extras | Generators | Picture Model** option creates mosaic-like models out of a file. In Figure 5.81, a very low-resolution picture of the champ minifig (shown in the Picture Preview box) is being converted into a model. Colors can be used to create height; the lighter the color, the higher/taller the model. If you do not want to use that feature, set the height slider bar to the left (the lowest height), as shown in Figure 5.81. As with the Fractal Landscape generator, we can also specify the maximum part size to use (as well as choose between plates and bricks for its construction).

Figure 5.81 The Create Picture Model Generator

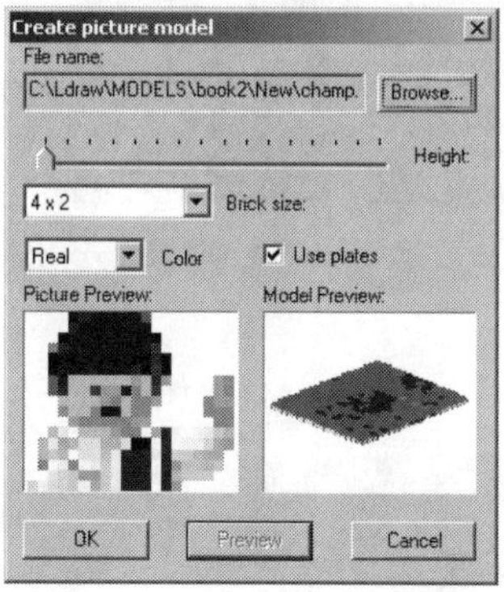

The color pull-down menu allows us to further use the color coding for height, in effect generating a landscape from a picture, similar to those generated automatically by the Fractal Landscape generator. If you just want to use the colors in the picture for the model (for a flat mosaic-like result, for instance), use the Real option. If you use the Fake color option, the resulting model will be coded like the Landscape generator we just saw: low bricks in blue, higher bricks in brown, then green, and finally the top ones in white. The None option generates a model with no colors (but with volume if we activate the Height option).

Figure 5.82 shows the model generated by the settings shown in Figure 5.81.

Figure 5.82 A Mosaic Model

Designing & Planning…

Virtual LEGO Mosaics

LEGO mosaics are becoming increasingly popular, particularly because LEGO has started offering a service via which you can provide a photograph and LEGO sells you the bricks and instruction sets to make a mosaic out of it. This idea is simply awesome, and some people have taken the concept even further, creating gigantic LEGO mosaics. Several other programs besides the picture model generator create LEGO mosaics out of pictures and photographs.

Of course, the larger the final mosaic, the more detail it will have. In the image used in Figure 5.81, the minifig's eyes are simple black boxes. However, large mosaics are cumbersome to work with both in real life and inside MLCad. A handy use of mosaic generator programs is to create LEGO lettering, which is a specialty of many LEGO fans.

Rotation Model

Finally, the **Extras | Generators | Rotation Model** menu option helps us create models that are made by rotating parts around a center axis, such as vases. As shown in Figure 5.83, we specify a profile for the object by drawing it in the grid at right. We can choose maximum brick (or plate) size and the grid size and width. We can also use different colors for the parts. Each box of the grid represents one LEGO brick (one stud).

Figure 5.83 The Rotation Model Generator

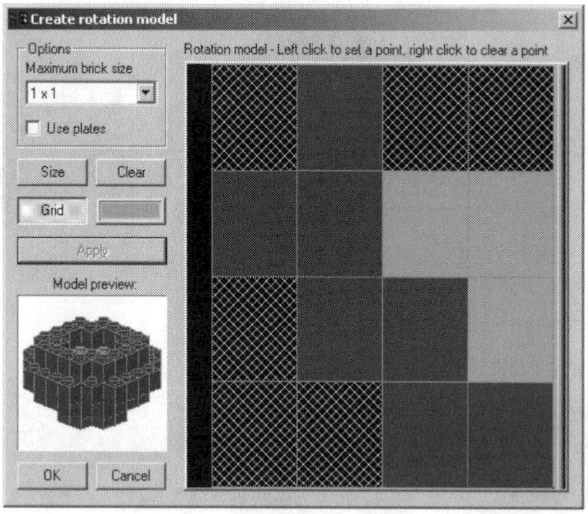

In Figure 5.84, we can see the end result: a plant pot generated by rotating the profile around an imaginary axis on the right side of the grid shown in Figure 5.83.

As you can see, all these tools generate models automatically for us. Other applications, such as LSynth, do a similar job. They run outside MLCad but generate files that can be loaded into the program or be further used with other programs. Figure 5.85 shows all the models we have created so far (and a few extra elements) integrated into one big scene and rendered in POV. See if you can put it together yourself!

Figure 5.84 A Plant Pot

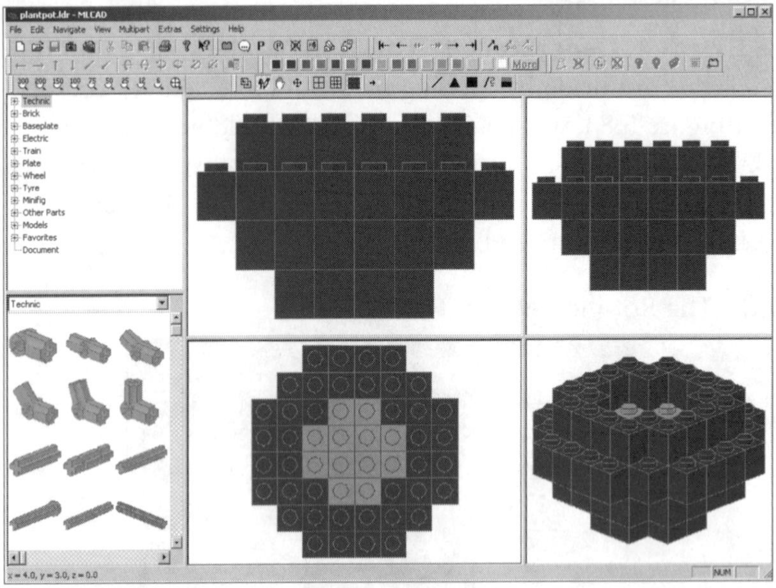

Figure 5.85 A Scene with All the Models

Summary

In this chapter, we covered in full detail all tools and techniques related to building LEGO-based model inside the MLCad application. This nifty program offers high-level support for the relatively basic modeling functions used in MLCad: adding parts to the model, changing their attributes, and moving and rotating them.

To make full use of the extended support for these functions, you must first be aware of the XYZ coordinate system that maps the interior of the virtual playground where we put our models together inside MLCad. Through the commonsensical approach and the built-in reference system of the LEGO parts, mastering this coordinate system is not a difficult task, yet it makes us much better modelers.

The abstract coordinate system becomes familiar through the use of the advanced options for the modeling tools. Even though we can move the parts in the modeling panes by dragging them with the mouse, often the preferred approach involves using the keyboard or toolbar buttons. In any event, the modeling panes always give us immediate visual feedback based on our actions.

Once we know how to use all the modeling tools, the next step is to learn how to combine more than one model into a single file. If this is a basic characteristic of the LDraw system, MLCad has expanded it by allowing new ways of using models as parts. Not only can models be inserted in other models as regular parts; MLCad also offers the possibility of grouping parts already sharing a model or, for high efficiency, Multi Part Project files, which store both sub models and the main model in one single file.

In the next chapter, you will explore the material of which the MLCad models are made. LDraw's file format is fairly accessible to the user (which brings many practical benefits). Furthermore, what starts as a MLCad file will often be used *outside* MLCad, whether to add to the model features not supported by MLCad or to use the model as part of a larger project. In any event, we finished the chapter by showing you some model generators included in MLCad; these tools generate entire models according to user-defined specifications.

Solutions Fast Track

Our Virtual Playground

☑ The modeling panes that we use to look "into" MLCad's virtual space can be set to different view angles: preset "flat" ones and 3D perspectives.

☑ Preset view angles also define the virtual space in which we built our models: They act as cardinal points.

☑ Models inside the virtual space are not necessarily aligned correctly with respect to the preset view angles. Thus we need a more precise set of coordinates to build models in it.

☑ The imaginary lines that connect opposite pre-set flat view angles define the three axes of the virtual space, XYZ.

☑ The XYZ axes cross at the center, where the coordinates are 0,0,0. From there, depending on the direction and the axle, we move in positive or negative units.

☑ The Statusbar displays the XYZ coordinates for the mouse pointer's position inside the modeling panes.

MLCad's Precision Modeling Tools

☑ Two quick methods to copy parts are moving them while pressing the **Ctrl** button and duplicating them via the **Ctrl + D** shortcut.

☑ The duplicate method is especially useful with parts that share a common part origin, such as the specialized parts of the minifig.

☑ Just as cities expand beyond the longitude and latitude points given for their position, parts expand beyond the exact XYZ point used for positioning them.

☑ The part origin is the point where the part "locks" with the grid and is defined in the part file.

☑ Parts that share a common origin often combine together easily.

- ☑ MLCad's support for color goes far beyond the needs of a large majority of its users. The program allows us to use and define a wide variety of colors, including transparent colors.
- ☑ Parts can be moved directly with the mouse in the modeling panes, but precise adjustments might require other tools, such as the cursor buttons of the keyboard.
- ☑ Part movement is always locked to an invisible grid, but the spacing between the grid steps can be set to fine, medium, or coarse steps.
- ☑ The default grid spacing can be further customized. Wider grid settings allow parts to "lock" together easily, but some operations such as rotations might require small adjustments.
- ☑ For a variety of reasons, a part can become misaligned with the grid steps, preventing it from combining with others. The Snap to Grid function repositions a part so that it is aligned with the grid once again.
- ☑ The part movement buttons of the Elementbar can be used in conjunction with the grid settings to move parts in very precise ways.

Advanced Modeling Techniques

- ☑ Rotations are the most complex modeling tools used in MLCad. They require three inputs from us: rotation axis, rotation point, and rotation angle. Rotations are performed using the buttons of the Elementbar or the keyboard shortcuts.
- ☑ The rotation axis is parallel to one of the three axes of the XYZ coordinate system. Finding it can take some (trial and error) time.
- ☑ The rotation angle is the distance covered in the rotation, measured in degrees; a complete circle has 360 degrees.
- ☑ The rotation point sets the locations for the local rotation axle (called *rotation vector*), which is parallel to the rotation axis.
- ☑ Rotation points often change, and MLCad offers great support for their management.
- ☑ Rotations require us to stay alert, and multiple part rotations are even more demanding. Setting the correct rotation point is key.

- ☑ MLCad also allows us to define the position and orientation of the parts in a model by entering them directly into the program.
- ☑ Both movement and rotation can be accomplished this way; in both cases, we can either enter the absolute coordinates or the values that the part should be offset by.
- ☑ Performing rotations this way requires a practical understanding of the mathematics behind rotations, but it enables us to do sophisticated part gyrations in seconds.
- ☑ The Hide and Unhide commands allow us to turn a part invisible (and visible again) and are used in models in which parts are placed in ways that obscure other parts.

Working with Submodels

- ☑ We can select individual parts by directly clicking on them in the modeling panes and select multiple parts by pressing the Ctrl key while clicking the parts. We can also select parts by color or type. Finally, we can select groups of parts by selecting an area of the modeling pane with our mouse.
- ☑ Grouping parts allows us to work with selections of parts as though they were a single object.
- ☑ MLCad offers solid support for groups, including the possibility of saving them as such in the model file.
- ☑ The advantage of using groups for submodels is the possibility of editing the submodel while we work with it.
- ☑ The LDraw system also allows models to be used as parts in other models, which in turn allows us to use the same submodel in different models.
- ☑ However, models inserted as parts are impossible to edit within the model file where they are inserted.
- ☑ Multipart project files are a variation of the standard LDraw files, which use models as parts while retaining the ability to use edit them individually. They are extremely efficient in many ways.

An Introduction to Automatic Modeling

- ☑ Model generators are programs that provide support for modeling functions not supported by MLCad.
- ☑ MLCad includes several model generators: Fractal Landscape, Picture Model, and Rotation Model.
- ☑ There are also other independent programs that will process MLCad's output to further open its possibilities.

Frequently Asked Questions

The following Frequently Asked Questions, answered by the authors of this book, are designed to both measure your understanding of the concepts presented in this chapter and to assist you with real-life implementation of these concepts. To have your questions about this chapter answered by the author, browse to **www.syngress.com/solutions** and click on the **"Ask the Author"** form.

Q: Why is knowing the XYZ coordinate system necessary for putting together LEGO bricks?

A: First, not all the parts in the library are simple bricks. More important, however, is the fact that most of MLCad's advanced modeling functions use it.

Q: Maybe building with virtual LEGO is harder than I thought?

A: Not really. The XYZ system is about as abstract as it gets for the common user. This system, like all new coordinate systems, seems confusing at first, but as you use it, it all makes a lot more sense. Think of when you move into a new building (home, office, or school). Finding your way around is strange at first, but soon it doesn't require a second thought.

Q: How can I achieve specific part movements or rotations?

A: Use the grid settings (see Chapter 6) or the Enter Position & Orientation dialog box.

Q: Isn't the grid annoying at times?

A: Not really. It's more like a safeguard. To undo a rotation, simply rotate the part(s) in the opposite direction using the same rotation point and rotation axis. Since LEGO parts connect in very specific ways, the grid always helps. If it seems annoying, try changing the setting.

Q: Will using complex rotations ever be useful for models other than minifigs?

A: Since most LEGO parts are not symmetrical, rotations are routine modeling tasks. Many specialized parts, including wheels, can require non-90-degree rotations to be included in models.

Q: I don't "get" automatic modeling. Should I be worried about this?

A: Don't worry about it. Model generators are extremely specific applications that support functions not provided by MLCad. Until you need those functions, the generators will do nothing for you. However, to learn more, read on to Chapter 7, which covers LSynth. Once you are familiar with that, you might want to come back to MLCad and try generating some landscapes and mosaics. Be aware that they tend to be cumbersome, so unless you have a clear objective in mind, you might see the results of different settings more clearly rendering the resulting models in POV via L3P or LPub. Keep mosaics and landscapes small or they will overwhelm you; but to fully take advantage of the Rotation Model generator, try generating large models first.

Q: Where do I go from here?

A: LDraw and MLCad files are efficient and user friendly. When we're working on models, not only can we select and work with parts in the modeling panes, we can also edit the model file directly. Read the next chapter to find out more about this and other file format capabilities.

Chapter 6

MLCad's Output

Solutions in this Chapter:

- A Peek into the Files
- Comments and Meta Commands
- Using Non-LEGO 3D Elements
- Adapting MLCad to Our Needs

☑ Summary

☑ Solutions Fast Track

☑ Frequently Asked Questions

Introduction

Once you are familiar with the modeling tools available in MLCad and are capable of creating LEGO models using your computer, it is time to take a look inside the LDraw file format. For many readers, the prospect of peeking into files and program architecture might not sound like a very user- friendly concept, but in fact it is. The fact that most programs do not allow direct access to their files works *against* the user, who is *forced* to use the programs' interfaces to achieve any and all his or her goals.

How much should we know about what is going on "behind the scenes" in the program as we happily build along? When we build with real LEGO, we might not explore the chemical composition of the plastic it is made of, but there are more subtle building rules that we acquire as we gain experience as modelers. The same happens with LDraw-based software: The very high-quality approach that James Jessiman took when designing the LDraw file format allows us to explore behind the scenes quite freely, without fear of breaking anything. LEGO parts are fun to build with because their design is extremely intelligent. Mirroring this trait, the LDraw file format is also extremely well designed and thought out.

When Jessiman created the software for storing LEGO models in a PC, he devised a file format and a program that would read it. Model files had to be written by hand, entering their part types and coordinates in a text editor according to the format Jessiman predefined. Things have, of course, changed quite a bit. However, all the software tools that came after LDraw have been built around these two core concepts: The general format is easy to understand (and thus users in theory can "type" a model into a text editor), and it is robust and flexible enough to support several very intricate layers of finely tuned details. The same easy-to-use file format can hold information on a model, a part from the library, and several other items, whether step-by-step instructions or pointers for other programs (an independent model generator such as LSynth, for example). The key words here are *friendly* but *powerful*.

MLCad actually lets us take a peek at the model file as we are working on it via the Model Parts List window. Apart from offering several other advantages, this is a great *modeling* tool—very helpful to use for selecting parts, for instance. In this chapter, we take a closer look at the Parts List window. That discussion is followed by an in-depth exploration of LDraw's file format. This is interesting from a purely practical perspective for several reasons. If we intend to create instruction steps for our models, basic knowledge of the file structure is necessary. If we want to include non-LEGO elements in our models, we are also required to have a working knowledge of the format. And if we want to create our own parts for the library, we most definitely need in-depth knowledge of the format and the 3D computer graphics field in general.

Just as LDraw model files carry coordinate information that, when combined with the parts library, generate models, they can also carry the instruction steps for those models or data for new parts. These might seem like secondary functions to some modelers, but in fact using this knowledge can further expand our modeling capabilities. For instance, MLCad's support for creating instruction steps is extraordinary—to the point where some of the tools can be used for other purposes, such as animation. In Chapter 9, we take a look at LPub, Kevin Clague's excellent application that allows you to generate photorealistic building steps, using many of the MLCad features we discuss in this chapter.

Once you are aware of the file format and all the possibilities MLCad offers you, we will discuss other ways to get the most out of our model files. Exporting models, printing model files, and generating "snapshots" of the step-by-step instructions are among some of the topics we will cover. The chapter ends with a description of MLCad's customization options.

A Peek into the Files

To make things simpler for rookie modelers, in the past two chapters we have completely ignored a very handy modeling support tool: the Model Parts List window. In this part of the chapter we talk extensively about the Model Parts List window and about program functions (such as model instruction steps) that require its use. Since the Model Parts List window shows the model file in an almost *raw* state, as it is being created, becoming familiar with it will also help you become familiar with the LDraw file format.

The Parts List Window

As you already saw in Chapter 4, the Parts List window is located above the modeling panes in the MLCad screen. Open MLCad and load the **champ.ldr** minifig we created in Chapter 5. Set the program to **Place** mode and adjust the screen so that the Parts List window is visible, as shown in Figure 6.1.

As you can see in Figure 6.1, the Model Parts List window lists all the parts used in the champ.ldr model. The first thing you should know about LDraw's file format (which is virtually the same as MLCad's) is that model files look like lists. These lists resemble the purchase receipts that you get at stores, which generally feature one item type per line, with additional information such as quantity or price. LDraw's format is exactly the same, only instead of quantity or price, it specifies part color, position, orientation, and type. Each part in the model has its own line in the model file.

Figure 6.1 Location of the Parts List Window

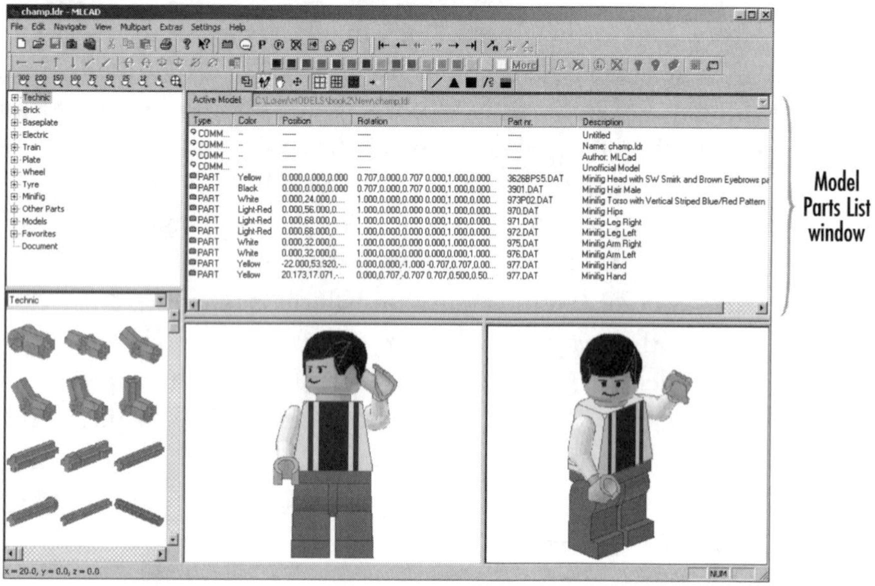

NOTE

Notice that the first line of the Model Parts List window shows the *active model*. This pull-down menu is only active when we are using multipart project files, and it is another way of navigating between the different submodels.

Figure 6.2 focuses on the contents of the Model Parts List window. At the top of the window are column headings for the contents of each item listed in the file. These heading are:

- Type
- Color
- Position
- Rotation

- Part nr (part number)
- Description.

Figure 6.2 The Parts List Window for champ.ldr

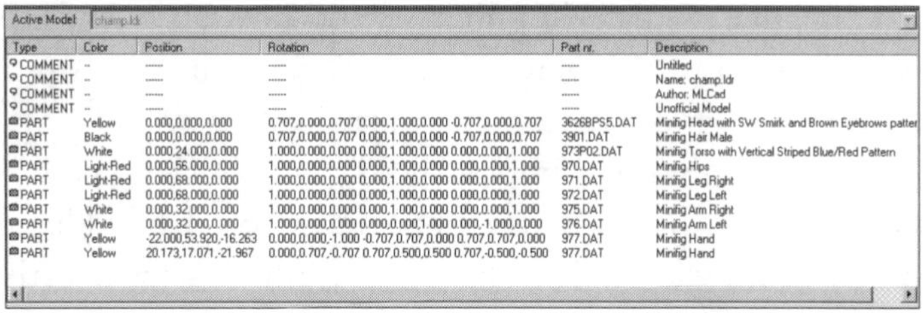

Active Model: champ.ldr

Type	Color	Position	Rotation	Part nr.	Description
COMMENT	--	------	------	------	Untitled
COMMENT	--	------	------	------	Name: champ.ldr
COMMENT	--	------	------	------	Author: MLCad
COMMENT	--	------	------	------	Unofficial Model
PART	Yellow	0.000,0.000,0.000	0.707,0.000,0.707 0.000,1.000,0.000 -0.707,0.000,0.707	3626BPS5.DAT	Minifig Head with SW Smirk and Brown Eyebrows patter
PART	Black	0.000,0.000,0.000	0.707,0.000,0.707 0.000,1.000,0.000 -0.707,0.000,0.707	3901.DAT	Minifig Hair Male
PART	White	0.000,24.000,0.000	1.000,0.000,0.000 0.000,1.000,0.000 0.000,0.000,1.000	973P02.DAT	Minifig Torso with Vertical Striped Blue/Red Pattern
PART	Light-Red	0.000,56.000,0.000	1.000,0.000,0.000 0.000,1.000,0.000 0.000,0.000,1.000	970.DAT	Minifig Hips
PART	Light-Red	0.000,68.000,0.000	1.000,0.000,0.000 0.000,1.000,0.000 0.000,0.000,1.000	971.DAT	Minifig Leg Right
PART	Light-Red	0.000,68.000,0.000	1.000,0.000,0.000 0.000,1.000,0.000 0.000,0.000,1.000	972.DAT	Minifig Leg Left
PART	White	0.000,32.000,0.000	1.000,0.000,0.000 0.000,1.000,0.000 0.000,0.000,1.000	975.DAT	Minifig Arm Right
PART	White	0.000,32.000,0.000	1.000,0.000,0.000 0.000,0.000,1.000 0.000,-1.000,0.000	976.DAT	Minifig Arm Left
PART	Yellow	-22.000,53.920,-16.263	0.000,0.000,-1.000 -0.707,0.707,0.000 0.707,0.707,0.000	977.DAT	Minifig Hand
PART	Yellow	20.173,17.071,-21.967	0.000,0.707,-0.707 0.707,0.500,0.500 0.707,-0.500,-0.500	977.DAT	Minifig Hand

Since the Model Parts List window is an organized representation of the file, let's take a look at the file itself to fully understand its general structure. As we said in the introduction, LDraw files were meant to be accessible, and accessible they are. Open the **Windows Notepad** accessory, and using the **File | Open** menu option in that application, load the **champ.ldr** file (see Figure 6.3). The file can be open in MLCad and Notepad at the same time.

Figure 6.3 The Champ File Opened in Notepad

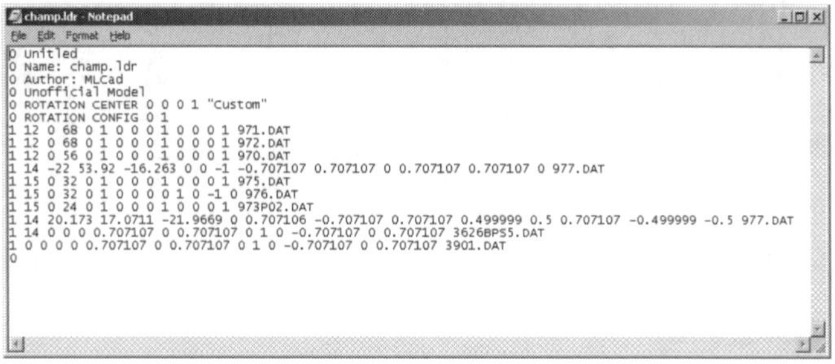

champ.ldr - Notepad

```
0 Unitled
0 Name: champ.ldr
0 Author: MLCad
0 Unofficial Model
0 ROTATION CENTER 0 0 0 1 "Custom"
0 ROTATION CONFIG 0 1
1 12 0 68 0 1 0 0 0 1 0 0 0 1 971.DAT
1 12 0 68 0 1 0 0 0 1 0 0 0 1 972.DAT
1 12 0 56 0 1 0 0 0 1 0 0 0 1 970.DAT
1 14 -22 53.92 -16.263 0 0 -1 -0.707107 0.707107 0 0.707107 0.707107 0 977.DAT
1 15 0 32 0 1 0 0 0 1 0 0 0 1 975.DAT
1 15 0 32 0 1 0 0 0 0 1 0 -1 0 976.DAT
1 15 0 24 0 1 0 0 0 1 0 0 0 1 973P02.DAT
1 14 20.173 17.0711 -21.9669 0 0.707106 -0.707107 0.707107 0.499999 0.5 0.707107 -0.499999 -0.5 977.DAT
1 14 0 0 0 0.707107 0 0.707107 0 1 0 -0.707107 0 0.707107 3626BPS5.DAT
1 0 0 0 0 0.707107 0 0.707107 0 1 0 -0.707107 0 0.707107 3901.DAT
0
```

Apart from the fact that MLCad formats the information neatly into columns in the Model Parts List window, the general structure is very similar to that of the file as it appears in the Parts List window. Each item is presented on one line, and each line contains roughly the same types of information.

There are some differences, however. Apart from the different spacing, the beginnings and ends of the lines differ between the two. The difference at the ends is trivial: MLCad adds the name of the parts to the end of the row, but to keep the size to a minimum, the actual file as seen in Notepad doesn't include them.

The lines also start differently, but this is not trivial at all. The lines in the file open in Notepad start with a single digit, which in the Model Parts List window is translated and listed as *type* instead. This does not specify *part* type (listed later in the row as part number), but rather the type of *list item* (also called *commands*) in that line. Remember the receipt analogy we just discussed? Not all the data in those receipts is used to show just the items purchased. There might be lines that include the store's name, the date of purchase, and sales tax calculations. The same thing happens with the lists of LDraw files. Parts generally take up the bulk of most models, but there is room for other information as well.

In the champ.ldr file, the first six lines hold items of the type *comment*. As you can see in Figure 6.2, the Model Parts List window for champ.ldr only shows four such comments. We will explain that (powerful) mystery shortly, but let's first consider the larger picture.

After the first six lines (or four in the MLCad window), the line items start with the number 1 and are listed as of the type *Parts* in the MLCad window of Figure 6.2. When MLCad "translates" those initial digits into the Model Parts List window as "types," it lists lines starting with 0 as of the *type* COMMENT and the ones starting with 1 as of the *type* PART. Are there any more numbers—more *types* of list items? Yes, and a large part of this chapter deals with them. The types are:

- **0** Comment or meta command
- **1** Part (or model) usage
- **2** Line
- **3** Triangle
- **4** Quadrilateral
- **5** Optional line

Is this too much detail already? Here is a handy rule of thumb to help you remember: Lines of types 0 and 1 are used for models (typically); types 2 through 5 (and higher) are used to define LDraw parts. As LDraw-based modelers, we start by building with virtual LEGO and then maybe proceed to the non-LEGO-based 3D world—defining new parts for the library, for instance. In the same way, in this chapter we first examine the two types incumbent to LEGO modelers, and then we proceed to the other types.

For the time being, think of the file as a list of generic items. Let's first find out how using that textual representation of the model directly influences our modeling in MLCad. Parts (line items of type 1) are the simplest and most familiar type for the beginner, so let's start there.

Working with the Model Parts List Window

The Model Parts List window is most useful when it's used with the modeling panes. That way, the whole modeling area behaves a bit like the Library windows. The Model Parts List window offers more precision, but the modeling panes offer more immediacy via the visual display of the model.

Let's see how this works in practical terms. If you select a part in a modeling pane, it is also displayed as selected in all the other modeling panes (except those set to 3D view) *and* in the Model Parts List window. Select the champ's torso part in one of the modeling panes and you will see that it is selected in both the modeling panes and in the Model Parts List window (see Figure 6.4).

Figure 6.4 Part Selected in the Preview Window and the Modeling Panes

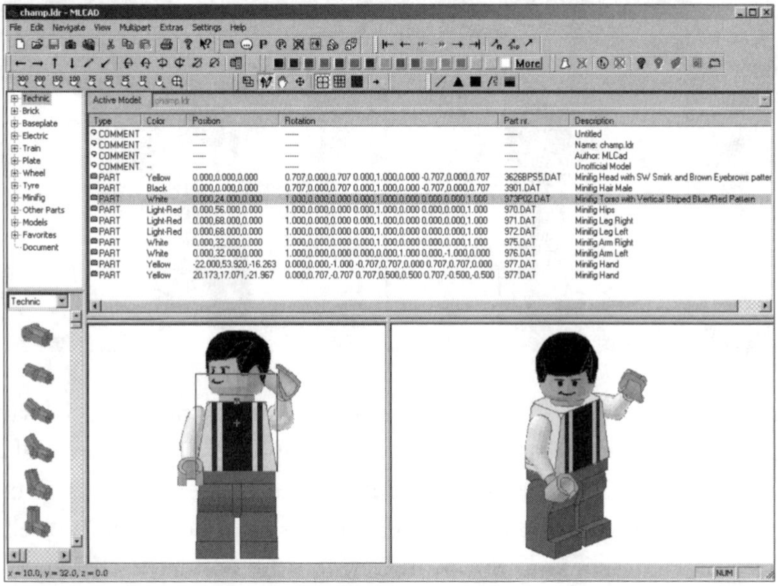

This is a great way to identify parts in models. Of course, the champ model is very simple, and locating parts in it via the modeling panes is not a big problem. However, most models contain a lot more than 10 parts. When we are working with large models with many parts, this way of selecting items in the Model Parts List window becomes very useful.

Selecting Parts

We can also select parts directly from the Model Parts List, just as we do in the modeling panes, by clicking anywhere on a particular part's line in the Modeling

Parts List window. To select several parts, press the **Ctrl** key on your keyboard while you continue clicking more items with the mouse. To select adjacent lines, select the first one and then hold down the **Shift** key while clicking the last item. The whole range in between will be selected, as shown in Figure 6.5. You can also use Ctrl and a mouse click to deselect particular parts.

Figure 6.5 Selecting Parts from the Model Parts List Window

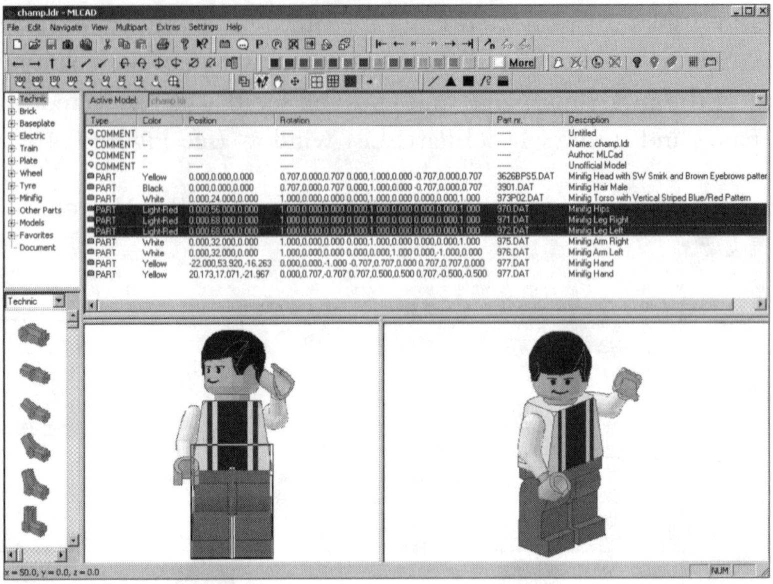

Once you have selected the parts you want, you can perform the same standard modeling functions described in the last chapter. The changes will be reflected in the appropriate fields (columns) of the Model Parts List window as well as in the modeling panes. You can also access many of these modeling functions directly by right-clicking the mouse inside the Model Parts Preview window and using the menu that pops up. In Figure 6.6, you see that this menu offers support for changing the part color, its type (via the Modify command), and its visibility. The menu also allows you to place a new part (or other list item) into the model file via the Add option.

Figure 6.6 The Right-Click Menu for the Model Parts List Window

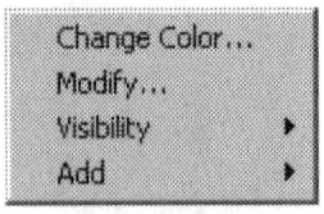

Sorting Parts

What other benefits does working directly with the file bring? For one thing, we can influence the order of the items as they appear in the file, which is important for tasks such as step-by-step instructions and many other neat file functions. Unless your model uses these *meta commands*, described in the next section of the chapter, the sequence of the items in the list does not matter. The model's parts can be moved up or down the list without modifying the model's appearance in the modeling panes in any way, but it can be helpful if we want our model files to be organized internally.

To move a part or a group of parts up or down the Model Parts List window sequence, in the Model Parts List window simply select the part or parts you want to move and drag it or them with your mouse to the new location. In Figure 6.7, we have moved the minifig's legs to the bottom of the Model Parts List window. Notice that there is no change in the appearance of the model as it is displayed in the modeling panes.

Figure 6.7 Altering the Order of the Parts List Sequence

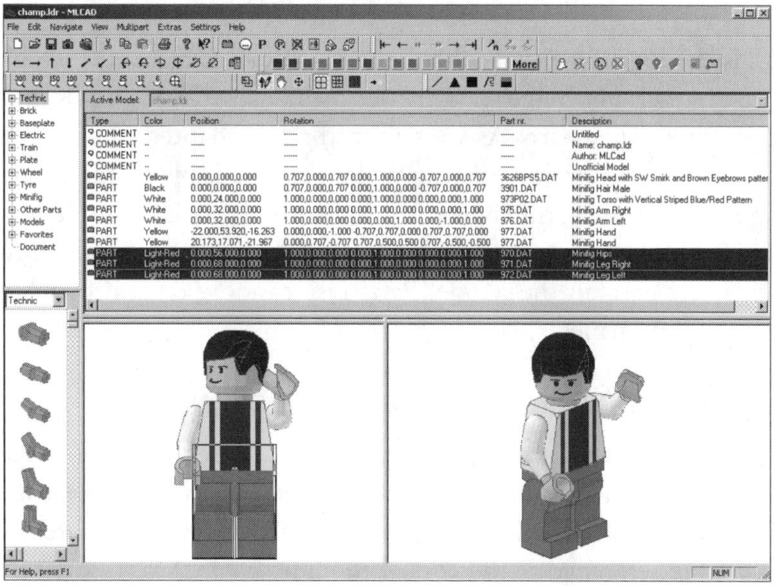

MLCad takes further advantage of this characteristic by offering sorting functions for the parts in a model. The **Edit | Sort** menu option activates the Sorting Parameters dialog window, shown in Figure 6.8.

Figure 6.8 The Sorting Parameters Dialog Window

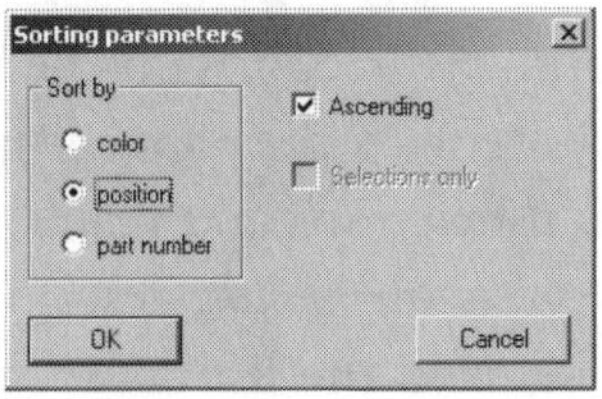

If we choose to sort by color, the parts are ordered according to the position of the color of the part in MLCad's palette: black parts first, followed by blue, green, dark cyan, and so on. Sorting the parameters by position orders them by placing them on the list according to their coordinates from bottom to top, back to front, and left to right (or, easier to remember, *according to their XYZ coordinates*). Finally, we can also sort automatically by part reference number. The Ascending option reverses the sorting order (for example, when we're sorting by color, black parts come last in the list). The Selections Only option allows us to sort selected parts of the model only. The Sort By box is a mutually exclusive list, as indicated by the use of standard Windows radio buttons. You can only sort by one criterion at a time.

Sorting is the kind of activity that becomes more attractive as we gain experience and our models get bigger and harder to manage. Those kinds of models might also require *hiding* parts. As we saw in the last chapter, using the Edit | Visibility | Hide menu option, we can make parts invisible in the modeling panes. However, they will still appear in the Model Parts List window, with their Description field specifying their hidden state, as Figure 6.9 shows. This tool allows us to select parts in the Model Parts List window even if they are not visible in the modeling panes.

Working with Submodels and Groups

What happens if we insert a model as a part into a larger model, or as a submodel? For all intents and purposes, the LDraw file format will treat its line exactly the same as it treats a line of type 1, part: It stores coordinate information about it and a file reference at the end of the line. Each part is stored in a single file that is in a format completely compatible with model files. Thus, there is no need to make any special fuss to accommodate a part or a submodel in the same line item definition. At the end, they both reference files with geometric data in them.

Figure 6.9 A Hidden Part Displayed in the Model Parts List Window

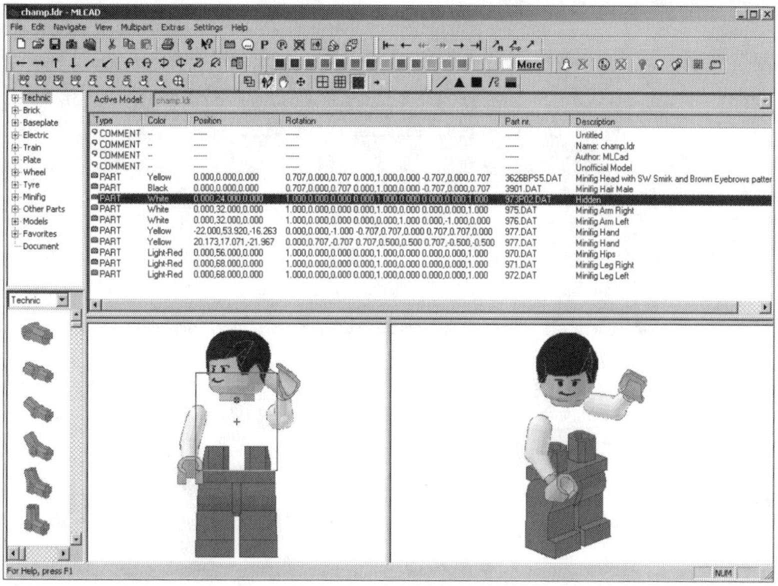

Let's see how it works in practical terms. Figure 6.10 shows the podium model inserted as a part in the champ.ldr file. In the Model Parts List window, the only difference with other parts is that the *Part nr.* column displays the submodel filename, while the description shows "Part unknown!" This can be fixed in the submodel's original file, as is explained in the next section of the chapter. The important information for the program, the part's position and orientation, is stored in exactly the same way. The part color information is irrelevant when the part is a submodel.

Working with groups is quite different from working with submodels inserted as parts into a file. In Figure 6.11, the two leg parts and the hip part have been grouped and located at the bottom of the Model Parts List window. When we group parts using the **Edit | Group | Create** menu option, the Model Parts List window displays a lot less information for the group, displaying just the Group name in the Description column. More telling, unlike the submodel of Figure 6.10, the line is no longer of the type PART. Now it is of the type GROUP. As far as the file format is concerned, Group is not really a type unto itself. It is a meta command. Let's take a closer look at what this means.

Figure 6.10 A Model Inserted as a Part Displayed in the Model Parts List Window

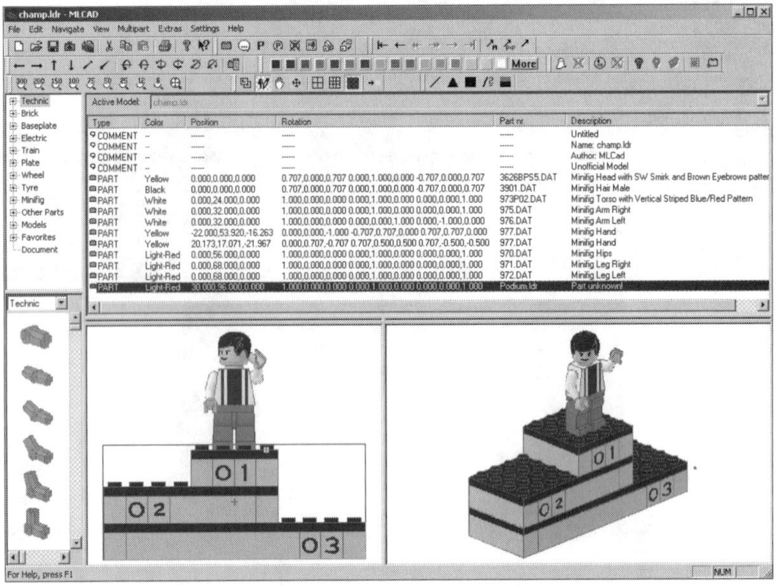

Figure 6.11 A Group Displayed in the Model Parts List Window

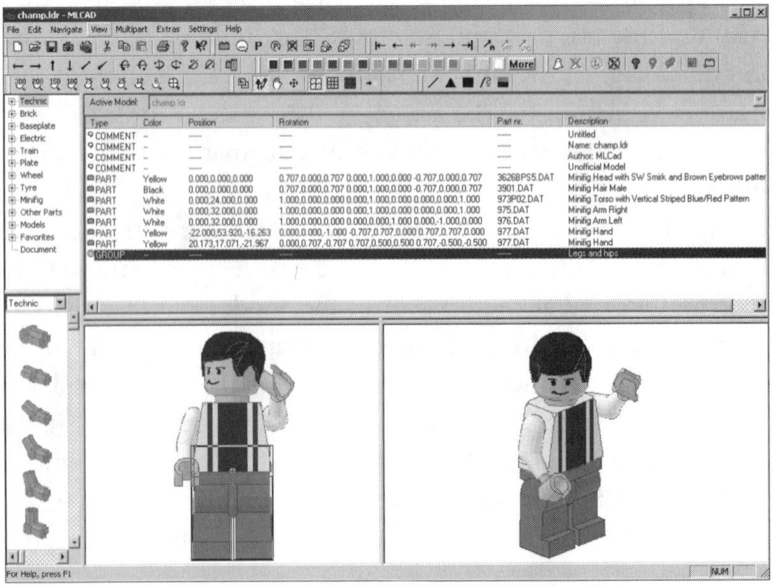

Comments and Meta Commands

To understand meta commands, we have to take a look at another level of detail in the file format. Doing so will ultimately also explain the mystery of the missing lines 5 and 6 of the file from the Model Parts window.

In the overall diagram of the file format, parts are the basic type of file item. To the virtual-LEGO modeler, a part is as tangible as the objects bought in a store and listed in the purchase receipt. Internally, a part works in a similar fashion: to the program, it references a specific file (an arguably "tangible" object) from the parts library. When models are used as parts of other models, the type is still the same. The line will still carry at the end a filename for the part or sub-model.

What happens if something goes wrong? Most software reading the model file would simply skip the erroneous line (the faulty part) and continue on. Whatever went wrong is isolated, preventing it from causing major trouble in the rest of the file. Unlike real LEGO, parts do not depend on each other, and thus we can choose to try and fix the problem or delete the part and substitute it with another part. Needless to say, this is a very elegant solution—very simple, yet completely effective. So effective, in fact, that it lends itself to further customization. Which is exactly what the COMMENT line item type (and the lines 5 and 6 "vanishing" mystery) is all about.

Let's look at the larger picture again. As we said earlier, types 0 and 1 are typically used for model files. If lines of type 1 are *always* parts or models used as parts, that leaves a lot for the 0 type. Why? Because when we say types 0 and 1 hold model information, we mean they hold *all* the other model information, such as instruction steps or special commands for other programs such as LSynth to complete the model. And those are only a few examples.

Why not assign other type numbers to file lines containing that kind of information? After all, types 2 and above hold information for parts. Surely there must be space for more types of line items, such as those dedicated to groups or instructions steps. In fact, it could be argued that James Jessiman actually wanted to cram into the type 0 file lines as much power as possible. If that was his original intent, he most definitely succeeded. The lines that start with a 0 hold a great variety of options for our models. These options are called *meta commands*.

What Jessiman did is allow this type of file item to be "open," to allow users to customize it for their own benefit. This allows for plenty of cool things to happen. But let's observe the worst-case scenario again. What happens when one of these fabulous meta commands that anybody can tweak and even invent goes completely berserk (which *can* happen)? Well, once again, nothing serious occurs. The problem will always be isolated to a specific line in the file, and somehow or other, it can be fixed. You can always load the file into Notepad and delete the offending line.

NOTE

Always make a backup copy of the files with weird effects; they might come in handy one day!

Using Comments

By allowing the customization of the comment line item type, Jessiman planted the seed for an evolution to happen. And it did—users and developers (including Michael Lachmann, the creator of MLCad, and, of course, James Jessiman himself) jumped at the opportunity and boldly created very neat additions to the file format. Shortly we will we explain some pretty amazing tricks you can easily perform with these additions, but let's start at the beginning of that evolution.

In their most basic form, comments are even simpler than parts. Comments are often used in computer files for the benefit of human users. Since not everybody is equally fluent in computer language, comments are text helpers included between the lines of computer code to make it more readable for the human reader. The same is true for LDraw files; since not everybody can read them, comments are there to add extra information.

Adding a Comment

Let's see how commenting works in practice. In the champ.ldr file, use **Edit | Sort** to reorder the parts' sequence by position. The legs and hips, being the lowest parts, come at the end of the file. On the Model Parts List window, select the part immediately above the line for the Hips part, as shown in Figure 6.12.

Now use the **Edit | Add | Comment** menu option to insert a comment. This action triggers a New Comment dialog box like that of Figure 6.13. In it, we have entered a description for the lower part of the file.

Once you click **OK**, the comment will be inserted into the list, as shown in Figure 6.14.

Figure 6.12 Selecting the Place to Insert a Comment

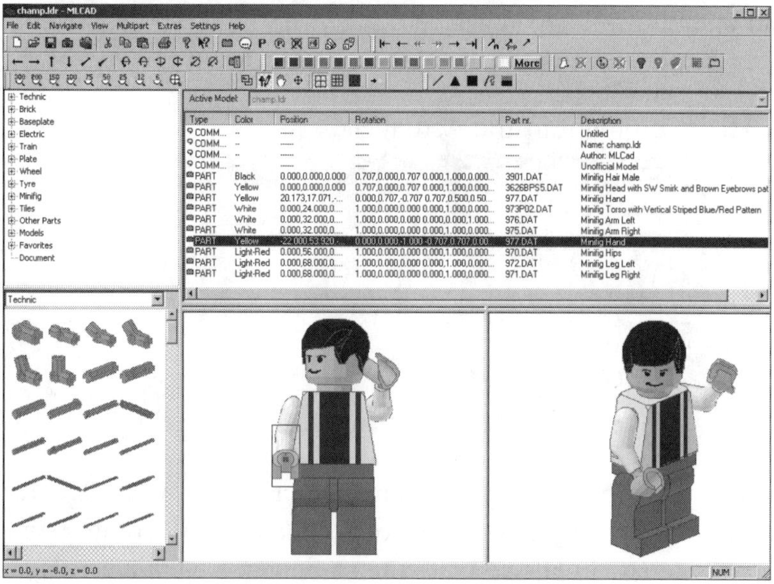

Figure 6.13 The New Comment Dialog Box

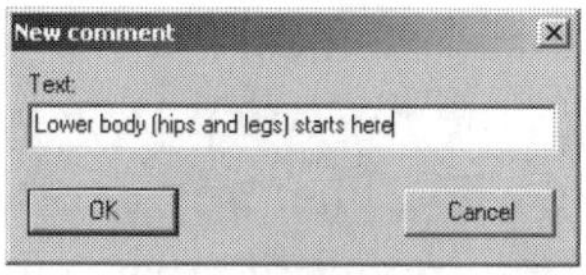

Figure 6.14 A New Comment Inserted into the Model Parts List Window

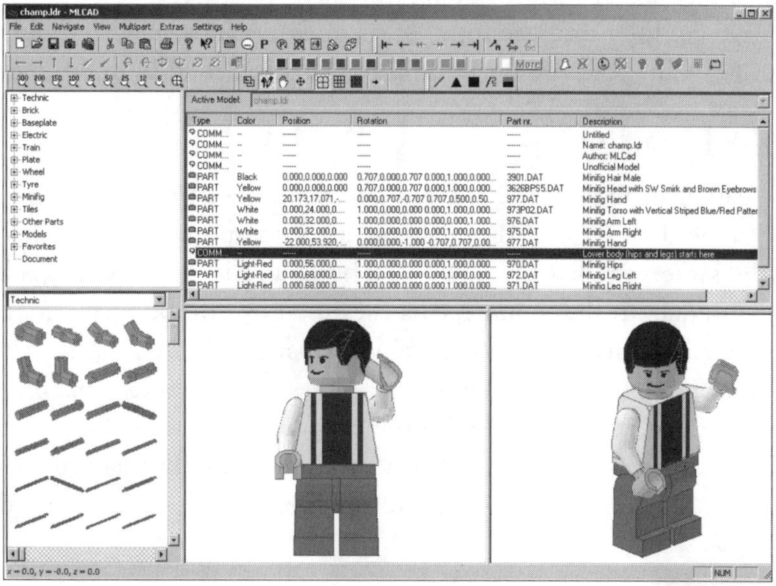

In MLCad, this insertion does not have a profound effect. However, if we save the file and open it in Notepad again, as shown in Figure 6.15, the comment does help us locate the parts for the hips and legs toward the end of the file.

Figure 6.15 A Comment in the Actual File

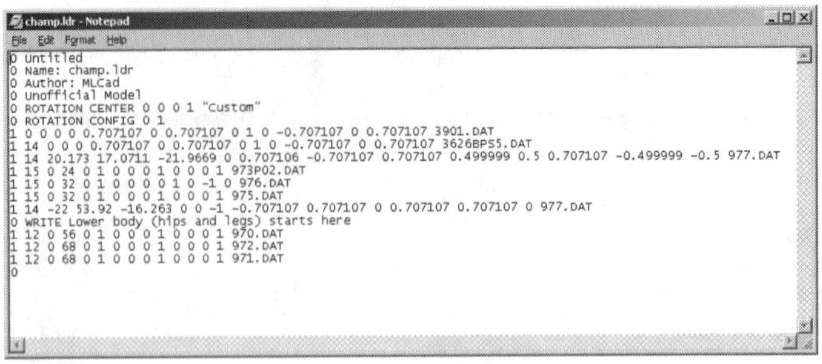

This is a comment in its most basic form. In principle, we can assume that when a program (for example, MLCad) reads the file, it will completely ignore the lines that start with a 0. In effect, 0 tells the computer, *Skip this line altogether—it's for humans to read.* However, nothing stops a developer from writing a program that in fact reads beyond the initial 0. When a program does this, we say it uses *meta commands*. This means it uses commands "hidden" inside the file as items of type 0—innocent comments. This makes the file compatible with programs that use meta commands and programs that do not. The programs that do not use meta commands simply skip the lines in the model file that start with 0. The programs that use meta commands explore beyond the 0, checking to see if there are any meta commands hidden inside *before* skipping the line.

Before we get fully into the topic of meta commands, let's see a simple example of how lines of type 0 can hold information used by the programs.

The File Header

As we saw earlier in Figures 6.2 and 6.3, the first lines of the champ.ldr model file were of type 0, or COMMENT. These lines compose what is called a *file header*. In the same way that the cover of a book and its first few pages hold general information about the volume, data files used by computer programs often use file headers like this to show general information about what is stored in the file. Computer programs often check the header of a file before opening it, to see if it is of a type that's compatible with the program.

When we create a new file in MLCad, the program automatically inserts a file header that forms the first six lines of the file. The first line carries the model's name, and the second one holds the filename. If this sounds confusing, think of how the parts have filenames based on their LEGO serial numbers, and they also have part names. For instance, part 977.DAT is also a "Minifig Hand" part. The program automatically adds the filename to the second line of the file header, but we have to add the model name in the first line ourselves.

Let's do just that. Select the first line of the file, as shown in Figure 6.16.

Figure 6.16 The First Line of the File Header Is Selected

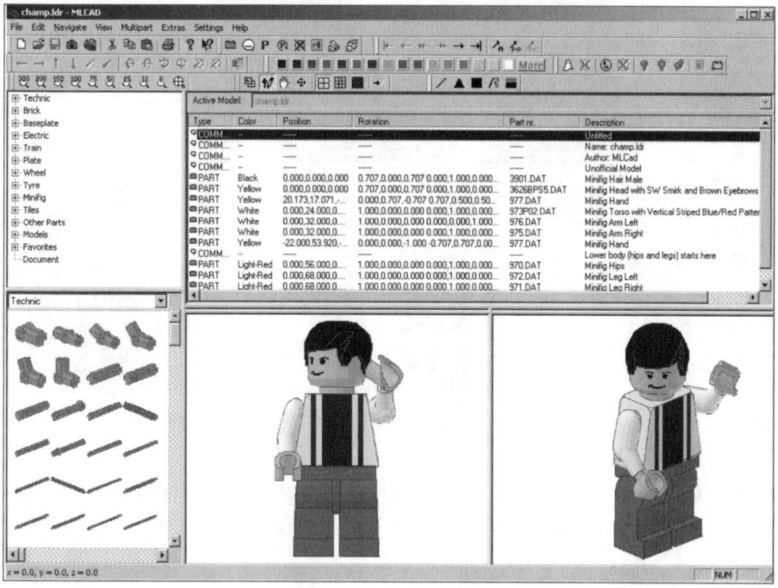

Any comment can be edited, either directly in the file via Notepad or another text editor or, in a more limited fashion, in MLCad. To add a name to the minifig model, we actually have to change the text in this section from *Untitled* to something else. To edit the comment you selected, use the **Edit | Modify** menu option. A Change Comment dialog window like the one shown in Figure 6.17 will appear.

Figure 6.17 The Change Comment Dialog Window

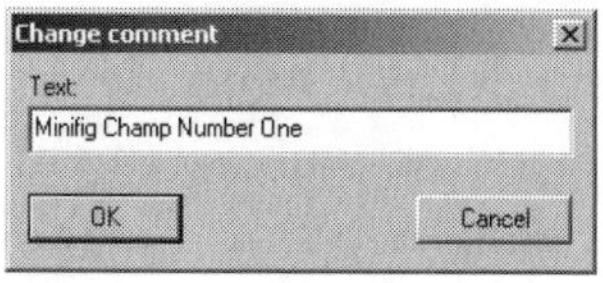

Enter a new name (we've used **Minifig Champ Number One**) and click **OK**. The Model Parts List window now displays a different Description for the

line item—the text you just entered in the Change Comment window, as shown in Figure 6.18.

Figure 6.18 The Model Name Is Changed

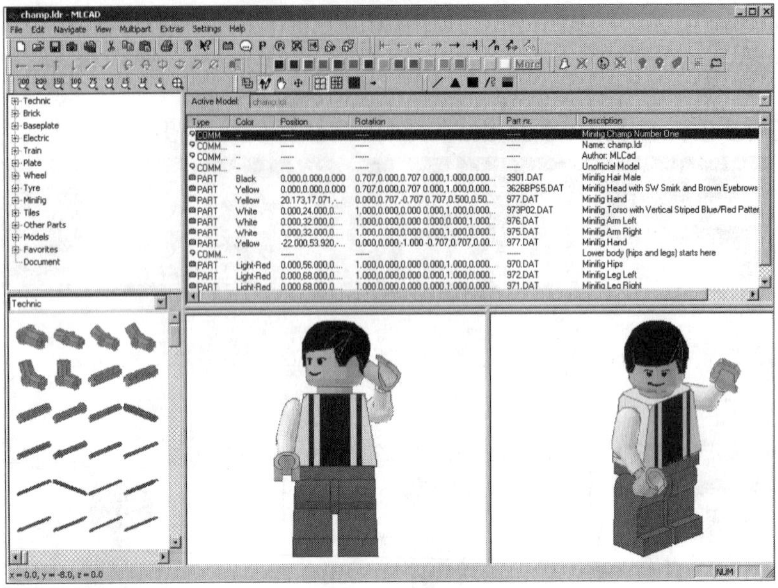

If you save the file and open it in Notepad again, the first line of the file will reflect the change as well, as you can see in Figure 6.19.

Figure 6.19 The Model Name in the File

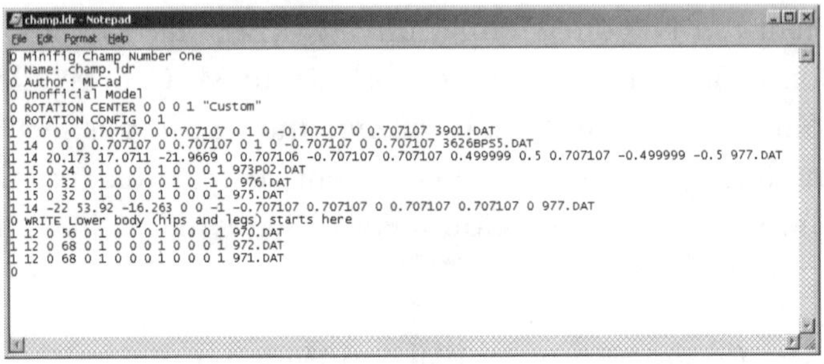

```
0 Minifig Champ Number One
0 Name: champ.ldr
0 Author: MLCad
0 Unofficial Model
0 ROTATION CENTER 0 0 0 1 "Custom"
0 ROTATION CONFIG 0 1
1 0 0 0 0 0.707107 0 0.707107 0 1 0 -0.707107 0 0.707107 3901.DAT
1 14 0 0 0 0.707107 0 0.707107 0 1 0 -0.707107 0 0.707107 3626BPS5.DAT
1 14 20.173 17.0711 -21.9669 0 0.707106 -0.707107 0.707107 0.499999 0.5 0.707107 -0.499999 -0.5 977.DAT
1 15 0 24 0 1 0 0 0 1 0 0 0 1 973P02.DAT
1 15 0 32 0 1 0 0 0 0 1 0 -1 0 976.DAT
1 15 0 32 0 1 0 0 0 1 0 0 0 1 975.DAT
1 14 -22 53.92 -16.263 0 0 -1 -0.707107 0.707107 0 0.707107 0.707107 0 977.DAT
0 WRITE Lower body (hips and legs) starts here
1 12 0 56 0 1 0 0 0 1 0 0 0 1 970.DAT
1 12 0 68 0 1 0 0 0 1 0 0 0 1 972.DAT
1 12 0 68 0 1 0 0 0 1 0 0 0 1 971.DAT
0
```

Now, when you use the model as a submodel in another file, the Description field of the Model Parts List window will use the model's name instead of the "Part Unknown!" shown in Figure 6.10 for the podium submodel.

NOTE

You will also have to save the file in the LDraw\models directory (or in the LDraw\parts directory). In MLCad, use the **File | Scan models** menu option to introduce your new model to MLCad. The program will look in the models directories to check for any new models, but it will not look into subdirectories of the LDraw\models subfolder. This is yet another reason to use subfolders and keep the base LDraw\models directory clear. It is better to use that disk space only for these types of functions. Remember that that models placed in this directory are also accessible via the Models group of the parts library windows.

The third line of the file header indicates the author of the model. This setting can be changed. The section "Customizing MLCad" at the end of this chapter explains how to set MLCad to automatically add names to all model files. The fourth line simply indicates that this is an unofficial model. This is especially useful with part files; it lets us (and, potentially, programs) identify which parts are officially approved by the LDraw community.

The two extra lines of the header that do not appear in the Model Parts List window store custom rotation points for the file. They are available in MLCad, not via the Model Parts List window but through the Rotation Point Definition dialog window, described in the last chapter. Why is that? Because they are meta commands.

Meta Commands

The file header MLCad inserts in all files shows us some excellent examples of meta commands in action. Despite the fact that some lines start with a 0, the program still extracts information from them (such as the model's name or its rotation points). In fact, the case of the file headers is a bit special. There is no official header for LDraw files; in fact, they do not need one. MLCad takes advantage of the position of the first line: If it starts with a 0, whatever text forms the rest of the line is the name of the model.

However, meta commands generally do not work this way. Meta commands are included *inside* lines of type 0. As we said earlier, programs that do not take advantage of meta commands simply skip those lines. Programs that do take advantage read the 0 and then the first word after the 0. If the program recognizes the word, it has found a meta command and will act accordingly.

The fifth and sixth lines of the automatic file header are a great example of typical meta commands. When MLCad reads a line that starts with 0 ROTATION (and not just 0), it knows it is dealing with meta commands that store information for rotation point definitions. This meta command then has another key word, which can be CENTER or CONFIG. 0 ROTATION CENTER sets custom points (with the name at the end of the line). 0 ROTATION CONFIG stores the rotation point the user was using when the file was last saved.

Meta commands come in many flavors. Indeed, we can almost foresee a future in which meta commands are located in a "library" of the size of the parts library. The finer points of the ROTATION meta commands are not very interesting, though. Instead of discussing them in detail, in the next sections we focus first on a family of meta commands that allow us to add instruction steps to our models. Then we will see some advanced meta commands that can be used for both sophisticated instruction steps and animation. In chapter 7, Kevin Clague introduces us to his LSynth application, which uses its own meta commands (approved by LDraw.org).

Before you begin to use meta commands in earnest, it is important that you understand that comments and parts coexist peacefully inside files. Do not lose sight of the fact that any and all of the effects that you will see in the next two sections correspond to specific lines in the file. You can locate them via the Model Parts List window or directly in the file via Notepad.

Designing & Planning …

The Meta Command Meta List

Since meta commands are highly customizable and task oriented, there is not a single place where you can get a definitive list of meta commands for LDraw-compatible software. Three important places to keep an eye on are the specification pages of www.ldraw.org, the CAD forums of www.lugnet.com, and the MLCad extension sheet found on www.lm-software.com (the complete URL is www.lm-software.com/mlcad/Extensions.pdf). You can find more information on LUGNET and LDraw.org in Chapter 10.

Adding Instruction Steps to Our Models

As we mentioned at the beginning of the chapter, our overall objective is to familiarize you with LDraw's powerful file format and its underlying fabric. Indeed, meta commands form a very important part of that detailed fabric. They are so important that if our virtual models can be thought of as three-dimensional, we can almost say that meta commands add a *fourth* dimension into the model.

What does this fourth dimension add to our models? It means that not only we can store a model in a file, we can also add information on how the model *works*; a scheme of how the model is built; and any other data you might find relevant. All this is achieved via different meta commands that we insert into the file or via the line item types 2 through 5.

In this section, we first examine some simple meta commands that allow us to structure the file in such a way that it not only contains a model editable in Place mode, it also allows us to view the step-by-step building instructions in View program mode. And all is kept in one file.

Instruction Steps Meta Commands

Both the LDraw file format and MLCad provide fabulous support for adding and viewing step-by-step instructions for our models. Although the end product—the effect of the instruction steps meta commands—is seen in View program mode, the actual creation of the steps is done in Place program mode. Figure 6.20 shows the Editbar's buttons, which add building instruction meta commands to our models. The corresponding meta commands that actually go into the file are at right in the figure.

Figure 6.20 The Editbar Buttons

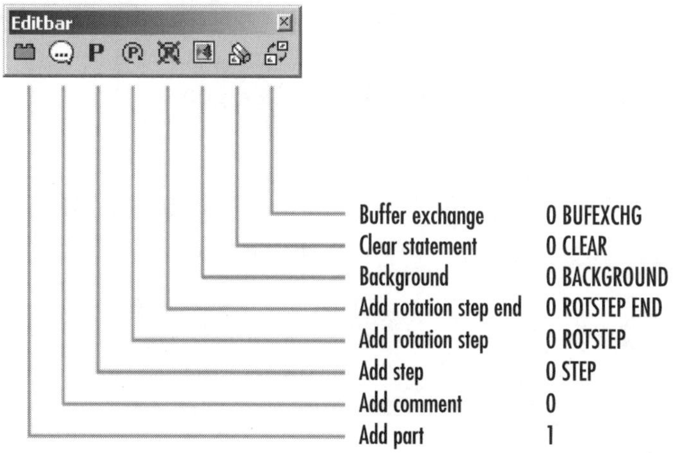

Designing & Planning…

Why Add Instruction Steps to Our Model?

Apart from the advantage of learning how to use meta commands by actually *using* them, there are other reasons you should consider adding instruction steps to your virtual LEGO models. For one thing, it allows you to share your creations without necessarily releasing the file. The instruction steps can be seen in the screen but also exported into images from within MLCad or via LPub for photorealistic effects.

Instruction steps also add quality to your models. Adding instruction steps is like adding order to the models, which makes your files more efficient if you ever decide to modify them in the future. Additionally, adding steps will force you to revisit your model, increasing your chances of spotting mistakes.

The strongest argument for using instruction steps is that it makes you a better modeler. Not only are you forced to be more organized in your modeling, you are also required to *think* more about your model: how it looks from different angles or how a potential audience would react to it (when using the instruction steps to build it with real LEGO, for instance).

Adding Steps Manually

To show you how to add instructions steps to the models, we will use the podium.ldr file. As we mentioned earlier, the champ model has too few parts to make adding steps to it worthwhile. The podium model, clocking at 20 parts and with a more "architectural" feel to it, is much better suited to showcase the basic methods and techniques of adding instruction steps.

Go ahead and open the **podium.ldr** file. As Figure 6.21 shows, the complete parts list for the model does not fit in the Model Parts List window; we need to use the scrollbar to navigate it. This, of course, makes it harder to locate parts in the list. If we do not want to include steps in the file, we can always document large models by adding comments to them, as we saw earlier with the minifig.

Figure 6.21 The Podium's Long Parts List

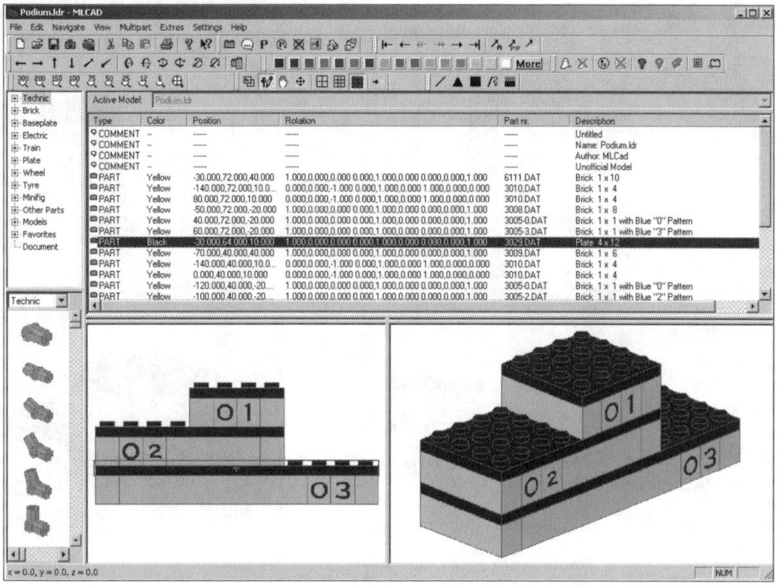

NOTE

Since you built the podium model from the ground up, the resulting sequence of parts should be useful for building its instruction steps. You might want to make sure that this is the case. **Edit | Sort** its parts by **Position**, as we have done with the podium in Figure 6.21. This method is not universal, but it works well as a first step with "architectural" models built from the ground up.

Adding a step meta command to a file in MLCad is possibly the easiest thing to do in the program. Select the place in the file where you want the step, as you did earlier when adding a comment to the champ.ldr file, and use the **Edit | Add | Step** menu option (or click the **Step** button of the Editbar). Figure 6.22 shows the podium file with a step added to it.

What is the effect of this meta command? As we said earlier, we create the steps in Place program mode, but we see their effect in View program mode. If we switch to View program mode by pressing the **F2** key, we will be presented with a screen similar to the one in Figure 6.23.

Figure 6.22 Step Inserted Manually

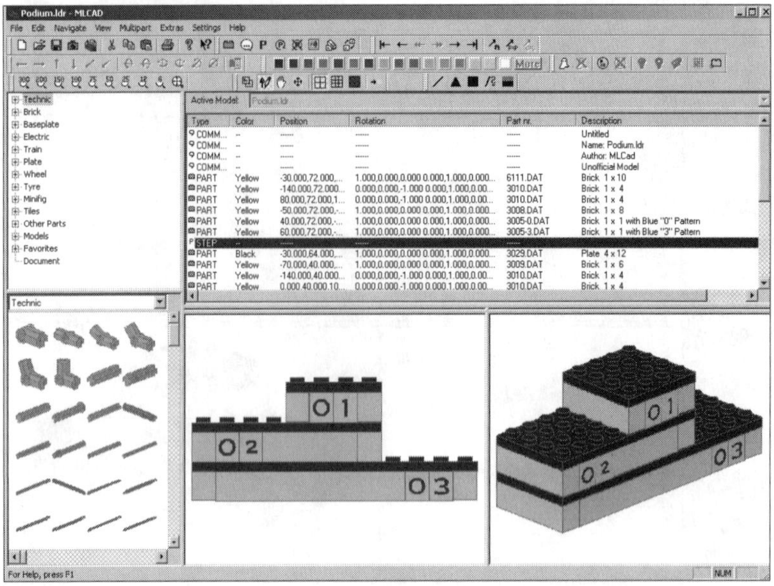

Figure 6.23 Podium.ldr in View Program Mode

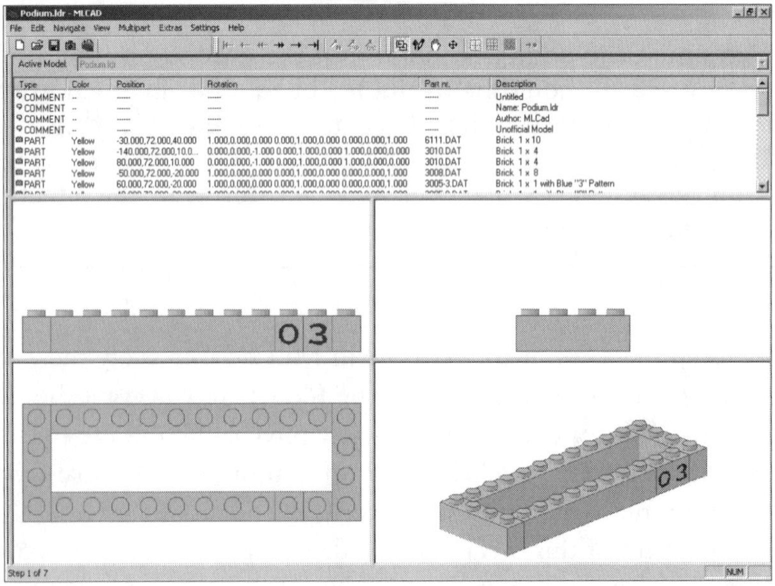

What you see in Figure 6.23 is the first step of your model. In View program mode, the STEP meta command makes MLCAD stop drawing the model when it encounters a 0 STEP line. Thus, any parts located in the list sequence *before* the STEP meta command make up the step.

NOTE

View and Place program modes share the same screen elements, but their layout can be configured independently. In Figure 6.23, we have minimized the library windows and have turned off many of the toolbars not used in View program mode.

If we press the **Enter** key, the model will advance one step forward. Since there is only one step in the file, the program includes in the second step all the parts located in the list after the first step, as shown in Figure 6.24.

Figure 6.24 The Second Manual Step

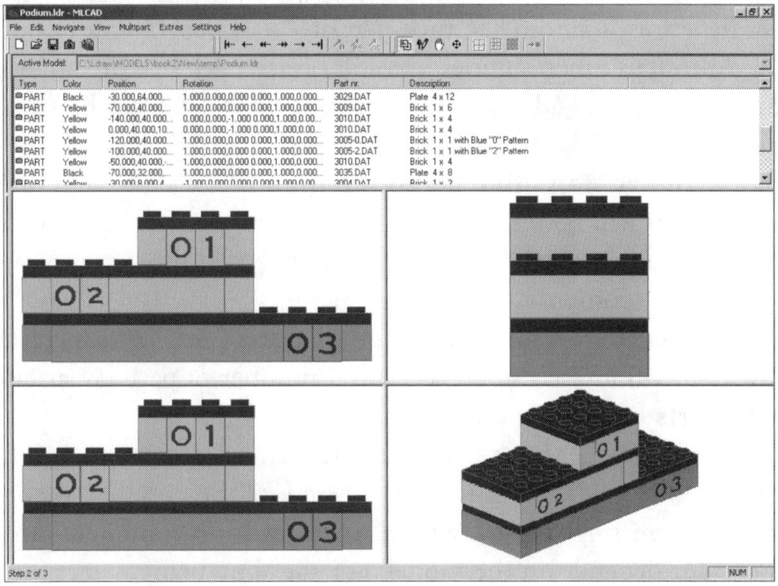

We could add more manual steps and further refine the instructions that way. Instead, let MLCad do the hard work while you provide finesse. Go back to Place program mode by pressing the **F3** key. In the Model Parts List window, select the step you have just added and delete it by pressing the **Del** key on your keyboard. The file should again look like Figure 6.21.

Adding Instruction Steps Automatically

Let's take a look at MLCad's automatic instruction step functions. Open the Autosteps dialog window (see Figure 6.25) using the **Edit | Add | Autosteps...** menu option.

Figure 6.25 The Autosteps Dialog Window

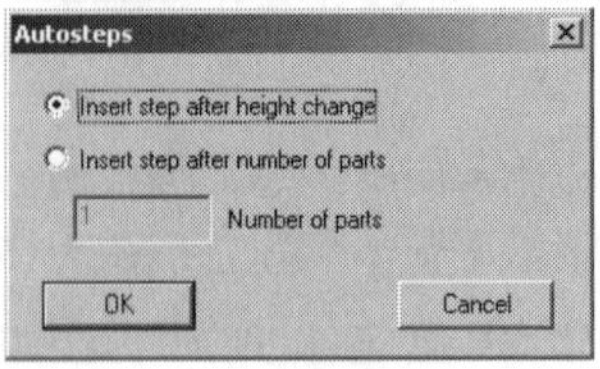

The Autosteps feature works a bit like the model generators we saw in the last chapter. It performs automated model-related tasks according to our specifications. The Autosteps function does not add or arrange parts; it simply goes through the parts lists and adds step meta commands to it according to one of two criteria.

If we select **Insert step after height change** in the Autosteps dialog window, a new step will be inserted before each new part whose value for the Y axis coordinate is higher than the previous part (we will show you how this works in detail in a moment).

If we select **Insert step after number of parts** in the Autosteps dialog window, MLCad will insert a step every fixed number of parts, as specified in the **Number of parts** box below this option.

The seemingly innocent fact that there are two major criteria to add steps automatically to our model actually hints at some of the complex theoretical issues involved in creating step-by-step instructions for our models. Which method is best? It depends on the model. For the time being, select the **Insert step after height change** option and click **OK**.

As we see in Figure 6.26, there are now STEP items scattered throughout the Model Parts List window. Specifically, a STEP item is inserted after the first tier of bricks, one right after the first black plate, one after all the bricks in the second tier, one after the second black plate, one after all the bricks in the third tier, and the last one after the final black plate on top of the podium.

Let's see how this all looks in View program mode.

Figure 6.26 The Podium File with Automatic Steps

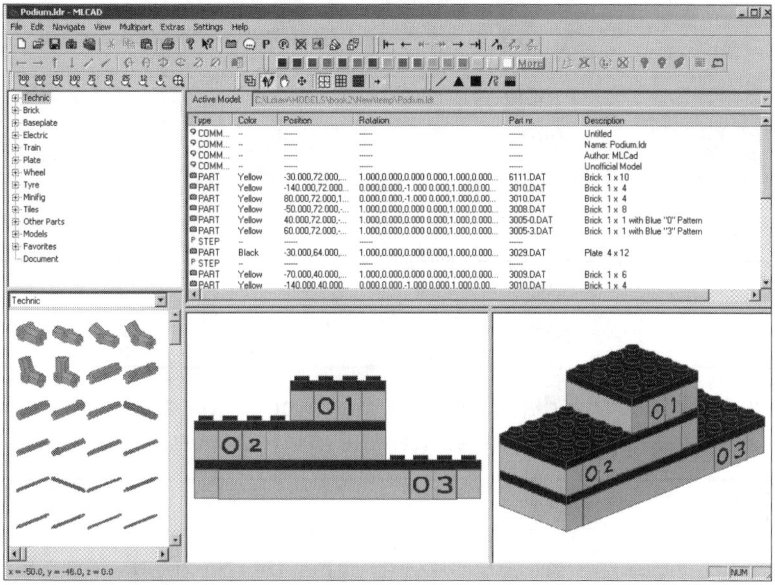

Viewing Instruction Steps

To view the automatically generated step-by-step building instructions, switch again to **View** program mode. You will see a screen similar to the one in Figure 6.23, since the manual step we added and the first automatically added step are identical in this case. In View program mode, browse the instruction steps through the Navigate menu, using the Movementbar buttons shown in Figure 6.27, or the keyboard shortcuts listed in Table 6.1.

Figure 6.27 The Navigation Buttons of the Movementbar

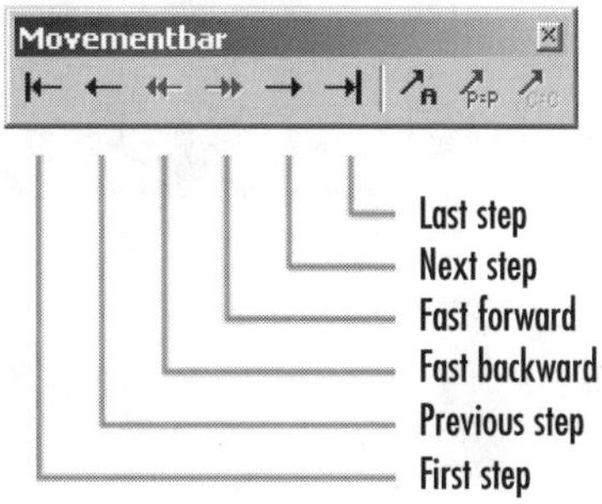

Table 6.1 Navigation Shortcuts

Key	Action
PgUp	Previous
PgDn	Next
Ctrl + PgUp	First
Ctrl + PgDn	Last

NOTE

In View program mode, if we click the modeling panes set to non-3D view angles, we will also advance one step forward.

If you advance to the next step (using either the Next Step button on the Movementbar or the PgDn shortcut), you should see something like the screen shown in Figure 6.28. As shown in the Model Parts List window in Figure 6.26, this step contains only one part, the 4X12 black plate.

Figure 6.28 Step 2 of the Podium

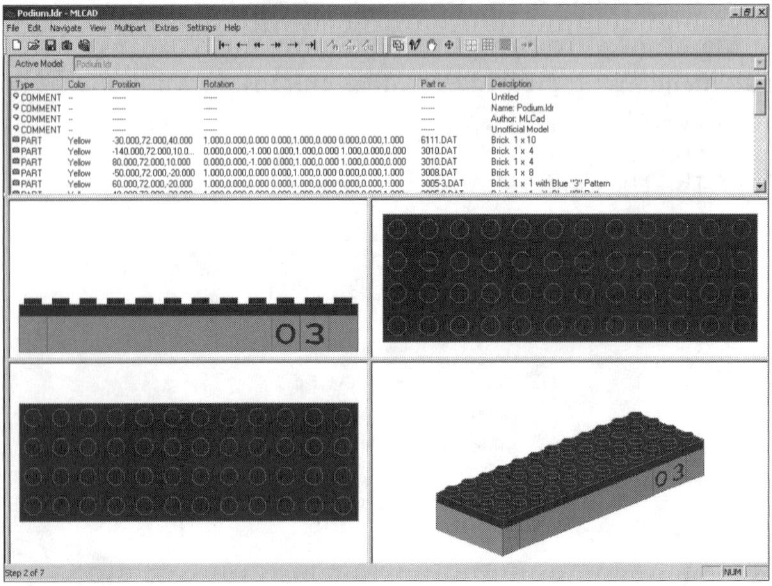

NOTE

Notice how in View program mode, the Statusbar displays the current step number and the total number of steps.

Advance to Step 3. It should look like Figure 6.29.

Figure 6.29 Step 3 of the Podium

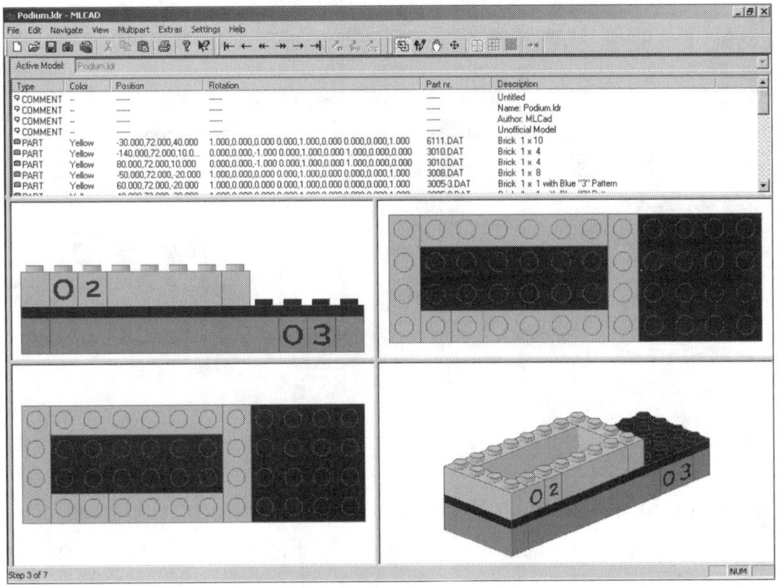

As you can see in Figure 6.29, MLCad automatically makes the parts from previous steps darker, thus highlighting the parts being added in the current step.

You can continue browsing through the rest of the steps, including going back to previous ones. However, if you look at Figures 6.23, 6.28, and 6.29, you will realize that there is no need for three different steps: The parts added in Figures 6.28 (Step 2) and 6.29 (Step 3) could very well be merged into one step. In fact, several of the steps that MLCad creates automatically could be combined in this manner to make for fewer overall instruction steps. Let's take a look at how to do this.

Fine-Tuning Instruction Steps

The first thing we need to do is delete the unnecessary steps. To delete steps, you have to go back to Place program mode; you cannot edit model items in View program mode. Select the steps right after each of the black plates in the Parts List Model window, as shown in Figure 6.30, by pressing **Ctrl** and clicking each STEP statement, then press the **Delete** key on the keyboard or use the **Edit | Erase** menu option.

Figure 6.30 Selected Steps

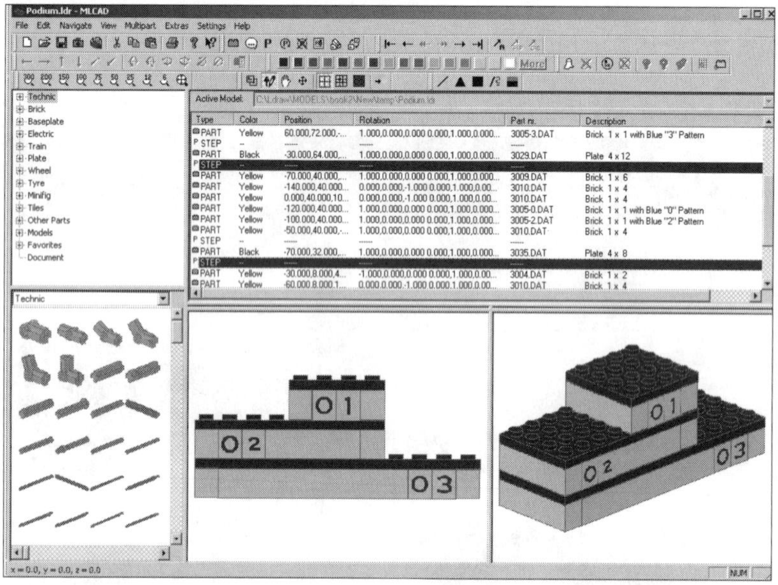

Once you delete the steps and go back to View mode, the instruction sequence will be shorter and more efficient. In Figure 6.31, we compare the sequence MLCad created automatically on the left with the edited sequence we just created on the right.

The example shown in Figure 6.31 pretty much sums up the art and science of creating step-by-step instructions. The set of steps at the right (which we tweaked ourselves) is more compact, but the set of steps at the left (which MLCad created automatically) is a bit easier to follow. Getting the instructions right is as complex as getting the model right!

In general we should aim for the simplest way to present the instruction steps, but many models made out of LEGOs do require intricate couplings between parts. If you get stuck when creating instructions steps for a complex model, the best thing to do is take a look at LEGO's official instruction booklets. They use a long list of handy tricks, including color-coded subassembly steps, multiple-angle views, and the like. Still, the best recipe is always the simplest one:

Avoid using too many parts in one single step, avoid switching between view angles, and throw in an arrow or two if doing so helps you avoid a step or two.

Figure 6.31 Instruction Step Sequences

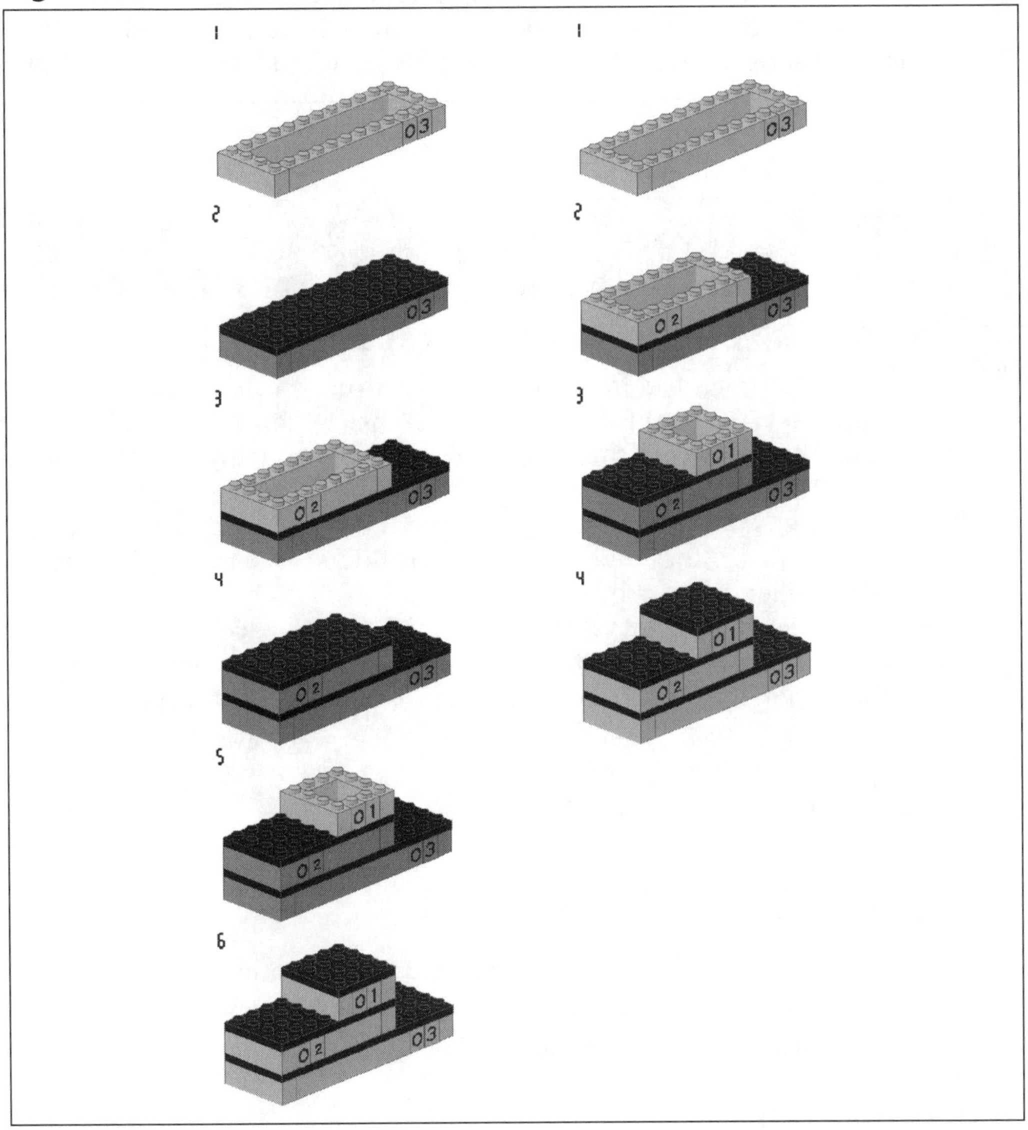

NOTE

Arrows and lines might make it easier for the viewer of the instruction steps but not for the author of the file. See the section on primitives later in this chapter to learn how to add lines to models. You will need to use *buffer exchange* statements, too; they're also covered later in this chapter.

Designing & Planning...

Brickshelf's LEGO Instruction Scans

The best way to get a feel for official LEGO instruction steps (apart from buying official LEGO kits) is to visit the Brickshelf Archive at www.brickshelf.com. This unofficial site offers free and public viewing scans of almost all LEGO instruction booklets, including not only those with instruction steps but also the catalogs. LEGO allows this dissemination of information with the understanding that Brickshelf can only do this for models no longer sold by LEGO.

The archive holds a lot of material, which you can access by year or model reference. Other Web sites, which we talk about later in the book, allow for even more specialized searches (for example, by part number). Use these sites to gain inspiration for your own creations—many LEGO fans have done so for years!

More Instruction Steps Tools

This section presents another five meta commands related to instruction steps that we can add to our models via the Editbar. They all perform very specific tasks, some of which are also supported by LPub.

Rotation Steps

The **Edit | Add | Rotation step** menu option rotates the model (and also acts as a regular step). For instance, it allows us to create the sequence of steps shown in Figure 6.32. Notice how new parts have been added to the model *and* how the model's orientation has changed. We generally use rotation steps to show parts or building sequences that are obscured by other parts of the model.

Figure 6.32 Rotation Step Sequence

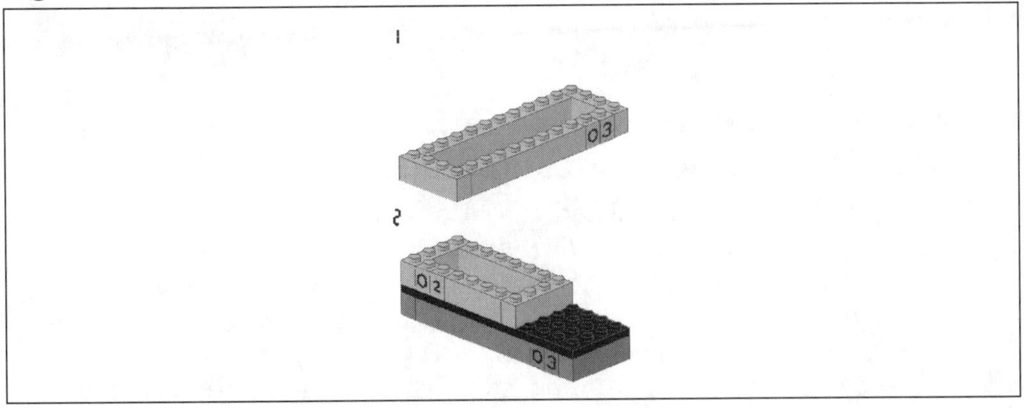

To make rotation steps less confusing, you might want to consider using one step to show the part rotation only and then add the parts in the next step. Following that optional rule, the sequence in Figure 6.33 uses one more step than the one in Figure 6.32, but it is easier to follow.

Figure 6.33 Improved Rotation Step Sequence

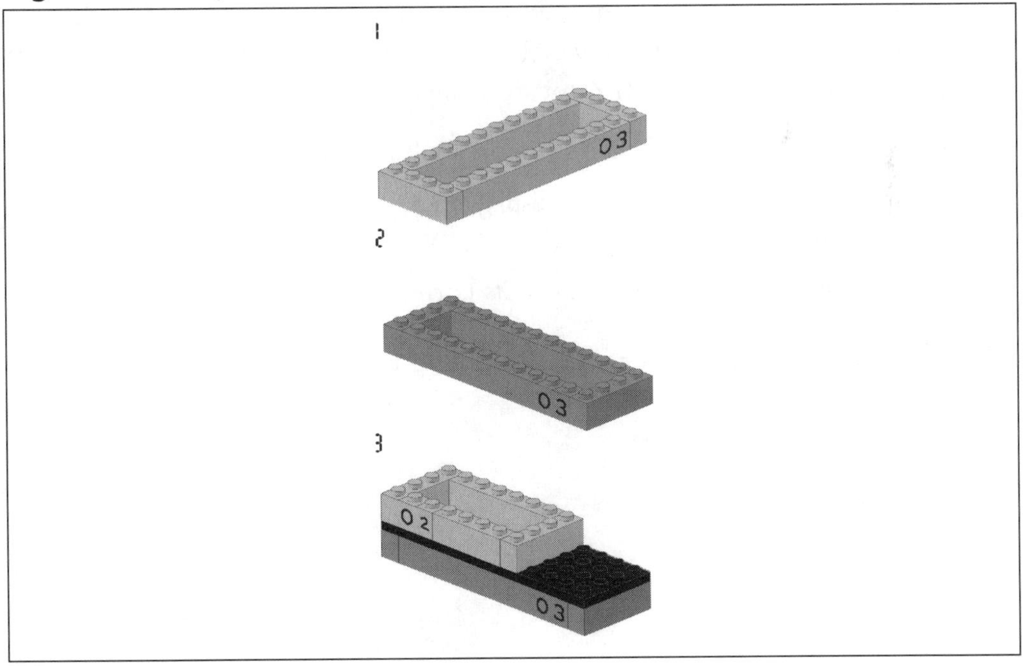

Designing & Planning...

The Importance of the 3D View Angle in Creating Modeling Steps

As we will learn shortly, MLCad can also generate image files of each step and even print them. In View program mode, we have four modeling panes in which to see the model from any view angle we choose, but when we generate these step images and printouts, the program will only render them using the default 3D view angle.

In complex models, rotation steps can become critical and will require all our attention. Whole instruction steps sequences might have to be adapted to reduce the rotations of the model, which are always distracting. In other cases, we might even require the help of some other instruction step meta commands that we will see shortly.

When we insert a rotation step, we activate the Enter Rotation Angles dialog window, shown in Figure 6.34. In it, we can enter rotation angles for the X,Y, and Z boxes at the top. The Relative, Absolute, and Additive options refer to how the program interprets those XYZ rotation angles:

- *Relative* rotations add their rotation angles to the default 3D angles.
- *Absolute* replaces the default display angles with the ones defined by the XYZ rotation angles.
- *Additive* rotates the model from its current display angle.

Figure 6.34 The Enter Rotation Angles Dialog Window

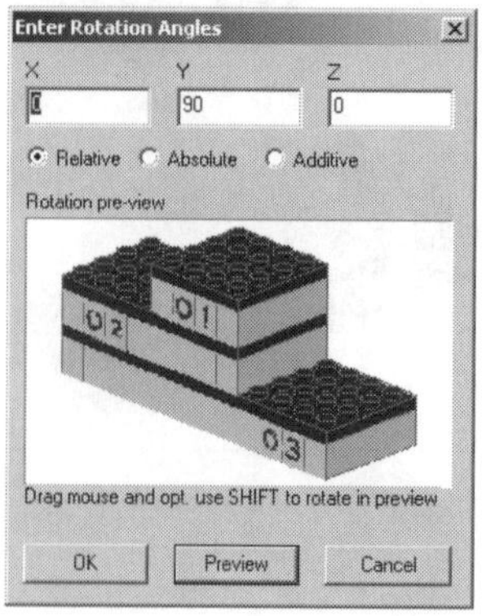

The rotations defined by rotation steps affect all parts listed after the rotation step, until a rotation end (0 ROTEND) or another rotation step meta command is encountered in the file. Rotation end causes the model to be displayed back in the default 3D rotation angles. Additive rotations modify the effects of the most recent rotation step. If no rotation step is in effect, additive rotation calculates the display angle by adding the additive rotation angle to the default 3D display angles.

The **Edit | Add | End rotation step** menu adds a 0 ROTEND meta command that rotates the model back to the default 3D display angles (and acts as a regular step).

Adding a Background

The **Edit | Add | Background** menu option inserts an image file as a background for the building step(s). Once we add a background image file, we are presented with the Background Image dialog window, pictured in Figure 6.35.

Figure 6.35 The Background Image Dialog Window

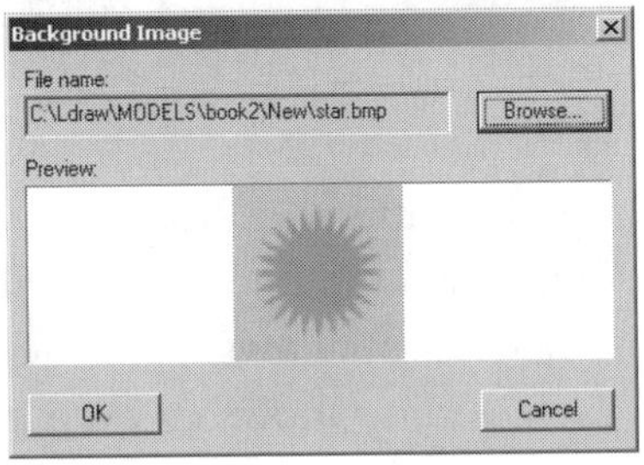

The inserted image appears on the background of the modeling panes only in View program mode. In Figure 6.36, the model has a Background list item at the beginning of the file (visible in the Model Parts List window). This way, the background image is shown in all the modeling steps, unless you specify another image at a later step. Each step can have one different background image.

NOTE

Unlike rotation steps, Background meta commands do not act as steps. The program will continue adding parts until it finds a Step or Rotation Step item.

Figure 6.36 Modeling Panes with Background Image

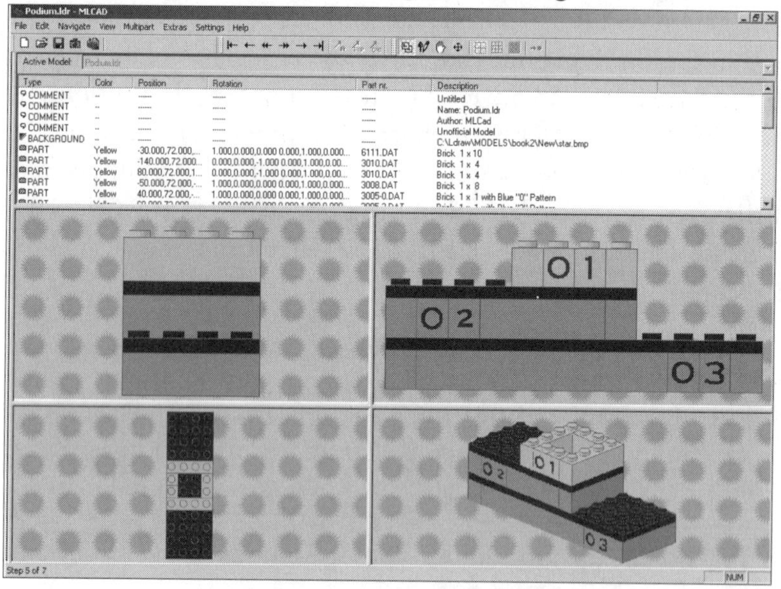

The *Clear* Statement

The Edit | Add | Clear Statement menu option clears the screen of all content from previous steps. This is actually an incredibly neat feature, not necessarily useful only for building instructions. It allows us to use the instructions step commands for other purposes, such as slide shows or even animations. The sequence shown in Figure 6.37 has been stored in a single file. Each step contains one complete minifig image, preceded by a *clear* statement. Thus each step just shows one minifig image (with a different stance in each step).

Designing & Planning…

Using Different Meta Commands with Different Versions of a Model

The animation sequence shown in Figure 6.37 looks like a weird-looking model when it's seen in Place program mode. Since MLCad ignores STEP meta commands in Place program mode, the three minifigs will be merged together into one. The result will be an image of a minifig with

Continued

many different limbs! Meta commands are often used *in context*; we can't have a model with all sorts of meta commands that looks great in all instances. Often several versions of the same model file are used for different things. One file might be perfect to distribute to other fans. Another will render just right in POV. A third one might contain different positions or views of the model and its internal parts. Think of these "versions" as using the same base file to create different minifig champs, as we did in the last chapter to populate the podium.

Figure 6.37 An Animation Sequence Using Clear Statements

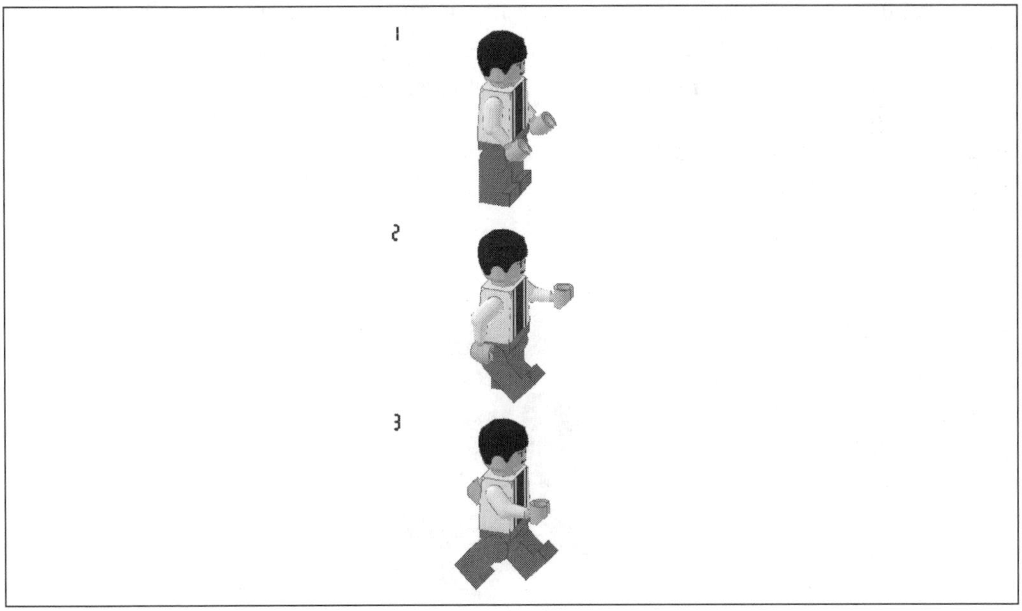

Buffer Exchange

The **Edit | Add | Buffer Exchange** menu option offers a much more sophisticated approach to managing the modeling panes' content while you're in View mode. When we add a *buffer exchange* meta command to a model, we activate the Buffer Exchange dialog window. This dialog window, shown in Figure 6.38, allows us to manage up to eight memory buffers (A–H). These buffers each store the display in the state it is in when it is activated. When you use the Retrieve option, the program redisplays the buffer's content.

Figure 6.38 The Buffer Exchange Dialog Window

This feature allows for all sort of very sophisticated tricks. For instance, in Figure 6.39, the *buffer exchange* meta command is used to insert a driver into a finished car. At the end of Step 1, we capture the seat and plate into Buffer A. In Step 2, we add the rest of the parts of the car and capture the display of the whole car into Buffer B. At the beginning of Step 3, we recall Buffer A (which contained Step 1) and show how to add the minifig driver to it. Finally, in Step 4, we recall the contents of Step 2 and add the minifig driver again. This can also be achieved with *clear* statements, but instead of copying the minifig twice, we would copy the minifig *and* the whole car twice.

Figure 6.39 The Buffer Exchange Sequence

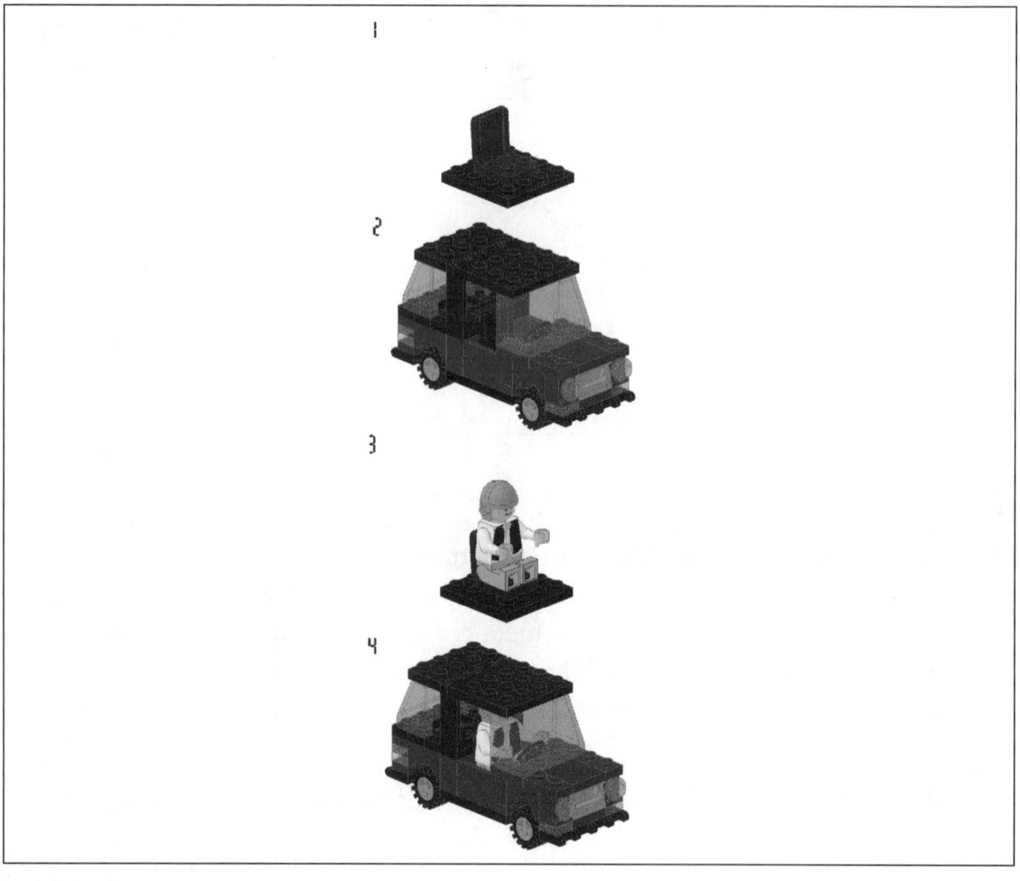

NOTE

In a complete sequence, there would probably be several instruction steps between 1 and 2, showing how the car is built.

Ghosted Parts

Through the use of ghosted parts and multipart project files, we can stretch the possibilities of buffer exchanges even further. We can add instruction steps to both the main model and the submodels. By ghosting a part in a submodel, we prevent it from being shown in the instruction steps for the main model, even though it will still be visible in the instructions for the submodel. This allows us to use *clear* and *buffer exchange* statements in all the models of a multipart project file—which is to say that it allows us to create extremely clear instructions steps and "automated" animation sequences.

Designing & Planning...

Michael Lachmann's LEGO Trains

Generally speaking, all LEGO fans have one or two selected LEGO themes to which they devote most of their LEGO time. In the case of MLCad's creator, Michael Lachmann, the theme seems to be trains. His secondary theme could be considered to be instruction steps. In the official MLCad documentation, Michael advises users to go to his Web site for examples of how to use these advanced instruction step techniques.

Michael's site is indeed worth visiting. First, trains are a great LEGO theme for all ages and interest groups. They combine technical aspects with aesthetic ones, giving modelers of every ilk something to chew on.

But more important in terms of the content of this chapter are some of the transport cars that Michael offers for download as *sample* MPD files. The instructions Michael has added to them showcase techniques closer to the ones official LEGO catalogs display than the typical MLCad model file with some steps added to it. Michael uses lines and arrows, "before and after" images, and other neat tricks to show the

Continued

user how to build his liquid gas "wagons." If after looking them over you have still questions, who better to ask than the guy who wrote the program? Contact Michael via the www.lugnet.com CAD forum.

Draw to Selection: Faking It in Place Program Mode

The meta commands we just looked at are used in creating steps. Although they are added in Place program mode, their effects are only visible in View program mode. This is because when MLCad is in Place program mode, it ignores meta commands related to instruction steps. However, if we ever need to "see" a step in Placemode, we can use a relatively new addition to the MLCad toolset: the **Draw to Selection** button of the Viewbar (see Figure 6.40).

Figure 6.40 The Viewbar's Draw to Selection Button

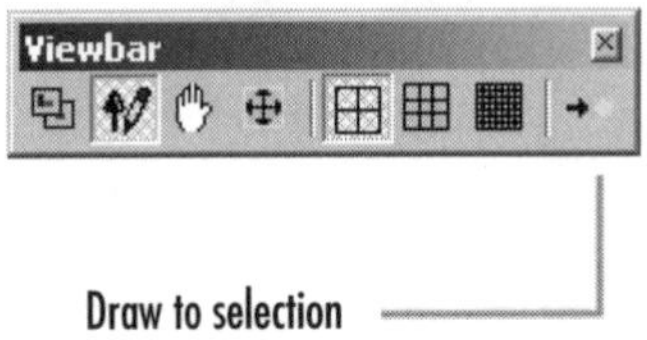

With this function activated in Place program mode, MLCad will draw to the modeling panes only the parts in the model parts list up to the currently selected one. This feature allows us to gauge how our model is progressing without leaving Place program mode. Figure 6.41 shows the podium model with **Draw to selection** activated and the first numbered brick of the second brick tier selected. The modeling panes only render the model up to that part, even though the model contains more parts.

NOTE

Don't forget to deactivate **Draw to Selection** once you are done with it. Leaving it activated drains your computer's resources and can also confuse you!

Figure 6.41 The Podium File with Draw to Selection Activated

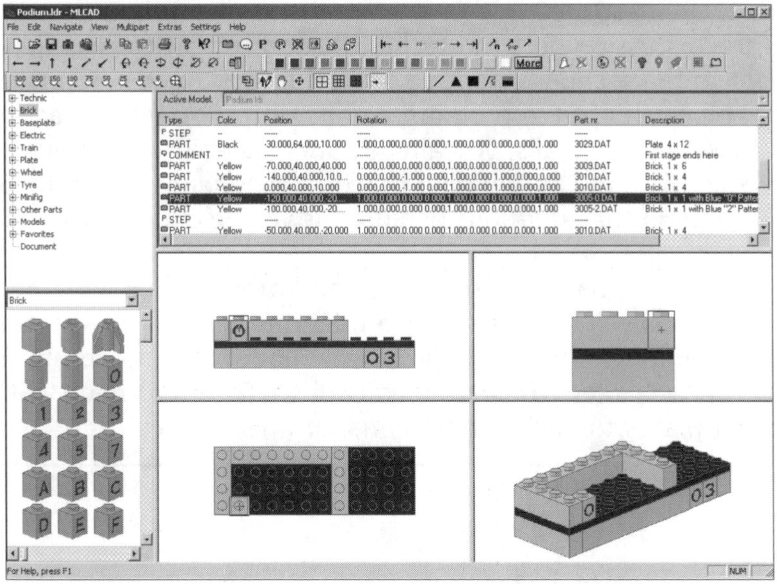

Using Non-LEGO 3D Elements

This is the last section of this chapter actually dedicated to "building" techniques. First we cover line items of type 2 and above. These are called *primitives* because they are basic (primitive) geometric elements. The more complex and sophisticated part and model files, composed of many primitive elements, contain literally thousands of these primitives.

A file formatted according to the LDraw file standard can hold models made out of library parts *and* primitives. The primitive elements can then be used for all sorts of functions—for example, adding lines (or arrows) to the models' instruction steps.

We then proceed to look at other primitives (or *meta primitives*) saved as files and used in developing parts. We end this section with an in-depth exploration of rotation matrices, a particular and essential mathematical aspect of the way part orientations and scales are defined inside the files.

NOTE

This is, to a large extent, an optional section of the book. In principle, you owe it to yourself to know as much as possible about the tools you use. On the other hand, the elements and techniques that we cover in this section are highly specialized. You will not need them at all to use the software described in the other chapters of this book.

If the discussion gets too technical and abstract for you, jump straight to the section "Adapting MLCad to Our Needs" to make sure you know how to adapt the program and its output to your needs. Other than these minor details, you already are a full-blown MLCad modeler. Your skills will grow as you gain practical experience using the program; understanding the equations behind it is very useful but by no means essential.

On the other hand, programmers and other similarly mathematically inclined minds will find here only a brief introduction to some of the methods and techniques used in LDraw models *not* based on LEGO parts. The Web sites we mentioned earlier (LDraw.org and LUGNET.com) are great places to find more material on this subject.

Working with Primitives

As you have probably realized by now, LDraw's and MLCad's file formats are quite flexible: Not only do they hold models made of the LEGO-like parts from the library; they also allow models to be used as parts in other models in a variety of ways, and they include some pretty sophisticated instruction step commands. In fact, the file format also offers support for 3D models *not* based on LEGO parts.

As we mentioned earlier, the parts in the library are standard LDraw model files. However, they are not made of LEGO-like parts. Instead, they are made of *primitives*. Primitives are basic geometric shapes such as lines and triangles that can be used to create *any* 3D model, not just those built using existing LEGO elements.

Primitives can also be used for other purposes. One of these purposes is to clarify building steps. In the sequence pictured in Figure 6.42, the line in Step 2 is used to show how to place the plate.

To use this kind of element in your instruction steps correctly, you must be familiar with the usage of *clear* and *buffer exchange* meta commands. Line primitives are just another type of file item (type 2). In a model file, they occupy a line, just like every other item. They behave more like parts than like meta commands. Primitives are also "tangible" objects rather than "triggers" for functions.

Figure 6.42 Using Line Primitives in Instruction Steps

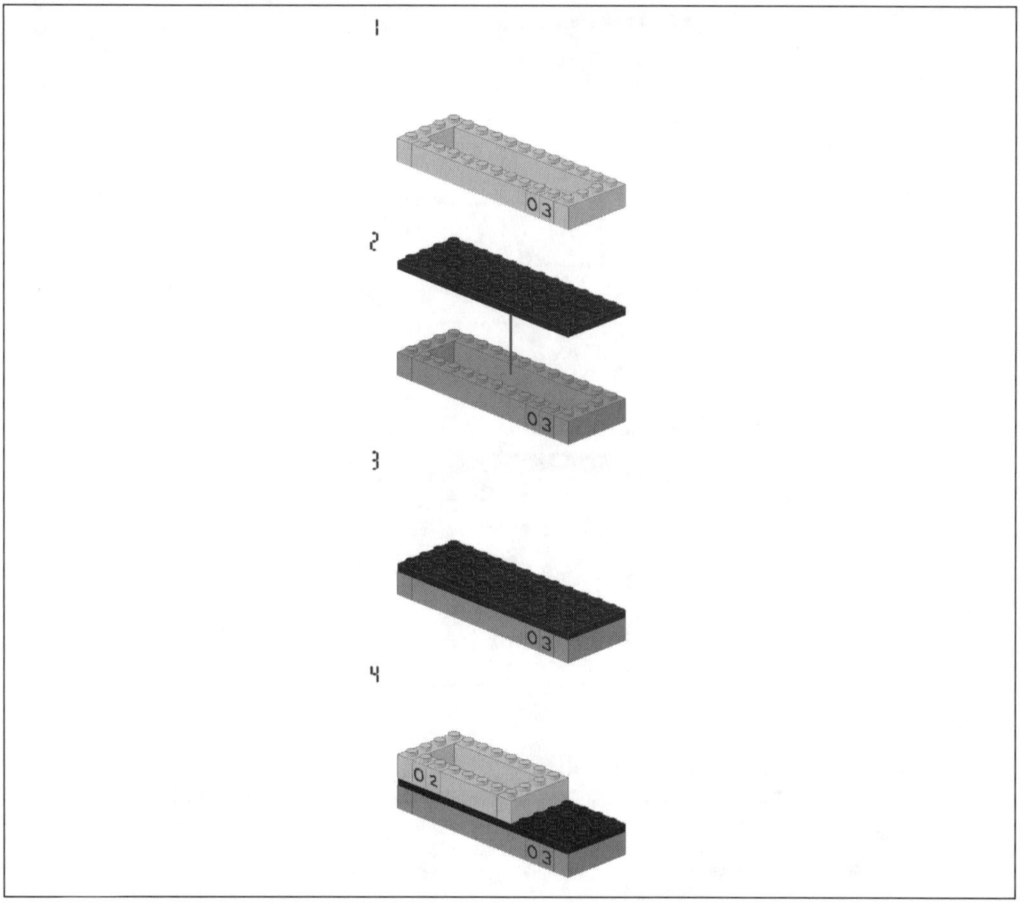

Adding a Line Primitive to a Model

To add primitives such as lines to a model, we use the Edit | Add | Primitive command options. They include Line, Triangle, Quad, Optional Line, and BFC Statement. These commands are also available via the buttons on the Expertbar, shown in Figure 6.43.

If we insert a line primitive into the model, we will be presented with the Add Line dialog window, pictured in Figure 6.44.

Any line can be plotted in 3D space using only two sets of XYZ coordinates. Enter those coordinates in the boxes of the dialog box and click **OK**. To create the line shown in Figure 6.45, we used coordinates 0,0,0 and 0,50,0.

Figure 6.43 The Expertbar

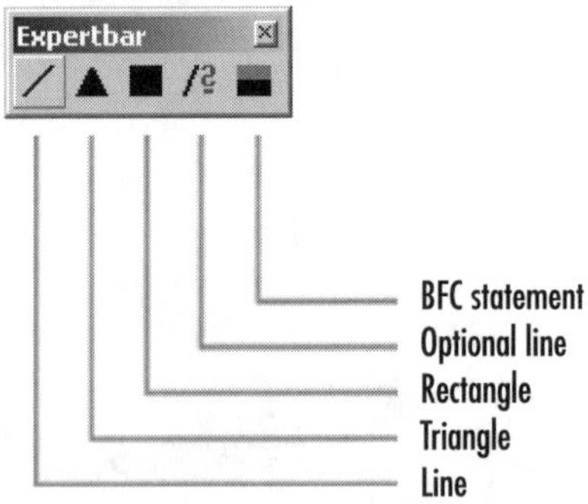

Figure 6.44 The Add Line Dialog Window

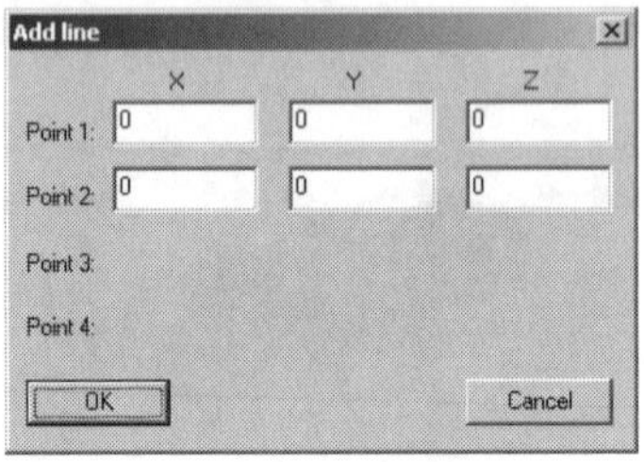

Figure 6.45 A Line Item in MLCad

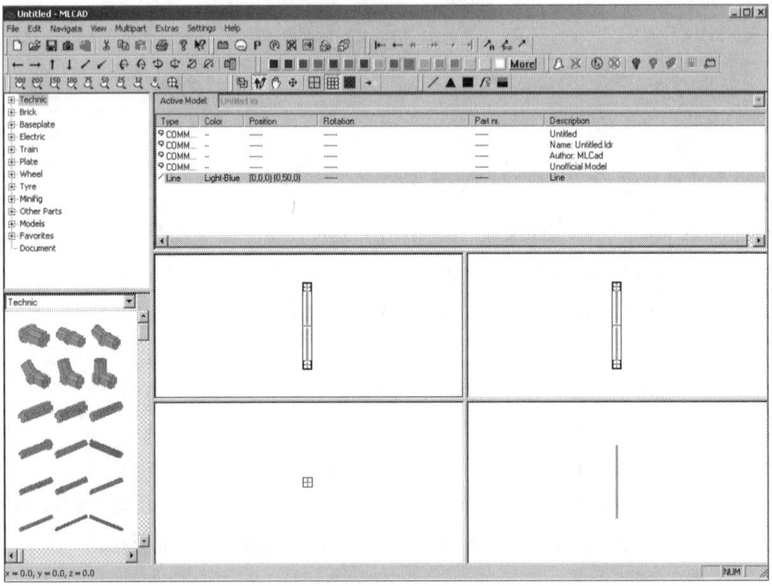

When a line item is selected, we can use the modeling functions with it. In fact, there are also some other ways we can model it. For instance, while we can move the line item in the modeling ports using the same methods as we use to move a regular part, we can also alter the line's shape by clicking and dragging with the mouse on any of its two end points (called *vertices*). In Figure 6.46, we have converted the line from vertical to oblique.

Figure 6.46 Editing Lines Using the Mouse

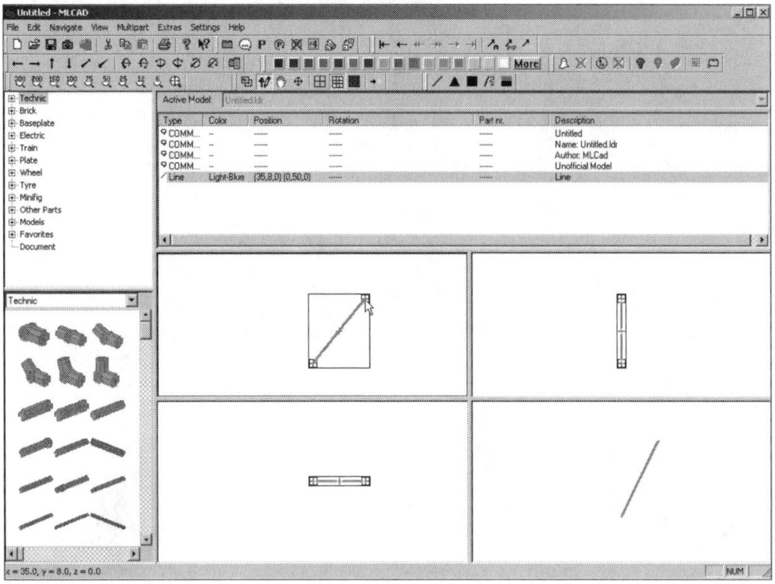

The Edit | Add | Primitive Triangle and Quad commands behave similarly, only instead of lines, they create surfaces. Conditional Line and BFC Statements are very specialized commands useful in very specific circumstances beyond the scope of this book.

Creating Custom Parts

Why create custom parts? It is generally not a question of *why* but of *when*. Once modelers become proficient with MLCad and other modeling tools, there might come a time when they need a specific LEGO part that has not yet been added to the library. Since the task of creating a new part for the library is much more complex than creating models out of virtual LEGO, this need by itself is generally not enough to motivate a modeler. But if we are extremely comfortable building models with MLCad and know of a part that seems to need only some minor modifying to convert it into the one we are looking for, we might bite the hook and go for it.

The Parts Library files are regular LDraw files, and thus we can open a part file in MLCad as though it were a model. In Figure 6.47, one of the simplest parts from the library (3070.DAT / Tile 1 x 1) has been loaded in such a way into MLCad.

Figure 6.47 Tile 1 x 1 Loaded into MLCad

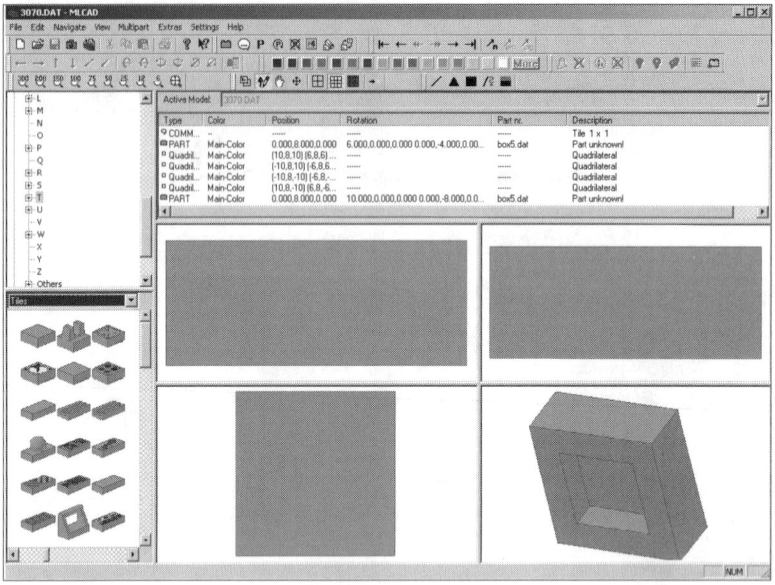

This is a part (or a model) that experienced MLCad modelers could put together in a few hours using Line and Quad elements, even if they had not used the primitive items before. However, notice in the Model Parts List window how the model is using Quads … but also submodels such as box5.DAT!

Why use submodels for basic parts? The answer is, just as MLCad allows us to use both individual parts and submodels as parts for efficiency and flexibility reasons, the LDraw file format allows us to build submodels made of primitive items for the same reasons. The files for these primitive submodels, some of which are shown in Figure 6.48, are stored in the LDraw\P directory.

Knowing when to use particular methods is what creating any model boils down to—but the jump from building computer models based on LEGO parts to building computer files for those parts is the same, if not wider, than going from constructing things out of LEGO to building them with other materials in real life. Not only do the rules change dramatically, but most of our cozy references are gone and our frequent mistakes will no longer be as easy to spot—or as easy to fix!.

Figure 6.48 Some Primitive Submodels

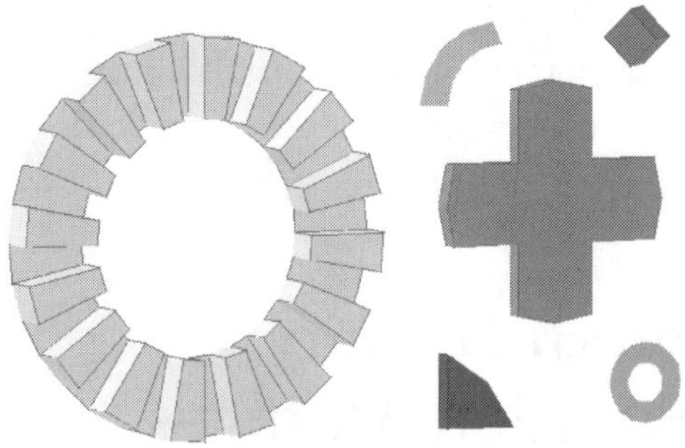

Designing & Planning…

Creating Official Parts for the Library

As LEGO adds new parts to its official kits, they are incorporated into the library as users model them according to LDraw standards. If you can't find a part in the library and you are sure that it is produced by LEGO in real life, the first thing you should do is make sure you have the latest (and most complete) release of the library installed. New releases come out regularly, incorporating new and updated parts. They can be found at the LDraw.org Web site.

If the part has not yet been officially included in the library, visit the parts tracker in the LDraw.org Web site. The Parts Tracker lists the ongoing new parts projects and their status. Often, developers release unofficial versions of parts that are not yet finalized. If the part is not in this list, you can inquire about its existence in the CAD forums of LUGNET.

If you decide to create the part yourself, besides gathering documentation and support from the sources listed, you should also know that to include the part in an official library release, you have to submit it to a peer-review process, also based in LDraw.org.

Explaining in a practical way how to build 3D models with non-LEGO elements would take another book at least as thick as this one. We don't mean to imply that creating them is not as fun and engaging as building LEGO-based models, but it is certainly not nearly as easy. A user whose only experience in 3D computer graphics is via MLCad would do well in exploring fully all the other LEGO-related software available on the Internet and maybe even doing some 3D modeling in other environments before attempting to create custom parts. This is especially true if you want to create parts that you want to become official additions to the library.

An Explanation of Location Vectors and Rotation Matrices

The rotation matrix is a key mathematical element of the LDraw file format. In Chapter 5, we bumped into it in the Enter Position & Orientation dialog window. Expressing the rotation of a part using a rotation vector is much easier than using a rotation matrix. Explaining how to use rotation vectors is also much easier than explaining how to use a rotation matrix. The rest of this section describes how to calculate and use rotation matrices. We get pretty heavily into advanced mathematics that we attempt to explain as simply as possible. Skim over or ignore the parts you do not understand, or skip this section entirely, and find comfort in the fact that you can always just use a rotation vector instead of a rotation matrix for performing complex rotations in MLCad.

You are probably asking yourself, "If rotation matrices are so hard to use, why does LDraw use them?" The answer is that the matrices can be used to make parts larger and smaller as well as for rotation. We'll see how this works after we get through the nasty math part.

The LDraw file format describes a 3D world, so it takes three numbers (X,Y, and Z) to describe a given location. In advanced mathematics this list of three numbers is called a *vector*. Thinking of a location as single thing (in other words, a vector with three parts, x,y,z) rather than three things (x, y, and z) can make it much easier to write things down mathematically.

We often use the vector concept to give directions to someone who is familiar with a town. An example name for a vector might be *the public library*" a location that many people know as a reference point. For people who don't know where the library is, we must explain the contents of the vector by saying "The library is at the corner of Broadway and Grove Street." Broadway and Grove Street can each be considered a number in a two-element vector, *the public library*.

In math, you can add and subtract vectors, which is to say, add the *x* components of each vector together, then the *y* components, and then the *z* components. Continuing our real-world example, we can describe the location of the post office as "Two blocks south of the public library," where *two blocks south* is one vector and *the public library* is the other vector.

The LDraw file format also describes how each part is oriented. This description includes the rotations about the X,Y, and Z axes that are needed to orient our parts correctly with respect to each other. Rather than describing the rotations of a given part as three numbers (rotation angle about the X,Y, and Z axes), the LDraw format describes the rotation amounts using nine numbers. The nine numbers are organized in a three-by-three configuration, called a *matrix*. The matrix is three numbers high and three numbers wide because LDraw describes a 3D space.

Here is the mathematical description for a rotation matrix *R*:

```
R = ( r11, r12, r13,
      r21, r22, r23,
      r31, r32, r33)
```

The next question is, what are the values for *r11* through *r33*? The answers lie in a mathematical topic called *trigonometry*, or the study of angles. We fill in the rotation matrix with sines and cosines of the rotation angles in our rotation vector. To fill in the rotation matrix, we need a sine-and-cosine pair for rotation about each axis (X,Y, and Z). The sine and cosine are a measure of how long a line looks.

Figure 6.49 shows three Technic axles that help us visualize sine and cosine.

The horizontal transparent axle represents the maximum value for cosine. The vertical transparent axle represents the maximum value for sine. The black axle is the one of interest. In the upper-left view pane, we can see the black axle at an angle, so it is easy to see the axle's *true length*. In the upper-right view pane, we can see how tall the rotated axle looks. This is the axle's *apparent height*. In the lower-left view pane, we can see how long the rotated axle looks. This is the axle's *apparent length*.

The cosine is defined as the *apparent length* divided by the *true length*, which is a number that can range from -1 to 1. The sine is defined as the *apparent height* divided by the *true length*, which can also range from -1 to 1. In trigonometry, the *true length* is called the *radius*. Trigonometry lets us change an angle to sine/cosine and back.

Figure 6.49 Three Technic Axles

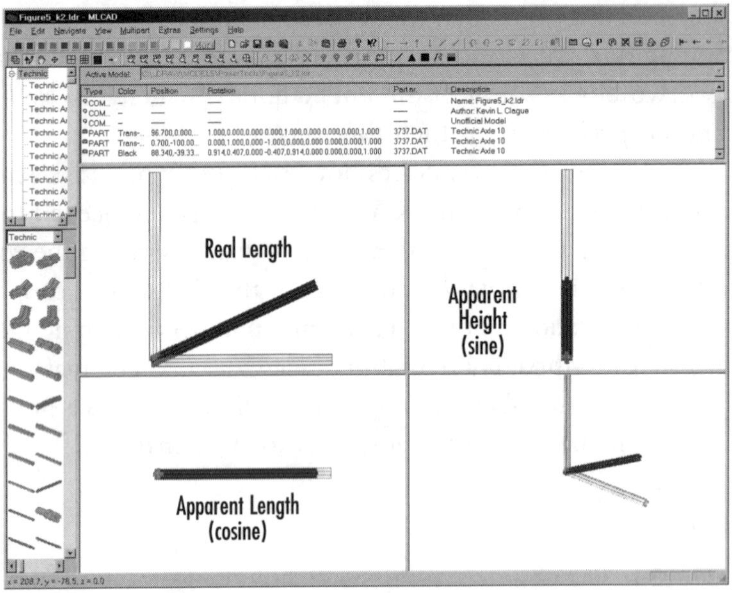

The rotation matrix used in the LDraw file format is filled with sines and cosines of the rotation angles about the X, Y, and Z axes. Here is how these sines and cosines are filled into the rotation matrix for rotation angles *xa*, *ya*, and *za*:

R = (cos(ya)*cos(za), -sin(za), sin(ya),
sin(za), cos(xa)*cos(za), -sin(xa),
-sin(ya), sin(xa), cos(xa)*cos(ya))

To rotate a location using a rotation matrix, we multiply the original location vector (described by *x*, *y*, and *z*) by the rotation matrix and get a new location vector (described by *X*, *Y*, and *Z*). Here is the full-blown equation to do this:

X = x*cos(ya)*xos(za) + y*-sin(za) + z*sin(ya)
Y = x*sin(za) + y*cos(xa)*cos(za) + z*-sin(xa)
Z = x*-sin(ya) + y*sin(xa) + z*cos(xa)*cos(ya)

If you are like we were in math class when we were at school, about now your eyes are crossed, the room is getting blurry, and suddenly that freckle on the back of your hand is very interesting. Having said that, let's try to bring this discussion back to something easy to grasp.

Figure 6.49 shows three examples of calculated rotation matrices, one for each axle used in the example. You can see the rotation matrices for these three axles in the Rotation column of the Model Parts List window. The first axle listed is horizontal and transparent white. This axle is unrotated and therefore has a rotation matrix of (1,0,0, 0,1,0, 0,0,1).

The second axle is transparent and vertical. In the LDraw part definition, the axle looks horizontal, so we rotated it to look vertical. MLCad figured out the rotation matrix for this axle as (0, 1, 0, -1, 0, 0, 0, 0, 1).

The back axle is last, and it's rotation matrix is (0.914, 0.407,0.0, -0.407, 0.914,0, 0, 0, 1).

Making Parts Smaller or Larger

One wonderful aspect of the LDraw file format is that it can be used to describe large, complicated LEGO designs. All the LEGO parts in the LDraw parts library are described in LDraw format using things called *part primitives*. The LDraw rotation matrices can also be used to make parts bigger or smaller (a process called *scaling* in mathematics). These scaling capabilities are most often used to create LDraw parts out of LDraw primitives.

To make a scaling/rotation matrix, we multiply a scaling matrix by a rotation matrix. The definition of a scaling matrix is:

```
S = (SX, 0, 0,
     0, SY, 0,
     0,  0, SZ)
```

Here *SX* is the scaling factor for the *X* dimension, *SY* is the scaling factor for the *Y* dimension, and *SZ* is the scaling factor for the *SZ* dimension. Setting *SX*, *SY*, and *SZ* to 1 performs no scaling. Setting a scaling factor greater than 1 makes parts bigger. Setting scaling factors greater than 0 but less than 1 makes parts smaller.

Since we know that *R* and *S* are matrices, we can describe the scaling rotation matrix *r* like this:

```
r = R * S
```

This simple formula hides a great deal of complexity because *r*, *R*, and *S* are matrices. This is what makes mathematics so powerful, yet sometimes hard to understand.

Expanding *r*, *R*, and *S* into their matrices, we get:

```
r(r11, r12, r13,      R(R11, R12, R13,     S(S11, S12, S13,
  r21, r22, r23,   =    R21, R22, R23,  *    S21, S22, S23,
  r31, r32, r33)        R31, R32, R33)       S31, S32, S33)
```

Now to calculate the matrix *r*, we need to perform calculations for each element of matrix *r*:

r11 = R11*S11 + R12*S21 + R13*S31

r12 = R11*S12 + R12*S22 + R13*S32

r13 = R11*S13 + R12*S23 + R13*S33

r21 = R21*S11 + R21*S21 + R23*S31

r22 = R21*S12 + R22*S22 + R23*S32

r33 = R21*S13 + R22*S23 + R23*S33

r31 = R31*S11 + R31*S21 + R33*S31

r32 = R31*S12 + R32*S22 + R33*S32

r33 = R31*S13 + R32*S23 + R33*S33

Wow! That is a lot of calculations. Thank goodness for computers. After your vision has cleared, take a look at Figure 6.50, which shows the primitive that is used to define Technic axles.

Figure 6.50 The Primitive Used to Define Technic Axles

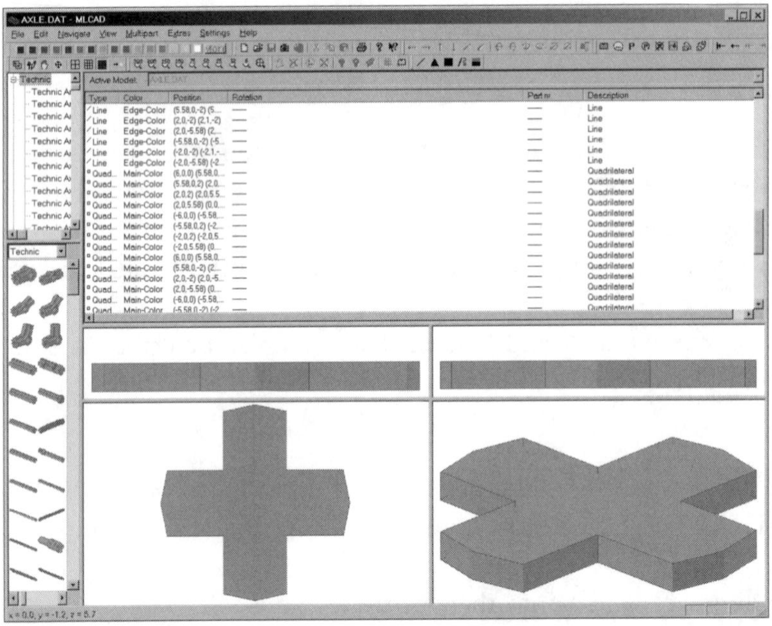

The Model Parts List window shows that the axle primitive is made up of lines and rectangles (called *quadrilaterals*). Figure 6.51 shows the definition for a number 10 axle. Notice that it contains only one part, with a scaling rotation matrix that uses an *X* scaling factor of 200 on axle.dat. In short, this tool is powerful and simple to use but complex to understand.

Figure 6.51 The Definition for a Number 10 Axle

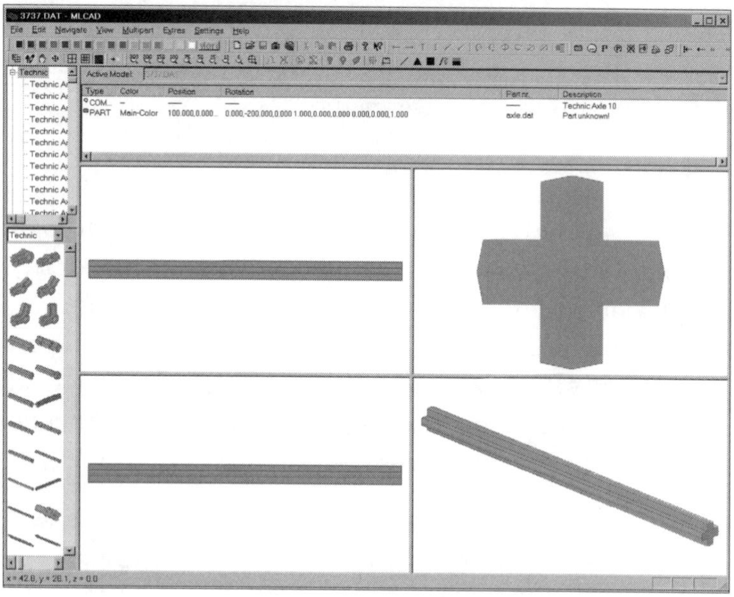

Adapting MLCad to Our Needs

This last section dedicated to MLCad shows you how to customize both the output of the MLCad program and many of its internal functions.

Extracting Data from MLCad

As you will soon learn, there is plenty of life for your LDraw files beyond MLCad. Other programs will add elements to your models or translate them into other formats used for photorealistic rendering, to name just a few of the options. Most of the LDraw-related software reads LDraw files directly, so there is no need to translate the files to other file formats (such as those used by professional 3D software) for them to remain useful. However, since a file can contain all sorts of information, from model data to animation steps, we can also ask the program to "filter" a file to remove all unnecessary information for a given task. We do this by *exporting* the resulting file under a different name.

Clearly, now that you know about the advantages and general structure of the LDraw file format (or MLCad's variants), it becomes obvious why it is such an excellent way to save model data created in MLCad. If you know what the model looks like and how it is stored, you will have plenty of knowledge to make the best decisions as to how to use the file, whether with other programs, releasing it as a model with built-in instruction steps, or something else. However, precisely because MLCad offers such wide support for the format, there are several other ways to extract model data out of MLCad apart from saving model files. Let's take a look at some of these methods.

Exporting Files

The **File | Export** menu function opens the Export Model dialog window. The Export Configuration tab, shown in Figure 6.52, allows us to extract selected sections of the model and save them as regular LDraw files. For instance, we can save the submodels in a multipart project model as individual files. In regular files with instruction steps, we can save *each step* (or a range of them) as an individual LDraw file. The **Ignore rotation steps** option causes files to be ignored when you're generating files from individual steps.

Figure 6.52 The Export Model Dialog Window

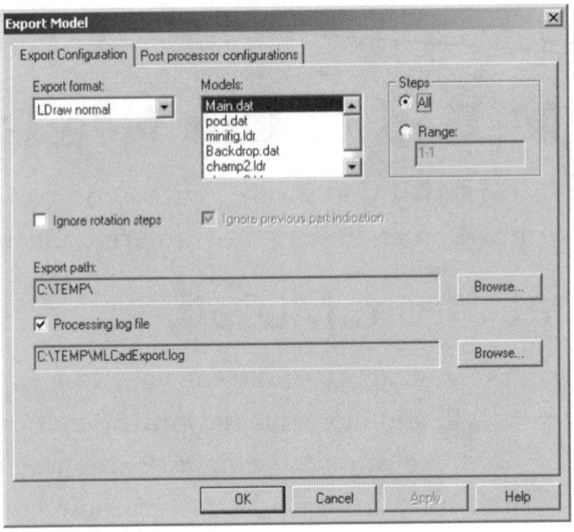

The Export Format option is also an interesting feature. We can direct MLCad to export the data as regular part-based files, or we can ask it to do a second pass and dissect the parts into their primitive elements. As you already know, both formats can be displayed and edited with MLCad, but this export feature also allows developers to create programs that read LDraw files without necessarily referencing the parts in the library. This could allow us to use our models with other non-LEGO-based 3D applications.

The Export Path field allows us choose the output directory where the file or files will be placed. Beneath it, we can also ask for a processing log file (useful when something goes wrong or for software developers).

A second tab in the Export Model dialog window, called Post Processor Configurations, is shown in Figure 6.53.

Figure 6.53 The Post Processor Configurations Tab

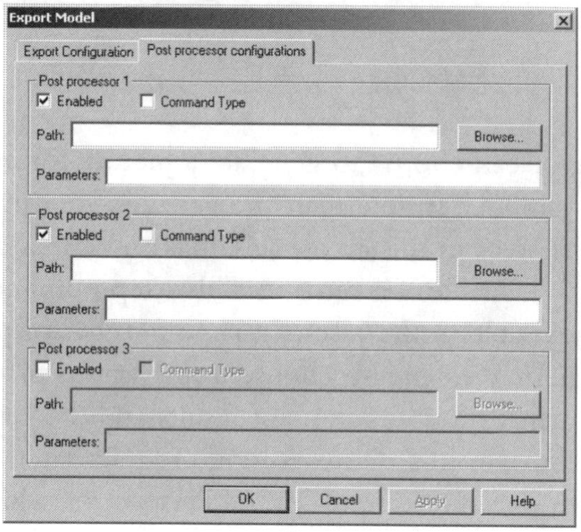

This tab allows us to manage the processes to which we will submit the file after it has left MLCad. For instance, we could instruct it to save (export) our model and run it through a program such as LSynth. We can specify up to three postprocessors (LSynth, L3P, and POV, for instance). For each, we can also include line commands (or *parameters*). Finally, the Command Type option lets us use MS-DOS programs. Uncheck this option for native Win32 applications.

This might not be the most practical way to perform these operations for one single file. However, when using multipart project files, it becomes a very useful tool for batch processes, where the computer crunches away the numbers while the user goes on with his or her life.

Generating Pictures

Apart from viewing the instruction steps using our computer (for example, via the View program mode in MLCad), we can also generate image files. This is done via the **File | Save Pictures** menu option, which opens the Save Picture Options dialog window pictured in Figure 6.54.

Figure 6.54 The Save Picture Options Dialog Window

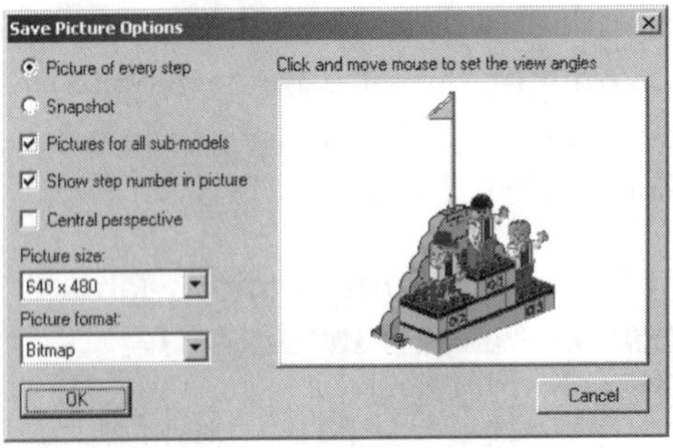

In this window, we can choose to generate a picture for every step or a snapshot of the complete model. In multipart project files, we can also generate an image for each of the steps in the submodels, making *clear* and *buffer exchange* statements even more valuable. We can also add a step number in every picture, which is probably advisable in the case of a large number of pictures, even if we later have to erase them. We can also choose from various picture sizes and image file formats. BMP format produces the best results and the largest files; use it if you plan to edit the images in a photo-editing program later.

One recently introduced feature is Central Perspective, which renders the models in non-isometric perspective. In Figure 6.55, the podium has been rendered in isometric perspective at left and with central perspective at right. In isometric perspective, all projected lines are parallel (in other words, the lines at the front and the back of the podium are parallel). In central perspective, the lines converge to one point. This produces more dramatic effects, especially at certain angles. For Figure 6.55, we have used a low X angle to remain close to the ground.

Figure 6.55 Isometric and Central Perspectives

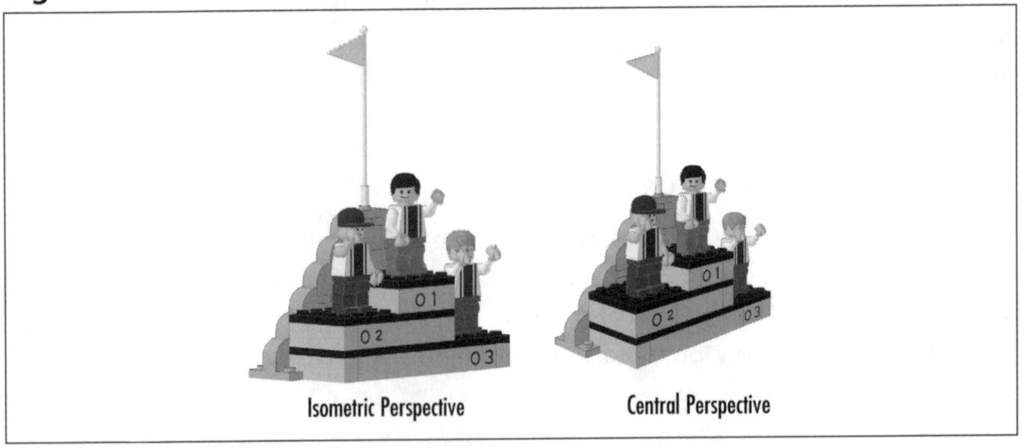

NOTE

You probably do not want to use central perspective with instruction steps, since the "coolness" factor should come second to clarity.

The **File | Save Part List** menu option allows us to inspect the contents of our model from a purely parts quantity perspective. Through the dialog window pictured in Figure 6.56, we will immediately be able to know how many total parts the current model has, listed by type and color. We can also sort the part lists by different criteria. More important, this data can be saved as a text file.

Figure 6.56 The Part Lists Dialog Window

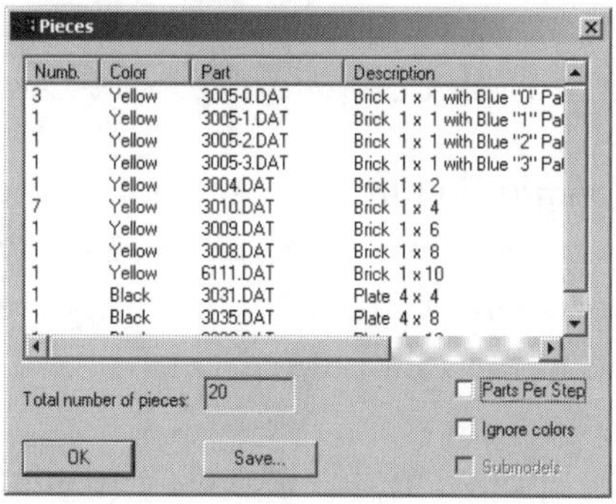

This dialog window is also available using **Extras | Reports | Pieces**. The **Extras | Reports | Dimensions** menu option gives us the dimensions of the model in studs, centimeters, and inches. **Extras | Reports | Comments** simply lists the comments, which can be useful to locate a specific comment when you're editing large documented files.

Using the print options of the File menu, we can also output our models and instructions step directly to paper through the print commands. They allow us to specify printer and page options, and there is a Print Preview menu option. However, we tell the program what to print via the Settings menu, which we cover in the next section.

Customizing MLCad

This last section shows you how to change some of MLCad's default program settings. To access the customization options, use the **Settings | General |**

Change menu option. This activates the MLCad Options dialog window, pictured in Figure 6.57. This dialog window has six tabs through which we can change the program's default settings. The **Settings | General | Default** menu option restores the default settings.

Figure 6.57 MLCad Options

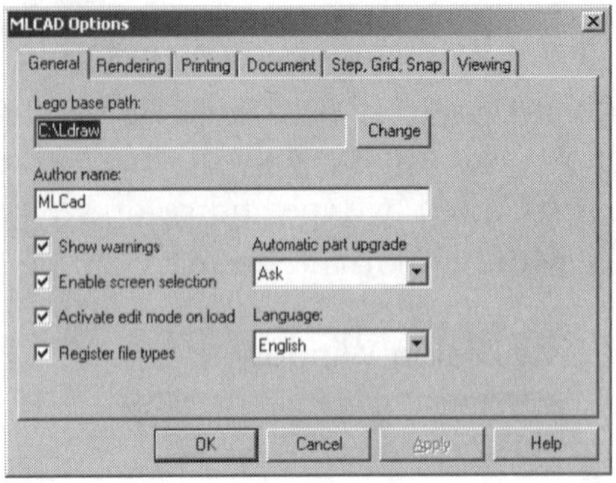

The General tab, shown in Figure 6.57, sets some of the program's administrative functions: the directory where MLCad and LDraw are installed, whether we want to see general program warnings, which language to display the program menus in, and whether Windows should register MLCad files. These are all pretty standard options.

However, three other options in this tab relate to some of MLCad's specific functions. Via the **Author name** option, we can instruct the program to insert our name (or whatever we type in this field) in every new file we create. It's well worth the minimal effort of entering our ID here and having it tagged to our files automatically. Using the **Automatic part upgrade** option, we can instruct the program to check to see if the model being loaded is using the latest version of the library parts that it needs. Some part files receive further treatment and are released with new library updates, so this is always a handy tool to keep our models "sanitized" with the most up-to-date part files. Finally, if **Activate edit mode on load** is turned on, our models will load directly in Place program mode instead of View program mode, which is the default.

The MLCad Options Rendering tab, pictured in Figure 6.58, allows control over the modeling pane settings. **Shading** uses hues of the color to which the part is set to give the model an impression of volume. Deactivating it makes MLCad render our parts in flat colors (perhaps desirable for certain effects). **Optimisations** allows us to take full advantage of our computer's display resources while using MLCad. **Line width** sets the width of all the lines in the program, including those that define the parts. We can also select the definition used in the studs, which, being numerous in most models, tend to eat up computer power. We can also activate the option **Draw to selected parts only**, which we saw at the end of the building steps section. Finally, through the **Preview options** we can set the color of the background of the view ports and the color used in the Parts Preview library window, as well as choosing between having small or large preview images for that window.

Figure 6.58 MLCad Rendering Options

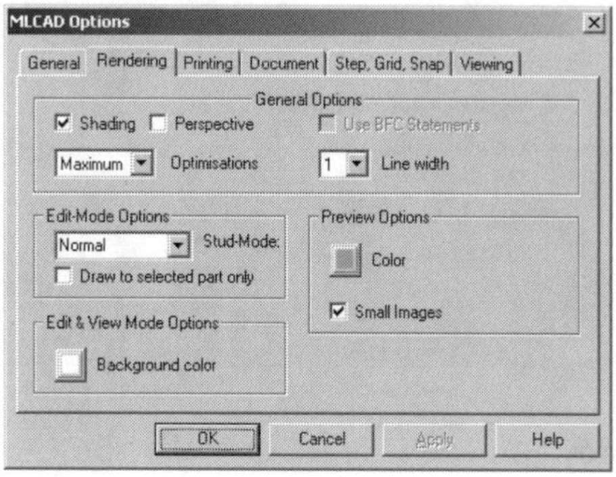

The Printing MLCad options tab, shown in Figure 6.59, offers control over the content and format of print jobs. Through it we tell the program what and how to print. We can set whether to print the finished model or the instruction steps; whether to print one or two steps per page; how to format the parts lists, and whether printouts of multipart project files should include the submodels as well. Finally, we can set the format of the page and whether to include file information on it and where.

Figure 6.59 MLCad Printing Options

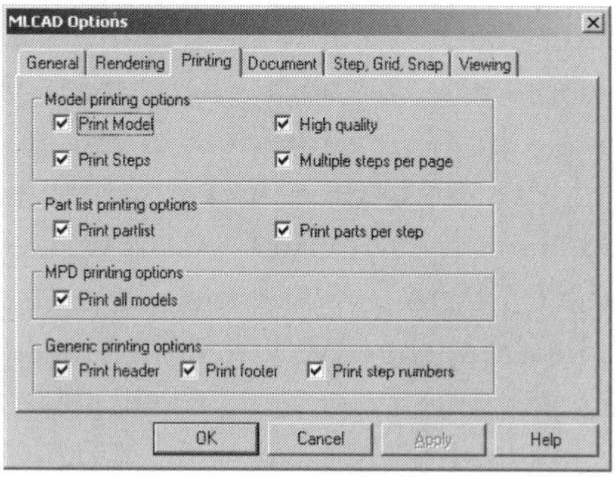

The Document tab in the MLCad Options window, shown in Figure 6.60, allows us to set four fairly important program options. The three boxes at top set the default 3D rotation angle. This angle applies to the modeling panes when set to 3D view angle and to the pictures and printouts of models that we generate. This default angle should also be taken into account for rotation steps.

Figure 6.60 MLCad Document Options

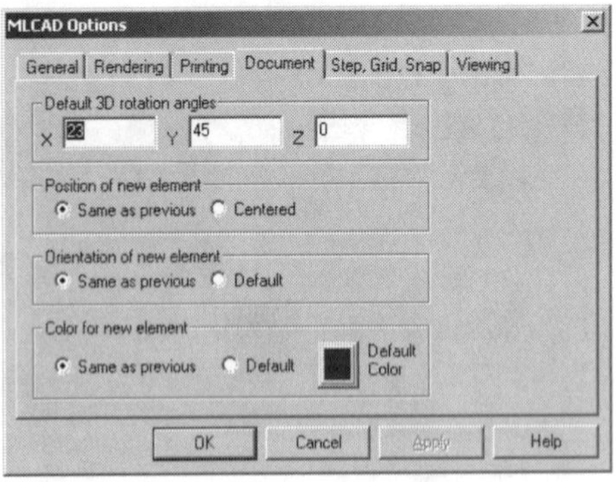

The **Position of new element** and especially the **Orientation of new element** boxes have a more direct impact on the way we use the program to model. The position of new elements is not as key, but we should decide whether we want the new elements that we add to appear in the center of the virtual space or in the same position as the last part. We tend to think that the best setting is centered, since it will make our models centered by default in the virtual space. The orientation of new elements is more important because it relates to rotations, and we know that rotations are tricky. In this case, we tend to think that **Same as previous** is probably the best setting because it helps the build-by-copy process while we're working on whole rotated sections of our model. In any event, a part can easily be aligned with the default orientation using the Enter Position & Rotation dialog window we saw in the last chapter. Use a rotation vector with XYZ coordinates (rotation angles) of 0,0,0 and a rotation angle of value 0 as well. Through the **Color for new element** option, we can also decide which default color to use in the new parts that we insert into a model.

The Step, Grid, Snap tab in the MLCad Options window, shown in Figure 6.61, allows us to alter the grid settings to set the spacing for any of the three grid modes in the three axes independently. We can also change the rotation angle defaults. This makes the grid a very precise and useful ally, even when we employ nonstandard dimensions in our models. Each grid setting (Coarse, Medium, and Grid off) has its own line in the dialog box. In each line, we can separately specify

MLCad units for the spacing of the grid in the three axes as well as the angle spacing for rotations.

Figure 6.61 MLCad Grid Options

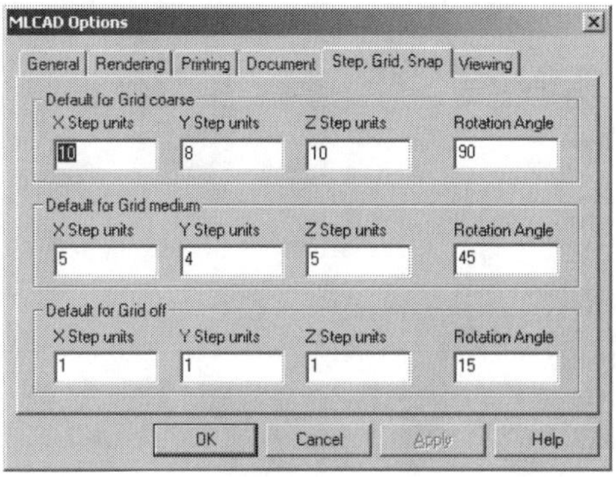

The Viewing tab of the MLCad Options window, shown in Figure 6.62, controls some settings of the View program mode. We can set how and if the parts for the current step will be highlighted. We can also set whether clicking with the mouse on a 3D modeling pane in View program mode will advance a step or rotate the model. Clicking the mouse in a modeling pane set to the 3D view angle will always rotate the model in Place program mode. However, while you're viewing the instruction steps, that function might not be as handy. Finally, we can also set the step count for the fast-forward and fast-reverse instruction step navigation buttons.

Figure 6.62 MLCad Viewing Options

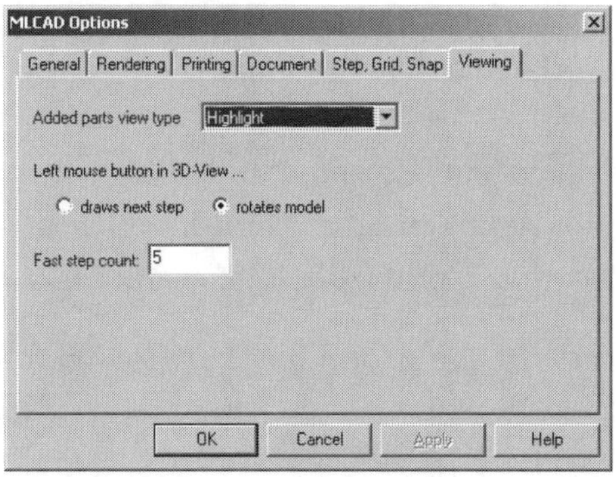

Customizing the Library Groups

Through the **Setting | Groups | Group Configuration...** menu option brings up the Parts List Configuration dialog window shown in Figure 6.63.

From here we can manage how the local copy of the library is organized in the library windows of MLCad's interface.

Figure 6.63 The Parts List Configuration Dialog Window

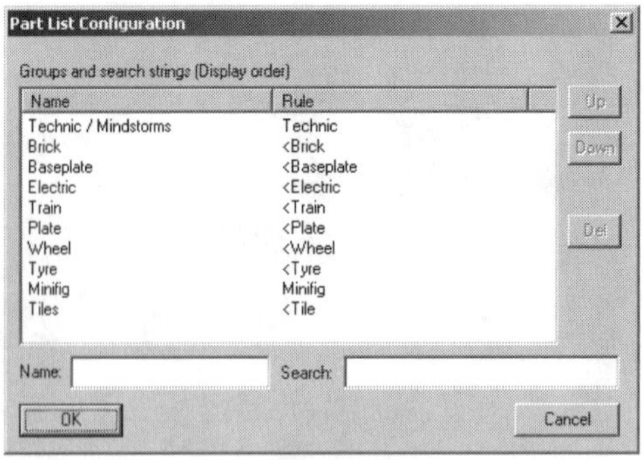

This window shows a list of the current library groups (and thus does not include the Favourites, Models, and Document groups). In the list window, we can select a group and edit it in several ways. First, we can assign it another place in the list by highlighting it and moving it with the Up and Down buttons, to the right in the window. The Del button deletes the selected group. We can also customize existing groups or create new ones.

Each group has a name and a rule. The name can be whatever we like. The rule tells MLCad which parts to include in that group. Basically, MLCad searches for parts in the library according to the criteria set in the rule. The first group in Figure 6.63 is Technic/Mindstorms and its rule is Technic. That means that under the heading Technic/Mindstorms, MLCad will group all the parts that carry *Technic* somewhere on their names. The second group is Brick, and the rule is <Brick. The < symbol indicates that the *initial word* of the name of a part has to be *Brick*.

To edit a group, select it first. Its name and rule will appear in the fields below the group list. Now you can edit them directly. To create a new group, simply enter a new name and a new rule. The searching rules are:

- **WORD** Include in the group any part for which the name *contains* this word.
- **<WORD** Include in the group any part for which the name *starts* with this word.
- **WORD>** Include in the group any part for which the name *ends* with this word.

- **WORD1&WORD2** Include in the group any part for which the name includes these *two words*.
- **WORD1 | WORD2** Include in the group any part for which the name includes *any* of these *two words*.
- **!WORD** Include in the group any part for which the name *does not include* this word.
- **WORD1&!WORD** Include in the group any part for which the name *includes the first word but does not include the second.*

Summary

Knowing how our models are stored opens new possibilities for them—and for us. Understanding the file format not only makes us more precise in our modeling, it also enables us to use new tools and techniques. Models not only carry geometric data or references to files, they also contain instruction steps and other information.

The Model Parts List window lets us view the file in which the program stores the model. In this window and in the real file, the model's parts are listed vertically, one per line. Each part line includes information such as color or position. Just as we can edit parts in the modeling panes, we can also use the Model Parts List window to work with them directly. In fact, we can type our models into a text file using Notepad or a similar text editor. While this method is not as user-friendly as using MLCad or other modeling programs, it is useful to remember that we can always access the file content this way.

Model files typically contain parts and comments. Comments are not meant to be used by the software, but early on James Jessiman saw the possibility to include an "opening" from independent programmers to use custom commands *hidden* inside comment files. These hidden commands are called *meta commands* and are very useful for a wide variety of circumstances, such as adding instruction steps to our models or flexible parts using external model generators like LSynth.

Part files use *primitives*, which are another kind of file list item (types 2 through 5). They hold the basic geometric data and shapes used to define the LEGO parts in the library. Primitives can be inserted into any MLCad model—for instance, to use in instruction steps. We can also load and edit the library parts files in MLCad using the same techniques as with models—and some new ones. Primitives can also be used to create new custom parts for the library, although this is an activity that requires serious knowledge of the LDraw file format and general computer 3D skills.

We ended the chapter exploring the ways in which MLCad can export, save, and print our computer models and how to customize the software for our needs.

Solutions Fast Track

A Peek into the Files

☑ The Model Parts List window displays the model file in a way similar to the way it is stored by the computer.

- ☑ The model file has one part or list item per line, which includes all the information for it, such as coordinates and colors.
- ☑ List items can be selected directly in the Model Parts List window.

Comments and Meta Commands

- ☑ Apart from parts, model files can also include comments and meta commands.
- ☑ Through them, MLCad offers plenty of interesting tools related to instruction steps that can also be used in other areas, such as animation.
- ☑ The format allows for developers to use meta commands inside files to trigger actions when they are read by the applications.

Using Non-LEGO 3D Elements

- ☑ Just as models use parts and comments (and meta commands), the LDraw files that define the parts in the library use primitive geometric elements to define their shape.
- ☑ These primitive elements can also be used inside regular models, using the same editing tools and a few extra options.
- ☑ Primitive elements can also be used to define complex 3D geometries, but this is far from being a trivial task.
- ☑ Rotation matrices are a mathematical concept that holds the key to much of the apparent simplicity and ease of use of LDraw-based software.

Adapting MLCad to Our Needs

- ☑ Through the Export option, MLCad allows us to extract data from our models in a variety of ways.
- ☑ Additionally, via the post processor window, we can control up to three applications besides MLCad that will further work with our exported data.
- ☑ MLCad can also generate and/or print image files for our models or instruction steps.
- ☑ Through the Settings menu options, we can customize MLCad's functions to make our modeling an easier task.

Frequently Asked Questions

The following Frequently Asked Questions, answered by the authors of this book, are designed to both measure your understanding of the concepts presented in this chapter and to assist you with real-life implementation of these concepts. To have your questions about this chapter answered by the author, browse to **www.syngress.com/solutions** and click on the **"Ask the Author"** form.

Q: Why is it important to learn about the file format defined for LDraw and used by MLCad?

A: The file format holds the key to the success of these LEGO CAD programs. Through it we can both describe parts and complete models and can add all sorts of items to it, from simple comments to special effects. If you do not have a general knowledge of the file format, this important area is locked to you.

Q: What if I am not very technically inclined?

A: The file format is designed to be accessible to everyone, even people with no knowledge or desire to learn math and computer programming. Its design is commonsensical and structured in several layers of detail. Dig into it until you start feeling lost (or bored!). That is about as far as you need to get; you can always come back when you need more information.

Q: Why are meta commands *hidden* inside comments, which confuses the user?

A: Indeed, this is a bit confusing *at first*. However, once you become more familiar with meta commands, you will come to understand the full implications of having a customizable file item.

Q: How many meta commands are there?

A: Just as with the number of parts in the parts library, we can assume there are plenty. Meta commands are linked to specific programs and specific functions, so thinking in terms of a complete list of them is completely irrelevant.

Q: *Buffer exchange* meta commands are confusing. Why should I bother with them?

A: Learn how to use *clear* commands first; these are pretty straightforward. Once you understand their function and possibilities, you will begin to feel annoyed by certain side effects of their use, such as having to duplicate the same parts in different steps. *Buffer exchange* meta commands help quite a bit in easing this annoyance.

Chapter 7

LSynth: A Bendable-Part Synthesizer

Solutions in this Chapter:

- Running LSynth
- Hose-Type Synthesis
- Band-Type Synthesis
- LSynthcp

☑ Summary

☑ Solutions Fast Track

☑ Frequently Asked Questions

Introduction

In the numerous lines of LEGO models, many flexible parts are available to builders. These parts include hoses, flexible axles, electrical cables, and rubber bands. The LDraw file format, unfortunately, does not provide the capabilities to handle parts that change shape. It would seem then that you cannot have bendable parts in your LDraw files. The solution to this dilemma is to create hoses and other bendable parts out of lots of little unbendable parts. For example, you can make a bent hose out of a bunch of little straight hose pieces. Placing all these little hose pieces just right is a tedious and tiresome job for a human, but is pretty straightforward for a computer. The process of having a computer create something like this for you is often referred to as *synthesis.*

As the author of LPub, the program that Syngress Publishing used to generate the building instructions for its *10 Cool LEGO Mindstorms* book series, I saw many of the authors struggle to generate these bendable parts and get them into their LDraw files. They went on the Internet and found some bendable-part synthesizers, including:

- **ldraw-mode** Created by Fredrik Glockner (www.math.uio.no/~fredrigl/TECHNIC/ldraw-mode), ldraw-mode adds LDraw editing to the extendable emacs text editor. It has a built-in hose generator.
- **Spring2DAT** Created by Marc Klein (http://marc.klein.free.fr/spring2d), Spring2DAT creates springs in LDraw format.
- **Rubber Belt Generator** Created by Philippe Hurbain (http://philohome.free.fr/rubberbelt/rubberbelts), Rubber Belt Generator creates rubber bands in LDraw format.

I wanted one program that could synthesize all the LEGO bendable parts, so I wrote LSynth. LSynth synthesizes these parts:

- TECHNIC ribbed hoses
- TECHNIC Flex-System hoses
- TECHNIC pneumatic hoses
- Electric cables
- TECHNIC fiber optic cables
- TECHNIC rubber bands

- TECHNIC rubber treads
- TECHNIC plastic tread
- TECHNIC chains

Like MLCad, LSynth is a work in progress. My next improvement will be to add spring synthesis to LSynth, so keep an eye out for it. You can keep up to date on the developments in LSynth by visiting www.users.qwest.net/~kclague/LSynth.

Extendable LDraw File Format

The LDraw file format allows users to add new features to the format as we create new programs compatible with LDraw. The LDraw format provides two basic kinds of records: those that describe the model in terms of parts, lines, triangles, or quadrilaterals and those that describe the model as comment lines. The LDraw program uses comment lines in two ways: as *meta-commands* (magic words that LDraw recognizes as something it should react to) and *comments,* which it ignores. Comment lines are defined to start with the digit 0. A meta-command that LDraw supports is STEP, which is discussed in Chapter 6. In the LDraw file, this looks like *0 STEP*. A true LDraw comment is any line in the file that starts with a 0, where the first word is not recognized by LDraw. For example, *0 THIS IS A COMMENT* is a comment because LDraw does not recognize the word *THIS* as something it should worry about.

Programmers can add new meta-commands to the LDraw file format by making their programs recognize the first word of comment records as a command. I wrote LSynth to recognize the word *SYNTH* as a meta-command, yet MLCad just thinks of 0 SYNTH records as comments.

Synthesis Specifications

You use SYNTH meta-commands combined with LDraw parts to describe the shape of part you want synthesized. This combination of SYNTH meta-command and parts is what I call a *synthesis specification.*

You put synthesis specifications directly in your LDraw design files. You can use the MLCad Add Comment dialog box to add SYNTH meta-commands and MLCad's Add Part mechanism to add the parts that describe the shape of the synthesized parts.

You use the *LSynth* command to synthesize your bendable parts. *LSynth* reads in your LDraw file that contains synthesis specification and creates a new LDraw file with the bendable parts synthesized.

To understand synthesis better, let's start out with an example of a LEGO TECHNIC ribbed hose. Figure 7.1 shows a photograph of a TECHNIC ribbed hose. You can see that it is made up of a regular sequence of ribs. Figure 7.2 shows a picture of LDraw part 80.DAT, a single TECHNIC hose segment consisting of a single rib.

Figure 7.1 A TECHNIC Ribbed Hose

Figure 7.2 A Single TECHNIC Ribbed-Hose Segment

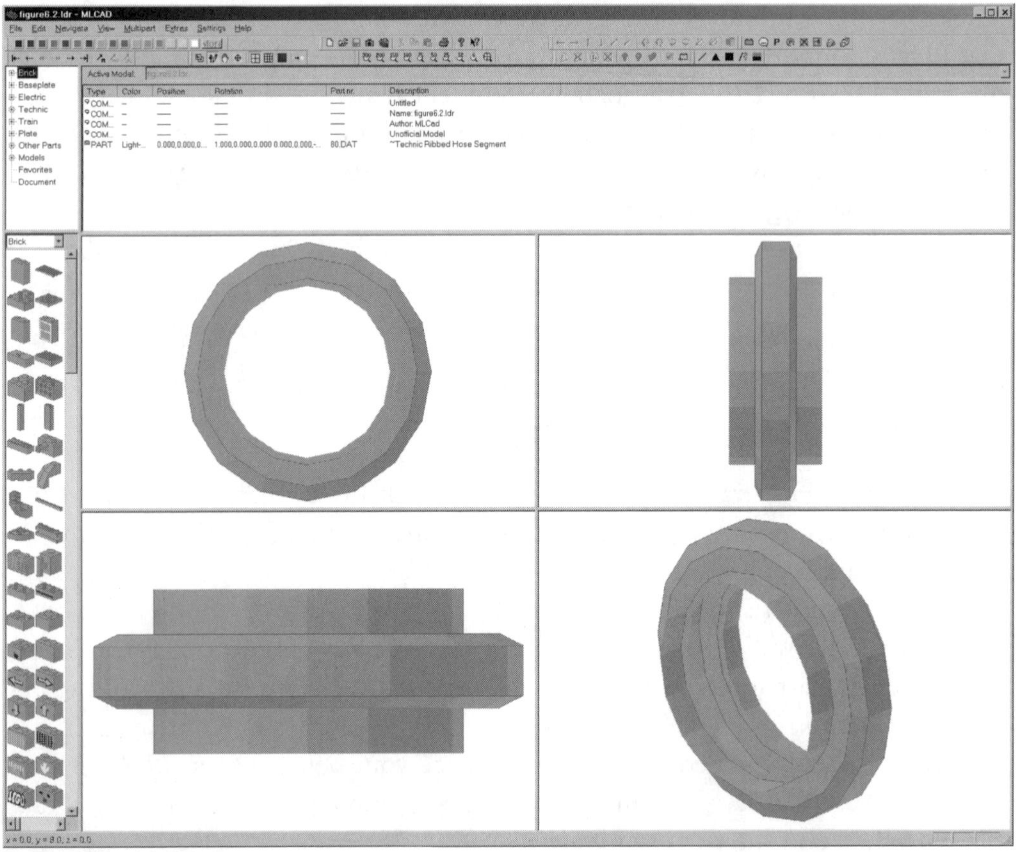

Building a straight hose rib by rib manually in MLCad would not be difficult, but it would be tedious. Imagine trying to lay out the same hose in a design where the hose bends. That would be much more difficult and time consuming. LSynth can perform this task for you quite easily. Figure 7.3 shows the image of a straight ribbed hose synthesized by LSynth.

Figure 7.3 A Synthesized TECHNIC Ribbed Hose

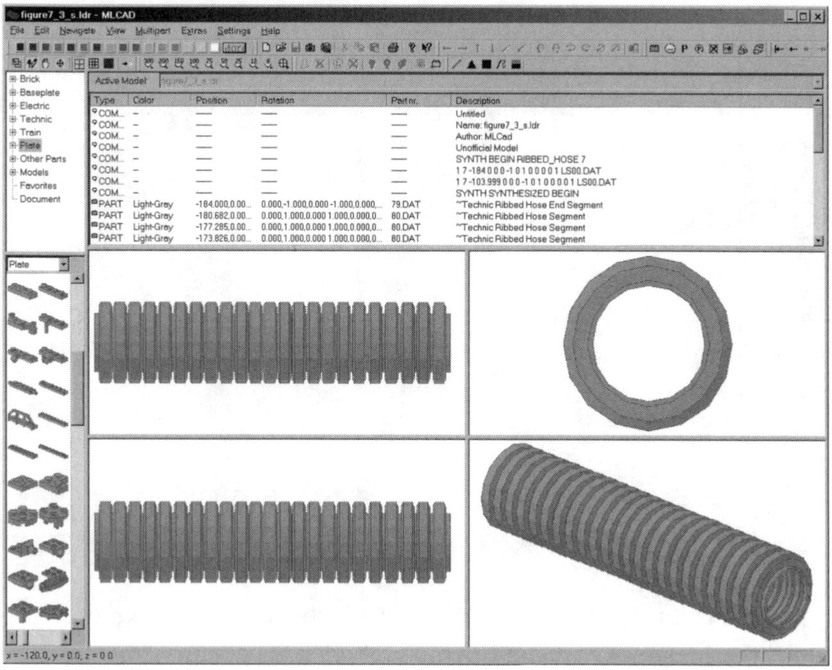

Figure 7.4 shows the synthesis specification of our ribbed hose in MLCad. You can see the uses of the *SYNTH* meta command in the Description column of the Model Parts List. The synthesis specification starts with a *SYNTH BEGIN* meta-command and ends with a *SYNTH END* meta-command. Between these two commands are LDraw parts that describe the shape of the part you want synthesized. For tubes, we use a special LDraw part I created to describe where the hose starts and ends. The parts that describe the shape of our synthesized parts are called *synthesis constraints* or simply *constraints*.

Let's look at the *SYNTH BEGIN* line more carefully. *SYNTH BEGIN* is followed by *RIBBED_HOSE* and then the number 7. *RIBBED_HOSE* tells LSynth what type of LEGO part I want to synthesize. The 7 is the LDraw number for the color light gray. The two white parts that look like arrows are the constraints that tell us where the hose starts (the arrow on the left) and where the hose ends (the arrow on the right). These arrows are both uses of the unofficial LDraw parts LS00.DAT that I created from the official LDraw part 755.DAT.

Figure 7.4 The LSynth Specification for a Ribbed Hose

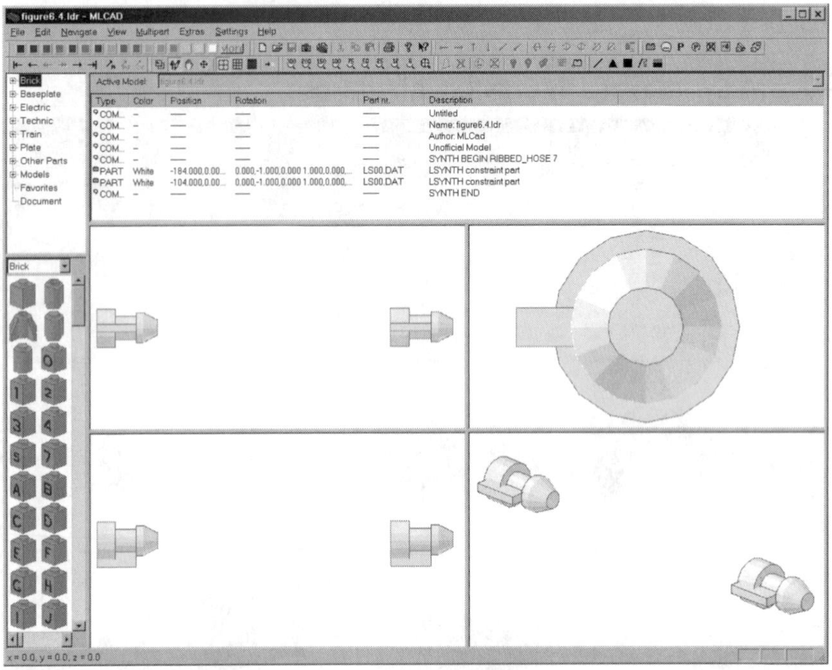

This specification describes everything LSynth needs to make our ribbed hose. From here it figures out how many ribbed-hose segments it needs to create the hose, where they need to be placed, and their orientation.

As you can see in this simple example, the initial specification is much simpler than the final synthesized hose. It's clear that synthesis can save you a lot of time and effort.

Running LSynth

LSynth is a Windows program that you installed in Chapter 2. Use Window's **Start | Programs | LSynth** menu to start LSynth.

Figure 7.5 shows the simple LSynth user interface. To function, LSynth requires two filenames. The first filename (called the *input file*) is the name of your file with synthesis specifications in it. When you click the **Browse** button (to the right of the Input File field), a familiar Windows Open File dialog box pops up. When you select an input file, LSynth automatically provides a default value for the output filename. The *output file* is the name of the LDraw file where LSynth puts your synthesized results. You can modify the output filename manually by clicking and typing in the Output File field. You can also click the **Browse** button to the right of the Output File field to pop up a Windows Save File dialog box, which you can use to choose an output filename. You can see in Figure 7.5 that I chose the file that I created using MLCad in Figure 7.4.

Figure 7.5 The LSynth User Interface

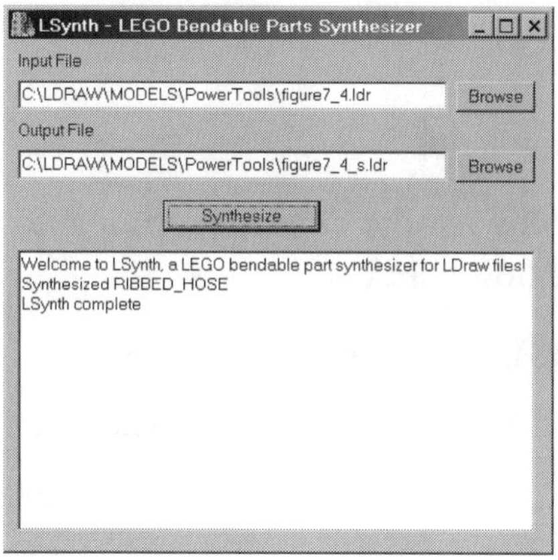

To synthesize figure7_4.ldr into figure7_4_s.ldr, click the **Synthesize** button. LSynth displays the types of parts synthesized in the memo window below the Synthesize button. The results of the synthesis in Figure 7.5 are for the image that we saw in Figure 7.3. The LSynth program is easy to run; LSynth's power lies in the synthesis specifications you place in your LDraw files and the synthesis algorithms inside the LSynth program.

LSynth uses two different synthesis algorithms: hoselike things that start in one place and end in another and closed-loop things such as rubber bands. The synthesis specification format for these two classes of bendable parts are different. First let's look at hose-type bendable parts.

Hose-Type Synthesis

TECHNIC ribbed hoses, Flex-System hoses, pneumatic hoses, flexible axles, fiber optic cables, and electrical cables all fall under the "hose-type" synthesis class because they all start in one place and end in another. (In other words, they are not loops.)

The general format for hose-type synthesis specifications is:

```
0 SYNTH BEGIN type color
constraint
constraint
0 SYNTH END
```

Notice that we need to supply several fields here:

- The *type* field
- The *color* field
- The two *constraint* fields

The *type* field controls what the cross section of a hose segment looks like. LSynth synthesizes these hose types:

- **RIBBED_HOSE** The corrugated hoses used in some TECHNIC sets.
- **FLEX_SYSTEM_HOSE** Standard Flex-System hoses.
- **PNEUMATIC_HOSE** Pneumatic hoses used to hook together pneumatic devices.
- **FLEXIBLE_AXLE** Bendable TECHNIC axles.
- **FIBER_OPTIC_CABLE** Fiber optic cables found in the Robot Invention System's Extreme Creatures Expansion Set as well as other LEGO sets.
- **ELECTRIC_CABLE** Standard electrical cables found in train sets, the Robot Invention System, and other LEGO sets.

The *color* field is a number that indicates the color of the synthesized hose. LSynth uses the standard LDraw color numbers, some of which are listed in Table 7.1.

Table 7.1 Some Common LDraw Color Numbers

Color	Number	Color	Number
Black	0	Dark gray	8
Blue	1	Light blue	9
Green	2	Light green	10
Dark cyan	3	Cyan	11

Continued

Table 7.1 Some Common LDraw Color Numbers

Color	Number	Color	Number
Red	4	Light red	12
Magenta	5	Pink	13
Brown	6	Yellow	14
Light gray	7	White	15

The *constraint* fields define where the synthesized part must exist as well as how the cross section is oriented. You must have at least two constraints per synthesis specification—the first for the beginning of the hose and the last for the end of the hose. If you need to route hoses around or through parts of your design, you can use intermediate constraints. Constraints must be listed in order from the beginning of the hose to the end of the hose. As we saw in our ribbed hose example, the unofficial LDraw part LS00.DAT is used to specify hose constraints.

Ribbed Hoses

Now that we've got the basics of hose synthesis, let's try a more complex ribbed hose synthesis to get a better feel for the power of hose synthesis. Figure 7.6 shows the podium we built in the earlier chapters, with a decorative ribbed hose arch at the back. Without LSynth, this example would be much harder to create than our straight-line ribbed hose because this hose is actually bent. Creating a bent hose like this by hand would be a tremendous amount of work.

Figure 7.7 shows the synthesis specification used to create this hose. The synthesis specification starts with the line *SYNTH BEGIN RIBBED_HOSE 14*, like our first example, only this time the hose is yellow. This time I have used three constraints: one for the start of hose, one for the top of the bend, and one for the end of the hose. In this example you can see how the directions in which the constraints are pointing play a big role in the final synthesis results.

Pneumatic Hoses

Some LEGO TECHNIC kits use pneumatic devices that run on air pressure. LEGO pneumatic devices include large and small pumps, large and small cylinders, air pressure tanks, T connectors, and valves. Pneumatic hoses are used to connect these devices. The LDraw library contains pneumatic parts but no support for their hoses. Entering a pneumatic design into MLCad without hoses contributes little to the understanding of the actual function of a pneumatic circuit.

Figure 7.6 A Podium with an Arch in MLCad

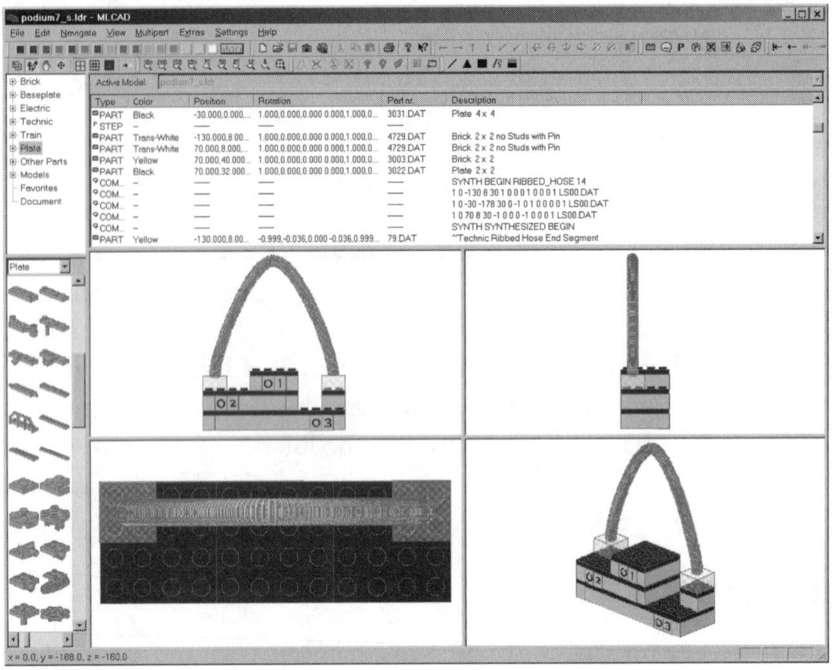

Figure 7.7 Synthesis Specification for the Podium Arch

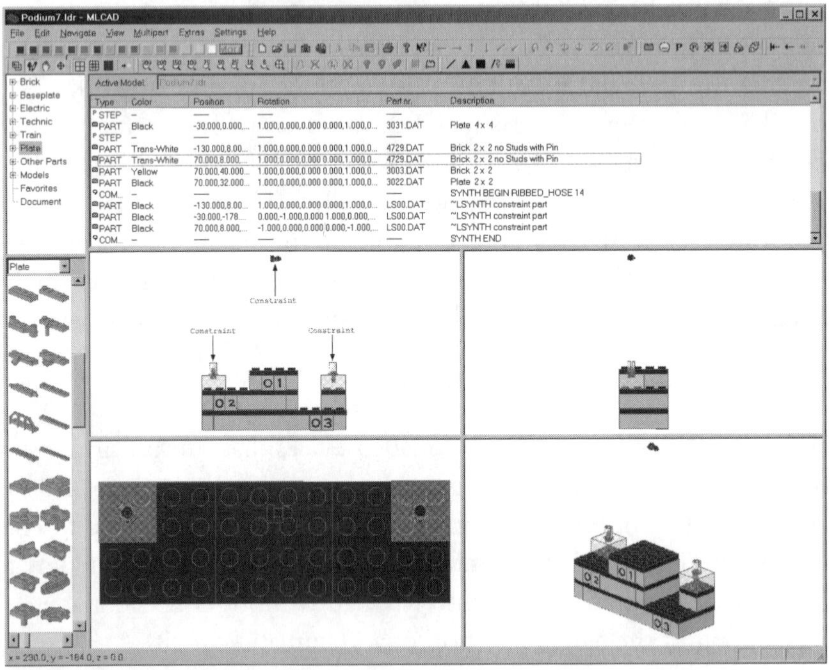

Figure 7.8 shows an LDraw file opened in MLCad that describes a pneumatic circuit made up of a large pump hooked to a pressure tank. The pressure tank feeds a pneumatic switch. The pneumatic switch controls a large piston. Notice that there are no hose images connecting any of the parts.

Figure 7.8 A Pneumatic Circuit with Hoses in MLCad

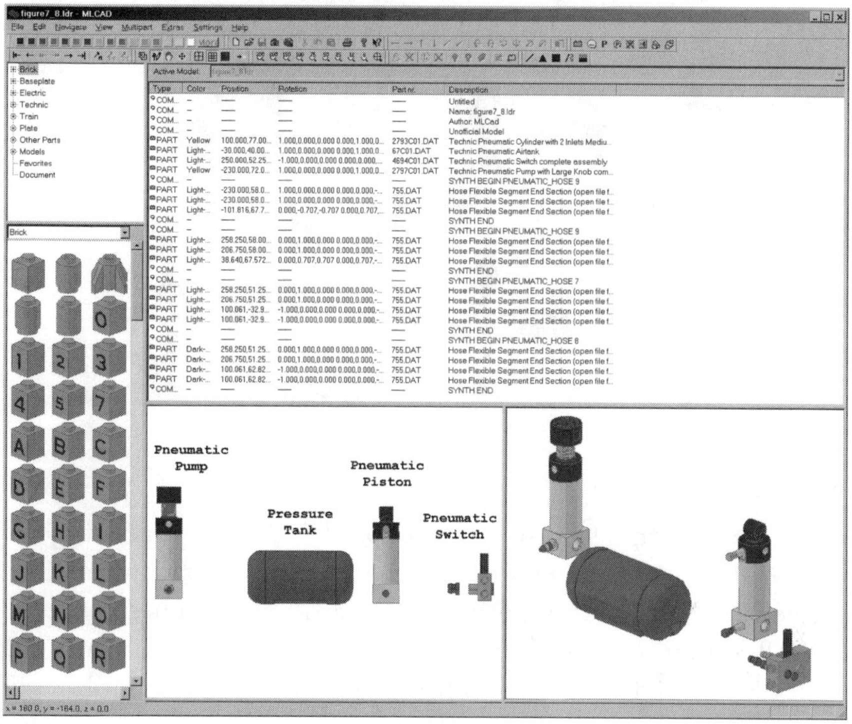

Figure 7.9 shows the synthesized results of this pneumatic setup after it has been run through LSynth. Quite a difference!

The thing to note in this LSynth example is that LSynth knows nothing about the pneumatic parts used. It only knows about the synthesis constraints you provide. This technique allows you to use any official or unofficial pneumatic parts you might desire.

When I was first designing the LSynth program, I thought about using the pneumatic parts as constraints, but I realized that this solution would not work. Most pneumatic parts have more than one place to hook hoses, so I would need a way to tell LSynth where to hook the hose onto the constraint. I then realized that it would be easier for LSynth to use a part like LS00.DAT to describe constraints for all hose-type synthesis specifications.

Figure 7.9 A Synthesized Pneumatic Circuit

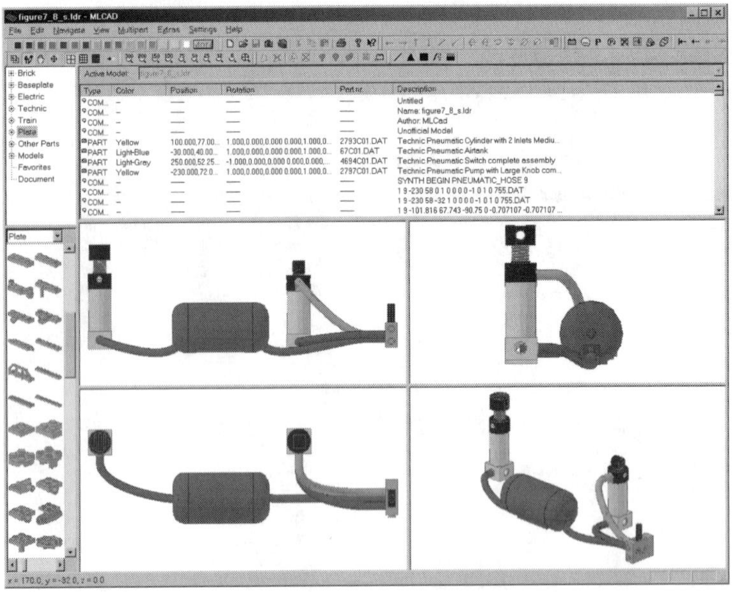

Electric Cables

The most common form of electric cable from LEGO is the standard cable with a 2x2 brick connector at each end. You can use the official LDraw part 5306.DAT, Electric Brick 2x 2x 2/3 with Wire End, as the end of the connector cable.

The small section of wire that juts out from the brick on this part (the wire end) makes 5306.DAT hard to use with LSynth. The cross-section shape of LSynth's wire is different from that provided in the 5306.DAT part. When you installed LSynth in Chapter 2, you also installed some unofficial parts created for use with LSynth. One part you installed was 5306A.DAT, Electric Brick 2x 2x 2/3, which has no wire end and which is easier to use with LSynth.

In addition to the standard electric cables, the Robot Invention System and some of the add-on packs for it provide three sensors that have electrical cables built in:

- Light sensor
- Rotation sensor
- Temperature sensor

The official LDraw part for the light sensor is 2982C01.DAT. It contains a wire end. The unofficial part 2982C02.DAT is a copy of the official part with the wire end removed.

The official LDraw part for the rotation sensor is 2977C01.DAT. This part also contains a wire end. The unofficial part with the wire end removed is 2977C02.DAT.

There is no official part for the LEGO temperature sensor.

Figure 7.10 shows a synthesis specification for the complete light sensor that comes with the Mindstorms Robot Invention System 2.0 set. This example contains an electric cable end connected to the light sensor by an electric cable. Note that the light sensor and cable end are inside the synthesis specification but are not constraints. LSynth ignores these parts inside the synthesis specification. Placing these ends inside the specification tells the LPub program that these are all one part.

Figure 7.10 Light Sensor Synthesis Specification

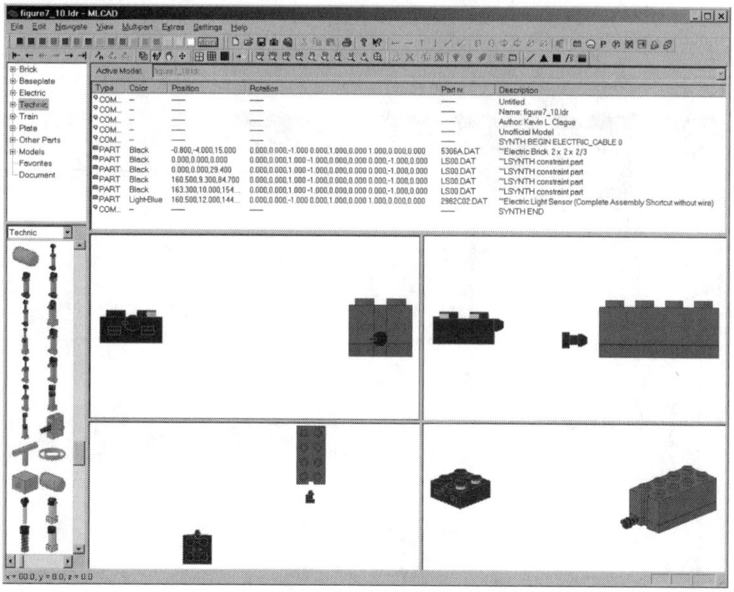

The synthesis constraints define where the cable goes to connect these two parts. There are two constraints for the sensor, one inside the sensor and one outside the sensor. This gives us a small amount of straight cable coming out of the sensor. I used the same technique at the connector end of the cable.

Figure 7.11 shows the synthesized results. This image is rendered with POV-Ray, which makes it much easier to see the cross-section shape of the cable.

Notice that an electric cable segment is made up of two long cylinders lying side by side. With electric cables we need to be concerned with how these cylinders are oriented relative to the electric cable ends. Figure 7.12 shows a synthesized electric cable with the cable segments oriented improperly (one on top of the other, as opposed to side by side) with respect to the cable ends. Figure 7.13, rendered with POV-Ray, shows the synthesis specification for this incorrect configuration. The difference between these two specifications is the rotation of the constraints. The constraints have a rectangular fin sticking out of the side. This fin is a visual cue to the orientation of your electric cable relative to the electric cable ends. This fin points to one of the wires in the electric cable.

Figure 7.11 Synthesize Light Sensor

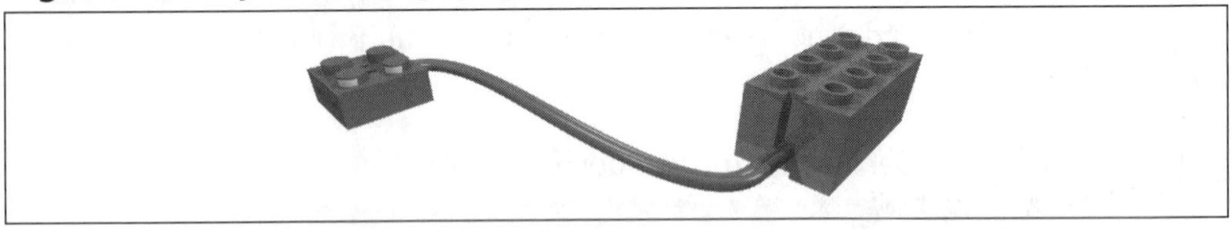

Figure 7.12 Incorrect Cable Orientation

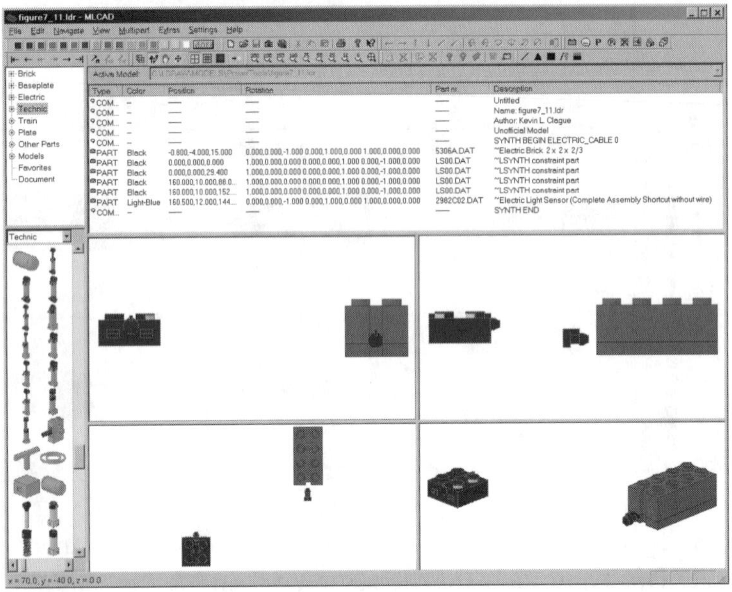

Figure 7.13 Synthesis Results of Incorrect Light Sensor Cable Orientation

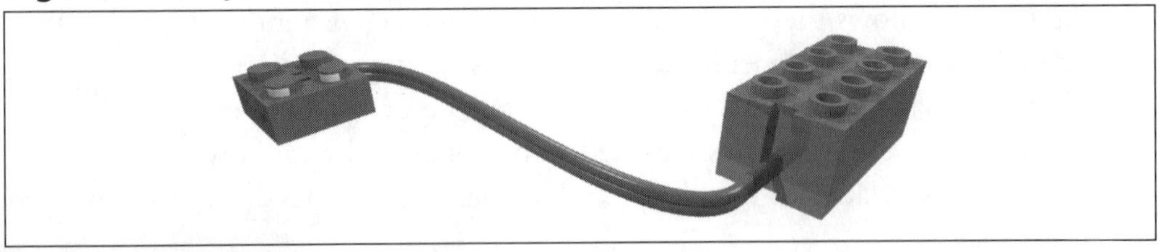

Fiber Optic Cables

The library of official LDraw parts does not contain parts for creating fiber optic cables. I created part LS30.DAT, which is the large end of a fiber optic cable. You use this part in conjunction with fiber optic cable synthesis to create a complete fiber optic cable.

Figure 7.14 shows a complete description of a fiber optic cable and the synthesized results. The individual segments of the fiber optic cable are listed after the line *0 SYNTH SYNHTESIZED BEGIN*. After the last fiber optic cable segment is a *0 SYNTH SYNTHESIZED END* line, followed by the *0 SYNTH END* line from the input synthesis specification. You cannot see these lines, because MLCad's model part list window is not large enough.

Figure 7.14 A Synthesized Fiber Optic Cable

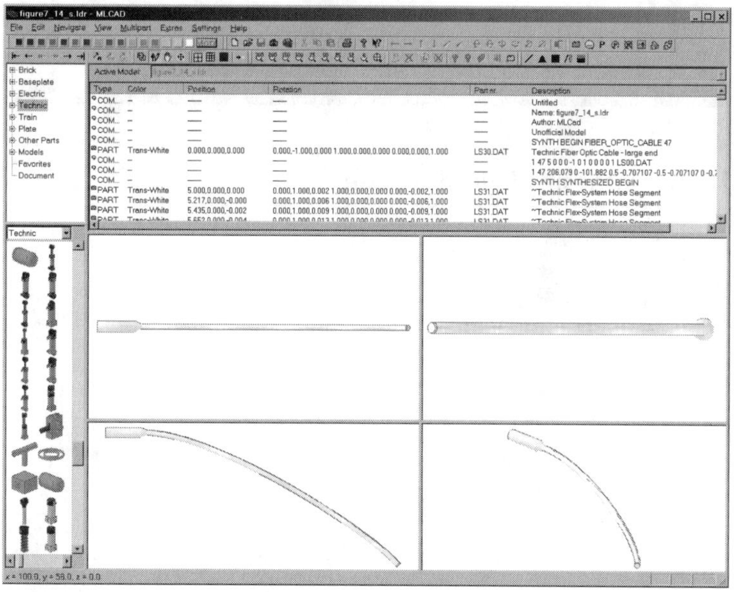

Flexible Axles

Five different lengths of flexible axle are provided in the official LDraw parts library. Unfortunately, LDraw and MLCad only allow you to use these flexible angles as straight, unbent parts (un-flexed). LSynth provides an unofficial part LS40.DAT that is the end piece for a flexible axle. Figure 7.15 shows an example of combining two flexible axle ends and an LSynth *FLEXIBLE_AXLE* specification to create a complete flexed axle. Figure 7.16 shows the synthesized flexible axle.

Figure 7.15 An Example of a Flexible Axle Specification

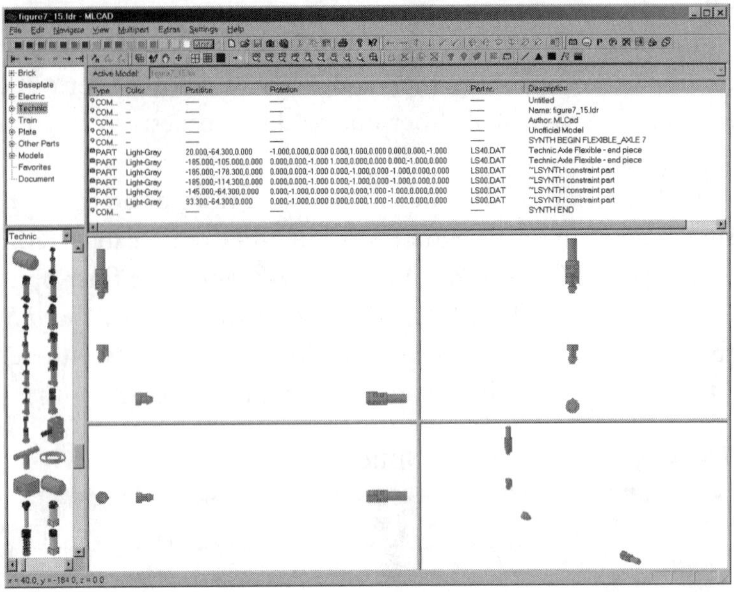

Figure 7.16 A Synthesized Flexible Axle

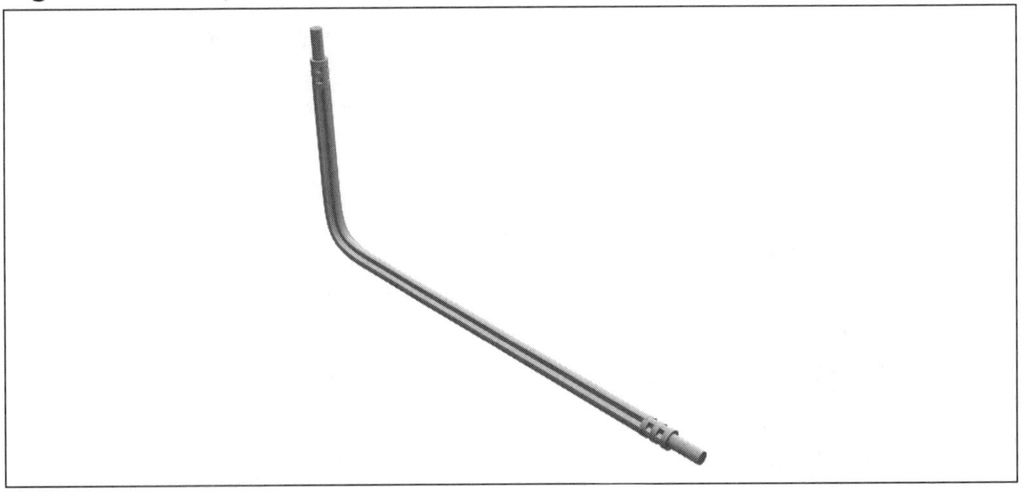

Advanced Issues

In LDraw, a *record* (a line in the file) that starts with the digit 0 is one of two things: a comment or a command. When LSynth is synthesizing your LDraw files, it copies hose constraints from your input LDraw file to your output LDraw file, but it puts a 0 in front. This turns your constraints into comments, so the constraints do not show up in your synthesized LDraw file. This is because normally you just want to see the hoses, not the constraints. If you want to see the constraints in the synthesized output file, you can use the *SYNTH SHOW* command before the list of constraints.

Band-Type Synthesis

Band types include rubber bands, rubber treads, chains, and plastic treads. These components are lumped together in the band types category because they all use LSynth's second synthesis algorithm for closed loops that go around circular things such as pulleys, gears, and wheels.

The general syntax for band-type synthesis specification is:

```
0 SYNTH BEGIN type color
constraint
constraint
0 SYNTH END
```

Again, we have the *type*, *color*, and two *constraint* fields to deal with here. We deal with color and constraint the same way we did for hose-type synthesis. The list of types is a bit different and includes the following:

- RUBBER BAND
- RUBBER_TREAD
- PLASTIC_TREAD
- CHAIN

Let's take a closer look at each of these types.

Rubber Bands

Figure 7.17 shows a synthesized example of two pulleys inside a rubber band. For rubber bands, we use official LDraw parts for constraints. Figure 7.18 shows a list of the LDraw parts that LSynth recognizes as rubber band constraints.

LSynth supports rubber bands that cross over on themselves. Figure 7.19 shows a rubber band that is inside two pulleys and crosses over on itself. The *SYNTH CROSS* lines tell LSynth that the rubber band crosses. *SYNTH CROSS* lines must always be used in pairs.

Figure 7.17 A Rubber Band Around Two Pulleys

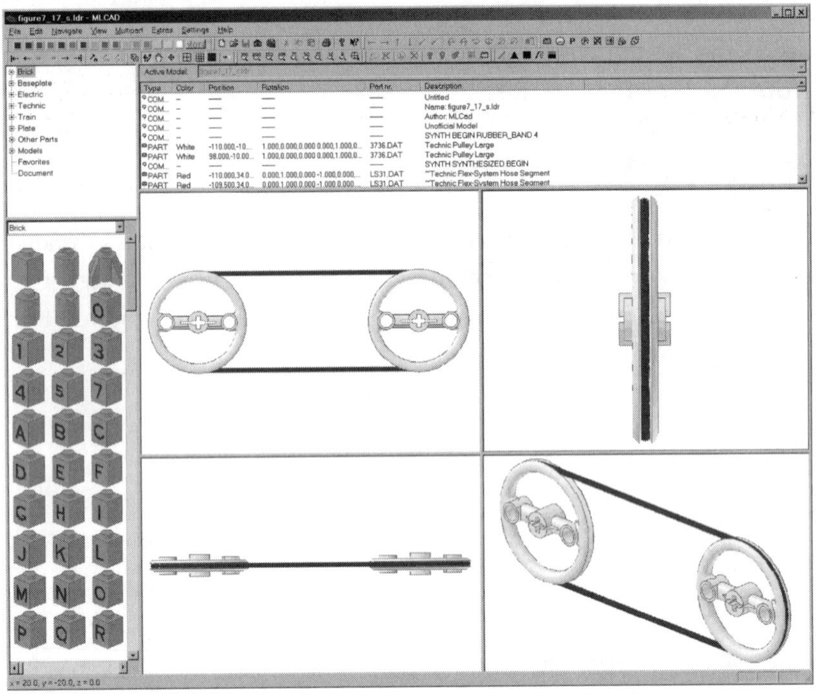

Figure 7.18 Supported Rubber Band Constraint Parts

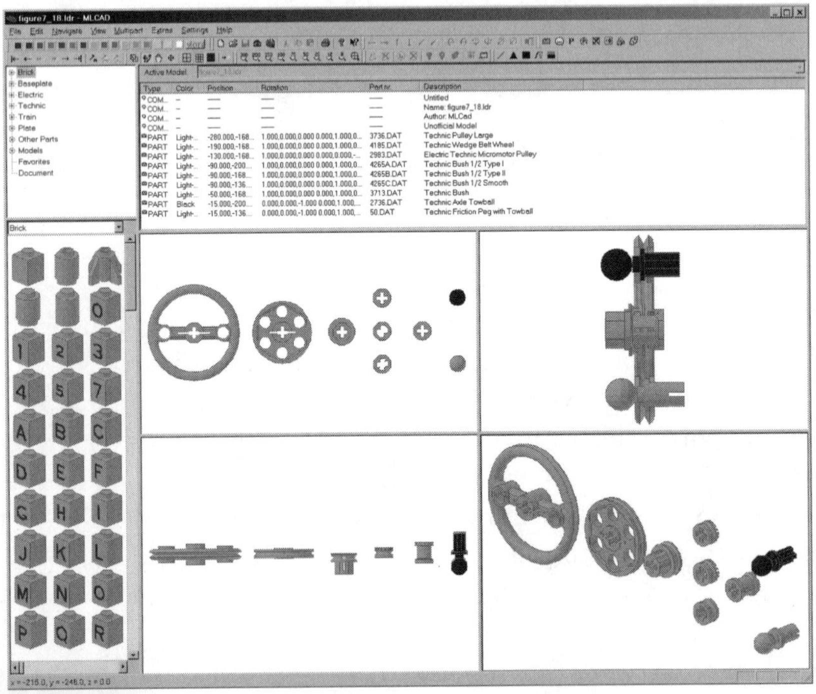

Figure 7.19 A Rubber Band Crossed Around Two Pulleys

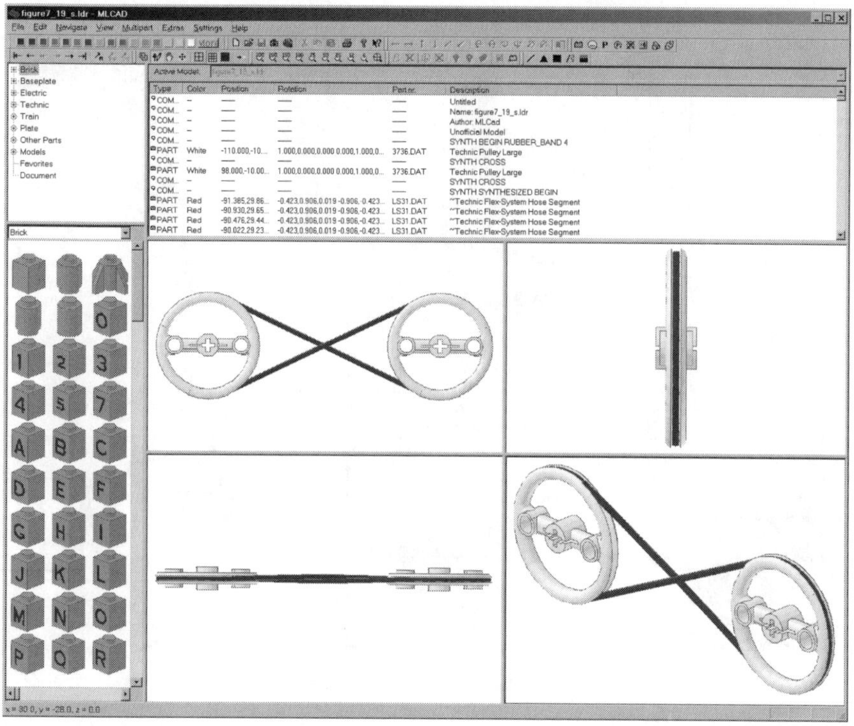

LSynth supports rubber band synthesis with multiple constraints (for example, multiple pulleys). You must have at least two constraints, but you can have more. For example, you can have a three-pulley synthesis or even more if you want. The order of the rubber band constraints is important. Figure 7.20 shows a four-pulley synthesis. LSynth assumes you list your constraints in the order needed to traverse the pulleys in a counter-clockwise fashion. In the example in Figure 7.20, the LSynth works its way from the white pulley (bottom left) to the light-gray pulley (bottom right) to the dark-gray pulley (top right) to the black pulley (top left), then back to the white pulley.

When using more than two pulleys, the extra pulleys (the third and beyond) can either be inside or outside the rubber band. Pulleys that are outside the rubber band press in on the normal shape of the stretched rubber band. Figure 7.21 shows a three-pulley synthesis with one pulley outside the band.

Figure 7.20 Pulley Traversal Order

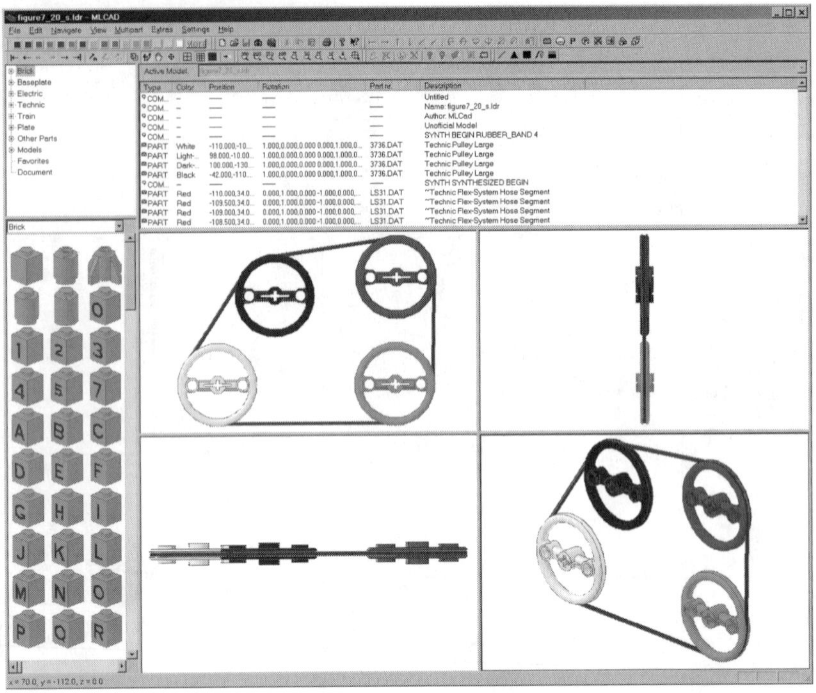

Figure 7.21 A Pulley Outside the Rubber Band

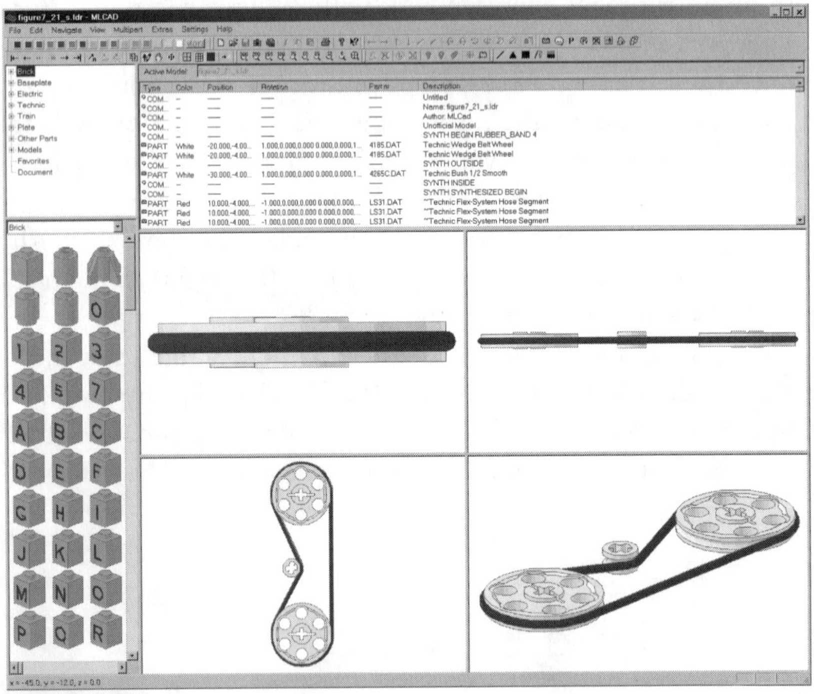

The *SYNTH OUTSIDE* line tells LSynth that all the constraints following the *SYNTH OUTSIDE* line are outside the rubber band. The *SYNTH INSIDE* line tells LSynth that all the constraints following the *SYNTH INSIDE* line are inside the rubber band. LSynth assumes that constraints are inside the rubber band until it sees a *SYNTH OUTSIDE* statement. For every *SYNTH OUTSIDE* statement, there must be a *SYNTH INSIDE* statement or else LSynth will not synthesize your rubber band correctly.

You can use *SYNTH CROSS* statements with more than two pulleys. Figure 7.22 shows a three-pulley synthesis with crossing.

Figure 7.22 A Crossing Rubber Band with Three Pulleys

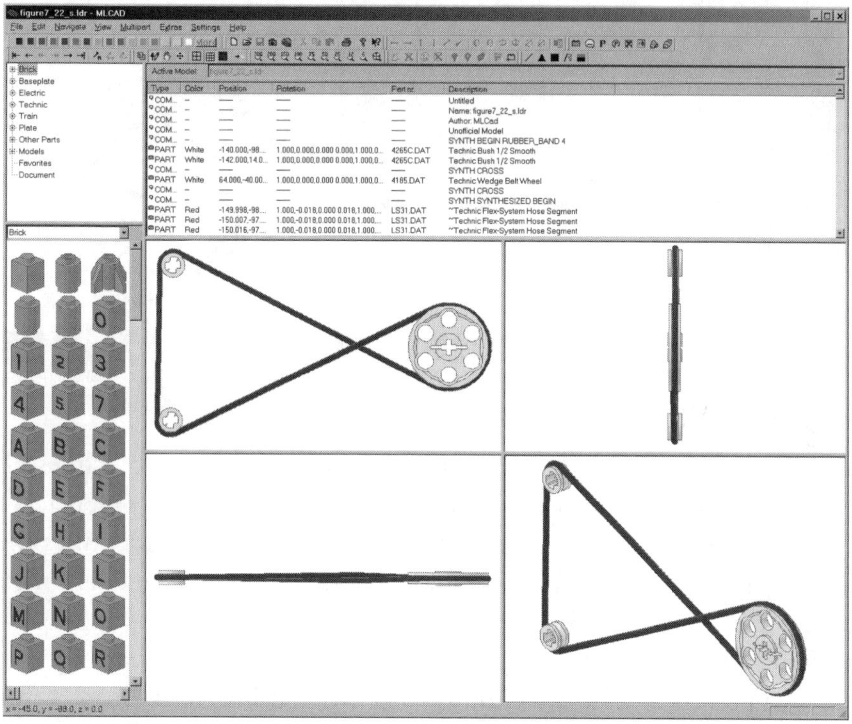

Figure 7.23 shows a four-pulley synthesis in which two pulleys are on one side of the cross and the other two pulleys are on the other side of the cross.

The next section explains the theory of how LSynth's band synthesis algorithm works. It is a bit mathematical, so if you don't get it, don't worry—when you need to make more complicated rubber band configurations, the theory will be easier to grasp.

Figure 7.23 A Crossing Rubber Band with Four Pulleys

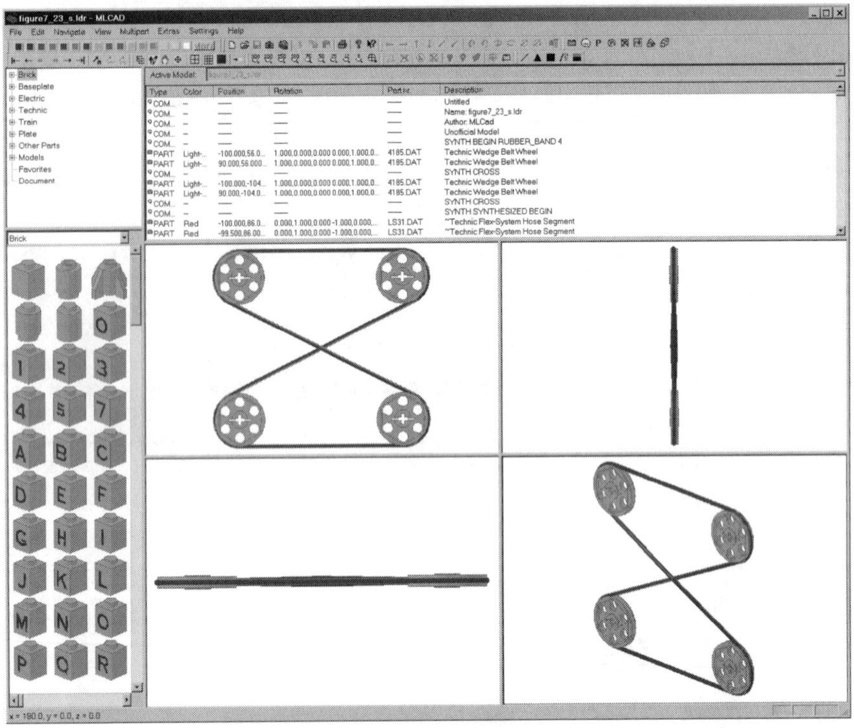

Band Synthesis Theory

When you have two circles that do not overlap, you can draw four lines between the two circles, where each line must touch each circle in only one place. In mathematical terms, these lines are called *tangents*. Figure 7.24 identifies the four possible tangent lines, given two example circles. The tangent lines are named T1, T2, T3, and T4.

As LSynth traverses your constraints synthesizing your rubber band, it must consider two pulleys at a time to know what tangent line to use to get from one pulley to the other. LSynth must consider whether each pulley is inside or outside the band and whether the rubber band is crossing when going between the two pulleys. LSynth chooses one of the four possible tangent lines depending on these conditions. Table 7.2 shows how LSynth decides which tangent line to use.

Figure 7.24 Lines That Are Tangent to Two Circles

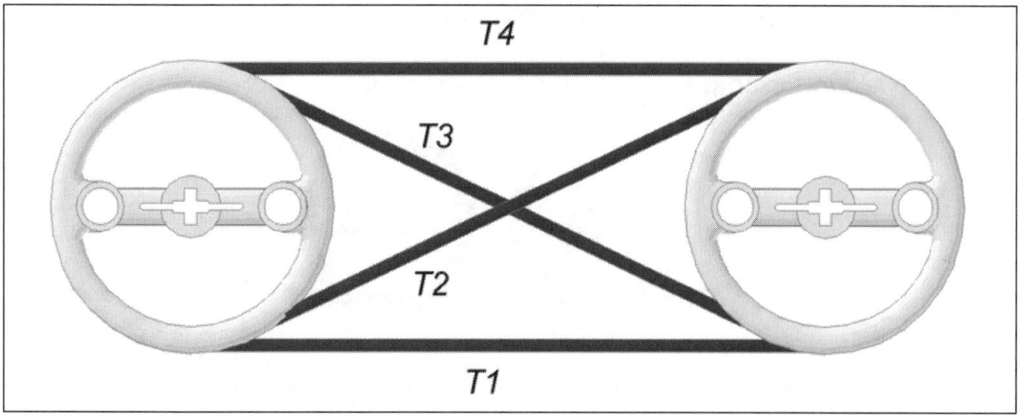

Table 7.2 Tangents Usage Chart

Case	First Pulley	Second Pulley	Cross	Tangent
1	Inside	Inside	No	T1
2	Inside	Inside	Yes	T2
3	Inside	Outside	No	T2
4	Outside	Outside	No	T4
5	Outside	Outside	Yes	T3
6	Outside	Inside	No	T3

Figure 7.25 shows a four-pulley synthesis where two of the pulleys are outside the rubber band. With four pulleys there are four places where the rubber band is stretched into straight lines. In Figure 7.25, when going from the white pulley (bottom left) to the light-gray pulley (bottom right), we're going from an inside pulley to an inside pulley, so LSynth uses tangent T1. When going from the light-gray pulley to the dark-gray pulley (top right), we're going from an inside pulley to an outside pulley, so LSynth uses tangent T2. When going from the dark-gray pulley to the black pulley (top left), we're going from an outside pulley to an outside pulley, so LSynth uses T4. (It looks like T1, but if you stand on your head it is clear that it is T4.) When going from the black pulley back to the white pulley, we're going from an outside pulley to an inside pulley, so LSynth uses T3.

The most interesting aspect of the information presented in Table 7.2 is that cases 2 and 3 are the same, as are cases 5 and 6. In fact, in all synthesis cases where there is an even number of pulleys, you can express your synthesis specification with either *the SYNTH CROSS* or the *SYNTH INSIDE* and *SYNTH OUTSIDE.*

Figure 7.25 An Example Using All Four Tangents

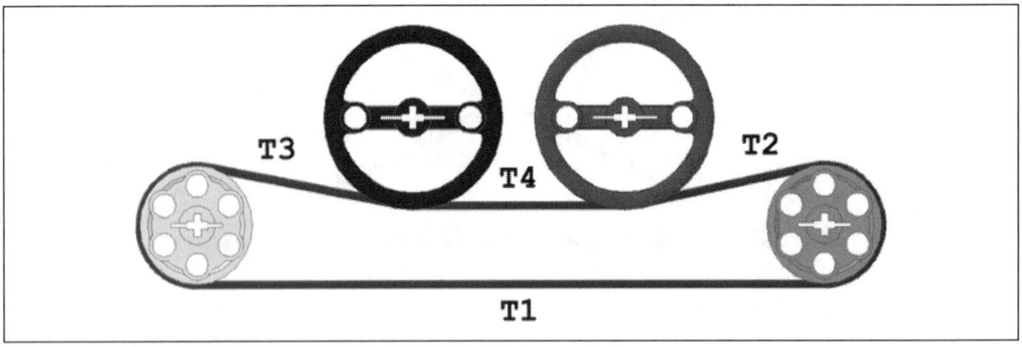

Figure 7.26 is a three-pulley case that violates the rule that says, for every *SYNTH OUTSIDE* you must have a corresponding *SYNTH INSIDE.* You cannot synthesize this case using *SYNTH CROSS.*

Figure 7.26 A Rubber Band with Three Pulleys and Three Crossings

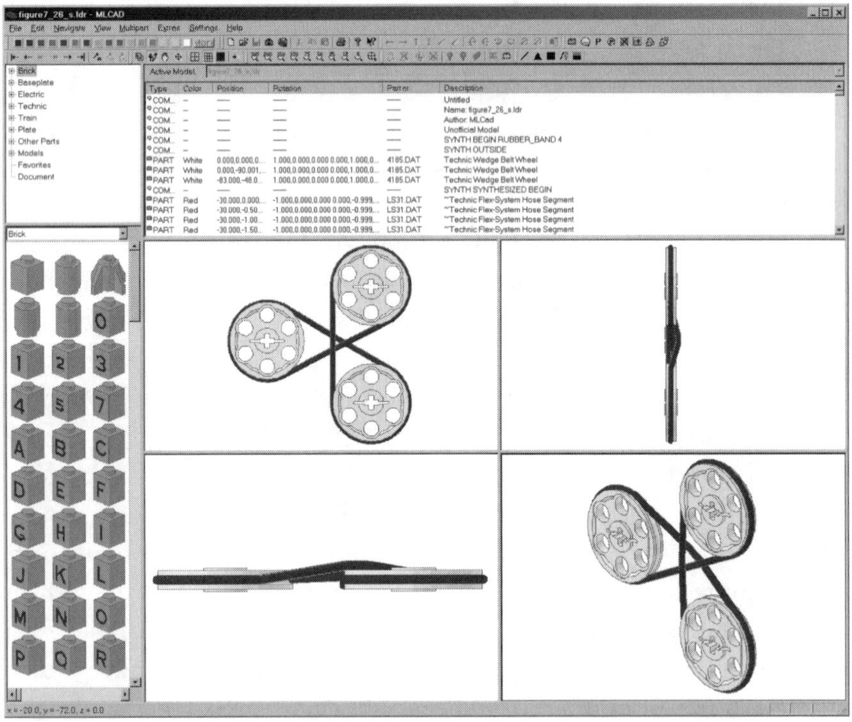

NOTE

The LSynth version that comes with this book requires that all the pulleys be in the same plane. In some situations, you might want pulleys to be at right angles to each other. I hope to add this capability in a future version of LSynth.

Rubber Tread

Rubber tread synthesis is similar to band synthesis except that treads use wheels as constraints instead of pulleys. Figure 7.27 shows a synthesized rubber tread using three wheels.

Figure 7.27 A Synthesized Rubber Tread

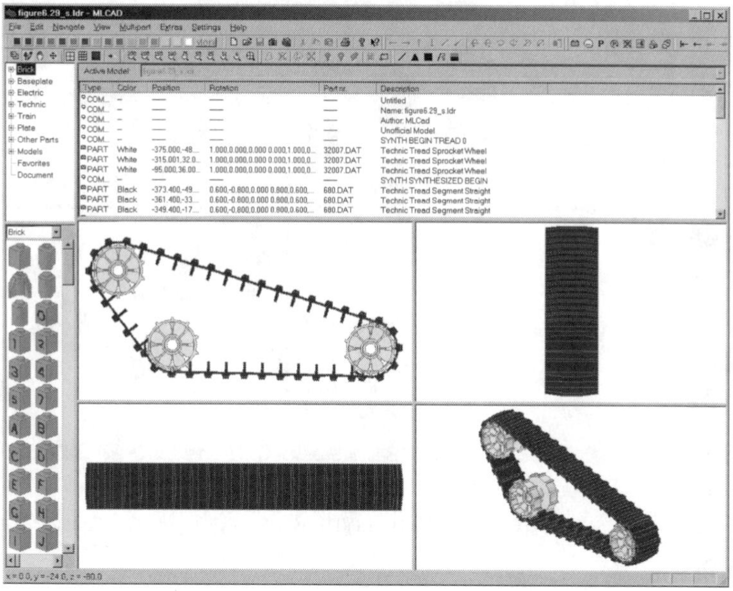

LDraw part 32007.DAT is the only wheel LEGO has designed to work with TECHNIC rubber tread. You can use *SYNTH INSIDE* and *SYNTH OUTSIDE* statements, but in the real world, rubber treads do not cross well.

TECHNIC Chain

LEGO TECHNIC chains are pretty uncommon in LEGO modeling, but they can be a wonderful replacement for gears, especially in instances where you need them to span long distances. LEGO chains are made up of a sequence of chain links. Part 3711.DAT is the LDraw part for LEGO chain links. LEGO TECHNIC gears are the correct parts to use for chain synthesis constraints. Figure 7.28 shows all the TECHNIC gears that LSynth supports.

Figure 7.28 Supported Chain and Plastic Tread Constraints

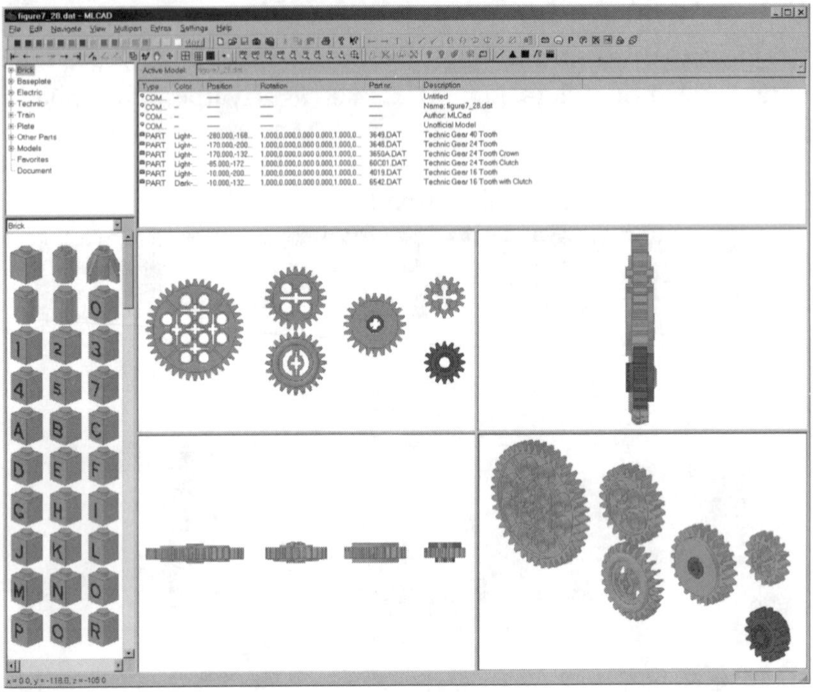

Figure 7.29 shows a four-gear chain synthesis with gears inside and outside the chain loop.

Plastic Tread

LEGO plastic treads are made up of individual plastic tread links, similar to chain links. LSynth lets you use the same list of gears that you can use as constraints for chains (Figure 7.28) as constraints for plastic tread. Plastic tread links are incompatible with outside gears, but LSynth does not stop you from using them for this purpose

Figure 7.30 shows a synthesis specification for a plastic tread around three gears.

Figure 7.29 An Example of Chain Synthesis

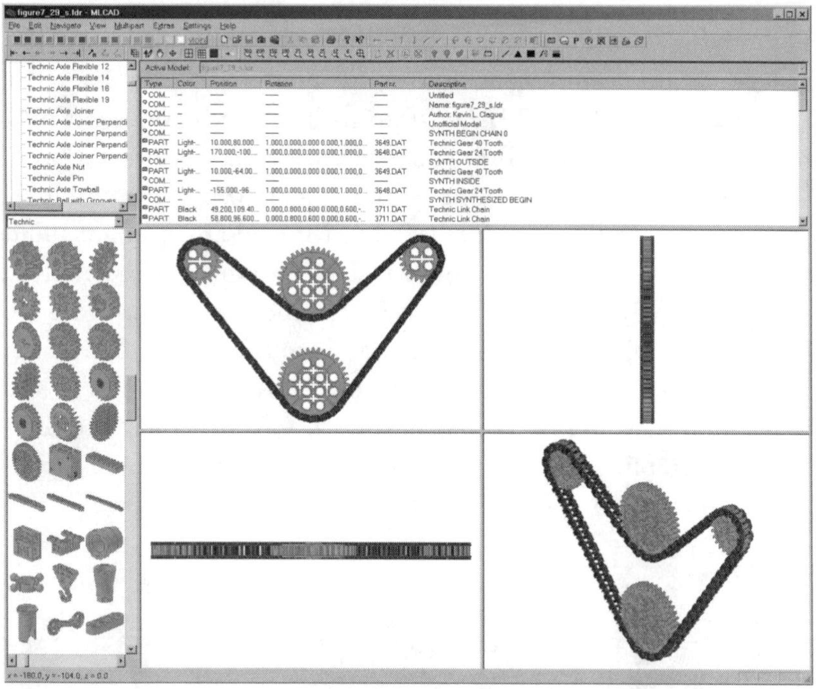

Figure 7.30 An Example of Plastic Tread

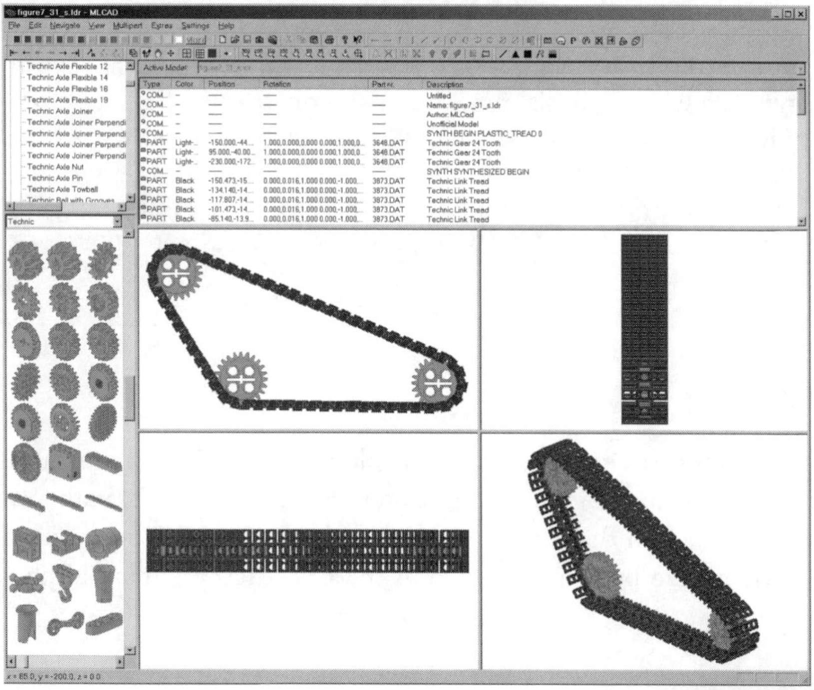

Advanced Issues

The band synthesis algorithm copies your band-type constraints into your output file unchanged. If you do not want to see your synthesis constraints in the output synthesis file, you can *use SYNTH HIDE* to make LSynth comment out your constraints. If you only want to hide some of your band constraints, you can also use *SYNTH SHOW* statements to make subsequent constraints visible.

Imagine you are creating an LDraw file in MLCad and you are adding *STEP* commands after every few parts so you can step through your assembly and see it being built. Imagine also that your design has a rubber band that goes around three pulleys. If the pulleys for the rubber band are added in different steps, you have to have two copies of the pulleys. One set is inside the synthesis specification (which cannot have *STEP* commands in it), and the other is set in the steps where you are assembling the design. It can be convenient to hide the pulleys in the synthesis constraints using *SYNTH HIDE* commands.

> **Designing & Planning…**
>
> **Lax Synthesis Rules**
>
> You might have noticed that LSynth lets you create synthesis specifications for configurations that do not exist in the real world. LSynth does not even check synthesis type against constraints that you use. For example, LSynth lets you use pulleys and wheels as constraints for chains and gears as constraints for rubber bands!
>
> People can be very inventive, so rather than not allowing them to "mismatch" constraints, LSynth allows it and hopes you know what you are doing.

LSynthcp

LSynth's capabilities are designed to be usable by other LDraw-compatible programs. LSynth is a simple graphics user interface program for LSynthcp, a command prompt program that actually performs the synthesis. When you press LSynth's **Synthesize** button, LSynth runs the LSynthcp program. LPub, my building instruction generator program, runs the LSynthcp program to synthesize parts as it is creating building steps. The LSynth program makes LSynthcp easy to run without having to use Windows command prompt mechanisms. I'm hoping that Michael Lachman, the author of MLCad, adds bendable part support to MLCad using the LSynthcp program.

Summary

LSynth is a powerful and easy-to-use tool for generating bendable parts. You add synthesis specifications into your LDraw design files using *SYNTH* commands and LDraw parts as constraints. Constraints tell LSynth the shape of the bendable parts you want synthesized. Defining synthesis specifications and synthesizing is much less work than creating these bendable parts by hand.

LSynth makes hoses, cables, rubber bands, treads, and chains very easy to create and work with. LSynth's tube synthesizer is more powerful than its predecessor because it allows for more than just the creation of hose starting points and ending points. LSynth is the first LDraw rubber band synthesizer that can have more than two pulleys, rubber bands that cross, or pulleys outside the rubber band. It is also the first LEGO chain and plastic tread synthesizer.

LSynth is actually just the graphical user interface program that runs the Windows command prompt program LSynthcp, which does all the actual synthesizing. Future versions of LSynth and LSynthcp will add more synthesis types and improve the synthesis algorithms already in place. Keep up to date on the latest developments in LSynth and LSynthcp by visiting my Web site at www.users.qwest.net/~kclague/LSynth.

Solutions Fast Track

Running LSynth

☑ Using LSynth is simple. To synthesize parts, follow these steps:

1. Use Windows' **Start | Programs | LSynth** menu to start the LSynth program.
2. Click the **Input File's Browse** button to pop up a Windows Open File dialog box.
3. Use the **Open File** dialog to select an LDraw file with synthesis specifications.
4. Modify the default **Output File** name if desired.
5. Click the **Synthesize** button to synthesize your bendable parts.

☑ Use MLCad to view your synthesized parts, just as you would for any other part.

Hose-Type Synthesis

☑ To synthesize hose-type parts, use the *SYNTH BEGIN* command entered into your LDraw file using MLCad's Add Comment dialog box.

☑ You can create RIBBED_HOSE, FLEX_SYSTEM_HOSE, PNEUMATIC_HOSE, FLEXIBLE_AXLE, FIBER_OPTIC_CABLE, and ELECTRIC_CABLE hose types.

☑ You must add two or more LS00.DAT parts to specify the shape of your hose.

☑ Use the *SYNTH END* command to end the synthesis specification.

Band-Type Synthesis

☑ To synthesize band-type parts, use the *SYNTH BEGIN* command in your LDraw file to identify a synthesis specification.

☑ You can create RUBBER_BAND, RUBBER_TREAD, PLASTIC_TREAD, and CHAIN hose types.

☑ Specify the shape of your band types using LDraw pulleys, wheels, and gears.

☑ Use the *SYNTH END* command to end the synthesis specification.

LSynthcp

☑ LSynthcp is the actual command prompt program that does all the synthesizing. It is accessed through LSynth's easy-to-use graphical user interface.

☑ For updates on both LSynthcp and LSynth, check out my Web site at www.users.qwest.net/~kclague/LSynth.

Frequently Asked Questions

The following Frequently Asked Questions, answered by the authors of this book, are designed to both measure your understanding of the concepts presented in this chapter and to assist you with real-life implementation of these concepts. To have your questions about this chapter answered by the author, browse to **www.syngress.com/solutions** and click on the **"Ask the Author"** form.

Q: Why did you create LSynth? Can't I build pretty much any model in LDraw and MLCad that I could in the real world with actual LEGO bricks?

A: LDraw does not have direct support for bendable parts. I created LSynth so that bendable parts would be more accessible to the LDraw community. I wanted to make LSynth easier to use and more powerful than the synthesizers already available.

Q: What format do the files I run through LSynth need to be in? What format will they be in when they are synthesized?

A: LSynth reads LDraw files that contain synthesis specifications; it writes out new LDraw files with synthesized parts.

Q: What are constraints? Are they actual LDraw parts or just place holders?

A: Constraints are LDraw parts that describe the shape of the bendable part you want synthesized.

Q: What is the difference between constraints in band-type synthesis and constraints in hose-type synthesis?

A: Hose-type constraints are written to the synthesis results file commented out, and band-type constraints are written to the results file unchanged.

Q: How do I choose where the finished synthesized file is saved to? Is this even possible?

A: You use the Output File field in the LSynth window to control where your synthesized results are saved.

Chapter 8

L3P and POV-Ray

Solutions in this Chapter:

- Running L3P
- Running POV-Ray
- L3P Options

☑ Summary

☑ Solutions Fast Track

☑ Frequently Asked Questions

Introduction

You might have noticed that some of the pictures of synthesized parts in Chapter 7 look much more realistic than those produced by MLCad. These photorealistic pictures are rendered (drawn) by a program called *POV-Ray*. POV-Ray uses a technique called *ray tracing* to make the pictures look so realistic. In order for standard LDraw files to be compatible with POV-Ray, they need to be run through a program known as L3P that translates them into a format that POV-Ray can work with.

Lars Hassing wrote L3P because he wanted to use the power of POV-Ray to make high-quality images of his LEGO designs. Lutz Uhlmann wrote the first program, known as L2P, that converted LDraw files to a format that's usable by POV-Ray. L2P requires a special parts library, LGEO, that contains much more complex versions of LDraw parts. It is a lot of work for one person to model all the LDraw parts, so Uhlmann was never able to get it all done. Hassing got impatient with the lack of LGEO parts and wrote L3P so that he could use POV-Ray with the existing LDraw parts library. Hassing also added automatic camera and light settings, enabling you to create a complete POV-Ray scene and produce a good picture of your model on your very first try.

Hassing is very well known in the LEGO CAD community for L3P and other powerful LDraw-compatible tools. You can read more about Hassing and his LDraw-compatible tools on his Web site at http://www/hassing.dk.

L3P is a powerful and easy-to-use tool that makes POV-Ray accessible to the LEGO community. When Jon Babock, our Syngress editor, first contacted us to coauthor their first LEGO-building instruction book, *10 Cool Dark Side Robots, Transports, and Creatures: Amazing Projects You Can Build in Under an Hour* (ISBN: 1-931836-59-0), we were curious how we, as the authors, were supposed to create the images for our robot chapters. I suggested that MLCad could be used for this purpose, but Jon had done his research and suggested using L3P and POV-Ray as a way to get much better-looking pictures. I could not imagine trying to lay out the artwork for chapters such as these by hand with a paint program or something similar, so I took Jon's suggestion and started working with L3P and POV-Ray.

I ended up writing a program called LPub (short for *LEGO Publisher*) that reads an LDraw design and automatically creates step-by-step building instructions. Chapter 9 is devoted to using LPub, but for now it is enough to say that I could not have created LPub without L3P. The LEGO community owes a lot to Hassing for L3P.

Running L3P

In Chapter 2, you installed onto your computer the Software Power Tools suite of applications described in this book. One of the things you installed was L3P. L3P is a command-line program that can be run using a Windows 98 DOS prompt, or a command prompt in newer versions of Windows. Since many readers, used to using a Windows graphical user interface, might be unfamiliar with using command-line programs like L3P, this chapter walks you through the steps involved.

In Windows 95/98, you use the **Start | Programs | DOS Prompt** menu to get a DOS prompt window. In Windows 2000 and Windows XP, you use the **Start | Programs | Accessories | Command Prompt** menu to get a command prompt window. Figure 8.1 shows a command prompt for Windows 2000 Professional. A command prompt represents a place where you type commands that tell the computer to do things. The line that starts with C:\> is the command prompt. You get a new command prompt every time you type a command and the command completes. Windows has a few built-in commands for changing directories (folders) and listing files within folders.

Figure 8.1 The Windows 2000 Professional Command Prompt Window

To run the examples used in this chapter, you need to change the directory to the directory c:\LDRAW\MODELS\PowerTools (where all the examples for this chapter were installed when you installed L3P and where you have been saving the models you've built in earlier chapters). To do this, type the command **cd c:\ldraw\models\powertools**, as shown in Figure 8.2. Don't worry about uppercase or lowercase characters in the name; the command prompt recognizes the directory name either way. Notice that after you typed the command and

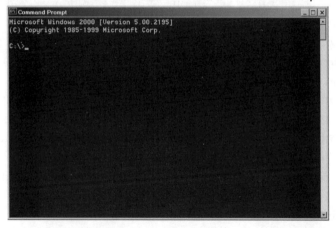

pressed **Enter**, the prompt changed from C:\> to C:\LDRAW\MODELS\PowerTools. The command prompt tells you what directory you are in. This is called the *current directory*.

Figure 8.2 Changing Directory

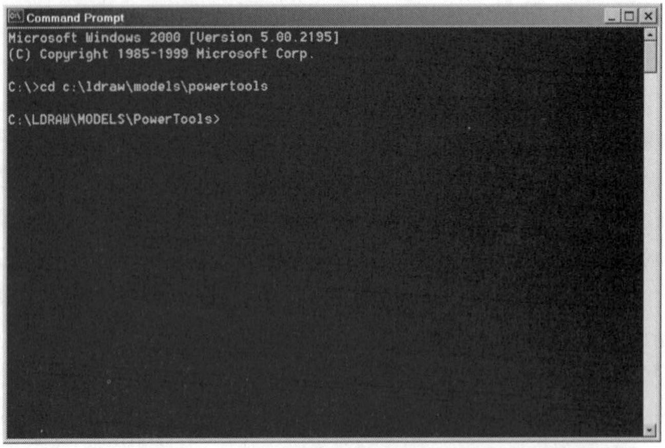

You can run programs at the command prompt by typing in the program's name and pressing the **Enter** key. In our case, the program we want to run is L3P. When you run L3P, it needs to know the name of the file you want translated from LDraw format to POV-Ray format. You provide this filename to L3P on the same line as you type the name **L3P**. The program name you type, combined with these parameters, is called the *command line*. In addition, some programs have optional parameters called *options*. In L3P, options are key words (think of them as special names that L3P recognizes) that start with a dash or hyphen (–). In Figure 8.3, we typed **l3p –o chapter8_podium.ldr**, which is made up of *L3P* (the program name); *-o,* an option that makes L3P overwrite the POV file you want if the file already exists; and *chapter8_podium.pov* (the name of the file we want translated to POV-Ray format. The file chapter_podium.ldr is the same podium that we worked with in Chapter 6.

Figure 8.4 shows you information that L3P printed after we pressed Enter (which made the L3P program run). The text printed after we entered the L3P command line is information that L3P printed as it translated chapter8_podium.ldr into chapter8_podium.pov. Since we didn't tell L3P a name for our translated POV file, it made up a name by substituting the .ldr suffix in our LDraw filename with the .pov suffix.

Now that we have a POV-Ray-compatible version of our podium file, let's render it in POV-Ray.

Figure 8.3 An L3P Command Line

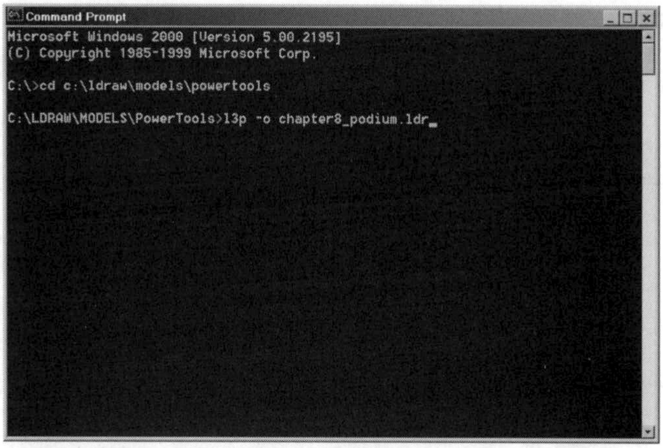

Figure 8.4 Results of Running L3P

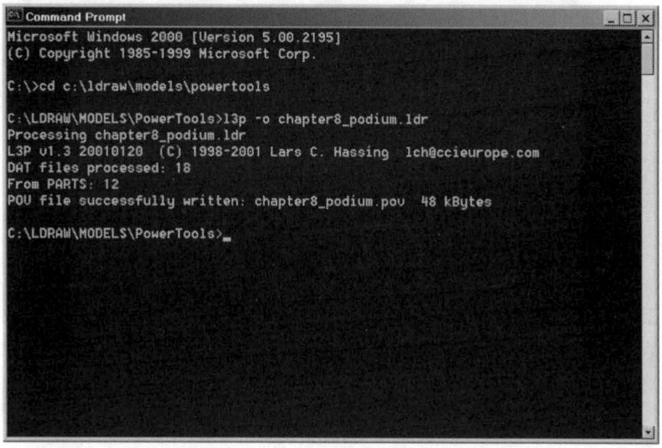

Running POV-Ray

You start POV-Ray using the **Start | Programs | POV-Ray for Windows V3.1 | POV-Ray for Windows** menu. POV-Ray for Windows is a graphical user interface (GUI) for the POV-Ray rendering engine. Figure 8.5 shows the POV-Ray for Windows GUI. This chapter explains just enough of the POV-Ray GUI to allow you to see the results of running L3P on your files.

As you can see from the Messages tab in the Tab Control section of the screen, a good many people have worked on POV-Ray. POV-Ray has a development history that goes back at least 20 years, with very capable computer programmers contributing their time and skills by adding new features to the program all the time. You can learn more about POV-Ray on the program's Web site (www.povray.org).

Figure 8.5 The POV-Ray for Windows Graphical User Interface

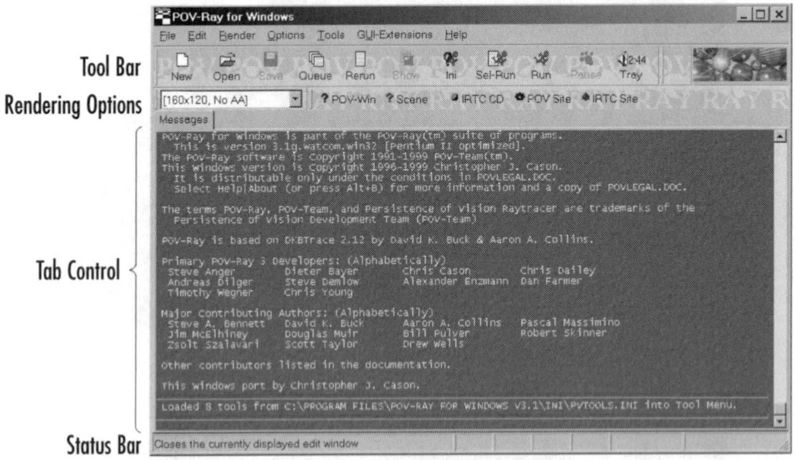

The first thing you will want to do is change the Rendering Options to a larger image size. This is done using the **Rendering Options** menu at the top left of the screen. The default is 160 pixels horizontally and 120 pixels vertically. We recommend selecting at least **640 x 480** for rendering your images. Figure 8.6 shows all the rendering options, with 640 x 480 selected.

Figure 8.6 POV-Ray Rendering Options

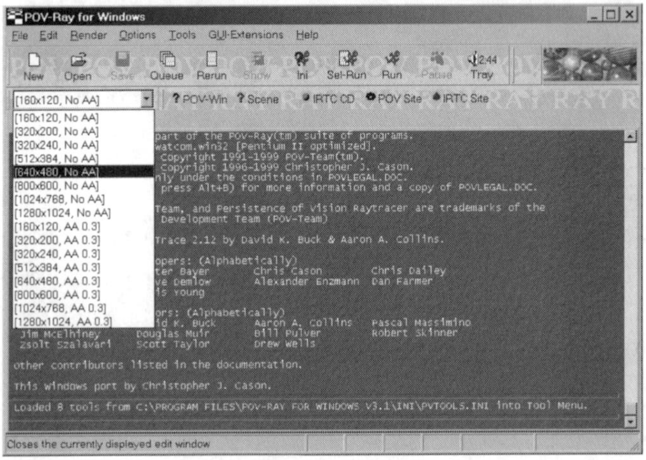

Use POV-Ray's **File | Open** menu to open our chapter8_podium.pov file in POV-Ray. POV-Ray reads in our file and displays it in POV-Ray as its own podium.pov tab in the Tab Control window, as shown in Figure 8.7.

Figure 8.7 chapter8_podium.pov Opened in Its Own Tab

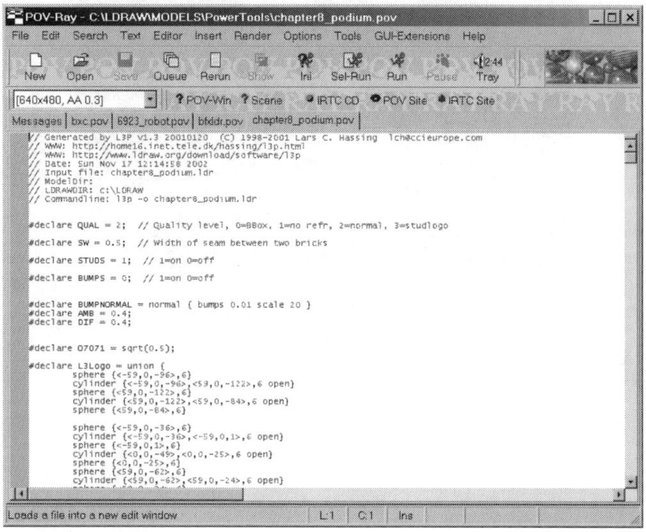

In the podium.pov tab, you see the POV scripting language, which we won't look at in detail here. POV-Ray is a professional-quality ray-tracing program that is available for free. It is worthy of its own book (or two), so we can't afford to go into POV-Ray in all the detail we'd like here. We look at it only enough to be able to use it to suit our MLCad modeling needs.

To render chapter8_podium.pov file, click the **Run** button in the Toolbar. This brings up the Rendering window, where you can see the image of the podium slowly begin to take shape. Figure 8.8 shows the POV-Ray rendered image of the chapter8_podium.pov file. Compare this image to Figure 8.9, which is the chapter8_podium.ldr file rendered by MLCad. MLCad cannot afford to spend as much time rendering our images as POV-Ray or MLCad's GUI would be unbearably slow, because rendering in POV-Ray can take minutes or even hours, instead of the seconds it takes MLCad.

POV-Ray requires, at a minimum, three things to render a picture like the one in Figure 8.8:

- Lights
- Camera
- Objects (our LDraw model, in this case)

Figure 8.8 POV-Ray's Rendering of Our Podium

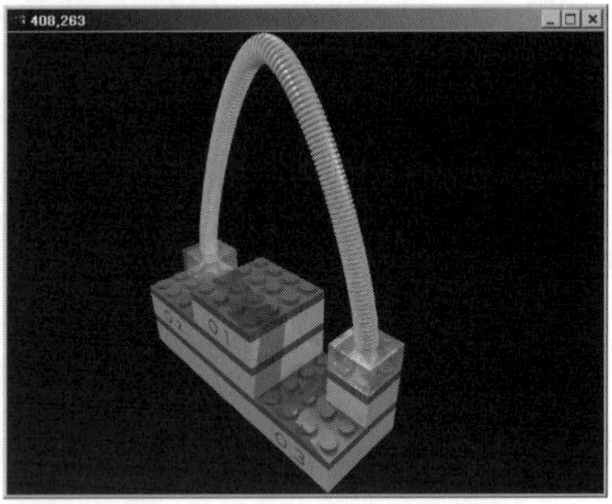

Figure 8.9 MLCad's Rendering of Our Podium

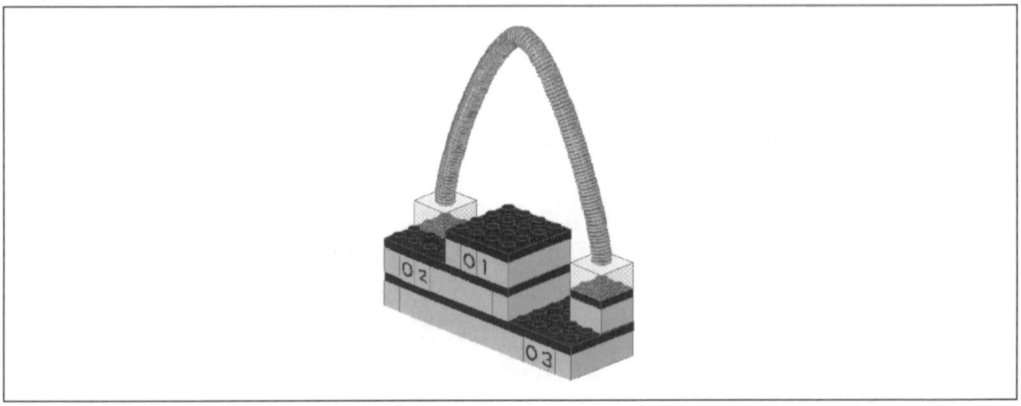

Collectively, POV-Ray calls these three elements a *scene*. L3P provides default lights, camera, and background when it translates our LDraw file to POV format. The Tab Control window in Figure 8.7 shows the beginnings of the description of our LEGO objects in the scene. Figure 8.10 shows chapter8_podium.pov scrolled to the bottom, where we can see the POV descriptions of the background, the camera, and the lights.

L3P has command-line options that let us change the default background, camera, and lights it creates when it translates our LDraw file to POV format. Let's go back to our discussion of L3P and take a look at how this is done.

Figure 8.10 chapter8_podium.pov's Lights, Camera, and Background

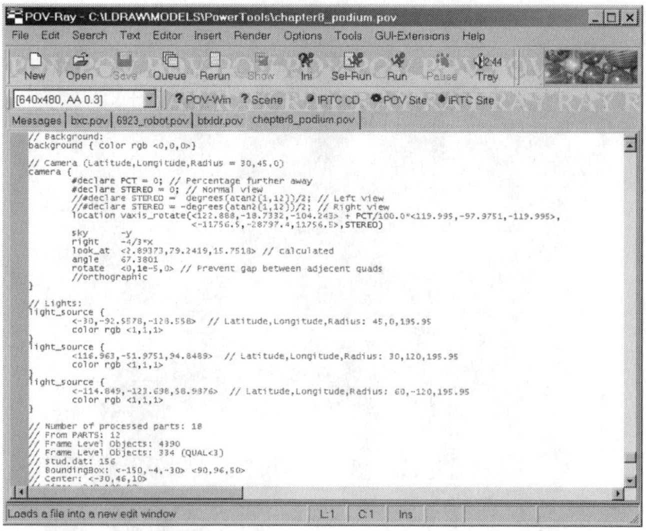

L3P Options

L3P provides with many command-line options that allow us to change the default LDraw-to-POV file format translation. Figure 8.11 shows L3P's own description of its command usage. To get this list, we typed the **L3P** command with no parameters or options, as you can see at the top of the screen.

This is a great reference for readers who understand L3P, but it is by no means a tutorial. Lars Hassing provides documentation for L3P in two other formats: a text file named l3p.txt and a Web page (http://home16.inet.tele.dk/hassing/l3p.html).

The topmost line of the L3P usage output in Figure 8.11 is Hassing's copyright notice. The next line is:

```
usage: l3p [options] modelpath [povpath] [options]
```

In English, this line means that you must provide L3P with the pathname of the file you want translated from LDraw format to POV format. This parameter is referred to as the *modelpath*. You can optionally provide a pathname for the newly created POV-Ray file. (You can tell this is optional because it is in square brackets.) The modelpath and povpath can be relative pathnames (for example, the names of the files in the current directory as shown by our command prompt) or absolute pathnames (which start with C:\ or D:\).

Figure 8.11 L3P Usage Output

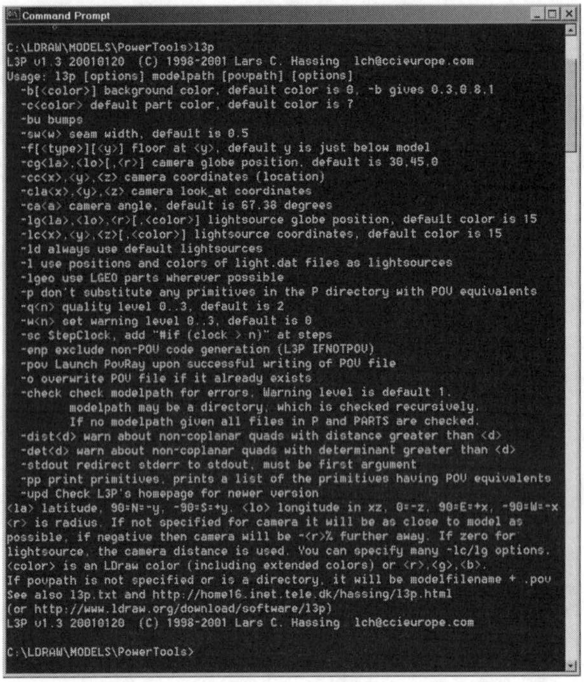

```
Command Prompt

C:\LDRAW\MODELS\PowerTools>l3p
L3P v1.3 20010120  (C) 1998-2001 Lars C. Hassing  lch@ccieurope.com
Usage: l3p [options] modelpath [povpath] [options]
  -b[<color>] background color, default color is 0, -b gives 0.3,0.8,1
  -c<color> default part color, default color is 7
  -bu bumps
  -sw<w> seam width, default is 0.5
  -f[<type>][<y>] floor at <y>, default y is just below model
  -cg<la>,<lo>[,<r>] camera globe position, default is 30,45,0
  -cc<x>,<y>,<z> camera coordinates (location)
  -cla<x>,<y>,<z> camera look_at coordinates
  -ca<a> camera angle, default is 67.38 degrees
  -lg<la>,<lo>,<r>[,<color>] lightsource globe position, default color is 15
  -lc<x>,<y>,<z>[,<color>] lightsource coordinates, default color is 15
  -ld always use default lightsources
  -l use positions and colors of light.dat files as lightsources
  -lgeo use LGEO parts wherever possible
  -p don't substitute any primitives in the P directory with POV equivalents
  -q<n> quality level 0..3, default is 2
  -w<n> set warning level 0..3, default is 0
  -sc StepClock, add "#if (clock > n)" at steps
  -enp exclude non-POV code generation (L3P IFNOTPOV)
  -pov Launch PovRay upon successful writing of POV file
  -o overwrite POV file if it already exists
  -check check modelpath for errors. Warning level is default 1.
         modelpath may be a directory, which is checked recursively.
         If no modelpath given all files in P and PARTS are checked.
  -dist<d> warn about non-coplanar quads with distance greater than <d>
  -det<d> warn about non-coplanar quads with determinant greater than <d>
  -stdout redirect stderr to stdout, must be first argument
  -pp print primitives, prints a list of the primitives having POV equivalents
  -upd Check L3P's homepage for newer version
<la> latitude, 90=N=-y, -90=S=+y. <lo> longitude in xz, 0=-z, 90=E=+x, -90=W=-x
<r> is radius. If not specified for camera it will be as close to model as
possible, if negative then camera will be -<r>% further away. If zero for
lightsource, the camera distance is used. You can specify many -lc/lg options.
<color> is an LDraw color (including extended colors) or <r>,<g>,<b>.
If povpath is not specified or is a directory, it will be modelfilename + .pov
See also l3p.txt and http://home16.inet.tele.dk/hassing/l3p.html
(or http://www.ldraw.org/download/software/l3p)
L3P v1.3 20010120  (C) 1998-2001 Lars C. Hassing  lch@ccieurope.com

C:\LDRAW\MODELS\PowerTools>
```

The povpath can be the pathname of a file or a directory. If no povpath is provided, L3P creates a povpath based on your modelpath name by replacing the modelpath suffix (.dat, .ldr, .mpd) with .pov. If you provide a modelpath filename that has no suffix, L3P adds a .pov suffix. If you specify a povpath that is the name of an existing directory, L3P creates the POV file using these naming conventions but creates the file in the directory indicated. Here are some examples of L3P commands and the locations of the generated POV files:

Command	Location of the generated POV file
`l3p car`	`car.pov`
`l3p car newcar`	`newcar.pov`
`l3p car car.txt`	`car.txt`
`l3p car scenes`	`scenes\car.pov`
`l3p car scenes\newcar`	`scenes\newcar.pov`
`l3p car scenes\car.txt`	`scenes\car.txt`
`l3p car c:\tmp`	`c:\tmp\car.pov`
`l3p car ..\tmp`	`..\tmp\car.pov`
`l3p smartcar.mpd`	`smartcar.pov`

L3P is safe to use, meaning that it will not overwrite any files. (You would use the *-o* option we used earlier to do that.) Don't be alarmed by the size of the POV file (typically 50 to 400KB). POV-Ray handles these large files very quickly.

Options (remember, these are keywords that begin with a dash) can be used either before the *modelpath* or after the *povpath* (if you provide one), but they must typed before you press the **Enter** key. We saw a list of all the L3P options in Figure 8.11. Some of the options require values (indicated by names surrounded by <and >). Some options have optional values (indicated by names surrounded by [< and >]).

The shortest form of L3P command line is:

```
l3p modelpath
```

Here, *modelpath* represents the filename of the LDraw file you want to translate into POV-Ray format.

Camera Options

The L3P camera options are the most important to learn, because they let you view your LDraw design from different angles. There are two important aspects to the camera: placement and orientation. These concepts should be obvious to anyone who has ever taken a photograph. Once you've decided you want to take a picture of something, you decide how close you want to stand from it (placement) and from what angle you want to photograph it (orientation), and then you point and focus the camera to get the picture you want.

Using the XYZ Coordinate System

In Chapter 4, the introduction to MLCad, we described the LDraw coordinate system that uses the X, Y, and Z coordinates to identify the locations of your LEGO parts in virtual space. L3P provides the *–cc* option to let you place your camera using X, Y, and Z coordinates. Unless you tell it otherwise, L3P assumes that the camera is looking at the center of your model. You can use the *–cla* option to make the camera look at the location of your choice instead. The L3P syntax for the *–cc* and *–cla* options are:

```
-cc<x>,<y>,<z>
-cla<x>,<y>,<z>
```

In both these options, <x>,<y>,<z> means fill in the values for X, Y, and Z.

Figure 8.12 is our podium translated to POV using this L3P command with the following *–cc* and *–cla* options:

```
l3p -o -cc110,-103,-91 -cla90,0,-30  chapter8_podium.ldr
```

To render chapter8_podium.pov, move your mouse back into the POV-Ray application. POV-Ray probably noticed that the chapter8_podium.pov file was changed by re-running L3P and will pop up the dialog window shown in Figure 8.12.

Figure 8.12 The File Changed Dialog Window

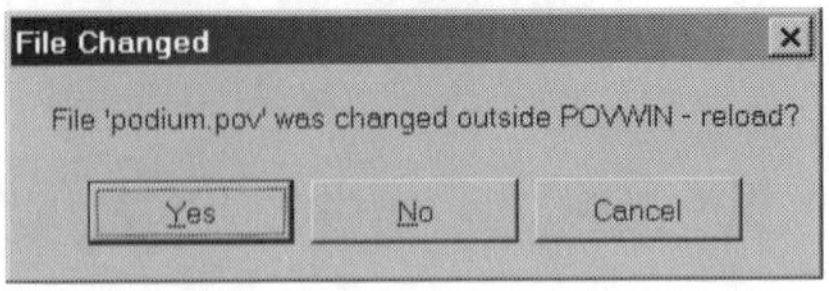

Click the **Yes** button. If this dialog does not show up, use POV-Ray's **File | Close** menu to get rid of the old version of chapter8_podium.pov, then use **File | Open** to open it again. Click POV-Ray's **Run** button in the toolbar to make POV-Ray render our podium. Figure 8.13 shows our podium with a closeup view of the third-place tier.

Figure 8.13 A Closeup of the Third-Place Tier on Our Podium

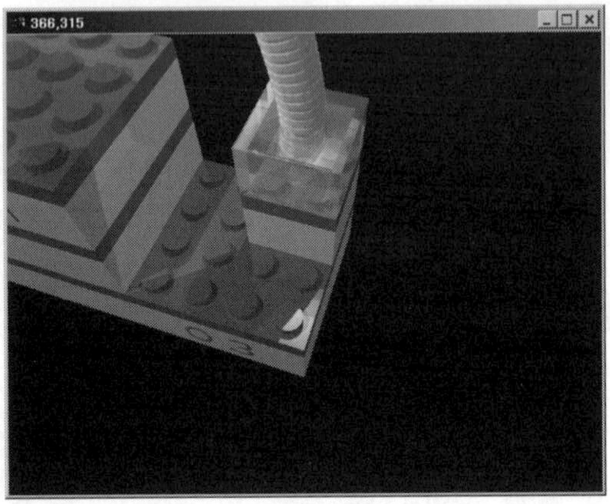

We used MLCad to find the coordinates of the center of the third-place level of the podium. We had to make an educated guess of the coordinates for the camera. When using the *–cc* option, you do not know how close to put the camera to the object without trial and error. If you guess too close, you could cut off part of your LEGO design that you want to see. If you guess too far, your

LEGO design might look very small against a lot of background. Hassing added globe coordinates, a way to place your camera using globe coordinates, to avoid the difficulty of using the LDraw XYZ coordinate system.

Globe Coordinates

Globe coordinates are used by sailing ships and aircraft to navigate around the earth using latitude and longitude. Every place on the earth's surface has a location that can be described in terms of latitude and longitude. *Latitude* is a measure of north and south. *Longitude* is a measure of east and west.

Latitude and longitude are angles that describe imaginary circles drawn on the earth's surface. Figure 8.14 shows a ball made of these circles. The equator is one of these circles and has a latitude angle of 0; it sweeps through all the possible longitude angles from 0 to 180 degrees east and 0 to 180 degrees west. All the other possible latitude circles are parallel to the equator and are either north or south of it. The extreme latitudes are at 90 degrees north (the North Pole) and 90 degrees south (the South Pole). The *diameter* (distance across) these circles get smaller the farther you get away from the equator.

Figure 8.14 Globe Coordinates

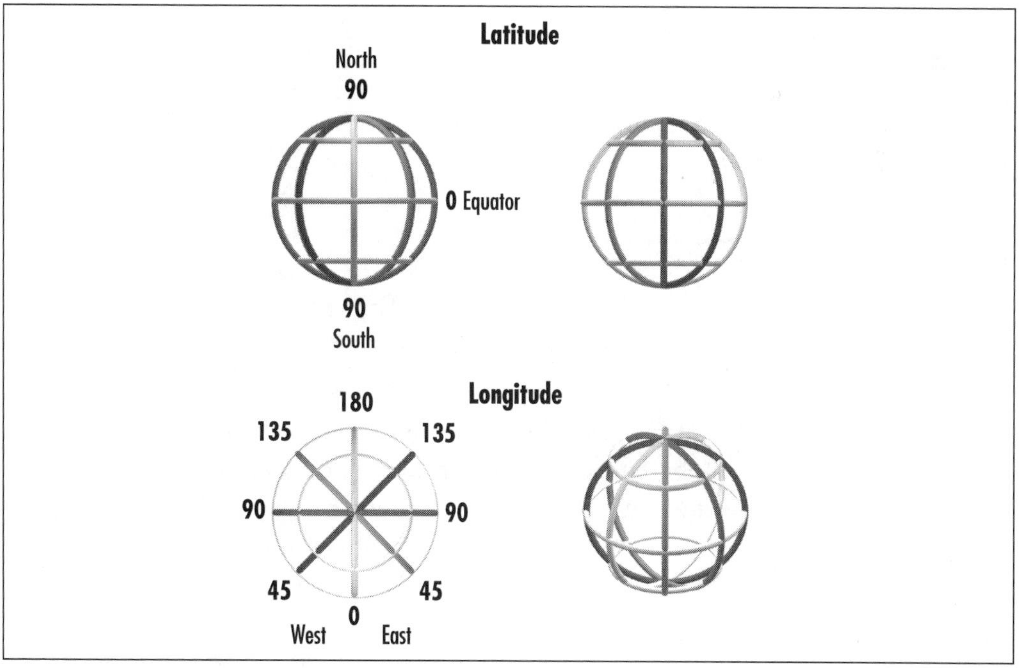

Longitude describes circles around the earth that go through both the north and south poles. All the longitude circles are the same size, the same size as the equator. On Earth, the longitude of 0 degrees is a circle that goes through the North Pole, the South Pole, and Greenwich, England (pronounced *Gren-itch* here in the United States), the place where the globe coordinates of latitude and longitude were invented. As the longitude angles get larger, you travel more west from Greenwich.

For L3P, Hassing uses positive longitude angles to represent east and negative longitude angles to represent west. In L3P, a positive latitude angle means north and a negative latitude angle means south, as shown in Figure 8.15.

Figure 8.15 L3P Globe Coordinates

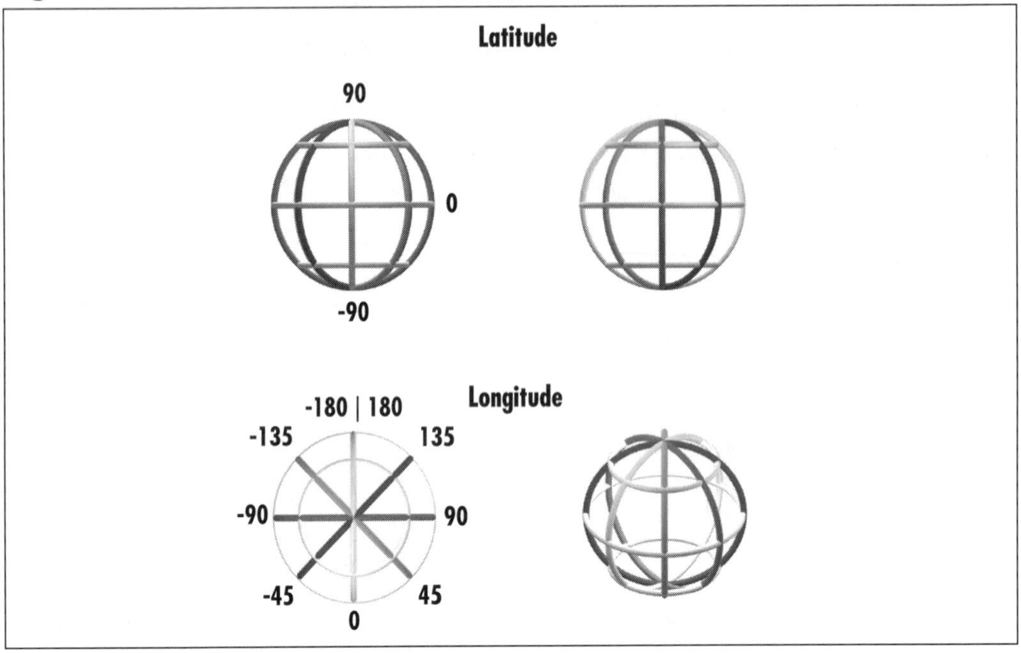

L3P treats the *camera look-at location* as the center of the globe and lets you place the camera with globe coordinates using the *–cg* option. If you do not specify any camera coordinates, L3P assumes the camera look-at location is the LDraw origin (X=0,Y=0, Z=0), a latitude of 30 degrees and a longitude of 90 degrees.

Figure 8.16 shows an example of using the *–cg* (camera globe coordinates) with a latitude of 30 degrees and a longitude of –45 degrees. Viewing the podium with L3P globe coordinates 30, -45 (*latitude*, then *longitude*) gives us a view of the podium with the second-place step closest to the camera. The syntax for the *–cg* option is:

```
-cg<la>,<lo>[,<r>]
```

Figure 8.16 Our Podium Viewed with the *-cg30,-45* Option

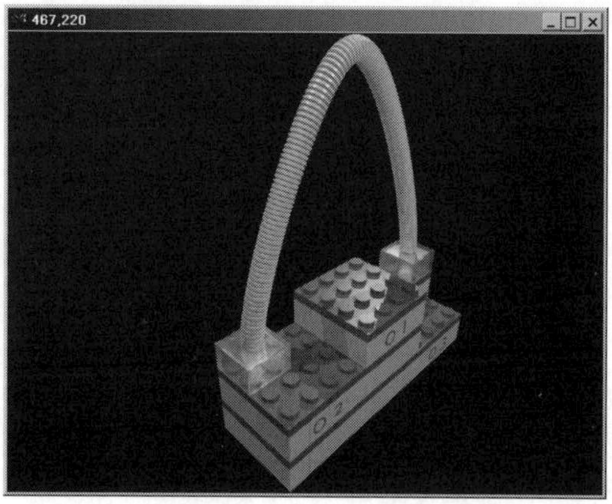

The *<r>* part of the *–cg* option is a *radius modifier*, which we explain next. The radius modifier is optional, as indicated the square brackets that enclose it. The following command gives us a good look at the second-place step on our podium:

```
l3p -o -cg30,-45  chapter8_podium.ldr
```

Radius Modifier

Let's run the podium through L3P again, but we'll use the radius modifier this time. In mathematics, the *radius* is defined as the distance from the center of a circle to its edge or the center of a sphere to its surface. To understand what *radius modifier* means, let's go back to talking about the latitude and longitude as used by sailors and pilots. Sailing ships float on the ocean surface. The distance from the center of the earth to the ocean surface is the *radius* of the earth, often referred to as *sea level*. Airplanes fly above *sea level*. The height the airplanes fly above sea level is referred to as *altitude*. To calculate the height of an airplane to the center of the earth, you need to add its *altitude* to the *radius* of the earth.

When L3P creates our POV file, it needs to know the distance between the camera and the camera look-at location. Hassing calls this distance the *camera radius*. When we do not provide a radius modifier, L3P calculates the distance between the camera and the camera look-at location so that the whole LDraw design can be seen in the rendered picture. We call this distance the *minimum camera radius.* The minimum camera radius acts kind of like sea level does for airline pilots.

By default, L3P uses this minimum camera radius as the camera radius. The radius modifier lets you change the camera radius. If the radius modifier is a positive number, it is used as a replacement for the camera radius. If the radius modifier is negative, it acts something like altitude does for airplane pilots in that it makes the camera move away from the camera look-at location. Using a negative radius modifier makes the camera radius larger. For example, using a camera radius of –10 moves the camera 10 percent further away by using a camera radius that is 110 percent of the minimum camera radius. In Figure 8.17, we use a negative radius modifier to back the camera away from the podium. The L3P command to do this is:

```
l3p -cg30,45,-50 -o chapter8_podium.ldr
```

Figure 8.17 Our Podium from a Greater Distance

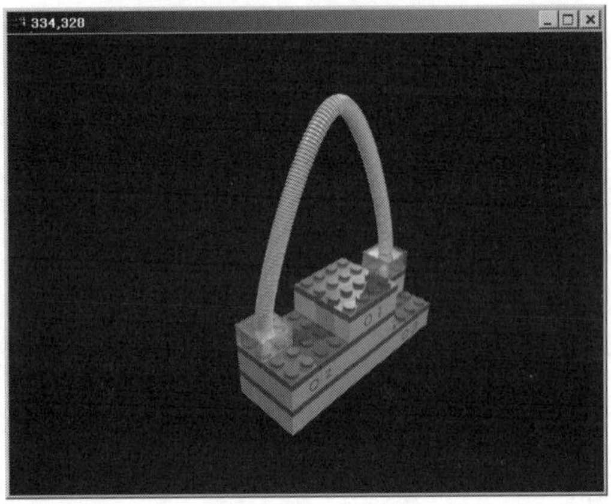

Moving the camera further away effectively gives you some space around the model, rather than the default "best fit."

Lighting Options

When we run L3P, it automatically creates three white lights for us. If it did not create lights, our picture would be all black; to avoid that fate, L3P creates these three lights by default. In Figure 8.18, you cannot see the lights themselves (they are invisible), but you can tell that there are three of them by the three shadows cast by the podium arch.

Figure 8.18 Podium Shadows from Default Lights

NOTE

Throughout this section on lights, we use L3P's *–f* option (explained in the Surroundings Options section later in this chapter) to create gray floors just below our model so that we can see the shadows cast by our lights and models.

L3P allows you to override the default light settings by letting you define your own lighting setup. You can specify lights using globe coordinates (latitude, longitude, radius) or Cartesian coordinates (X, Y, Z). The format for the option that defines a light using globe coordinates is:

```
-lg<la>,<lo>,<r>[,<color>]
```

We used the following command to create a single light that is directly above the podium but low enough to be inside the arch:

```
l3p -o -lg90,0,40 -fg chapter8_podium.ldr
```

You can see the results in Figure 8.19.

When using the *–lg* option, you must provide a latitude, longitude, and radius. The radius *<r>* is a radius modifier and works as we discussed. If you do not provide *<color>*, L3P assumes the light is white. The *<color>* can be in either LDraw color format or RGB.

You can define lights using XYZ coordinates as well the *–lc* option:

```
-lc<x>,<y>,<z>,[,<color>]
```

Figure 8.19 A Light Specified Using Globe Coordinates

We used the following command to create two light sources, one inside each podium arch base:

```
l3p -o -lc-130,32,30 -lc70,32,30 -fg chapter8_podium.ldr
```

You can see the results in Figure 8.20. You can use as many *–lg* and/or *–lc* options as you want to create as many lights as you desire.

Figure 8.20 Podium Arch Bases with Internal Lights

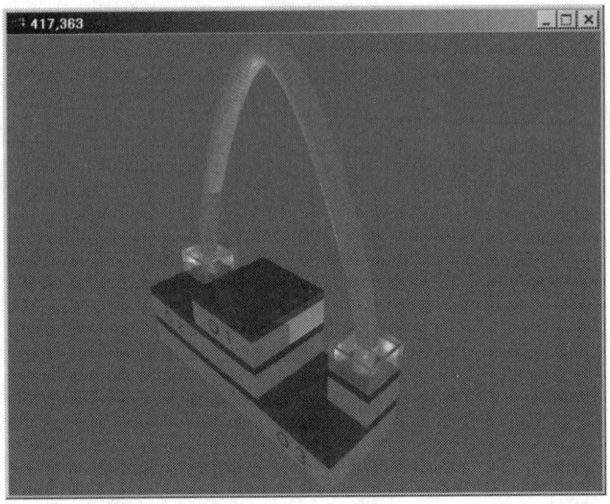

As an alternative to the *–lc* and *–lg* options, you can add light sources to your LDraw design using MLCad. You do this by adding light.dat parts to your design and using the *–l* option to L3P. The location and color of the lights are important, but the rotation is not. Figure 8.21 shows figure8_21.ldr file with four light sources added using MLCad. The light sources are the three plus (+) signs in front of the podium near the floor and the one + sign behind the podium.

Figure 8.21 Our Podium with Four light.dat Light Sources

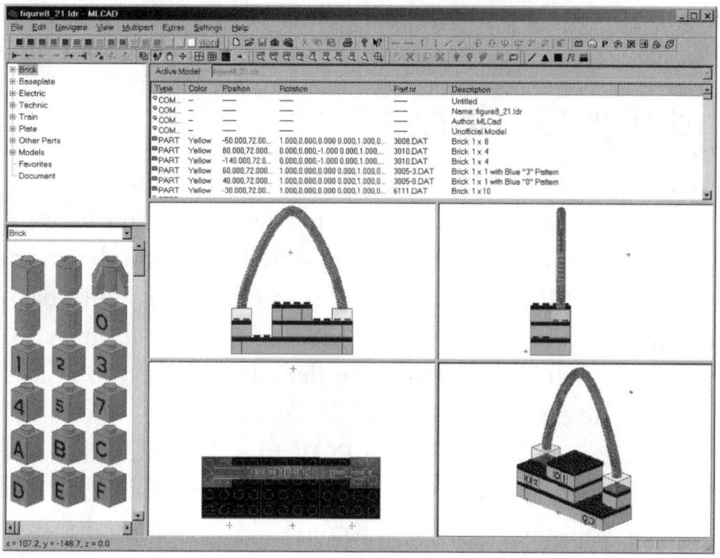

The *–ld* option can be used to force L3P to use its default lighting values, even if you use *–lc*, *-lg,* or *–l* options. We used the following command to convert figure8_21.ldr to POV format (including the light.dat usages) and use the L3P default lights. The results are put in the file figure8_22.pov.

```
l3p -o -ld -l -fg figure8_21.ldr figure8_22.pov
```

Figure 8.22 shows the figure8_22.pov file rendered using POV.

Figure 8.22 Using light.dats, *-ld* and *–l* Options

You can see the effects of the footlights at the base of the podium quite clearly. You can also see that the light source added behind the podium has the effect of wiping out the shadows of the default light sources near the base of the podium.

Surroundings Options

Near the beginning of this chapter, Figures 8.5 and 8.6 showed you an L3P/POV-Ray image and MLCad for comparison. There are many differences between these images, but the most striking difference is the background color. In MLCad, we see our LEGO designs drawn in front of a white background (MLCad's default background color). L3P does not know about MLCad's *Background* meta command to override the default background color, so setting a background in MLCad has no effect on L3P or POV-Ray.

When translating our LDraw files to POV files, L3P provides a black background by default, as shown back in Figure 8.5. L3P provides the *–b* option (for *background*) that lets you choose a background color to be used instead of black. The syntax for the *background* option is:

```
-b[<color>]
```

The less-than sign (<), combined with a greater-than sign (>), tells you that instead of typing *<color>* on the command line, you actually type a value for the color. You can think of the combinations of less-than and greater-than signs as meaning "fill in the blank." The square brackets around *<color>* mean that color is optional.

You specify a color in either LDraw format or by using POV-Ray's RGB format. RGB stands for *red, green, and blue*. By combining various amounts of red, green, and blue, you can make any color you want. The amount of color is a number from 0 to 1, indicating how much of that color you want. Table 8.1 shows some common colors and their POV-Ray encodings.

Table 8.1 Examples of POV-Ray RGB Color Encoding

Color Name	POV-Ray Format
Black	0.0,0.0,0.0
Gray	0.5,0.5,0.5
White	1.0,1.0,1.0
Red	1.0,0.0,0.0
Blue	0.0,0.0,1.0
Yellow	1.0,1.0,0.0

If you use the *–b* option without providing a color, L3P uses the POV-Ray color 0.3,0.8,1, which means a little red, a lot of green, and as much blue as you can get. This gives you a nice light-blue background. Type the following in your Command Prompt window and press **Enter** to translate our podium to POV-Ray with a white background:

```
l3p -o -b15 chapter8_podium.ldr
```

Figure 8.23 shows our podium with a white background.

Figure 8.23 Our Podium with a White Background

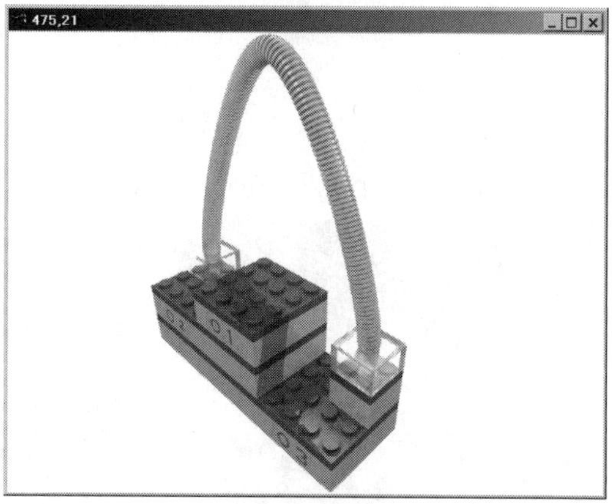

L3P also provides a floor option (*-f*) that creates one of two types of floor below your LDraw model. This option is accessed with the following command:

```
-f[<type>][<y>]
```

The floor *<type>* and *<y>* are optional. The *–fg* (or simply *–f*) form of the option gives you a solid light-gray floor, as shown in Figure 8.24. Adding a floor makes a stunning change to the look of our podium scene because of the shadows cast by the podium.

Providing a <y> value defines the altitude at which the floor should exist. If a <y> value is not provided, L3P places the floor directly below your model.

Using the *–fc* form of the floor option gets you a red-and-white checkerboard floor, as shown in Figure 8.25. The checkerboard floor makes it easy to see that the floor goes on into infinity.

Figure 8.24 Our Podium with a Gray Floor

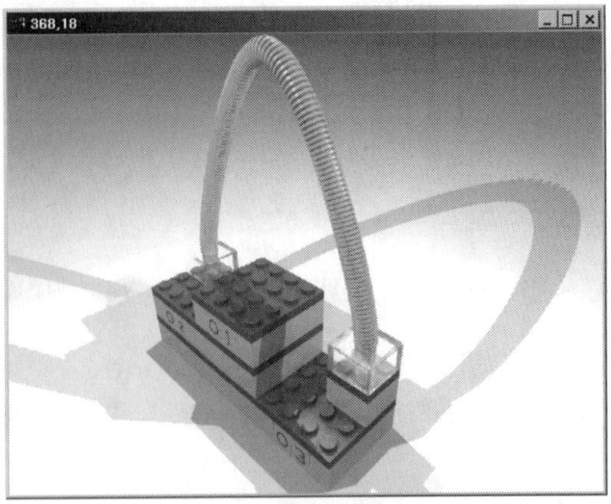

Figure 8.25 Our Podium with a Checkerboard Floor

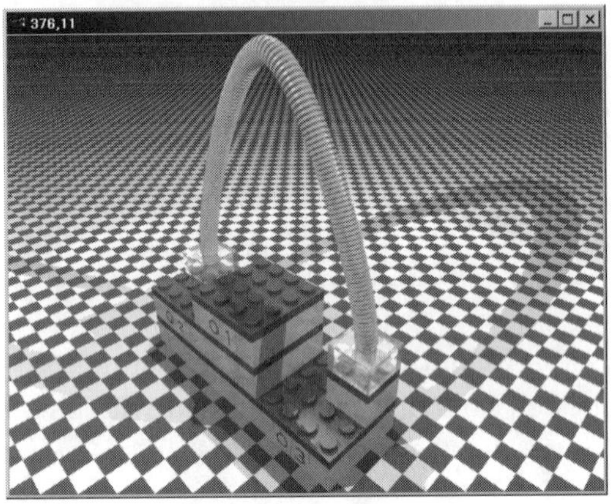

Model Options

L3P provides a number of options that affect the process of translating our design from LDraw format to POV-Ray format. Here we take a look at some of the more useful options.

Bumps and Seams

Adding or removing bumps and seams from your final POV-Ray image is an interesting way to customize the final image. Bumps can give your image a more realistic feel. To make bumpy the flat surfaces in the translated design, use the *–bu* command-line option.

In the perfect world of LDraw and POV-Ray coordinates, two bricks placed right next to each other can appear as one brick because the seam between the two bricks is zero width. When translating from LDraw to POV, L3P creates gaps between all the LDraw parts so that you can see the seams. You can change the seam width from the default of 0.5 LDU (in LDraw units, a 1 x 1 brick is 20 LDU wide) to a different value using the *–sw<w>* option. A seam width value *<w>* of 0.0 eliminates seams. We used this command at the command prompt to get the POV file rendered in Figure 8.26. It shows our podium with bumps and no seams.

```
l3p -o -fg -bu -sw0 chapter8_podium.ldr
```

Figure 8.26 A Seamless Bumpy Podium

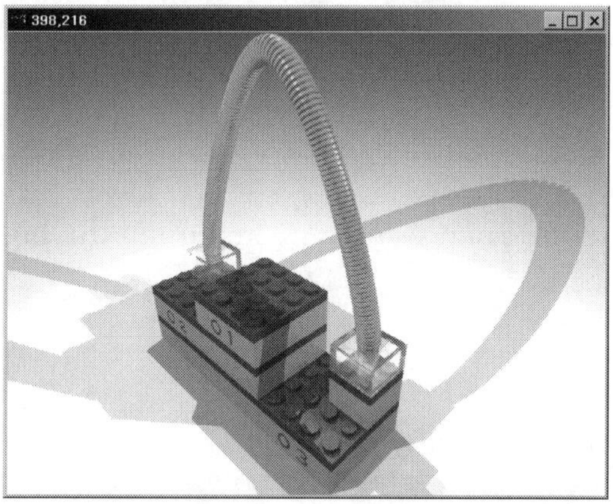

The Quality Option

The ray-tracing rendering technique that POV-Ray uses can be quite slow, especially on older computers, so L3P provides a quality option to control now much work POV-Ray must do to render your picture. L3P controls the amount of work your computer must do by controlling the number of triangles or rectangles it uses to create your POV file and the level of detail in which the surfaces should be rendered. The syntax for the quality option is:

```
-q<n>
```

An *<n>* value of 0 is the lowest quality, which replaces each LDraw part with a box the same size as the part it replaces. Quality 0 can be very handy when you are in a "trial and error" mode of designing a complicated scene. An *<n>* value of 3 is the highest quality level. Figure 8.27 shows our podium drawn using a quality of 0, which would look like this:

```
-q0
```

Figure 8.27 A Box-Quality Podium

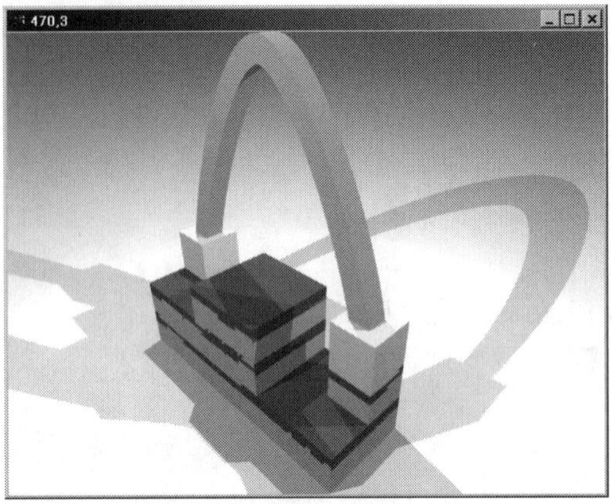

Figure 8.28 shows the same scene rendered using the highest quality of 3, which would look like this:

```
-q3
```

Figure 8.28 The Highest-Quality Podium

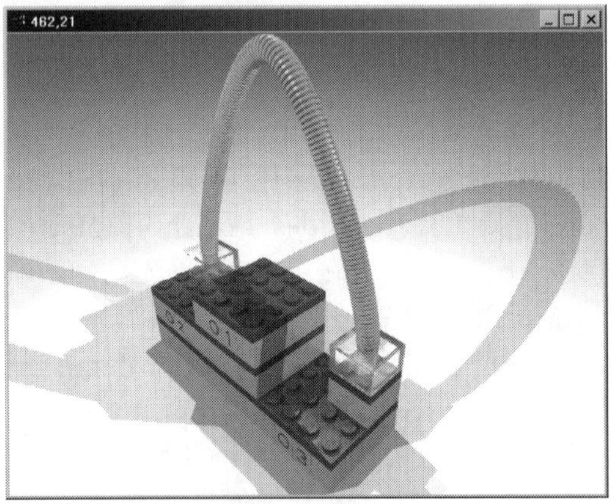

For more details on the quality option, visit www.hassings.dk/l3/l3p_q.html.

The Color Option

The *–c*<color> option is used when you want to directly render a part without having a model file referencing the part. Parts are by default colorless, and L3P defaults to gray (LDraw color 7). Use the *–c*<color> option to get another color.

Using LGEO Parts

The original LDraw-to-POV translator was L2P, meaning *LDraw to POV.* L2P uses a database of high-quality versions of LEGO parts that are much more real-looking than the LDraw versions of parts (mainly due to their rounded edges). This database, called LGEO parts, contains approximately 800 of the more than 2,000 parts that compose the LDraw library. By default, L3P only uses LDraw part shapes when translating from LDraw to POV.You can use the *–lgeo* option to have L3P use LGEO parts if they are available. Figure 8.29 shows two versions of a LEGO Technic bushing.The LDraw version is on the left, and the LGEO version is on the right.

Figure 8.29 LDraw and LGEO Versions of Technic Bushings

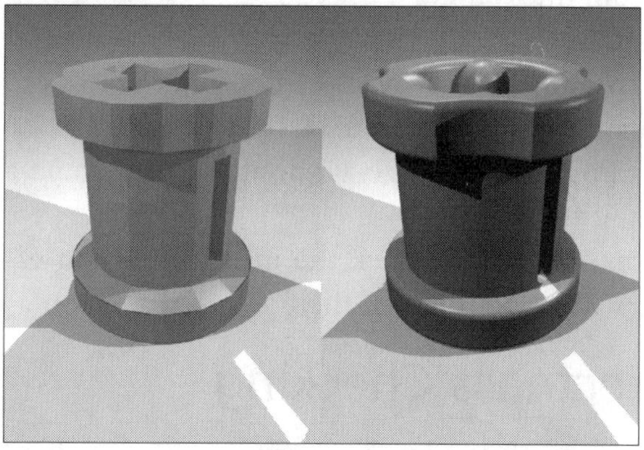

In Chapter 2, when you installed L3P onto your computer, the LGEO library was also installed. At that time, an environment variable called *LGEODIR* was set up in your AUTOEXEC.BAT file. LGEODIR tells L3P where the LGEO parts database is installed. To render the LGEO part in Figure 8.29 using POV-Ray, we had to tell POV-Ray where to find the LGEO parts. To do this, click the **Ini** button in the POV-Ray Toolbar. This will bring up the Render Settings dialog box, shown in Figure 8.30. You then type **Library_path-c:\ldraw\lgeo** into the **Command line options** field of the Render Settings dialog window.

Figure 8.30 POV-Ray's Render Settings Dialog Window

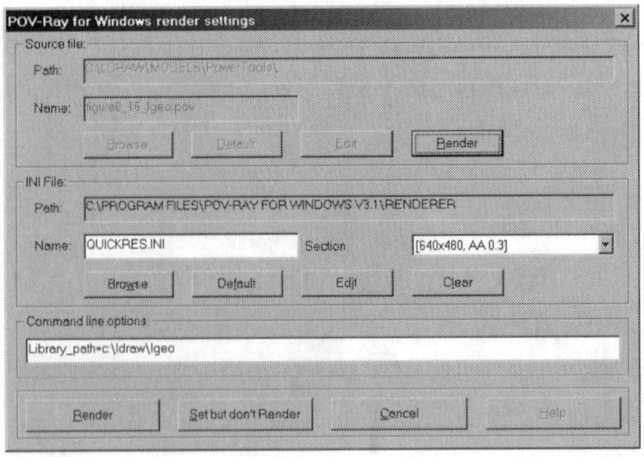

Miscellaneous Options

When L3P translates our LDraw files from LDraw format to POV format, it replaces some of the LDraw primitives with POV equivalents. This has two benefits: higher-quality images and faster rendering. You can prevent this translation using L3P's *–p* option. You can get a list of the POV primitives that L3P supports using L3P's *–pp* option.

The *–stdout* option makes L3P print errors and warnings to the command prompt using standard output instead of standard errors. This makes it possible for programs (such as LPub) that use L3P without the user's help to capture any errors that L3P might encounter.

The *–upd* option makes L3P check the installed version of L3P against the version on Hassing's Web site to see if there is a new version available.

Model Correctness Checking

The *–w<n>* option controls the level of warning outputs that L3P prints when processing your LDraw files. The default value for *<n>* is 0, which means L3P only prints errors. Values of *<n>* from 1 through 3 make L3P print increasing numbers of warnings. An *<n>* of 3 makes L3P the most picky, therefore printing the most warnings.

The *–check* option can be used to check the quality of LDraw parts. You can use the following commands to check a single LDraw file:

```
l3p -check <part_file>
```

To check all the files in a directory, use this command:

```
l3p -check <part_directory>
```

To check all the parts in the parts directory, use this command

```
l3p -check
```

Typically, the *–check* option is used by people making their own parts that they'd like checked for correctness. The *–dist<d>* and *–det<d>* options can be used in combination with the *–check* option to make checking even more thorough.

The POV Step Clock

L3P's *–sc* option and POV-Ray's *clock* feature can be used to make a sequence of images from your LDraw file. When you run an LDraw file with LDraw *Step* meta commands in it through L3P, and you use the *–sc* option, L3P replaces the *Step* meta statements with POV-Ray clock controls. When you run POV-Ray, you provide it with a clock value. The clock value controls how much of the model is rendered. This feature can be used to create animations of your LEGO designs or simple building instructions. To demonstrate the step clock, we chose a simple building instruction example. We recommend using LPub for building instructions because it has many advantages, instead of L3P's step clock feature, since it is short and simple. Figure 8.31 shows a simple LDraw design in MLCad with a total of three steps.

Figure 8.31 A Simple LDraw Design with Steps

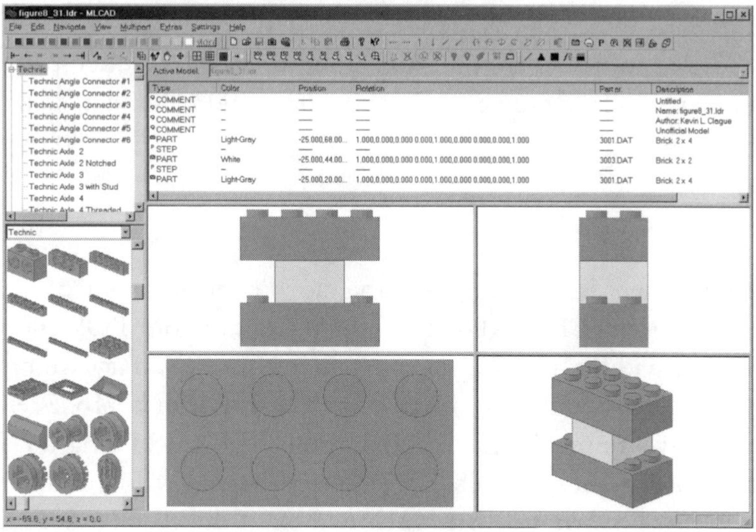

Using the following command, we created a POV file with step clock controls in it. You can see the step clock controls in Figure 8.32.

```
l3p -o -b15 -sc figure8_30.ldr figure8_31.pov
```

Figure 8.32 A POV File with Step Clock Controls

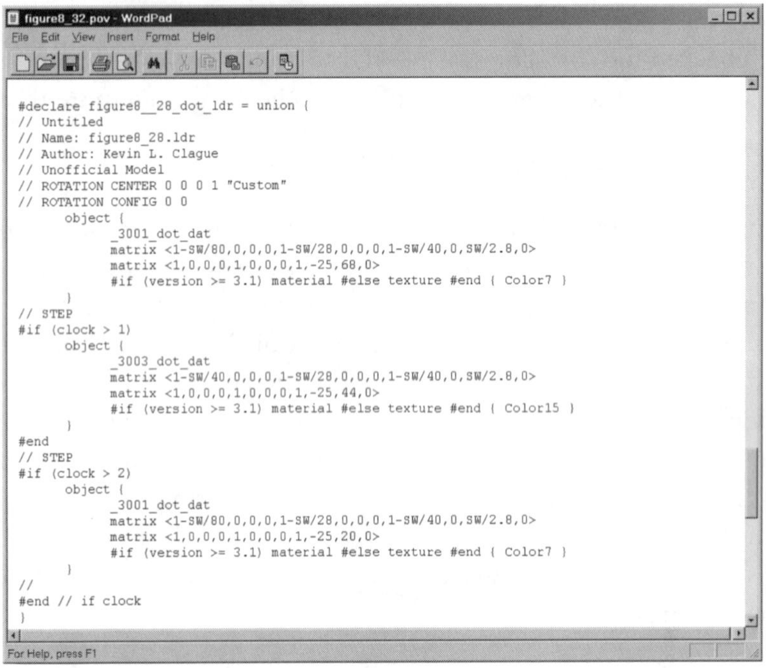


```
#declare figure8__28_dot_ldr = union {
// Untitled
// Name: figure8_28.ldr
// Author: Kevin L. Clague
// Unofficial Model
// ROTATION CENTER 0 0 0 1 "Custom"
// ROTATION CONFIG 0 0
      object {
            _3001_dot_dat
            matrix <1-SW/80,0,0,0,1-SW/28,0,0,0,1-SW/40,0,SW/2.8,0>
            matrix <1,0,0,0,1,0,0,0,1,-25,68,0>
            #if (version >= 3.1) material #else texture #end { Color7 }
      }
// STEP
#if (clock > 1)
      object {
            _3003_dot_dat
            matrix <1-SW/40,0,0,0,1-SW/28,0,0,0,1-SW/40,0,SW/2.8,0>
            matrix <1,0,0,0,1,0,0,0,1,-25,44,0>
            #if (version >= 3.1) material #else texture #end { Color15 }
      }
#end
// STEP
#if (clock > 2)
      object {
            _3001_dot_dat
            matrix <1-SW/80,0,0,0,1-SW/28,0,0,0,1-SW/40,0,SW/2.8,0>
            matrix <1,0,0,0,1,0,0,0,1,-25,20,0>
            #if (version >= 3.1) material #else texture #end { Color7 }
      }
//
#end // if clock
}
```

In Figure 8.32, three objects are listed: _3001dot_dat, _3003_dot_dat, and _3001_dot_dat again. These are the POV equivalents of the three parts we added in our LDraw design in Figure 8.31. Notice that _3003_dot_dat is enclosed in this code:

```
#if (clock > 1) {

#end
```

This *#if* statement is an example of the clock control statement that L3P inserted into our model. If we didn't use the *–sc* option, our POV file would look just like this, but without the *#if/#end* statements. In POV-Ray, we provide a *clock* value. As POV-Ray looks through our POV file, it only processes the sections of the file where the *#if (clock* comparisons hold true. Using the **Ini** command button on POV-Ray's Toolbar, we can set the *clock* value using POV-Ray's *+K* option in the **Command line options** box of the Render Settings window, as shown in Figure 8.33.

Figure 8.34 shows the result of rendering with a clock value of 2. Notice that only the parts from the first two steps of Figure 8.31 show up in the image.

Figure 8.33 Setting a Clock Value of 2 Using the *+K* Option

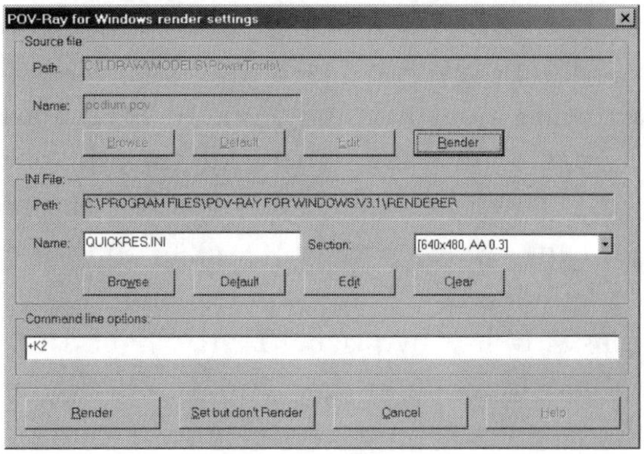

Figure 8.34 Rendering figure8_31.pov with Only Two of Three Steps

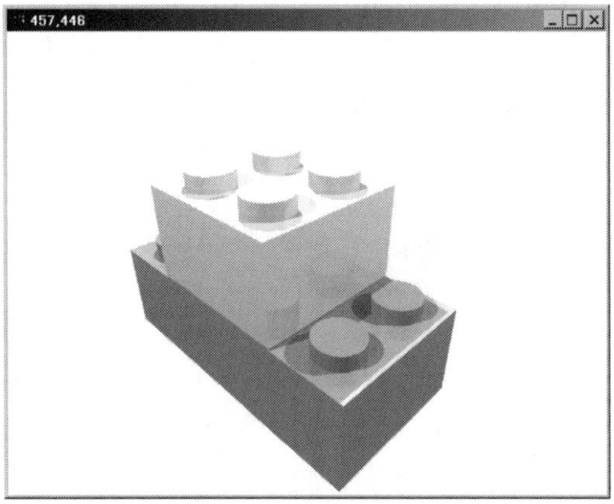

Summary

L3P is a powerful tool that marries the wonderful world of LEGO CAD to the advanced capabilities of the POV-Ray ray-tracing image-rendering tool.

L3P is a command-line tool run using a Windows command prompt. LP3 has many options that let you control the camera, lighting, surroundings, and model characteristics as your LDraw file is translated to POV-Ray scripting format. You can place camera and lights using Cartesian (XYZ) coordinates or globe coordinates (latitude and longitude). L3P allows you to define your lighting setup using command-line options as well as by placing LDraw light.dat parts directly into your LDraw file. L3P provides options for defining a background of whatever color you want. L3P also allows you to create either a solid-gray or red-and-white checkerboard floor under your LEGO model.

L3P was written by Lars Hassing as a replacement for the L2P program, the original LDraw-to-POV-Ray translator. L2P provided its own library of POV-Ray parts that are more accurate models of the real-world LEGO parts than the LDraw models. L2P's part library is called the *LGEO part library*. L3P has support for the LGEO part library and can use these parts in translating your LDraw files to POV-Ray.

POV-Ray is a world-class image-rendering tool that uses a photorealistic image-rendering technique called *ray tracing*. It can take from minutes to hours to render a POV-Ray image, depending on the complexity of the model being rendered and the quality level you choose when rendering your model. The most important key to faster rendering is a faster computer. This chapter barely scratches the surface of POV-Ray's capabilities and features.

Solutions Fast Track

Running L3P

☑ Use the **Start** menu to reach the Windows command prompt.

☑ Change the directory to the directory that contains the LDraw file you want translated to POV.

☑ Use the L3P command to translate your file.

Running POV-Ray

- ☑ Use the **Windows' Start | Programs | POV-Ray for Windows 3.1 | POV-Ray for Windows** menu to start POV-Ray.
- ☑ Use POV-Ray's **File | Open** menu to open your POV-Ray file created by L3P.
- ☑ Use POV-Ray's **Run** tool button to render your POV-Ray file.

L3P Options

- ☑ Camera options include the following:
  - The *-cla* option defines where the camera is pointed in LDraw XYZ space.
  - The *-cc* option defines where the camera is placed in the LDraw XYX space.
  - The *-cg* option lets you define a camera using globe coordinates.
- ☑ Lighting options include the following:
  - The *-lc* options can define light sources using LDraw XYZ coordinates.
  - The *-lg* options can define light sources using globe coordinates.
  - The *-l* option allows light sources to be defined directly in your LDraw file using light.dat parts.
  - The *-ld* option makes L3P add its normal default lights to the lights you define.
- ☑ Surroundings options include the following:
  - The *-b* option lets you define the color of a backdrop behind your model.
  - The *-f* option lets you create gray or checkerboard floors under your model.
- ☑ Model options include the following:
  - The *-bu* option adds surface imperfections to your models that otherwise look too perfect to be real.

- The *-sw* option lets you create seams between bricks to make it easier to see where one brick stops and the next one starts.
- The *-q* option lets you control the number of triangles and rectangles used to render your images and the detail of the surfaces. Higher quality means slower image calculation.
- The *-lgeo* option lets you use high-quality LGEO parts when they're available.

Frequently Asked Questions

The following Frequently Asked Questions, answered by the authors of this book, are designed to both measure your understanding of the concepts presented in this chapter and to assist you with real-life implementation of these concepts. To have your questions about this chapter answered by the author, browse to **www.syngress.com/solutions** and click on the **"Ask the Author"** form.

Q: What does the L3P program actually do?

A: It translates your LEGO design from LDraw file format to POV-Ray file format. The translated file can be used with POV-Ray to render photorealistic images of your LEGO designs.

Q: What exactly is a command-line program? How is it different from the standard Windows program I'm used to using?

A: A command-line program is a program that is run using a Windows command prompt. You run a command-line program by typing its name and possibly some parameters to tell the program what to do.

Q: When do I use the L3P command-line options—before, after, or during the translation of my files from .LDR/.DAT format to .POV format?

A: The L3P command-line options are provided on the command line following the program name *L3P*. The options can be used to make L3P modify its behavior from the default behaviors.

Q: What are the benefits and drawbacks of using the XYZ coordinate system to position the cameras? How about the globe positioning system?

A: You can easily use MLCad to figure out XYZ coordinates, but it could take many tries to get the camera at the right distance from your LEGO design so that the design just fits in the rendered picture. Using globe coordinates, you tell L3P the latitude and longitude angles you want to view the model from, and L3P automatically puts the camera at the right distance to nicely frame your LEGO design.

Q: What sort of errors or warnings might I expect to see when running my files through L3P?

A: The most common errors occur if you specify an LDraw filename that does not exist or if your LDraw file you try to translate uses LDraw parts you do not have installed in your LDraw parts library. Some options can be used by advanced users to check the geometries of their designs.

Chapter 9

LPub

Solutions in this Chapter:

- The LPub Main Window
- Single-Image Generation
- Creating Building Instructions
- Saving Option Configurations
- Instruction Philosophies and Tips

☑ Summary

☑ Solutions Fast Track

☑ Frequently Asked Questions

Introduction

LPub is a program I wrote as a graphical user interface (GUI) to L3P and POV-Ray that allows you to create still images and building instructions for your LEGO designs.

When Syngress first contacted us about authoring for the first of their series of LEGO MINDSTORMS project-based books, *10 Cool LEGO MINDSTORM Dark Side Robots, Transports, and Creatures*, we were curious about what exactly this project would entail. As authors, our job was to create original robots of our own design and then create artwork that showed the step-by-step building instructions. When we asked if Syngress had any methods in mind for creating the artwork for the step-by-step building instructions, they told us that they were still researching ways of creating the artwork but had little more in mind than a paint program of some type.

My first love when building with LEGO is two-legged walkers. I built my first successful biped before I even learned that there is a large community of Internet LEGO fans. Once I got my first walker to walk and turn, I decided to look on the Internet to see if anyone had made anything like it. I was surprised by two things: No one had made a walking design like his, and people had a CAD program to document their designs: MLCad. I had dreaded the thought of having to take my walker apart, so I was pleased to find that it would be possible to document my walker in MLCad before taking it apart. This was my first step into the wonderful world of LEGO CAD.

By the time Syngress contacted me about the book, I'd played around with MLCad quite a bit and had used it to document most of my designs. I knew about LDraw STEPs and MLCad ROT-STEPs (rotation steps). I proposed to Syngress that we use MLCad to document our robot designs and use MLCad picture sequences as the images for the book. Jon Babcock, our editor and primary contact at Syngress, did some homework on LEGO CAD on the Internet. He suggested that we look at using MLCad, L3P, and POV-Ray to make extremely realistic-looking pictures of our robots. This started me down the road of developing LPub, a program that automatically creates step-by-step building instructions, where each step has an image of the partially assembled design and an image that lists the parts added to the step.

I have programmed computers for 30 years and enjoy making handy computer programs. I jumped at the chance to write tools to automate the generation of building instructions. LPub (which stands for *LEGO Publishing*) started out as a clunky Perl script that was a bit cumbersome but served as a good starting point for development. I then rewrote it using the C++ programming language and gave it a GUI.

LPub as it exists today is a program that glues together the following programs and processes that already existed in the LEGO CAD community:

- **MLCad and the LDraw file format** A powerful duo of programs for building and rendering virtual LEGO designs.
- **L3P** A flexible and powerful tool that translates LDraw format files into POV-Ray scripts.
- **POV-Ray** A world-class ray-tracing rendering tool that makes photorealistic images from information in POV-Ray scripts.

After much effort and four robots created for *10 Cool LEGO MINDSTORMS Dark Side Robots, Transports and Creatures*, I got a break from authoring and fully overhauled the user interface for LPub. I realized that I could generalize LPub by providing a GUI to many of the L3P and POV options, eliminating the need to use the command prompt or switch between the command prompt and the POV-Ray application. This ability made LPub very similar to the L3P Add-On (L3PAO) program by Jeff Boen (Onyx) at www.flash.net/~onyx/L3PAO/L3PAO.html, with one huge advantage: the ability to make advanced, step-by-step building instructions.

The LPub Main Window

LPub is a Windows application that can be started using the **Start | Programs | LPub** menu. LPub's main window is shown in Figure 9.1. LPub's interface contains a menu of three options at the top of the screen:

- **File** Used to open and save files in LPub.
- **Generate** Used to actually generate the instruction steps using your LDraw files.
- **Help** Used to access documentation on using LPub.

Below the menu is a standard set of Windows tabs that let you access the option configurations for all the tools with which LPub interfaces. Each tool has many options that let you control the tool's behavior. The supported tools are:

- **L3P** The powerful tool that we learned about in Chapter 8, which translates LDraw files to POV-Ray script format.

- **POV** The world-class ray-tracing program that creates photorealistic images through computer-intensive techniques.
- **LSynth** The bendable part synthesizer that creates tubes, cables, and rubber bands for MLCad.
- **LPub** The core program that creates step-by-step building instructions.

Figure 9.1 LPub's Main Window

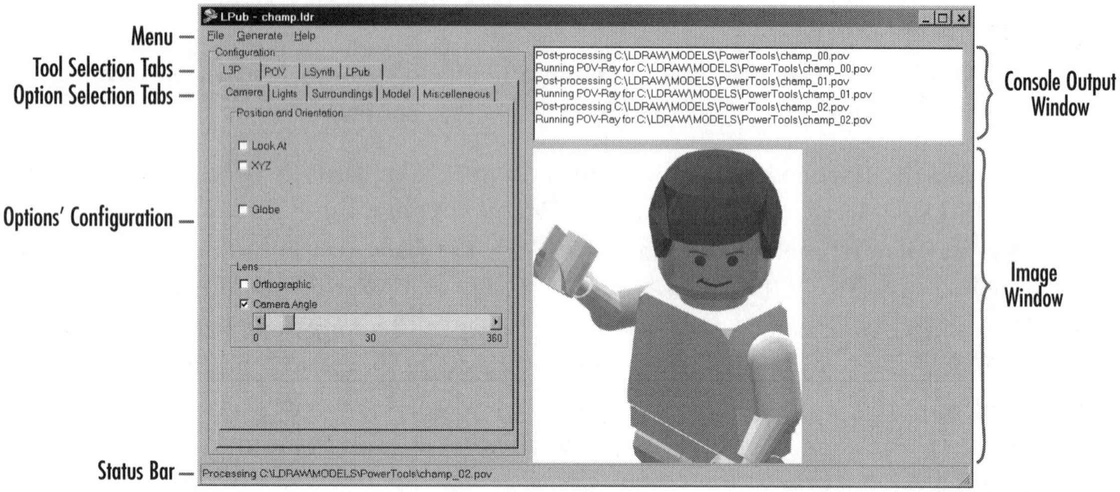

L3P, POV, and LPub have so many options that each has its own set of option selection tabs that provide access to its options by subject. Throughout the rest of this chapter, we walk you through each tool's option configuration tabs and explain how each option works.

The white Console Output window to the right of the main LPub screen is where LPub dumps out information as it is processing your files. It can take hours to generate all the building instruction images for a complicated LEGO design. Information is printed into this window as the work progresses. The Status Bar indicates which files are currently being processed. The Image Window displays the last generated image, so you can see what some of the results look like.

LPub provides two major capabilities: generating a single image of your completely assembled design and generating a sequence of images showing how to build your design step by step. When you choose to generate a sequence of building steps, LPub creates two images for each step:

- **A construction image** A partially assembled design image, showing the parts that have been added up to that point in the construction of the model.
- **A part list image** An image that lists all the parts added in that particular step.

Let's take a look at how to perform each of these steps.

Single-Image Generation

To generate a single image of our MLCad generated model, we first need to open its LDraw file using LPubs **File | Open LDRAW File** menu, shown in Figure 9.2. This action pops up a standard Windows **Open** dialog screen, as shown in Figure 9.3. The examples used in this chapter were installed when you went through the installation procedure described in Chapter 2. The examples are in a folder named PowerTools that is under the standard LDRAW\MODELS folder. Go ahead and select the **chapter_9_champ.ldr** minifig file you built in the earlier chapters, as shown in Figure 9.3.

Figure 9.2 The Open LDRAW File Menu

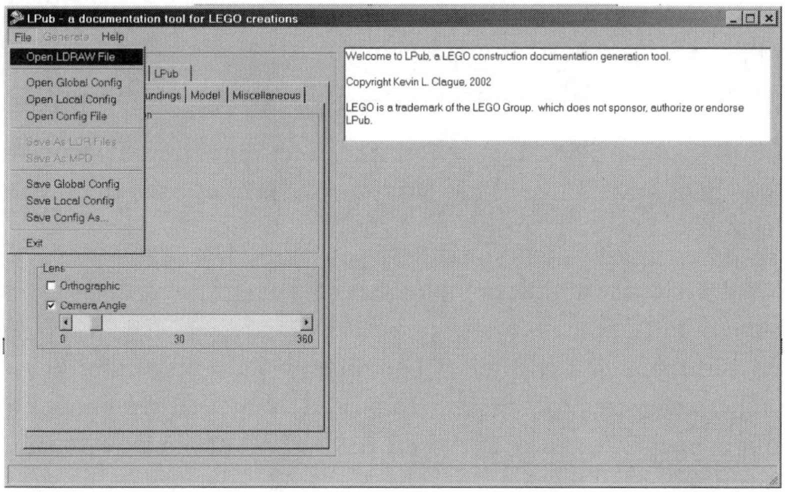

Figure 9.3 The Open Dialog Window

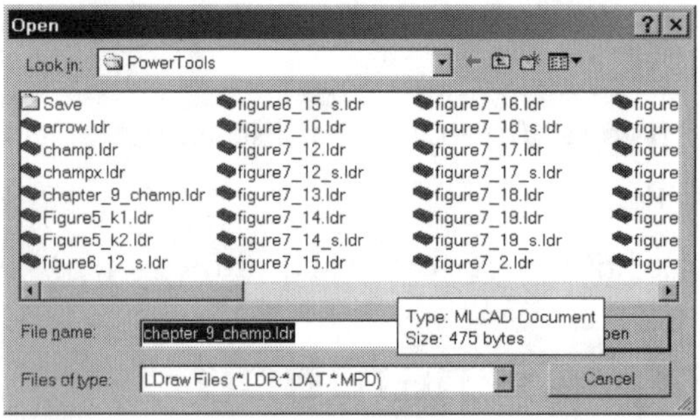

Click the **Open** button to open champ.ldr. To generate a picture of champ, use LPub's **Generate | Complete Assembly** menu. This makes a few things happen in quick succession:

1. LPub pops up a cancel dialog window that you can use to stop the generation of your image. The cancel dialog remains up until the drawing is complete or you click the Cancel button. If you see something that is not right in your image, you can click **Cancel** to stop POV-Ray rendering.
2. LPub runs L3P on chapter_9_champ.ldr, creating a new chapter_9_champ.pov file.
3. LPub starts the POV-Ray program.
4. LPub has POV-Ray render an image of the chapter_9_champ.pov file.
5. LPub iconifies POV-Ray when it is done rendering. This makes POV-Ray hide down in the Title Bar, out of the way, until the next time you need it.

Figure 9.4 shows the almost completed rendering of the chapter_9_champ.pov file in LPub and POV-Ray.

The final image, called chapter_9_champ.bmp, ends up in the PowerTools directory, where chapter_9_champ.ldr is located. Now that we know how to render a single image, let's look at the UI to the L3P options we learned about in Chapter 8.

Figure 9.4 LPub and POV-Ray

L3P Options in LPub

As we saw in Chapter 8, L3P is a powerful tool for translating LDraw files into POV-Ray scripts. Without L3P, there would be no LPub. LPub provides a GUI to L3P options to make it easy to use L3P without having to use a command prompt. In Figure 9.1, you can see that the L3P tool tab is selected, and the L3P Options tabs are visible. There are so many L3P options that it made sense to organize the options by subject, where each subject has its own tab. The L3P options are broken into five subjects:

- **Camera** Defines where the camera is located, where its looking, and the type of lens used.
- **Lights** Defines where lights are located and their color.
- **Surroundings** Defines the surroundings, including the floor and backdrops.
- **Model** Defines characteristics specific to the LDraw model.
- **Miscellaneous** A variety of options not covered by the first four tabs.

This organization of the L3P options might look familiar to those of you who read Chapter 8. They have been designed to match the L3P options very closely. This chapter shows you the LPub GUI that allows you to access these L3P options but does not rehash the significance of the options themselves. For more details on what these options actually do in L3P, refer back to Chapter 8.

The L3P Camera Tab

Figure 9.5 shows the Camera tab selected in the L3P tools tab and shows all the options associated with L3P/POV camera.

Figure 9.5 L3P's Camera Options

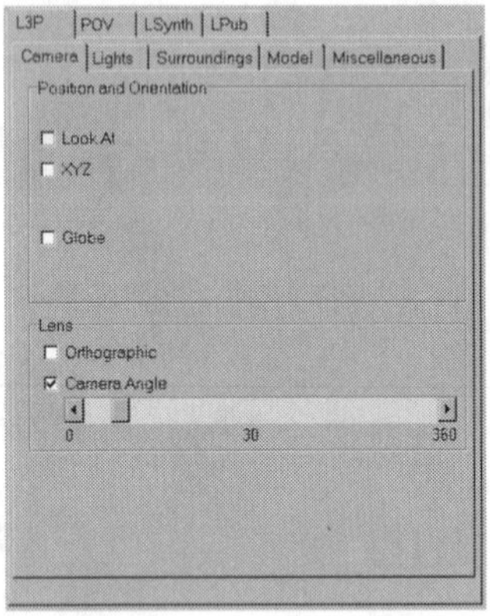

In the Position and Orientation pane at the top of the Camera tab, there are three check boxes:

- **Look At** Controls where the camera looks when generating your picture. This is gives you access to L3P's *–cla<x>,<y>,<z>* option.
- **XYZ** Lets you define where the camera is placed using X, Y, and Z coordinates. This choice gives you access to L3P's *–cc<x>,<y>,<z>* option.
- **Globe** Lets you define where the camera is placed using globe coordinates. This is gives you access to L3P's *–cg<la>,<lo>[,<r>]* option.

None of the boxes are checked in Figure 9.5, so if we were to generate a picture now, none of these options would be fed into L3P. Figure 9.6 shows champ.ldr drawn with Look At, Globe, and Radius checked. Once these are checked, the fields that provide values for the option become visible so you can use them. Note also that the Radius check box only appears if the Globe check box is checked.

Figure 9.6 Using Look At, Globe, and Radius

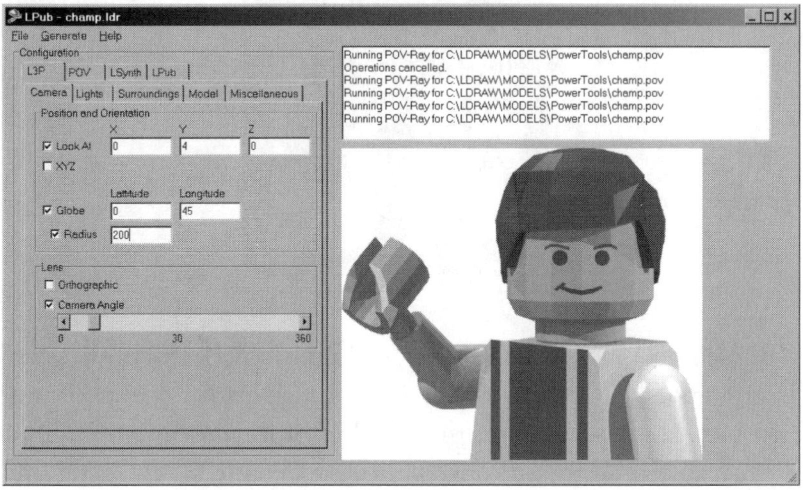

The Lens pane at the bottom of the Camera tab lets you choose from among L3P's default lens configuration (which is used if you have nothing checked in this pane), a POV-Ray Orthographic projection camera, or L3P's *–ca<a>* camera angle option. Although the Orthographic option provided here is not available through an L3P option, it makes the most sense to discuss this topic here (rather than in the POV-Ray section) because this is where all the other camera options are discussed. POV-Ray's normal camera provides perspective, where objects get smaller the farther they are away from the camera. With POV-Ray's orthographic projection lens, objects are the same size no matter how far they are from the camera. The orthographic lens can be good for making building instructions.

The Lights Tab

Figure 9.7 shows the L3P Lights tab. At the top, you can see two check boxes:

- **Always uses default light sources** Checking this box tells L3P to use the *-ld* option and use the default light sources.
- **Use light.dat as light sources** Checking this box tells L3P to use the *-l* option, which means L3P detects light.dat parts in your LDraw file and uses them as light sources.

Figure 9.7 L3P's Lights Tab Options

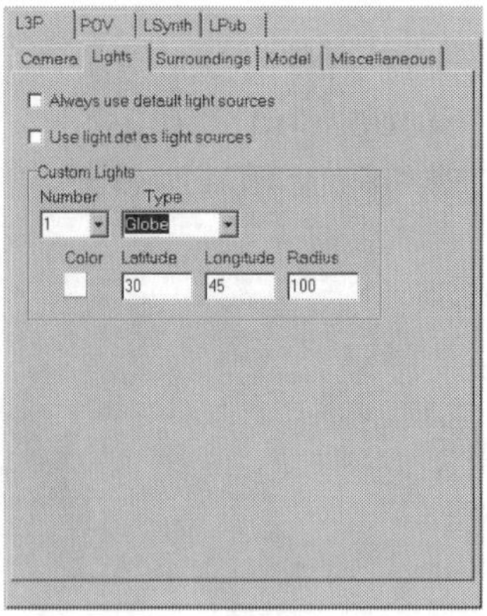

Below these check boxes is the Custom Lights pane, which has several options. In the upper-left corner of the Custom Lights pane is the Number selection list box. Here you can set the number of lights you would like LPub to use. LPub is capable of using up to 32 lights. By default, all the lights are turned off, so L3P provides you with default lights. To the left of the Number selection box is the Type selection box, where you can select either Globe or X,Y,Z. Depending on which type you choose, fields will appear below this selection box that allow you to type in the coordinates (Latitude, Longitude, and Radius coordinates if Globe is selected; X,Y, and Z coordinates if X,Y,Z is selected). This allows you to use the L3P options *-lc<x>,<y>,<z>, -lg<la>,<lo>,<r> -lc<x>, <y>,<z>, -lg<la>,<lo>,<r>*.

When a light is selected, the Color indicator button and coordinate fields become available for use. If you click the **Color** button, a color dialog box pops up that you can use to select a new color. Figure 9.8 shows the color selection dialog window. You select a color from the color dialog window by clicking one of the colored boxes and then clicking the **OK** button. The Color button for your light changes to the color you selected.

The Surroundings Tab

Figure 9.9 shows the Surroundings tab in the L3P tab. There are two panes on this tab:

- **Background** The only option here is checking the Background Color box. This choice lets you work with the L3P option *-b<color>*. Checking this box brings up a square button to the right of the Background Color box (shown in Figure 9.9). Clicking this button brings up a color dialog window where you can select the background color. By default, LPub has the Background Color checked with a background color of white. If you uncheck this box, you will get L3P's default background color of black.

- **Floor** There are two check boxes available in this pane. The top is the Enable check box. Checking this box brings up a Type selection box to the right of the Enable check box, which allows you to choose from a gray floor or a checkered floor. Below the Enable check box is the Altitude check box. Checking this box brings up the Y field to the right of it, where you can type an altitude value.

Figure 9.8 The Color Selection Dialog Window

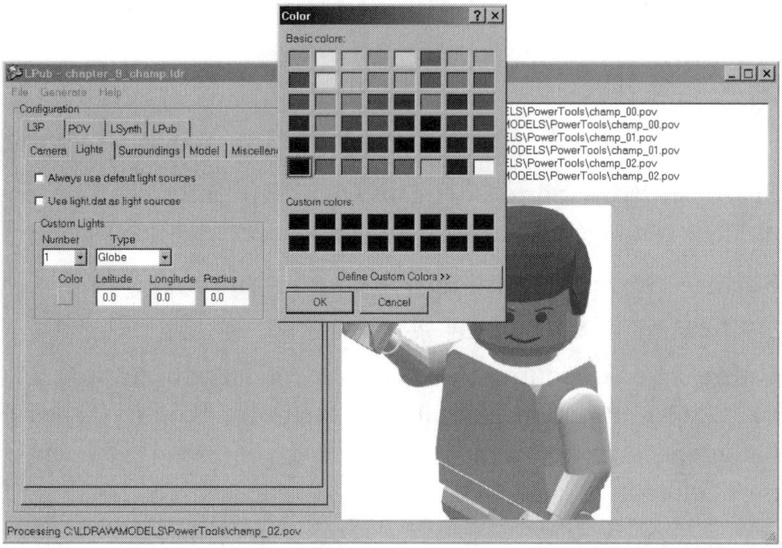

In Chapter 8, we learned that the floor option has this syntax: *-f[<type>][,<y>]*. Checking the **Floor Enable** check box is the same as using just the following portion of the floor options: *-f[<type>]*. Checking the **Altitude** check box lets you use the entire option format like this: *-f[<type>][<y>]*.

Figure 9.9 L3P's Surrounding Tab Options

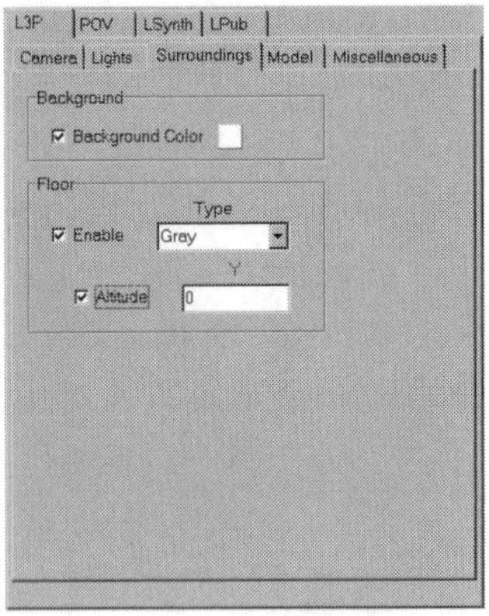

The Model Tab

Figure 9.10 shows the Model tab for L3P. There are five options on this tab:

- **Seam Width** Controls the gaps seen between bricks. The larger the number, the bigger the gap. (See L3P option *-sw<w>*.)
- **Surface Bumps** Adds imperfections to flat surfaces to make them more realistic. (See L3P option *–bu*.)
- **Color** Used to control the color of a model with just one LDraw part in it. (See L3P option *–c*.) Clicking the **color** button to the right of the Color check box pops up a color dialog window so you can select a new color.
- **Render Quality** Controls the amount of detail transferred from your LDraw file to the POV file. (See L3P option *–q<n>*.)
- **Use LGEO** Allows the use of LGEO parts that are more accurate models of LEGO parts than their LDraw part equivalent. (See L3P option *–lgeo*.)

Figure 9.10 L3P's Model Tab Options

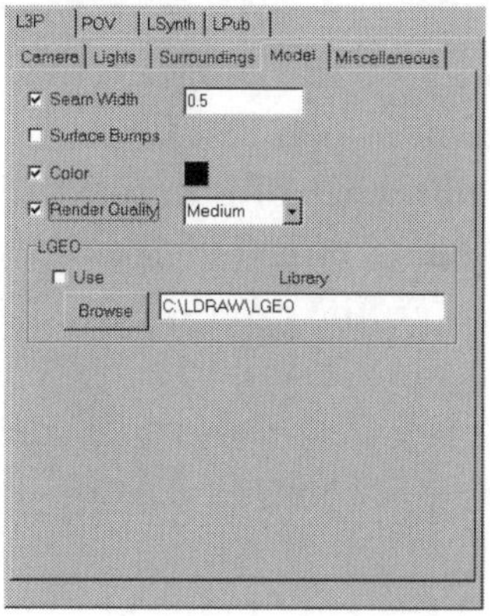

The Miscellaneous Tab

Figure 9.11 shows the Miscellaneous tab in the L3P tab. These options are for advanced users. Three check boxes are available here:

- **Don't Substitute Primitives** Prevents L3P from substituting its own drawing primitives for L3P drawing primitives. (See L3P option *–p.*)
- **Exclude Non-POV Code** Prevents inclusion of non-POV code from your LDraw file into your POV file. (See L3P option *–enp.*)
- **Enable Stepclock** This option can be used to create animations with POV-Ray. (See L3P option *–sc.*)

Figure 9.11 L3P's Miscellaneous Tab Options

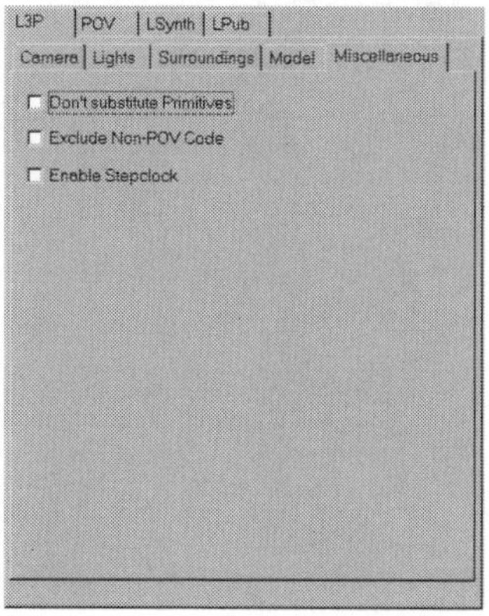

POV-Ray Options in LPub

POV-Ray is an advanced and powerful program that has lots of optional features. The creator of POV-Ray did not provide a GUI in LPub for all of POV-Ray's features, because it would simply make LPub too large and complex and cause it to lose its LEGO focus. Instead, Clague added access to the POV-Ray features that he felt are the most useful to the typical LEGO fan. As people have begun using LPub, they've asked for more features, and Clague has added many of them. He will probably add more POV-Ray features in future versions of LPub, so keep an eye on his site, www.users.qwest.net/~kclague/LPub/, for LPub updates and news of future releases.

Figure 9.12 shows the POV tab in LPub and all its subject tabs. These include:

- Rendering
- Output
- MEGA-POV

Figure 9.12 POV's Options Subject Tabs

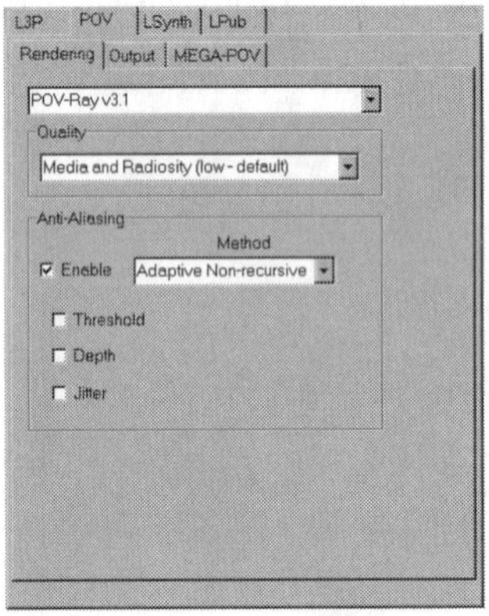

Let's take a look at each of these tabs in detail.

The Rendering Tab

The Rendering options are shown in Figure 9.12. There are three options in the Rendering tab:

- **Renderer** This list selection box at the top of the pane lets you choose from the renderers installed on your computer. Figure 9.13 shows the rendering options installed in Chapter 2. It shows POV-Ray version 3.1

as well as a program called MEGA-POV that works with POV-Ray. MEGA-POV provides the ability to outline each of the bricks in a design so that images look more like the artwork used in LEGO's official building instructions.

Figure 9.13 Renderer choices

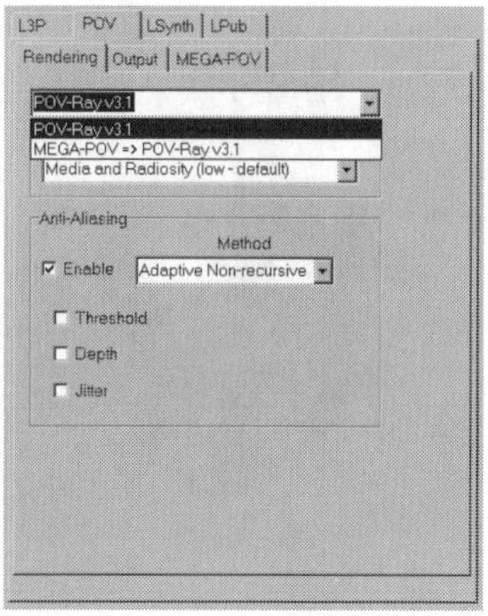

- **Quality** Like L3P, POV-Ray has image quality controls that allow you to trade off computing time against image quality. Figure 9.14 shows the list of POV-Ray quality choices. Figure 9.15 shows our LEGO champ minifig rendered in both the lowest (left) and highest (right) qualities. The lowest quality does not calculate shadows or shading, so the white torso and arms appear invisible against the white background. The torso and arms are very visible in the high-quality rendering on the right.

Figure 9.14 Quality Choices for POV-Ray

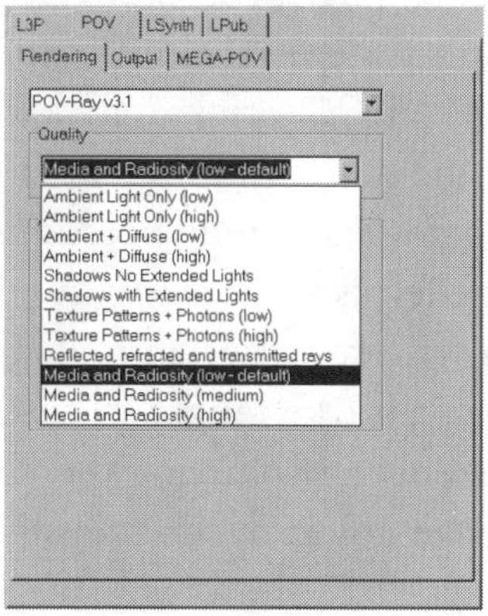

Figure 9.15 Our LEGO Champ Rendered with Lowest and Highest Qualities

- **Anti-Aliasing** Improves the look of sharp edges that are at angles in your image. Threshold, Depth and Jitter are options you use to control anti-aliasing if it is enabled.

Anti-aliasing is a technique to make your images look less computer generated. Your computer screen is made up of an array of picture elements known as *pixels*. These pixels are arranged in a grid of rows and columns. When you draw a horizontal line, you are drawing into pixels that are all in the same row. When you draw a vertical line, you are drawing into pixels that are all in the same column. When you draw a line at an angle, you draw into pixels where both the row and column change as you move from one end of the line to the other. Figure 9.16 shows a closeup view of a LEGO brick drawn without any anti-aliasing. You can see the aliasing effect in the jagged lines at the bottom of the brick. Figure 9.17 shows the same brick up close and rendered using POV-Ray's two forms of anti-aliasing. The brick on the left is rendered with nonrecursive anti-aliasing; the one on the right is rendered with recursive anti-aliasing. Recursive anti-aliasing is the superior technique, but it can take longer to compute.

Anti-aliasing has the effect of blurring the edges of your image by using lighter versions of the edge colors at the transitions from one pixel row or column to the next. Figure 9.18 shows you the Anti-Aliasing pane selections available to you on the Rendering tab. Checking the **Enable** check box at the top of the pane allows you to select either Adaptive Non-recursive or Adaptive Recursive from the Method selection box to the right of the Enable check box.

Figure 9.16 A LEGO Brick with Aliasing

Figure 9.17 Nonrecursive (Left) and Recursive Anti-Aliasing (Right)

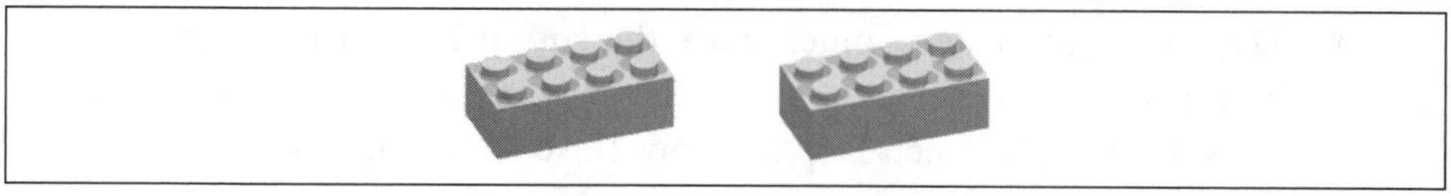

Figure 9.18 Anti-Aliasing Choices for POV-Ray

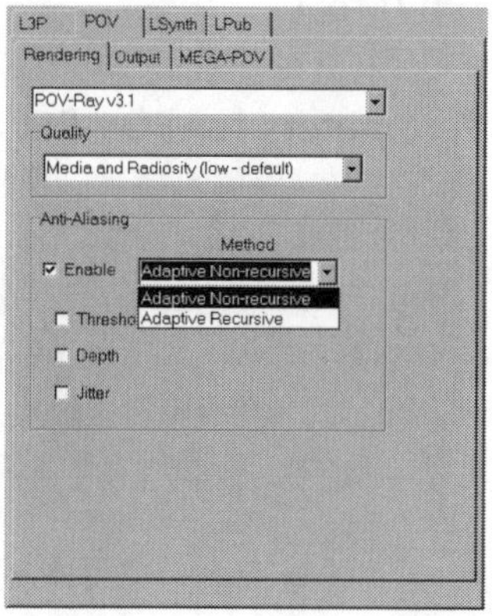

NOTE

Remember that even though selecting **Adaptive Recursive** will result in POV-Ray producing better-looking images, it will also take your computer longer to render the images.

Below the Enable check box in the Anti-Aliasing pane are three additional check boxes, all of which are shown checked in Figure 9.19. By checking the various check boxes, you make the value fields visible so you can change the default values. Without checking these, you get the default POV-Ray values for anti-aliasing, which are great for the average user:

- **Threshold** Controls when anti-aliasing is applied. The meaningful limits for threshold are from 0 to 3. A POV-Ray color has three components, red, green and blue, that can each range from 0 to 1. The largest possible color difference is from black (0,0,0) to white (1,1,1). The sum of the color differences in this case is 3. If the color difference between two neighboring pixels is greater than the threshold, anti-aliasing is applied. The lower the threshold, the more often anti-aliasing is applied.
- **Depth** Controls how much work the anti-alias algorithm performs for each pixel that needs anti-aliasing. The larger the number, the harder POV-Ray works. Keep Depth below 10 so as to not slow your rendering too much.
- **Jitter** Makes the edges blurry, helping hide the jagged edges. Jitter has a meaningful range from 0 to 1.

Figure 9.19 Anti-Aliasing Parameters for POV-Ray

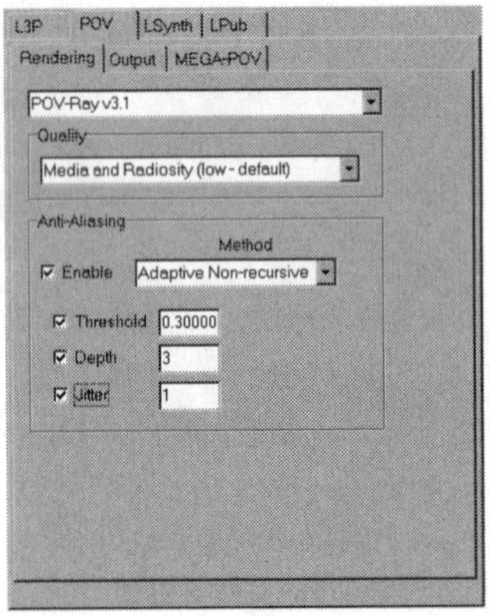

The Output Tab

Figure 9.20 shows the Output tab of the POV-Ray tab. It has three panes:

- **Dimensions** Lets you control the width and height of your rendered image (in pixels). A dimension of 800 by 600 is a standard image size for your computer. You can make this size larger or smaller by keying values into the **Width** and **Height** fields. Next time you render a picture, POV-Ray will use these values for its image size.

- **Format** POV-Ray supports four image formats: Windows bitmap, Portable Network Graphics (PNG), Targa-24 compressed, and Targa-24 uncompressed. LPub provides a fifth format by converting images to JPEG format. Figure 9.21 shows the Format selection box listing the format choices.
- **File** POV-Ray picks a default filename based on the name of the POV script provided. You can override this default by checking the **Specify File Name** check box and filling in the blank provided.
- **Buffering** POV-Ray writes out image calculations as it completes each line. You can make POV-Ray do this less often by checking the **Output Buffering** check box.

Figure 9.20 POV-Ray's Output Tab Options

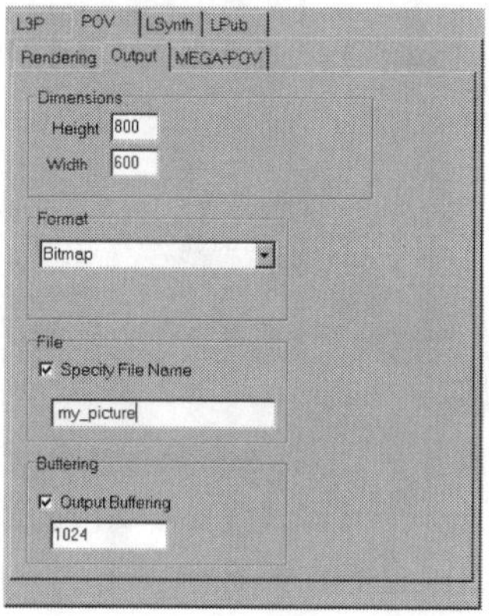

Figure 9.21 Single-Image Formats

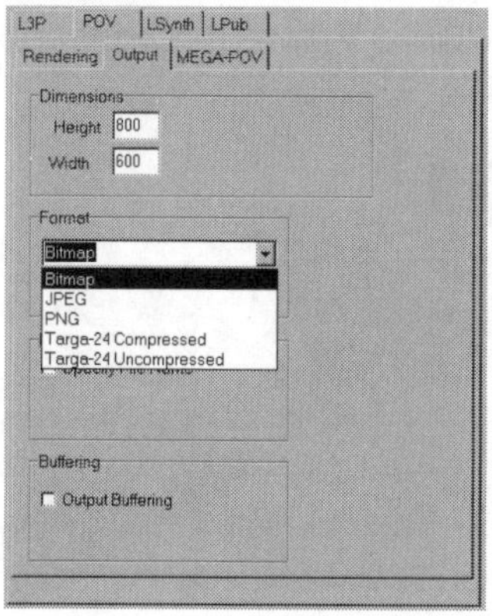

The MEGA-POV Tab

Figure 9.22 shows the MEGA-POV options tab. It contains two panes:

- **Edge Detection** Provides fields for which the values control how MEGA-POV identifies edges.
- **Line Format** Controls the color, sharpness, and width of the line drawn once an edge is detected.

Figure 9.22 POV's MEGA-POV Tab Options

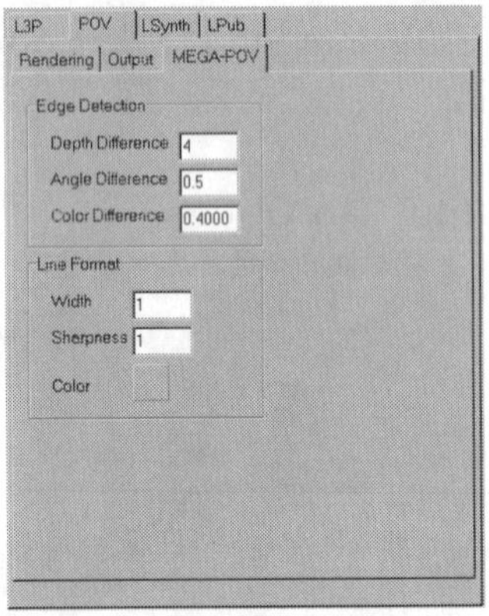

You can use MEGA-POV's edge mechanism to outline your LEGO parts as they are in LEGO building instructions. The white brick in Figure 9.23 shows one of the problems with white LDraw parts on a white background. It is hard to tell where the white brick ends and the background begins. Figure 9.24 shows the edges drawn in by MEGA-POV.

Figure 9.23 A White Brick on a White Background

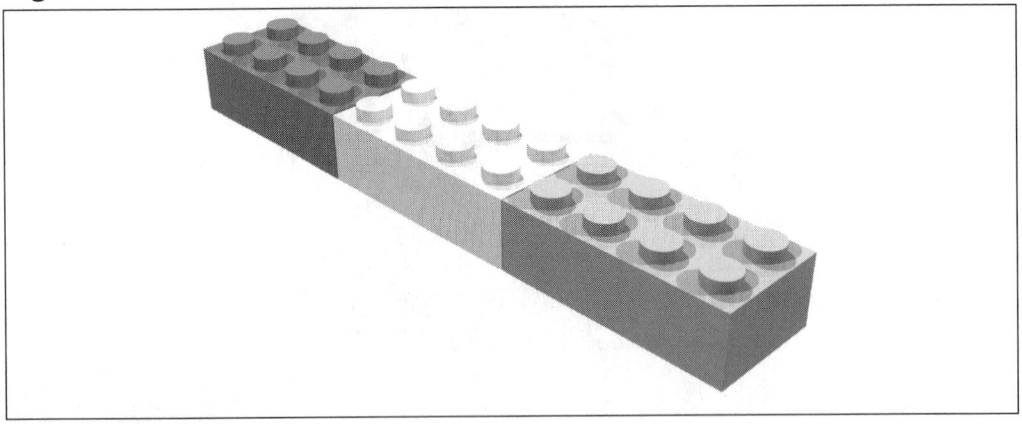

Figure 9.24 A White Brick Outlined by MEGA-POV

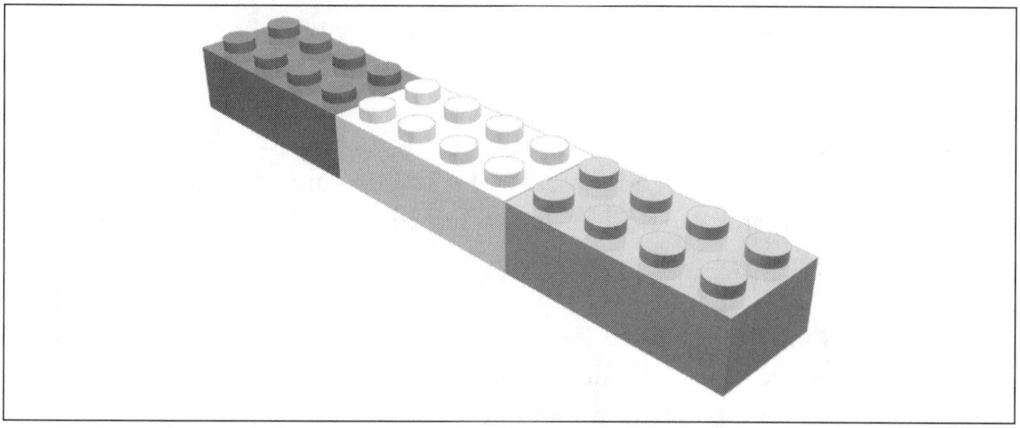

LPub's default values for edge detection and drawing are very subtle, but they work nicely with the black and white pictures.

LSynth Options in LPub

Figure 9.25 shows the LSynth tab, which contains only one option: Enable Flexible Part Synthesis. Check this option if you want to enable flexible part synthesis using LSynth.

Figure 9.25 LSynth Controls

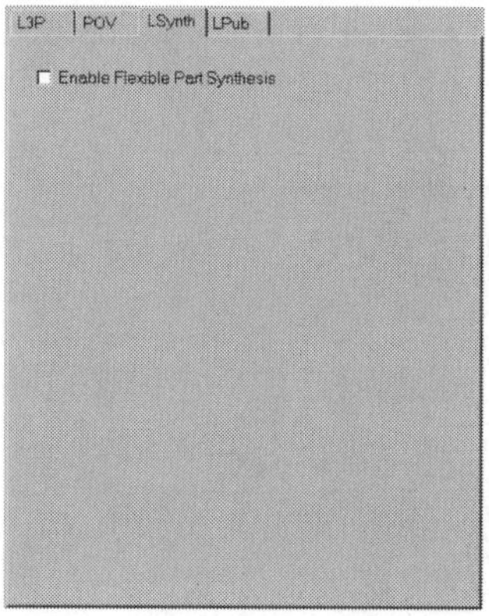

Creating Building Instructions

LPub's building instruction feature is powerful yet easy to use. You access LPub options using the LPub tab. With just a few mouse clicks, you can go from simple

LDraw files to book-quality building instruction images. A few more mouse clicks and you have Web pages that walk users through the building sequence. LPub combines concepts from MLCad and the capabilities of L3P and MLCad, plus some of its own wizardry, to make the creation of building instructions highly automated.

MLCad can produce step-by-step building instructions that honor both the LDraw step feature and MLCad's own ROT-STEP feature. MLCad creates a picture for each STEP and ROT-STEP. It also provides visual clues as to which parts in a STEP's picture are new to the current step by drawing parts from previous steps in darker colors, thus de-emphasizing them. This graying of previous steps' parts eliminates most of the need for arrows and picture inserts used to clarify where parts are added to the model in a particular step. This is very powerful stuff, but MLCad's image quality is quite low compared with POV-Ray's ray-traced images.

L3P's *stepclock* feature can be used to produce step-by-step building instructions using POV-Ray, but it has two major weaknesses. L3P does not support MLCad's ROT-STEP meta-command, making it impossible to see parts that are added on the backside of your LEGO model. L3P also provides no way of graying parts added in previous steps.

One feature available in LPub that neither MLCad nor L3P provide is the ability to produce an image for each *step* that shows all the parts added in that step. *Part-list images* are a normal part of LEGO building instructions; they show you the parts you need for a given step in a picture rather than a text list of part names.

LPub follows these steps to model MLCad's step image generation using L3P and POV-Ray:

1. LPub reads in the LDraw file.
2. LPub uses L3P and POV-Ray to create an image of the fully assembled design.
3. LPub then identifies each STEP and ROT-STEP in the LDraw file.
4. It then creates a sequence of new LDraw files, one per STEP or ROT-STEP. These steps contain all the original LDraw file's parts up to that step. We call these *step-DATs*.
5. If a ROT-STEP is in effect for this step-DAT, LPub rotates all the parts in the step-DAT according to the ROT-STEP specification.
6. It then runs the step-DATs through L3P to get a POV script.

7. LPub then modifies the POV script, graying out parts from all but the last step.
8. LPub runs POV-Ray to get an *assembly image* for the partially completed assembly up through that step.
9. It then makes a part-list image for the step that shows the parts added in that step.
10. If a part is encountered that is a user-designed LDraw model, LPub assumes it is a subassembly that is needed as part of the entire assembly. LPub processes this subassembly starting with Step 1. When all the images are created for the subassembly, LPub resumes processing the current file. This mechanism lets you to break your model into submodels, which can make building instructions easier to understand.

Once all the images are created, you can use LPub's Web page generation feature to create Web pages that organize all the images into step-by-step building instructions. Figure 9.26 shows our minifig athlete in MLCad with STEP and ROT-STEP meta-commands, making for extremely clear building instructions. To make building instructions for chapter_9_champ.ldr, follow these steps:

1. Use LPub's **File | Open LDRAW File** menu to open **chapter_9_champ.ldr**.
2. Use LPub's **Generate | Instruction Images** menu to make LPub generate the instruction images.
3. Use LPub's **Generate | Screen Web Pages** menu to create and display the building instructions.

Figure 9.27 shows the resulting Web pages generated by LPub.

The first Web page (top left) shows you the complete assembly you are going to build. The second Web page (top middle) shows you the parts needed for Step 1 and how they go together. Step 2 (top right) shows the parts from Step 1 ghosted as well as the parts needed for Step 2 and how they are added to the parts you have already assembled. Steps 1 and 2 show the partial assembly rotated, so you can see where the right arm and hand are added. In Step 3 (bottom left), the rotation step is no longer in effect and the last parts are added. This number of steps might seem like overkill for a simple minifig, but they illustrate most of LPub's building instruction features. Notice that you control all the contents of your building instructions using information in your LDraw file. Using MLCad, you use steps and rotation steps carefully to make the assembly sequence clear to the reader.

Figure 9.26 chapter_9_champ.ldr with Steps Added for Building Instructions

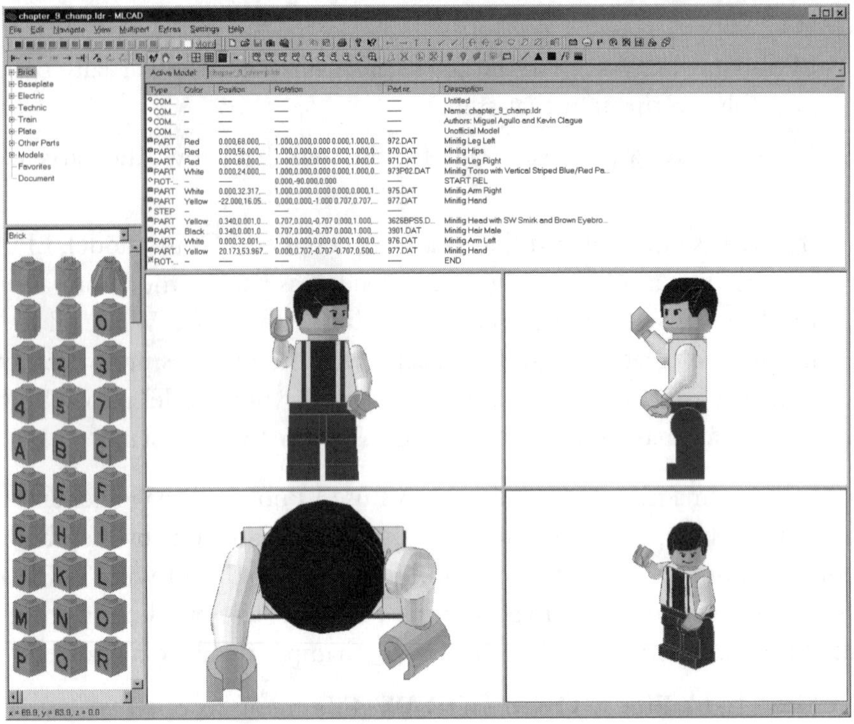

Figure 9.27 Web Page Building Instructions for Our Champ

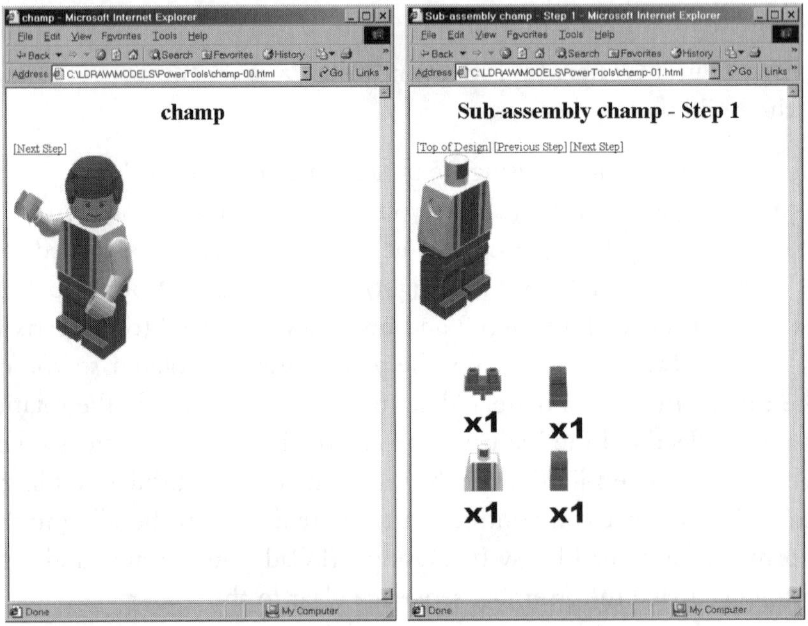

Figure 9.27 Continued

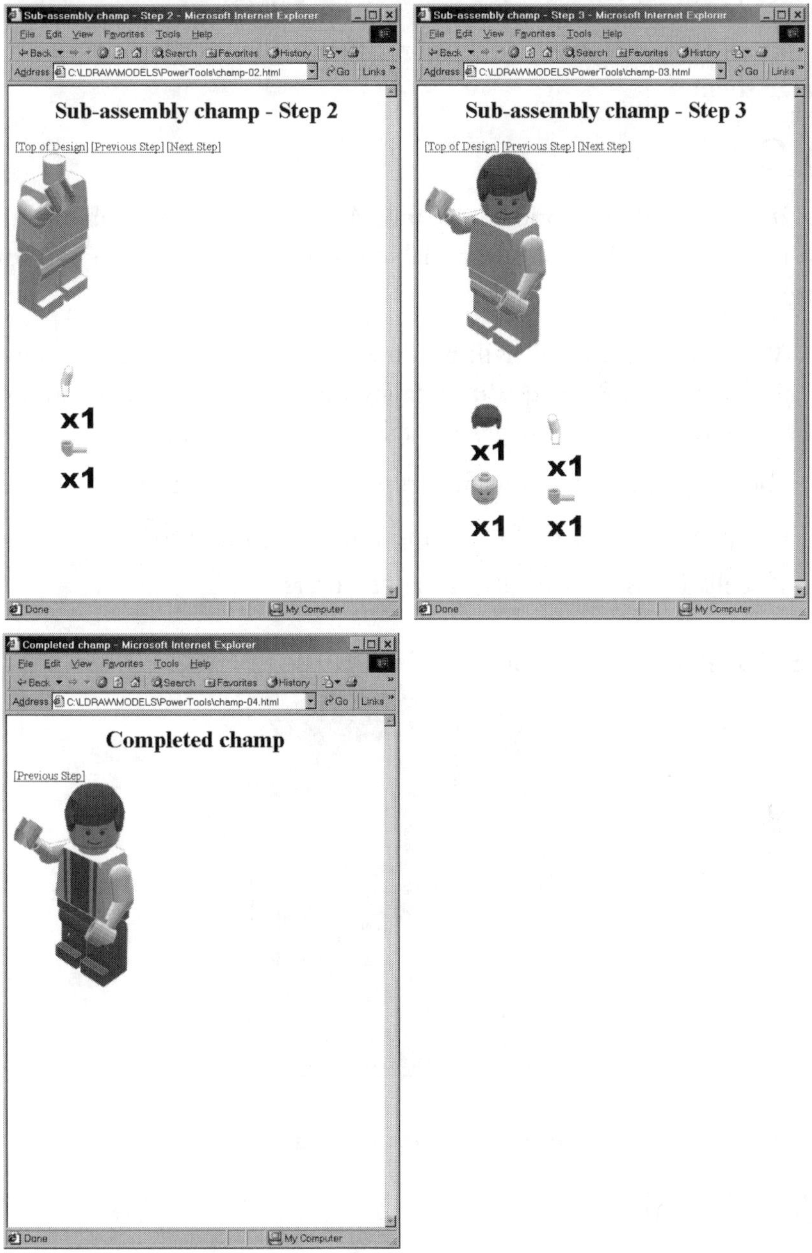

LPub provides a few options to control the look of the generated building instructions, but the contents of the building instructions are for the most part controlled by the information you put into your LDraw file. Once LPub starts generating your building instructions, no user intervention is required. This is important to realize, because for complex LEGO models, it can take POV-Ray hours to render all your instruction step images. LPub has a reasonable set of default options for generating building instructions, but it provides options to let you change them.

LPub Options

Figure 9.28 shows LPub's option tabs: the Mode tab, the Steps tab, and the Controls tab. Through the options available on these tabs, you can control the following:

- Whether or not LPub will produce *print-quality* images or *computer screen-quality* images. Print-quality images are higher resolution and look better than screen-quality images but take longer to render.
- The image format of the final results (Windows bitmap or JPEG).
- The previous Step Color Scaling (ghosting) controls.
- Miscellaneous controls for advanced users.

Figure 9.28 LPub's Options Subject Tabs

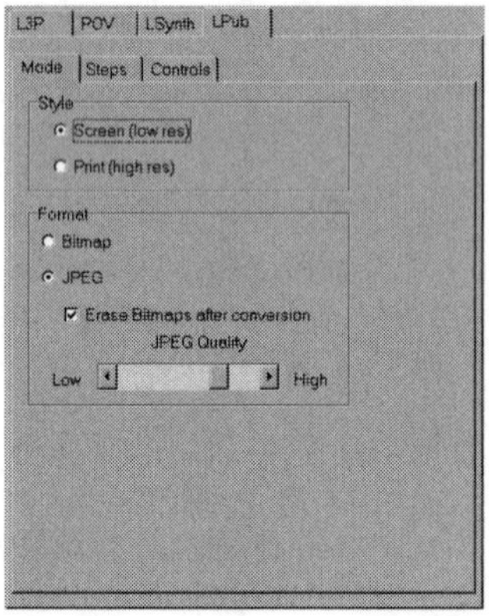

Let's take a look at each of these tabs in detail.

The Mode Tab

Figure 9.29 shows the LPub tab opened to the most important of its subject tabs, the Mode tab. It contains two panes:

- **Style** Lets you choose between two general styles: screen-style images (lower-resolution color) and print-style images (high-resolution back and white). Selecting **Screen** or **Print** styles is an easy way to select default values for the other LPub options, such as Format.

- **Format** Your step images can be in one of two formats: Bitmap or JPEG. Bitmap images take up more space on your hard disk but are crisper looking, with sharper edges. JPEG images take much less space but are not as crisp as bitmap images.

Figure 9.29 Color, JPEG Print Resolution Configuration

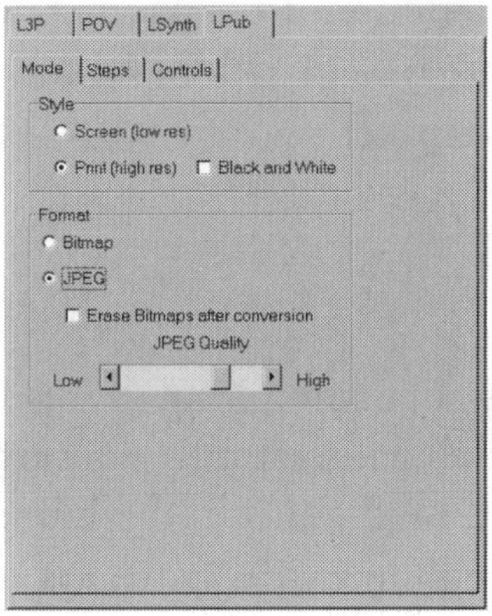

Selecting the **Screen (low res) Style** option gives your images the following characteristics:

- Color images
- JPEG file format
- Assembly image resolution 800 x 600

Selecting **Print (high res) Style** gives your images the following characteristics:

- Black and white images
- Windows bitmap format
- Assembly image resolution 2048 x 1536
- L3P camera angle of 30 degrees
- L3P surface bumps
- L3P background color of white

The idea behind the Style options was to make it easy for Syngress authors to easily set all the LPub settings to ensure that their results were consistent. Once you've chosen a style, you can you can override any of the defaults by going to the appropriate tool tabs (the L3P, POV, and LSynth tabs we just looked at) and modifying the option settings there. Figure 9.29 shows an LPub Style and Format configuration that produces color JPEG images at print resolution. The Black and White check box appears and is checked when Print Style is selected. We unchecked this box to create print-quality color images.

The Format pane at the bottom of the Mode tab in Figure 9.29 lets you choose the format of the images rendered for building instructions. There are two check boxes at the top of this pane:

- Bitmap
- JPEG

POV-Ray always generates building instructions in bitmap format, but if JPEG Format is checked, LPub automatically converts the bitmaps that POV-Ray renders to JPEGs. By default, LPub discards the bitmaps after conversion, but this can be avoided by unchecking the **Erase Bitmaps after conversion** check box.

When converting to JPEG format, you can choose the quality of the result using the JPEG Quality scroll bar at the bottom of the Format pane. The higher the image quality, the larger the image file size. The lower the image quality, the smaller the image file size.

Black-and-White Image Generation

The building instructions in Figure 9.27 were generated for computer screens using the Screen (low res) style. Figure 9.30 shows the same LDraw design generated in Print (high res) style. We used a paint program to merge all the pictures together into Figure 9.30.

For black-and-white image generation, LPub translates the colors of the parts in your LDraw file to grayscale (used in the production of this book). This option was created to make it easier to publish Syngress's line of grayscale LEGO books. With the color information converted to grayscale, it is difficult to know the color of the parts to be added to the models. To preserve this color information for the reader, LPub tags the parts in the part-list images with color labels. Parts that are white, black, or gray are not labeled. Table 9.1 shows a list of colors and their color labels. In Figure 9.30, you can see the color labeling above the minifig's head, hands, hips, and legs in the part-list image. In this case they are labeled *R* (hips and legs), which stands for *red*, and *Y* (head and hands) for *yellow*.

Figure 9.30 Our Champ in Black and White

Table 9.1 Part-List Image Color Labels

Label	Color	Label	Color
B	Blue	TB	Transparent blue
G	Green	TG	Transparent green
DC	Dark cyan	TDC	Transparent dark cyan
R	Red	TR	Transparent red
M	Magenta	TM	Transparent magenta
Br	Brown	TBr	Transparent brown
LB	Light blue	TLB	Transparent light blue
LG	Light green	TLG	Transparent light green
C	Cyan	TC	Transparent cyan
LR	Light red	TLR	Transparent light red
P	Pink	TP	Transparent pink
Y	Yellow	TY	Transparent yellow
Ppl	Purple	TPpl	Transparent purple
O	Orange	TO	Transparent orange

The Steps Tab

Figure 9.31 shows the LPub Steps tab, which gives you access to two panes:

- **Previous Step Color Scaling** In large LEGO designs, it can be hard to tell where new parts are added. To reduce this problem, LPub has the ability to modify the colors of parts added in previous steps to make it easier to see the parts added in each step and where they go.
- **Generate** Generating POV-Ray images can take a long time. You can use the Generate pane check boxes to avoid some of the work. Normally, you want all these options checked, but under certain circumstances, you might want to prevent some of the work from being done, in which case you would uncheck those options.

In the Previous Step Color Scaling pane, you define how much you want the color of parts added in previous steps to be changed. There are two aspects to this pane: *amount of scaling* and *color*. The scroll bar controls the amount of scaling. Setting it all the way to the left, to 0%, tells the program that previous parts' colors will be unchanged. Setting it all the way to the right, to 100%, tells the program that all the previous parts' colors should be lost and only the scaling color used. If you set the scroll bar to 50/50, previous parts' colors will be an even mixture of the original part color and the scaling color. You access the scaling color by clicking the square button to the right of the scroll bar. The default scaling color is white. You can change the color by clicking the color box, which pops up a color selection dialog window. MLCad tends to draw parts from previous steps darker than the parts added in the current step. You can make LPub do this by setting the color to black and setting the percentage to 50%.

Figure 9.31 LPub's Steps Tab Options

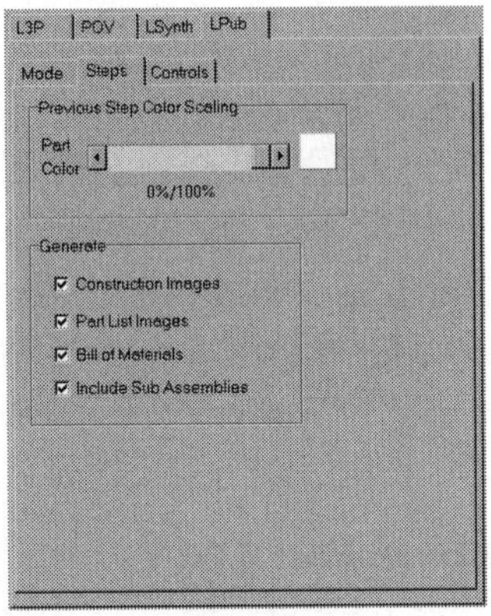

The Generate pane is used to control the pictures created by LPub as your building instructions are generated. Normally you want all three boxes checked. The three check boxes in this pane and what they control are as follows:

- **Construction Images** The images of the parts being added together for each step.
- **Part List Images** The images of the parts lists for each step.
- **Bill of Materials** A part-list image for all the parts in the design, like the ones LEGO has at the back of its constructopedias. You can see an example of this image in Chapter 11, where we have included a full set of building instructions for an AT-ST rendered in LPub.
- **Include Sub Assemblies** Since it is typical to have large LEGO designs described in multiple LDraw files with subassemblies, this box should usually be checked.

The Control Tab

Figure 9.32 shows LPub's Control tab. This tab consists of eight check boxes that allow users to control LPub's behaviors:

- **Erase generated DAT files** Normally, LPub erases the step-DAT files it generates to clean up after itself. If you want access to these step-DAT files after LPub is done using them, uncheck this box.
- **Erase generated POV files** Normally, LPub erases the POV files it generates after it is done with them. Uncheck this box if you do not want these files erased.
- **Run POV-Ray** This option is used to debug problems in the generated DAT or POV files. If all we want is to examine the contents of the DAT or POV files without running POV-Ray, we uncheck this check box.
- **Run POV-Ray only if needed** LPub can tell if changes have been made to your LDraw files since the last time building instructions were generated, and the program only regenerates images for files that have been changed. Unchecking this box makes LPub regenerate all images.
- **Check DAT(s) against Part List** When you try to generate building instructions from other people's designs, there is a chance that they used LDraw parts that do not exist in your parts database. This can cause processing problems for LPub, so LPub checks for this issue and won't process the files. You can uncheck this box to defeat the checking, but LPub will fail when creating part-list images.
- **Crop Images** Often the image of the model is smaller than the window it is generated in. LPub throws away the excess window in a process called *cropping*. If you do not want your images cropped, uncheck this box.
- **Auto load local configs** This topic is discussed in the "Saving Option Configurations" section later in this chapter.
- **Overwrite on MPD extract** When creating building instructions for a multipart DAT file, LPub separates the individual files that reside in the one multipart DAT file into individual files on your hard disk. If a file on your hard disk has a name that matches one of the files listed in the MPD file, LPub will not overwrite that file. Instead, it pops up an error dialog window describing the situation. You can make LPub overwrite the file by checking this check box, and then try generating building instructions again.

Figure 9.32 LPub's Control Options

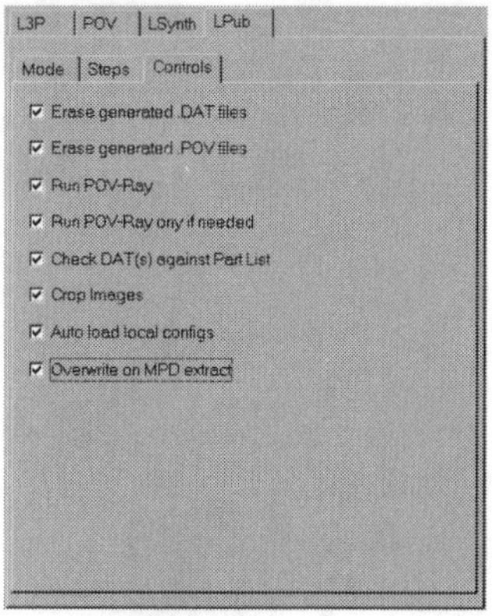

Now that you've seen all the options LPub has to offer, let's see what other features we have available to us.

Tabular Web Page

LPub can create a single Web page with a table showing the building instructions for a model in sequence. This feature is especially handy when you're generating high-resolution assembly images. Using Web pages created with LPub's Generate | Screen Web Pages menu with large images does not work well because when the two images for a given step are combined, they will not fit on your computer screen. Using LPub's Generate | Web Page Table menu, you can create a tabular Web page, like the one shown in Figure 9.33. Clicking the small images in the table lets you view the large images by themselves.

The table format was designed specifically for communication between authors and Syngress Publishing when assembling the company's *10 Cool LEGO MINDSTORMS* line of books. It is also very useful for creating your own instruction steps at home. In this concise format, we can add any text we want for each step as well as make notes for the folks that are laying out the book. If any special artwork images are needed, we can indicate them in the Callout Text column.

Figure 9.33 A Tabular Web Page

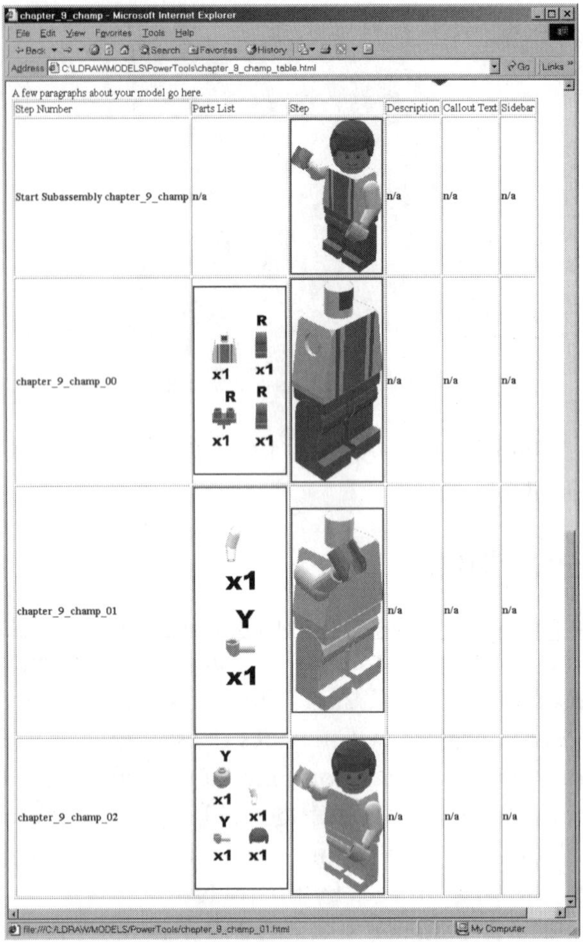

L3P, POV, and LSynth Options for Building Instructions

One thing that is not clear from the user interface is which of the L3P, POV, and LSynth options have an effect on the building image-generation process. These following options have an effect on building images:

- L3P Lighting Options
- L3P Camera Angle
- L3P Render Quality
- L3P Surface Bumps

- L3P Seam Width
- L3P Background Color
- L3P Floor
- L3P LGEO Parts
- POV-Ray Buffering
- POV-Ray Quality
- POV-Ray Anti-aliasing
- LSynth Enable

Saving Option Configurations

LPub allows you to save your option configurations using LPub's menu. You can save your configurations in three places:

- Local to the current directory with default name config.lpb
- Global for all directories with the default name config.lpb
- To a specific filename using the Save dialog window

Kevin Clague developed LPub while he was authoring a book for Syngress. He wanted to be able to have high-resolution settings in his directories for the book and screen settings for everything else. To do this, Clague created the local and global configuration save mechanisms. He typically creates a subdirectory for each robot. He created local configuration settings in the subdirectories that contained robots for the book. He set up the global configuration for screen settings. Later, he added the ability to save configurations to a specific file using the Save dialog window.

To save the current option configuration as global, use the **File | Save Global Config** menu. To save the current option configuration local to the current subdirectory (the directory where you opened your last LDraw file), use the **File | Save Local Config** menu. To save the current option configuration to a file of your choosing, use the **File | Save Config As...** menu.

When you start LPub, the program automatically opens the global configuration as the option settings. You can force LPub to restore the option configuration from the global configuration settings using the **File | Open Global Config** menu. If you open an LDraw file in a subdirectory that has a previously saved local configuration, the local configuration will automatically be read in when you open your LDraw file. This automatic loading is an option that you can disable by unchecking the **LPub | Controls | Auto load local configs** check box that we saw back in Figure 9.32. You can force LPub to restore the option configuration from the local configuration (in the current directory) using the **File | Open Local Config** menu. You can force LPub to restore option configurations from named configuration files using the **File | Open Config File** menu.

NOTE

LPub supports DAT files (the original type for LDraw files), LDR files (the preferred type of LDraw files), and multipart-DAT (MPD) files. You can create an MPD file from a group of individual DAT or LDR files using the **File | Open LDraw File** menu to open the top-most level LDraw design file, then using the **File | Save As MPD** menu. LPub pops up a Save dialog box so you can choose a location and name for your MPD file.

You can split a MPD file into individual DAT or LDR files by using the **File | Open LDraw File** menu to open an MPD file. Use the **File | Save As LDR** menu to have LPub save the MPD file as individual DAT or LDR files (using the filenames found within the MPD file).

Instruction Philosophies and Tips

When we started laying out our first book of building instructions, there was a lot of discussion among the authors regarding the techniques we could use to make good building instructions. We wanted to make the instructions easy to understand yet compact enough to fit into a book. Here are some of the concepts we came up with:

- Break the design into subassemblies that are easy to merge into the complete assembly.
- Break up the subassemblies and final assemblies using the STEP feature provided by MLCad.
- Build the assembly from back to front and bottom to top to reduce the possibility of new parts being hidden by parts from previous steps.
- Limit the number of parts added in a given step to a reasonable number (eight or fewer).

- Use MLCad's rotation steps so that added parts can always be seen in the assembly image.
- Keep the number of rotation steps to a minimum.
- For pins and short axles that hook parts together, put the pins into the design in one step, and put the part that they hook to in a different step, so you can see where the pins are added.
- Try to relate the building instructions on screen to what you would actually do with your hands.

Breaking the entire design into subassemblies gives the builder a sense of accomplishment on each subassembly and keeps the steps required to complete a task to a reasonable number. The back-to-front building strategy, combined with rotation steps, ensures that the builder can always tell where the parts in the design are added.

Let's now take a look at a few more tips you can use to make your instruction steps clearer and easier to understand.

Rotation Steps

Rotation steps can be visually confusing because the builder has to reorient him- or herself to see the model at a new angle. We try to use rotation steps that only rotate around one axis, either turning the subassembly over to see the underside or turning it around to see the back side. We also tried to limit ourselves to standard angles such as 45, 90, or 180 (and –45, –90, or –180) degrees. Rotating about more than one axis and using arbitrary angles makes it much harder for builders to reorient themselves when they come upon a rotation step.

Often rotation steps are needed so the builder can see how to assemble the back side of a design and then the front. For these situations, you can often get away with what appears to the builder as only one rotation of the design, by using a rotation step as your first step. This solution lets you build the back side of the design first. After the back side is complete, use a rotation end, which turns the model back to the front side, where you can add all the front side parts.

In the real world, the builder typically lays the parts out on a table, with the partially completed assembly resting on the table. Try to keep this in mind when laying out building instructions. You might want to have a subassembly's instructions oriented differently than the way it will be used in the final assembly. Keeping the real world in mind can make a big difference in the clarity of your building instructions.

The BUFEXCHG Meta-Command

Recently Kevin Clague found time to add MLCad's buffer exchange feature to LPub. Willy Tschager, an LPub user, sent Clague a cute little robot design (see Figure 9.34) to show him the power of buffer exchange. It shows how to use buffer exchange and an arrow (made of a triangle and a quadrilateral) to illustrate how to put the robot's hinge waist together. The lines in the Type column of MLCad's Model Part List window, with the word *BUFEXCHG,* are uses of buffer exchange. Figure 9.35 shows the building instructions using a Web page table. We've edited the Web page using WordPad to add description and sidebar comments describing the assembly as well as where and how buffer exchange was used.

Figure 9.34 A Buffer Exchange Example

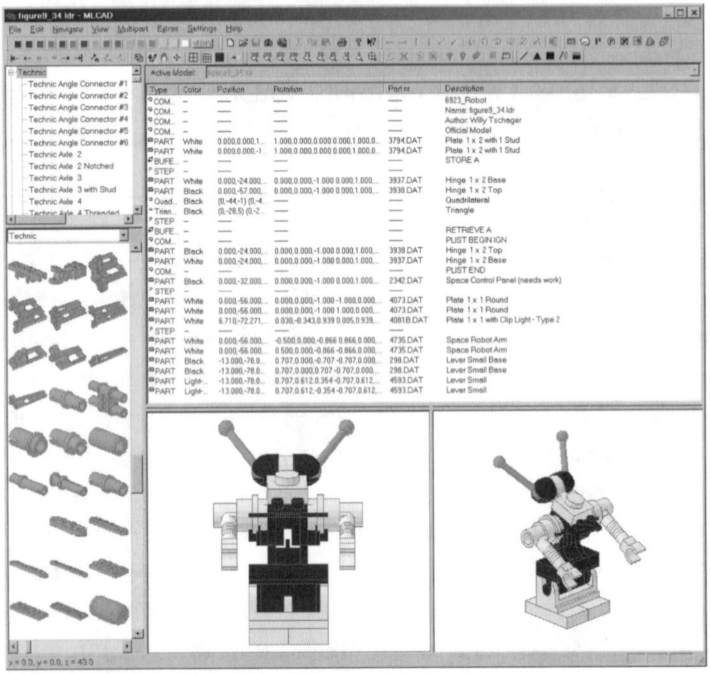

Tschager used buffer exchange and arrows to show how to assemble two pieces. In doing this, he added the hinge base and top twice in the LDraw file, even though the completed assembly only has one hinge base and top. Tschager used *BUFEXCHG STORE A* in Step 0 to remember the contents of the model before the first pair of hinge parts were added for illustration only in Step 1. *BUFEXCHG RETREIVE A* was used to forget the illustration-only hinge, and the complete hinge was added in Step 2. LPub ended up putting two sets of hinge parts into part list images, one in Step 1 and the other in Step 2.

Figure 9.35 Building Instructions Using Buffer Exchange

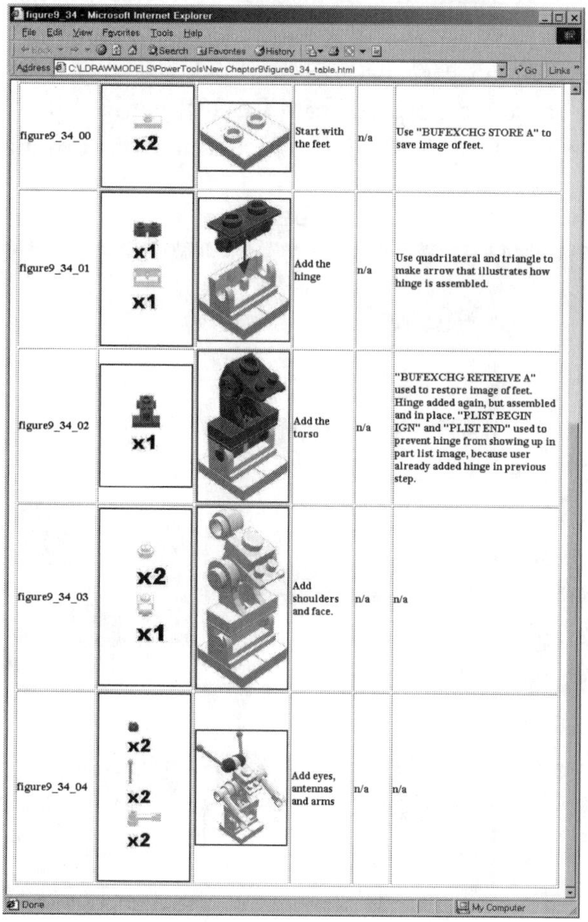

figure9_34_00	x2		Start with the feet	n/a	Use "BUFEXCHG STORE A" to save image of feet.
figure9_34_01	x1 x1		Add the hinge	n/a	Use quadrilateral and triangle to make arrow that illustrates how hinge is assembled.
figure9_34_02	x1		Add the torso	n/a	"BUFEXCHG RETREIVE A" used to restore image of feet. Hinge added again, but assembled and in place. "PLIST BEGIN IGN" and "PLIST END" used to prevent hinge from showing up in part list image, because user already added hinge in previous step.
figure9_34_03	x2 x1		Add shoulders and face.	n/a	n/a
figure9_34_04	x2 x2 x2		Add eyes, antennas and arms	n/a	n/a

The PLIST Meta-Command

Kevin Clague created a new meta-command for LPub called *PLIST* (short for *part list*) to let us control some aspects of part list images. To solve the "parts added twice" issue created by buffer exchange, Clague created *PLIST BEGIN IGN*, which means "when creating part list image for this step, start ignoring added parts." You use *PLIST END* to tell LPub to stop ignoring added parts.

Another issue with LDraw parts and LPub's part-list image feature is that some LEGO parts are represented by more than one LDraw part. Figure 9.36 shows a Technic shock absorber as an example of this case. The shock absorber is in two parts because you can show your LEGO designs with the shock absorber compressed as much as you want. In a construction image, this works great, but in a par-list image, you get each half of the shock absorber as its own part. This is confusing because there are no parts from LEGO that look like the parts in the part-list image. Step 1 of Figure 9.37 shows a confusing-looking part-list image.

Figure 9.36 A Two-Part Shock Absorber

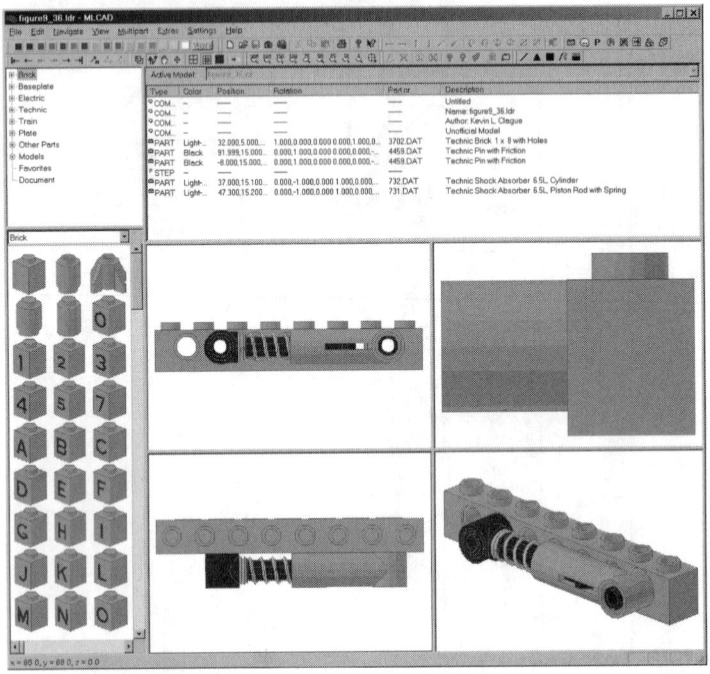

Figure 9.37 Building Instructions for the Shock Absorber

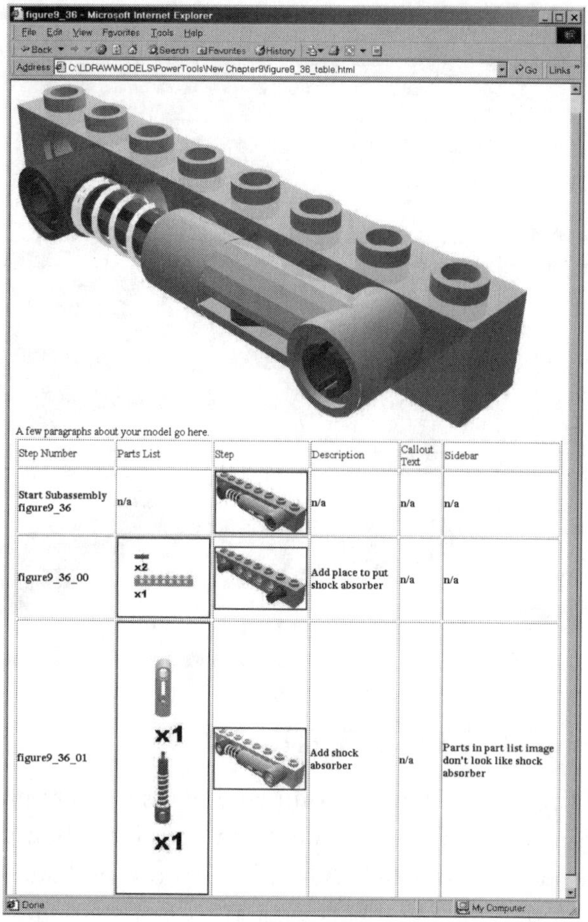

Clague created a second form of the *PLIST* meta-command, *PLIST BEGIN SUB*, which lets you substitute a rendered image of another LDraw file in place of images of the parts in the current LDraw file. Figure 9.38 shows use of a part-list image substitution in MLCad. Notice that the shock absorber parts are surrounded by *PLIST SUB BEGIN shock.ldr* and *PLIST END*. LPub ignores the shock absorber parts added in the LDraw file and instead generates an image of the shock.ldr file and puts it into the part list image. You can see the results in Figure 9.39.

Figure 9.38 Part-List Image Substitution

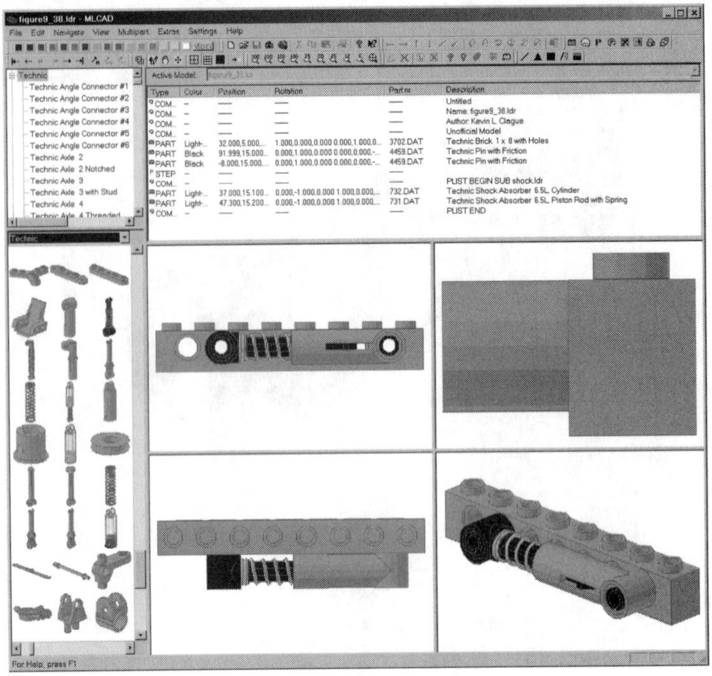

The GHOST Meta-Command

MLCad provides the *GHOST* meta-command. Parts that are ghosted are visible in the LDraw file in which they are added, but invisible when the LDraw design is used as a sub-model in another LDraw file. This can be handy for things that change shape when the subassembly is used as part of a larger assembly. The defining part in the LEGO Robot Invention System is the RCX, a programmable computer brick that has sensor inputs and motor outputs.

When creating building instructions for the RIX, it is common to have motors added in a subassembly and hooked up to the RCX in a different subassembly. More than likely, the cable that connects the motor and the RCX must be connected to the motor in the motor subassembly. Subsequent steps in that assembly bury the motor end of the cable under other bricks that make up the rest of the subassembly. In the end, you have a motor subassembly with a cable hooked in, but only at one end. LEGO's technique for this is to show a cable that just hangs out of the design. When you see the subassembly added to the main model and the motor cable hooked to the RCX, the cable has a different shape.

Figure 9.39 Shock Absorber Building Instructions Revisited

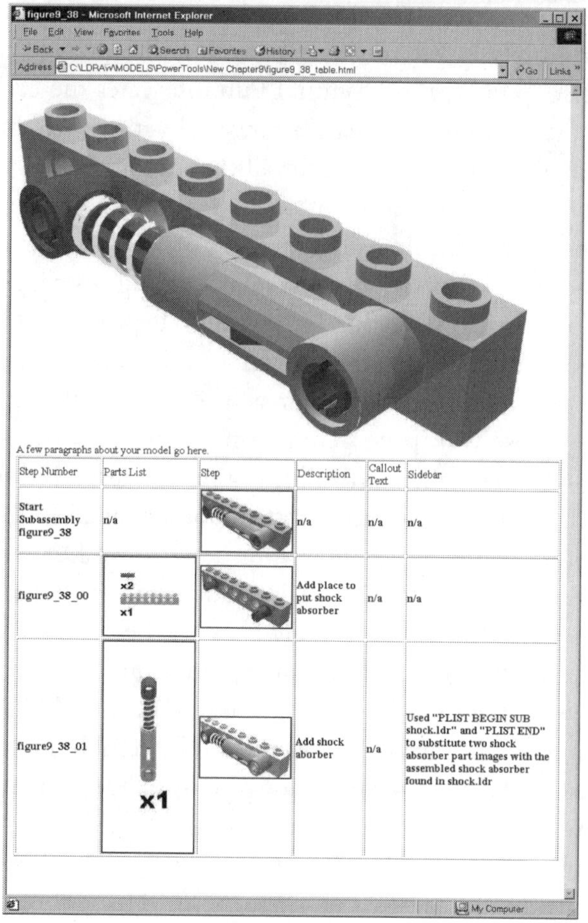

You can do the same thing using MLCad and LPub. In the motor subassembly, mark the cable as ghosted. When the motor subassembly is added into another LDraw design, the cable is invisible. You can use a new cable with a different shape to hook the motor to the RCX. You will want to bracket this new cable with *PLIST BEGIN IGN* and *PLIST END* so that the new cable does not show up in the part-list image (it was already added in the subassembly).

Summary

LPub is a tool that simplifies the process of creating photorealistic images using L3P, POV-Ray, MEGA-POV, and LSynth. LPub integrates the capabilities of these programs with its own features to create a powerful yet easy-to-use LEGO image renderer. LPub provides default settings for all these programs, yet provides a GUI that lets advanced users get the most out of the features these programs provide. LPub can help you create stunning single images of your LEGO creations.

LPub's most powerful feature, step-by-step building instruction generation, depends only on information in your LDraw files. It produces all the assembly and part-list images with no human intervention. Just open the LDraw file you want and pull down a menu, and LPub creates all your building images. LPub also provides easy-to-use Web page generation features for one-step-per-page format or a tabular format where the all the steps are displayed in one table.

LPub is a great companion for the most important of all LEGO CAD programs: MLCad. You use MLCad to enter your designs in LDraw format. You use MLCad's steps and rotation steps to describe building your design in small steps, turning the design if needed so you can always see where parts are added. You can use MLCad's buffer exchange feature to create visual annotations in one step and have them disappear in the next. You can also use MLCad's *GHOST* meta-command to make things disappear when you use a subassembly in another assembly.

LPub provides *PLIST* meta-commands to complement its support of MLCad's *BUFEXCHG* and *GHOST* meta-commands. You can use *PLIST BEGIN IGN* to ignore parts added in a step and not place them in the step's part-list image. You can use the *PLIST BEGIN SUB* to ignore parts added in a step and use the image of another LDraw file instead.

By combining the capabilities of L3P, POV-Ray, MEGA-POV, and LSynth with LPub features, you can create professional-quality building instructions that rival LEGO's own building instructions.

Solutions Fast Track

The LPub Main Window

☑ The LPub menu is used to make LPub perform actions such as opening files or generating images and Web pages.

- ☑ The LPub Options Configuration tabs provide a GUI to L3P, POV-Ray, MEGA-POV, LSynth, and LPub options.
- ☑ The LPub Status Bar keeps you updated on what processing step LPub is working on.
- ☑ The LPub Output console gives you a history of the things that LPub has worked on.
- ☑ The LPub Image window shows you the most recently generated image.

Single Image Generation

- ☑ Use the **File | Open LDraw File** menu, then the **Generate | Complete Assembly** menu, and voilà—LPub runs L3P and POV-Ray for you automatically.
- ☑ LPub provides a GUI to L3P, POV-Ray, MEGA-POV, and LSynth options that will affect your single-image generation.
- ☑ LPub offers support for DAT, LDR, and MPD file formats.
- ☑ LPub offers support for multilevel LDraw designs using LDraw files for subassemblies.

Creating Building Instructions

- ☑ LPub offers highly automated building-instruction creation by using the **File | Open LDraw File** menu, then the **Generate | Instruction Images** menu. You can then sit back and wait for LPub to create all your images.
- ☑ Support for color or black-and-white building instructions ready for publishing is offered by LPub, as is the creation of partial-assembly images and part-list images for each step in your LDraw design.
- ☑ LPub offers support for advanced building instruction meta-commands such as *BUGEXCH, GHOST,* and *PLIST.*
- ☑ Automatic generation of Web pages that glue the images together into building steps is also possible with LPub.

Saving Option Configurations

- ☑ LPub allows you to save your options configuration globally so that the option settings become the defaults when LPub is run next time.
- ☑ You can also save your options settings for a specific folder (typically, a specific LDraw design) or to a specific filename.
- ☑ You will have automatic restoration of global option configuration when the LPub program is started.
- ☑ You will have automatic restoration of local configuration when a file is opened in a specified directory.

Instruction Philosophies and Tips

- ☑ Take some time to think about what parts you will add in each step of your instructions. Don't use too many parts or it might be difficult to see where they all go. Keep rotation steps to a minimum, and break the finished design into several manageable subassemblies.
- ☑ LDraw's *STEP* meta-command is the cornerstone of making clear, step-by-step building instructions.
- ☑ MLCad's *ROT-STEP* meta-commands enable you to turn your model as needed so that you can always see added parts.
- ☑ MLCad's *BUFEXCHG* meta-command can be used to add visual annotations (such as arrows) in one step and make them disappear in the next.
- ☑ LPub's *PLIST* meta-command complements buffer exchanges by making part-list images match the assembly steps.

Frequently Asked Questions

The following Frequently Asked Questions, answered by the authors of this book, are designed to both measure your understanding of the concepts presented in this chapter and to assist you with real-life implementation of these concepts. To have your questions about this chapter answered by the author, browse to **www.syngress.com/solutions** and click on the **"Ask the Author"** form.

Q: What does LPub do?

A: Using L3P and POV-Ray, LPub creates photo-realistic images of completely assembled models, or step by step building instructions for your LDraw models.

Q: Can't I create instruction steps like this in MLCad? How does LPub differ?

A: Yes. You can create building instructions in MLCad, but there are many differences. MLCad's renderings are not photo-realistic like the POV-Ray images created by LPub. MLCad does not produce part list images, or Bill of Material images. MLCad does not produce web pages to glue all the images together. LPub offers all of these options.

Q: I noticed that I can configure the options for several of the other applications covered in the book through LPub. Does this mean I don't need to learn how to use these applications directly?

A: If you are only interested in having your images rendered by POV-Ray, then simply accessing it through LPub will be enough. If you want to use synthesis in conjunction with MLCad, you will need to learn how to use LSynth on it's own.

Q: I'm still not clear about the Saving Configuration Options. When would I want to save these to the current directory, to all of the directories, or to a specific file?

A: Good question. If you change options, then render images and quit LPub, your changes will be forgotten. If you want these changes to be saved and available next time you use LPub, you should save them using the global save mechanism. Saving option configurations locally is typically used by people who want to render some designs in black and white for print publishing, and some designs in color for web publishing. My global configuration settings (the ones I get when I fire up LPub) are set to web page publishing. For designs that are going into book form I switch the options to black and white, then save them locally. The saving and restoring of named configurations can be used if you have more than one favorite setting, and you want to be able to use them from lots of directories.

Q: How does MEGA POV differ from POV-Ray? Are they essentially the same program?

A: MEGA-POV is an extension to POV-Ray. MEGA-POV post-processes an image rendered by POV-Ray, finds sharp edges in the image, and draws them into the picture. This is popular because it makes building instructions look like modern LEGO building instructions with parts outlined in black.

Chapter 10

Going Beyond: LEGO Resources

Solutions in this Chapter:

- A Thriving Online Community
- Other Resources

☑ Summary

☑ Solutions Fast Track

☑ Frequently Asked Questions

Introduction

Now that you have mastered the applications we have looked at in this book (or are in the process of mastering them), you might ask yourself, "Now what?" In the previous chapters we have explored a single slice of the enormous 3D modeling cake awaiting the reader who wants to go further. The LDraw system has spawned a whole range of programs that not only make modeling in 3D accessible and free but also offer you plenty of possibilities to grow.

LDraw's success is due in a great part to the community that supports it. It turns out that LEGO modeling is a hobby that is perfectly suited to the Internet. In fact, it could almost be said that the LEGO community was waiting for the Internet to happen. The wealth of LEGO-related resources on the Internet is mind-boggling and, for the most part, free, so they are well worth exploring. The official LEGO Company Web resources are both excellent and varied, but the real treasure lies in the numerous fan-created and fan-oriented LEGO-related services. They contain all sorts of diverse information, from documents pertaining to the original LEGO patents to guides on how to participate in online auctions of LEGO parts. You will find original instructions from official kits as well as MOCs, short for *My Own Creations*, which are examples of LEGO builders' personal creations, proudly displayed in ever-growing numbers.

The emphasis in the online LEGO community is on participation and collaboration, making the sum larger than the individual parts. This mostly amateur community has rolled out some impressive successes and continues to do so at an increasing pace. Perhaps one of the community's greatest successes is that it is organized in a fashion similar to the LEGO line itself. We have clear starting points that then lead into an infinite galaxy of possibilities. In the real world, these starting points are the LEGO parts themselves; in the LEGO community, they are the key Web sites that allow us to explore the rest of the LEGO community very easily.

This chapter covers some of these key sites and what they offer the computer 3D software user and LEGO fan. We also explore two examples of personal LEGO-related Web sites—those of the co-authors of this book. This discussion will give you a very good idea of what's out there, how to find it, and how to use it to improve your LEGO 3D modeling skills.

A Thriving Online Community

The LEGO fan presence on the Internet has taken a tremendous advantage of the medium—in great part because the LEGO system adapts very well to it. After all, LEGO is a global brand. Since users all over the world employ the same parts and techniques, the system serves as a common language and starting point for those users. The Internet and the evolution of computers in general has brought LEGO fans the perfect tools to articulate their world.

These online resources bring to the hobby a system of exhaustive support for individual users so that no matter what their interests, users will find other people with similar likings. Newcomers find themselves immediately empowered with all that is available to them. Let's take a closer look at some of the major LEGO fan sites. This listing is by no means complete, and it is not our intention to trivialize by omission or brevity any of the numerous sites that we did not have the time or space to cover. We present only some of the sites that are both the most popular and the most useful; as you explore them, they will quickly lead you to an almost endless array of other LEGO-related sites.

LUGNET: The LEGO Users Group Network

The LEGO Users Group Network, located at www.LUGNET.com, is the unofficial central point of LEGO fandom. It serves as a one-stop center for all LEGO aficionados. Not only will you find all sorts of support at LUGNET, from specialized forums to parts, contests, and sets databases, but you'll also find resources for individual builders and groups as well as links to many other sites. Figure 10.1 shows LUGNET's main screen.

Since the nature of fan Web sites is to learn and share with the rest of the community, this main LUGNET screen is a truly interactive starting point. The left column below the site's logo allows visitors to browse the forums, which are open to anyone to browse and post questions and comments. These forums are extremely varied in nature and cover almost every possible aspect of the LEGO universe.

In the center column we find a wealth of options. At the top is a simple interface to several very powerful databases of LEGO sets and parts. There are entries for over 4,500 LEGO products, each of which contains exhaustive information. For instance, the entries for kits show a picture of the set, plus a rating fans have given the set, as well as those fans' comments on the set—including information on who owns, who wants to own, and who is selling the kit (all voluntarily submitted). Details such as the date of release, the official price, and number of parts are also included. Additionally, there are often links to inventories and even to commercial sites that sell the kits, with price comparisons.

Figure 10.1 The LUGNET Web Site

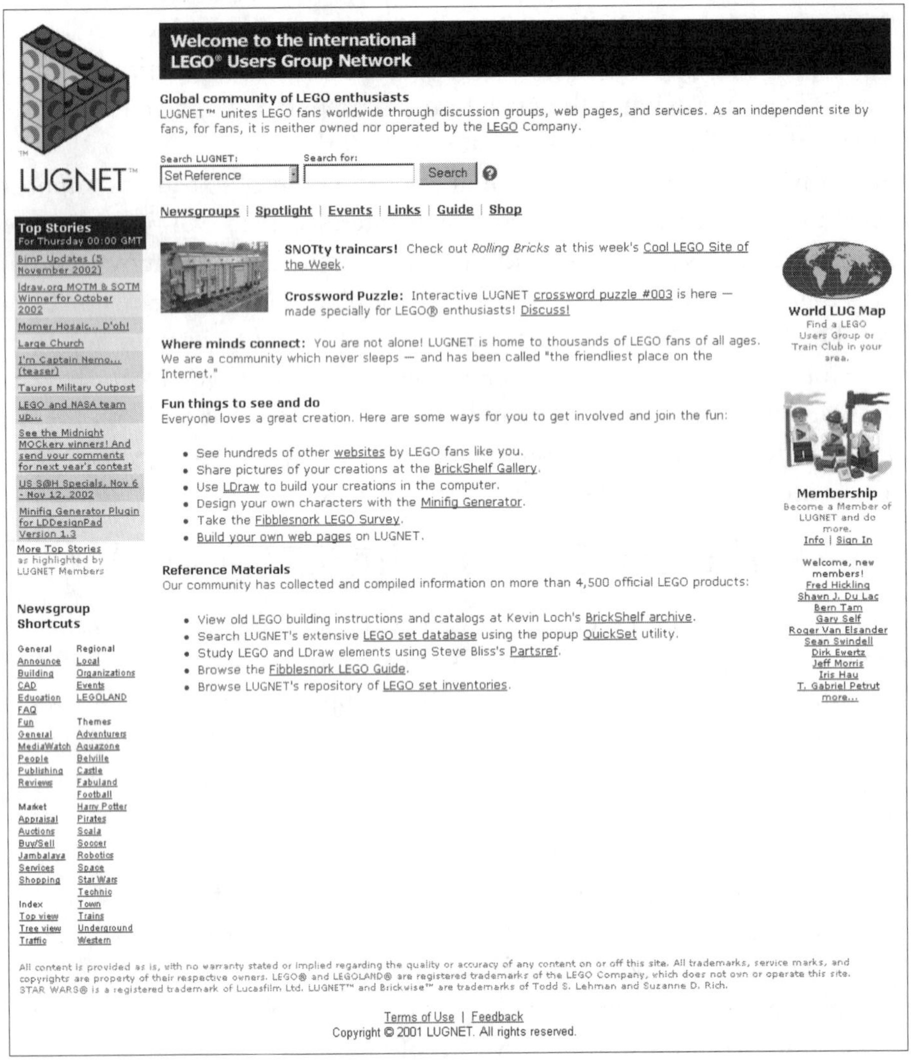

Below this feature are listings for all sorts of contests, events, and sites offering a variety of support and services. The space on the page's right side is dedicated to organizations; you can locate neighboring LEGO fans by navigating the world map to your local area. LUGNET is in fact also an organization in the sense that it accepts paying members, although that is not a requirement. The vast majority of the site's content and services are offered for free, but the membership fee is low and the effort on the part of the people who run the site is well worth some reward.

NOTE

The Massachusetts Institute of Technology (MIT), one of the most prestigious technical colleges in the world, and LEGO, one of the most prestigious toy companies in the world, have officially teamed up in the past to design a variety of LEGO MINDSTORMS robotics elements such as sensors and programmable bricks.

LUGNET is also the result of this collaboration, albeit an unofficial one, between LEGO and MIT. Suzanne D. Rich, the cofounder of LUGNET along with Todd S. Lehman, attended MIT's Media Lab and later received some funding from LEGO to start the LUGNET site. Although LEGO and LUGNET maintain their independence from each other, both enterprises gravitate around the same subject, and there is plenty of collaboration between them. After all, LUGNET *is* the voice of the LEGO fan.

Brickshelf

Brickshelf, located at www.brickshelf.com, is a site that visitors frequently access from the front page of the LUGNET site, pictured in Figure 10.1. At Brickshelf, the efforts of Kevin Loch have created a site that deserves individual recognition. Figure 10.2 shows Brickshelf's main screen.

Brickshelf offers three amazing and free services to the LEGO user. One is the Instructions and Catalogs Scan Library, which is a repository of a large part of the official printed material from the LEGO Company. This material is very useful to the LDraw modeler, who can find instructions to recreate official models, look at the way LEGO produces its instruction steps, and see how they treat the models in the artwork found in the catalogs and on the boxes. There is a wealth of inspiration to be found here.

The library is only one of the three services offered by Brickshelf. The other two are combined into one offering: the Brickshelf Gallery. Anyone is welcome to create an account and post images, animations, video footage, and even small text descriptions of anything LEGO related. The site has moderators that periodically check that everything is in order. This is a great place to post any LEGO-related image. Many people use it in combination with the text-only LUGNET forums by posting questions in the LUGNET forums and linking to pictures in the Brickshelf gallery.

Figure 10.2 The Brickshelf Web Site

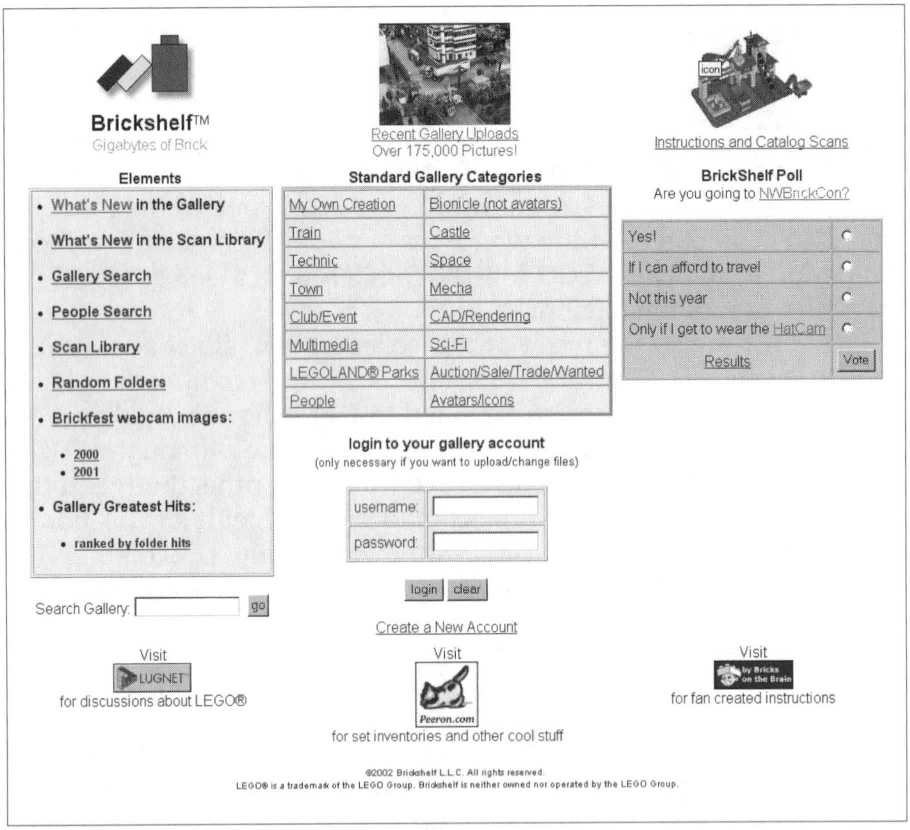

Designing & Planning…

Adult LEGO

Not without reason, parents are often alarmed (or appalled) by the material that their children can find on the Web. Refreshingly, very few people have chosen to use LEGO to portray themes that clearly belong in the adult realm. The overwhelming majority of LEGO fans sites are extremely conscious of the fact that LEGO is primarily a toy for children. In terms of moderation (Internet self-censure), these people tend to err on the very safe side. Large sites like LUGNET and Brickshelf can be considered absolutely safe for children of all ages. Indeed, some forums of LUGNET regularly run parent discussions on the subject.

This material is available for anyone to see, and new material is added *daily*. The Recent Gallery Uploads link at the top center of the main Brickshelf page provides constant fan-posted visual material for your perusal. Thus Brickshelf provides plenty of inspiration as well as space for exhibition of your creations.

LDraw.org

LDraw.org, located at www.ldraw.org, is the centralized resource center for all things related to the LDraw system, including other LEGO-related software. Figure 10.3 shows LDraw.org's main screen.

Figure 10.3 The LDraw.org Web Site

This site is hands down the best effort on the part of the LDraw fans, who are doing a fantastic job of articulating a software system based completely on voluntary contributions from different developers. There are contests, software updates, links, lots of other information, and, as usual, several ways to chip in and add something to the system centered around LDraw. If you develop software or parts for the library or simply create great renderings, definitely look it up.

One of the key aspects of the LDraw.org site is that it provides the user with many essential technical resources, such as white papers on LEGO CAD programming, file format information, and peer review for new parts. The dynamic nature of this great site and the LDraw.org team as a whole hold true to the high level of quality established by James Jessiman for his original software.

NOTE

After the tragic death of LDraw's creator, James Jessiman, in 1997, his program soon became *the* standard for LEGO CAD and took on a life of its own, refusing to disappear even with the lack of an authoritative figure to support it. Other programs have added new functionality to LDraw, and with LDraw.org, a group of devoted fans have put together a excellent centralized location for all things LDraw. The project is coordinated by LDraw.org's Webmaster, Tim Courtney, who confesses on the site to having a team of thousands of collaborators: all the users of the LDraw system.

Brickfilms

Brickfilms, located at www.brickfilms.com, is a relative newcomer to the LEGO fan site world. However, we chose to include it in this reference because its content holds a lot of interest for 3D computer modelers. The main Brickfilms screen is shown in Figure 10.4.

Brickfilms is a repository of LEGO animation links, resources, and films. Many of the films are actually stop-motion movies done using real LEGO parts and a digital camera. However, we found some completely digital animations as well as some very accomplished works that combine 3D CGI and real footage.

Even the animations that use real-life footage exclusively can be very valuable to the computer 3D modeler. This is because both real-life movies and 3D illustration are based on the same technique: creating visual messages made of light.

Many of the tricks involved in traditional photography and cinematography, such as the effect of lights, placement of the camera, and angles of perspective, can be directly applied to our 3D virtual images.

Figure 10.4 The Brickfilms Web Site

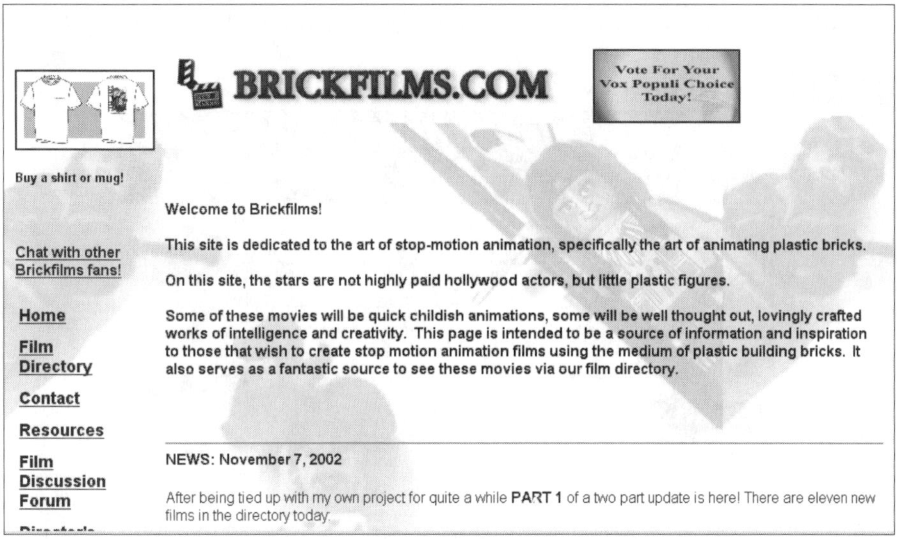

Designing & Planning…

LEGO Animation

In principle, computer animation seems less tedious a process than traditional stop-motion animation. We mentioned earlier in the book some basic animation techniques using the CLEAR and/or BUFFER EXCHANGE statement in LDraw files. However, making animated movies involves much more than making the parts move. To tell a story, we need to learn the language of film. Since animations are not bound by the laws of the real world, animated movies have yet another layer of techniques specific to them (including how things deform as they move). We encourage you to explore this engaging medium, but be aware of the challenges that lie ahead.

From Bricks to Bothans

From Bricks to Bothans, located at www.fbtb.net, is also a relative newcomer to the online LEGO community. Instead of focusing on specific modeling techniques, this site focuses on a specific theme—in this case, the *Star Wars* movie universe interpreted using LEGO. The site's main page is shown in Figure 10.5.

Figure 10.5 The From Bricks to Bothans Web Site

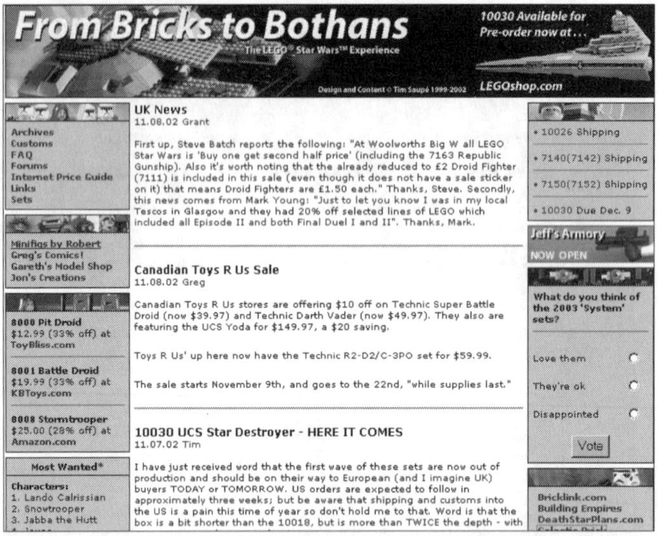

Apart from being one of the key sites dealing with the LEGO *Star Wars* theme, FBTB (as it is commonly referred to) highlights yet another type of fan site: one suited to specific tastes that might or might not coincide with official LEGO themes. Before LEGO partnered with Lucasfilm Ltd. to produce the current LEGO *Star Wars* kits, the company had already introduced several other space/science fiction lines, with great success. Today LEGO space/science fiction fans have developed other themes only loosely related to the official space kits as well as sites, building standards, and the usual myriad creations, support, and resources found in all LEGO fan Web site networks.

What does a site like FBTB offer to a purely digital modeler? It offers plenty of building ideas—and not only complete projects. Since the site is based on interpreting a concrete piece of visual fiction using plastic bricks, we can learn specific ways to translate into bricks any real object, starting with those of the *Star Wars* galaxy. Equally important is the fact we mentioned earlier—that movies and digital models have a lot in common. Studying how fans recreate complete scenes from the *Star Wars* movies can have a big influence on how we present our own creations.

The opposite also holds true: To the sci-fi LEGO fan, the tools described in this book offer the ability to achieve many building objectives. For instance, creating large spaceships is much cheaper if we do so using the computer. A digital model is not the same thing as a real model—then again, real models can't fly, but digital ones do, with amazing grace.

Designing & Planning…

Minifig-Scale Toilets and Other LEGO Fan Themes

You can be absolutely certain that there are plenty of fans for even the most specialized LEGO themes, which often break down into subthemes and sub-subthemes interwoven with each other. For example, architectural building could be a theme; its subthemes would include various scales, the nature of the buildings, and historical recreations, to name just a few. Sub-subthemes could include specific time periods, parts usage, or ways to create interior details. There is a fan page on the Brickshelf site with over 10 different minifig scale *toilet bowls* made of a combinations of regular LEGO parts. It would be impossible and futile to try to compile an approximate list of sites or even main themes, since new ones come out so often. If you are interested in a specific theme, try looking for information on LUGNET first.

Bricklink

Bricklink, located at www.bricklink.com, is a virtual LEGO mall. It is a collection of online stores devoted to LEGO, created and maintained by LEGO fans. Figure 10.6 shows Bricklink's main screen.

One of the important impacts of the Internet on the LEGO fan world has been in the commercialization of LEGO. Apart from the official company channels, LEGO has proven to be a product well suited for fan e-commerce. There are enough parts to create a wide interest, their quality is legendary, and they can be easily coded for remote transactions. These characteristics, combined with the new Internet e-commerce tools, has resulted in a dramatic increase in the number of ways LEGO can be traded—and thus valued. LEGO fans looking to expand their parts collections no longer have to depend exclusively on the LEGO Company or the local toy store to meet their needs. Hundreds of LEGO transactions are conducted on the Internet every day, and many involve fan-to-fan commerce.

Figure 10.6 The Bricklink Web Site

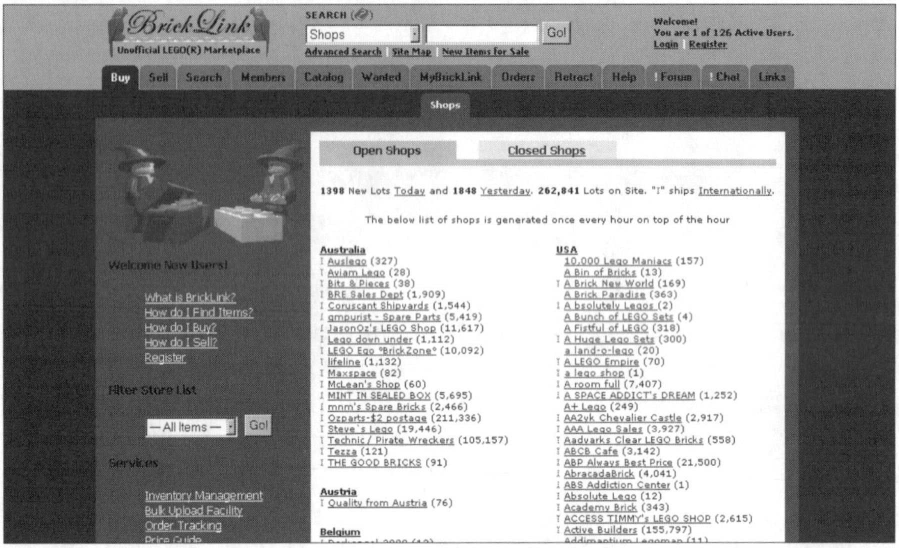

Designing & Planning…

Trading in LEGO

Since LEGO parts can be bought and sold on a daily basis, they behave a bit like stock market shares. We can't pin down an absolute price per part or per set, but we can assign a price range for them. If you want to plunge in and experience the economics of LEGO in action, here are some tips:

- **Stick to the LEGO brand** From a purely economical perspective, other brands of building bricks simply lack the fan support to make them a good investment. Most LEGO fans are *very* discriminating about quality. Never try to pass off bricks that are not official LEGO parts as official LEGO bricks. If you are not sure, be clear about it or you will earn yourself well-deserved scorn from the LEGO community.

Continued

- **Best deal** The best possible deal is buying new, sealed LEGO kits at a discounted price. You can get them directly from the LEGO Web site and LEGO Direct services or in some stores and outlets. Some fans buy discounted boxes and resell directly to other fans who might not be able to take advantage of the discounts locally. Others sell the parts individually (generally at a greater profit but understandably, since more effort is involved).
- **Secondhand LEGO** The cheapest way to acquire LEGO parts is to buy secondhand bulk inventories. The quality of LEGO does not end in its clever design. The plastic of which LEGO parts are made, acrylnitrile butadiene styrene (ABS), is a really cool material—highly resistant and completely recyclable. Most secondhand parts have multiple tiny scratches on their surface and thus lack the "luster" of new parts, but they tend to work flawlessly.
- **Two enemies** The two great foes of LEGO plastic are odors and children's strong teeth. LEGO parts can withstand great weights, several days on top of a hot radiator, and other domestic perils. However, the material absorbs odors such as cigarette smoke or perfume very easily. Washing parts is certainly possible, but doing so requires a certain amount of care. Teeth marks found in LEGO parts, often very profound, are the result of an instinctive movement found among many (hopefully younger) LEGO fans who use their *mouths* to disengage parts that are attached to each other. Please don't do this! The ABS plastic, although quite strong, has little chance against pointy human teeth. Obviously, this practice is also very unhygienic.
- **Catch of the day** To take the pulse of the market, visit the eBay LEGO auctions (search for LEGO from the main www.eBay.com screen). Several of these auctions occur every hour. Fans tend to code their listings very precisely, so you can focus on specific themes, parts, or kits that you are looking for. LEGO auctions are a universe unto themselves. Often, selected parts from a bulk lot are reintroduced into the market cleaned and well photographed within a matter of days, often fetching a much higher price than the lot they were sold in just a week before.

Bricklink is only one of several ways that LEGO modelers can acquire LEGO parts via the Internet. Bricklink's main value to the LEGO CAD user is that it enables us to purchase (or at least find) very specific parts. If we create a model using the computer and then decide to recreate it using real LEGO, we will probably find that locating the exact parts for our model might not be an easy task. Some parts are scarcer than others, and not all are available in every color we could hope for. Bricklink is not the cheapest place to get LEGO, but it is perhaps the ultimate real-world parts warehouse.

The First LEGO League

Yet another very important application of the LEGO building system is education. Figure 10.7 shows the site of the First LEGO League (FLL) International, located at www.firstlegoleague.org, an international LEGO robotics tournament designed for children and teenagers.

Figure 10.7 The First LEGO League International Web Site

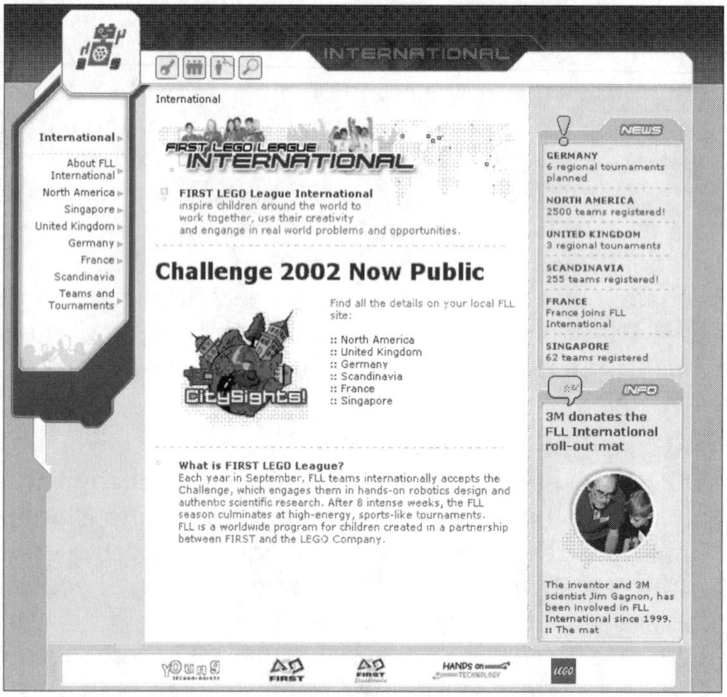

The LEGO Company has been involved in education for a long time with its Dacta product line, but perhaps its most important educational product is the LEGO MINDSTORMS robotic elements, introduced in 1998. Developed in

partnership with the prestigious Massachusetts Institute of Technology, these robotic sensors, "intelligent" bricks, and unique parts—fully compatible with the rest of the LEGO lines—have created quite a stir in the robotics field. In 1994, even before their commercial introduction, Fred G. Martin presented a Ph.D. thesis at MIT that outlined a new and better way to teach future engineering topics such as robotics using LEGO bricks.

Using Martin's blueprint, any teacher in any school in any country can now train students in the art and science of creating robots—from making them move to making them think and interact with their environment. The First LEGO League is not a machine contest. It is a *behavioral* robotics contest. Young kids are presented with an open challenge to design robots that will compete against the solutions adopted by other teams. In the end, these projects constitute an integral way to learn technology, from making reduction mechanisms with gearwheels to driving motors using onboard computers that react to the local environment.

Designing & Planning…

LEGO Contests

LEGO contests are always fun because winning or losing is a secondary priority. Any well-designed contest will generate enough different solutions to the initial challenge that everyone involved is bound to be both entertained and educated. Although robotics contests are perhaps the most well-known of these competitions, they come in all flavors, including purely aesthetic contests. A recent initiative awarded prizes to creations based around a Halloween theme, with the following subcategories: minifig characters, buildings, dioramas, and sculptures. Contests come in many shapes and sizes; some are online events, others are organized by local clubs. The LUGNET forums are a good place to find contests; don't forget to look in the LUGNET sections dedicated to local organizations!

The Official LEGO Web Site

As a global brand, LEGO has multiple official sites on the Internet. Most of them can be accessed via the main site, located at www.lego.com, the main screen of which is shown in Figure 10.8.

Figure 10.8 The Official LEGO Web Site

Designing & Planning…

Trademarks, Copyrights, and Patents

True to its unique character, LEGO has taken a largely unconventional approach to the defense of its intellectual property. Although many companies sternly discourage any and all user initiatives, LEGO has issued a set of general guidelines regarding the matter. These guidelines manage to encourage user experimentation while setting reasonable rules of fair play. The guidelines can be found at www.lego.com/eng/info/printpage.asp?page=fairplay and are of special importance to users who want to show their work to the public. The following is a quick overview of each concept and the position taken on it by the LEGO Company:

- **Trademarks** Trademarks are the names of companies or their products and themes. LEGO trademarks include LEGO, Duplo, Technic, and MINDSTORMS, to name just a few. LEGO has a specific disclaimer for fan-created Web sites that specifies that those trademarks belong to LEGO, which does not endorse these sites. Giving credit where credit is due is pretty much what the company is asking for, and that seems reasonable. Additionally, you should not include any LEGO trademarks in any publication title, whether in a Web site or

Continued

program. For example, MINDSTORMS CAD is not a good name for your new CAD program, because it creates confusion in the public's mind as to whether it is an official LEGO product or not. To be on the safe side, avoid any trademarks or logos in your illustrations (including the LEGO logos on the studs of LEGO bricks, which can be turned on or off in LP3). Trademarks do not expire as long as the owner renovates them periodically. When in doubt, ask LEGO.

- **Copyrights** Copyrights refer to published works. In LEGO's case, the most obvious examples are instructions sets and box art. LEGO, with reason, discourages the reproduction of currently available instruction sets. For discontinued sets, they might be more lenient, but before you publish old LEGO material on your Web site, be aware that there exists a central, nonofficial repository of all LEGO printed material accessible through LUGNET and Brickshelf. Generally speaking, always ask for permission from the copyright holder before reproducing their work. The fair-use doctrine, which you might have heard about, allows the owner of a copy of a work to do some reasonable archiving and citation. Don't sweat about including one specific instruction step to make a point, but be respectful of other's property. Copyrights do expire, but in the actual legal climate surrounding them, most of us will not be around by the time any LEGO copyright expires.
- **Patents** Patents are industrial copyrights, so to speak. They define and protect products and technologies. For instance, each LEGO piece has its individual patent, and there are patents on the connection systems and just about every specification relating to them you can think of. Legal matters pertaining to patents rarely affect the end user, unless he or she decides to manufacture a complementary line of pieces that connect with LEGO pieces. Even then, some of the patents will have already expired, which is why LEGO keeps some of its manufacturing processes secret. In part, it also explains why LEGO introduces new pieces and lines regularly. People who have "hacked" (deciphered the electronic workings) of LEGO electronic equipment, such as the RCX or MicroScout bricks, should actually be more concerned about copyright violation. In any event, LEGO has put out a specific, reasonable, and experiment-encouraging policy covering such issues. It can be found at http://mindstorms.lego.com/sdk2/SDK.asp.

The site is full of services to the LEGO user, with shops, forums, contests, Web hosting, articles, games—even an online LEGO CAD. The site also features special Internet services for the fans of specific sets or lines, such as instruction sets to build extra models from commercial kits, guidelines for programmers of the robotic LEGO elements, and game-like worlds based on LEGO products, such as Bionicle.

Just as LUGNET is primarily geared to adult LEGO users, the official LEGO site is tailored for the intended recipients of the company's product: young and imaginative people. Adult LEGO fans who sometimes feel ignored by the company's official products should perhaps remember that this wonderful toy and fantastic learning system is, after all, a product designed primarily for *children*.

3D Café

3D Café, located at www.3dcafe.com, is not a LEGO-related site, but it is one of the primary sources of information on 3D software. Its main screen is shown in Figure 10.9.

The site is oriented to all sorts of users, including both professionals and amateurs. It also offers a lot of free stuff, such as tutorials, programs, and even 3D models. This is definitely a site to keep an eye on if you want to explore the world of 3D illustration beyond LEGO-related software.

Just as the LDraw system is part the larger fan LEGO community, it is also a component of the larger 3D software world, which has its own community, Web sites, and software. As we mentioned earlier, mastering the software covered in this book gives you very important insight into the field of 3D design. And, like the LEGO community, the resources and support for 3D design are waiting to be tapped. It is up to you to make the best of them. 3D Café is an excellent place to begin your search for resources and inspiration.

Personal Web Sites

There are many other points of entry into the LEGO online community, but since it is impossible to cover all them, here we focus on what constitutes much of the underlying fabric of the community—the personal sites created by fans.

When we began working on this book for Syngress, our editor, Jon Babcock, proposed that we tell our individual LEGO stories in the book. We both replied, "OK, we'll put them at the very end." However, as the shape of this chapter took form, we became aware of the value of telling our stories here, if only because we have been fortunate enough to see much of our LEGO efforts quite literally paid back in full in a short period of time. We consider ourselves completely typical adult LEGO fans who perhaps happened to be at the right place at the right time.

Figure 10.9 The 3D Café Web Site

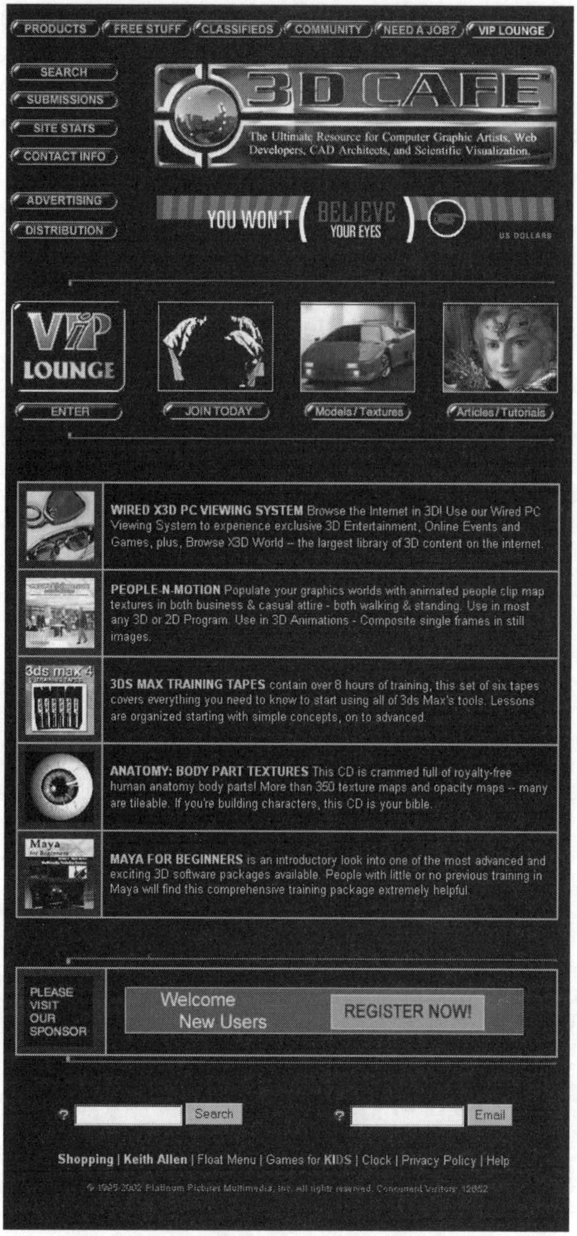

Like a large majority of LEGO fans who have taken up the hobby in the last two or three years, we quickly—and independently—made the connection to the LEGO online community. Seeing that some of our vague ideas could indeed be worked out with LEGO, each of us started a flurry of activity, asking questions on forums, purchasing LEGO kits and parts, learning about LEGO mechanics, inventing new ways of doing things, and, eventually, creating our own LEGO-related Web sites.

Kevin Clague's Web Site

Kevin Clague is one of the authors of this book and some of the software in it. His personal Web site, located at www.users.qwest.net/~kclague, is shown in Figure 10.10.

Figure 10.10 Kevin Clague's Personal Web Site

Kevin's Web site showcases his three major LEGO interests:

- Two-legged robotic walkers
- LEGO pneumatics
- CAD software

At the top of his site is an image of the first LEGO book he wrote for Syngress, *LEGO MINDSTORMS,* followed by links to his LEGO CAD software tools, LPub and LSynth. Notice how Kevin offers these programs to other fans for free. As we said earlier, the community is greatly based on the voluntary sharing of knowledge among fans.

Following the software section, Kevin highlights his own LEGO robots and models, or MOCs. Kevin's building interests primarily employ Technic and MINDSTORMS elements to create robots that walk on two legs. His site narrates the trajectory of his hobby. Robo-homind, the yellow walker, was the first biped he ever made. He shows a few variations of this type of robot, and then he moves on to some other bipedal techniques. Go-Rilla is the fastest biped on the site, but he cannot turn. Kevin has several more new bipeds in the works.

The site also includes a section on pneumatics. Kevin enjoys pneumatics quite a bit because they don't involve motors or gears, and they allow us to build complicated machines that require no electronics. Designing pneumatic circuits is very much like designing simple electronic circuits, which he did in his college electronics classes. Kevin also includes a section on custom sensors: homemade robotic sensors made by embedding electronic components inside regular LEGO parts.

Despite being very specialized, the stuff on Kevin's site is eclectic enough to potentially interest a wide variety of fans. This approach is common to most personal LEGO Web sites.

Miguel Agullo's Web Site

Miguel Agullo is the other author of this book. Kevin and Miguel have actually never met in person and did not know one another before they connected via their common LEGO interests. Since then, they have worked together on several LEGO-related projects. Miguel's Web site, Technic Puppy Journal, located at www.geocities.com/technicpuppy, is shown in Figure 10.11.

Miguel is a professional graphic designer who has found a lot of inspiration for his LEGO Technic works in the creations of Japanese LEGO fans. For that reason, the overall aesthetics of his site are meant to mimic modern Japanese popular fan publications.

Figure 10.11 The Technic Puppy Journal Web Site

As usual, part of Miguel's site is devoted to explaining and showcasing his own creations. Each model includes instructions and detailed explanations on how they work. There are links to pictures and videos, but the site is mostly illustrated with MLCad and POV images, one of Miguel's interests. Other parts of the site are devoted entirely to fans. Since Miguel has not yet created any software he feels is worth releasing, he created a list of links to LEGO bipeds, one of his building interests. He has also coded some unofficial parts for the MLCad library, also available through his site.

Designing & Planning...

Why Not Chip In?

The key word to remember about the online LEGO community is *collaboration*. Any and all contributions you can make are welcome. Not only that, but there are plenty of venues to explore for presenting your thoughts, techniques, and creations. Quick commentary is best pursued through the ever-active LUGNET forums. LEGO images can be posted in Brickshelf and in many other places, including official LEGO Web sites. If you want to create you own site, consider taking advantage of the generous gift offered by the New Zealand online store Ozbricks, which gives free hosting space to LEGO fan Web sites, with no strings attached. Check them out at www.ozbricks.com/.

Other Resources

We mentioned earlier that the Internet not only puts us in touch with other parts of the world, but it can also help us find LEGO fans near us. No amount of training or support can substitute for actual social interaction, for which LEGO constitutes a perfect excuse.

And, of course, no amount of digital LEGO will ever produce exactly the same sensations and experiences as visiting a LEGO convention or theme park. There you can admire the work of professional modelers … by the metric ton. Fans of all themes will find something that appeals to them specifically and maybe even develop a liking for new LEGO themes. Let's now take a look at some of the other resources available to the LEGO fan.

LEGO Clubs and Conventions

Just as the online community offers the LEGO fan a plethora of resources made up of many combined individual efforts, local gatherings of LEGO fans can also be a great resource for new ideas and contacts. LEGO clubs come in all sizes. They range from informal gatherings of friends who hold semi-private meetings in which they talk shop, show new models, and trade parts to large organizations capable of organizing entire conventions open to the general public.

Since LEGO is a modular toy, resources can easily be pooled to create, for example, large town or space dioramas composed of individual contributions from different members seamlessly integrated together. As long as the participants agree on some general guidelines, such as the size of the "real estate" (LEGO baseplates) to use, their individuals works will blend into something else. Few people are not impressed by such large setups.

LEGO clubs and conventions are also good places to give and receive training. It's hard to look for a LEGO modeling school elsewhere! This training does not need to be all introductory; contests are a very exciting way to explore advanced modeling concepts, using a setup similar to that of the First LEGO League Web site mentioned earlier.

Basically, attending conventions and joining LEGO clubs are great ways to meet people who *use* LEGO. Any hobbyist can appreciate talking to other people about a shared hobby without having to explain or justify it first. One easy way to find out about clubs in your area is to navigate the world map tool found on the LUGNET site. If you can't find one in your area, maybe it's time for *you* to get things going and start a LEGO club of your own.

LEGO Parks

Visiting a LEGO park will expose you to the work of professional modelers, and lots of it. This is actually pretty important for 3D software usage. Not only do these professionals achieve fabulous results using the same material you use, but they also engage in creations that are not typically available from the company as kits, such as LEGO sculptures or architectural models of real buildings.

Equally interesting is the act of observing how parts are used in different building scales. Larger models obviously allow for more detail, but each scale has specific ways to milk the best out of particular LEGO parts. Many of the models in the "town" sections of the parks are built at a scale several times larger than that of the minifig, which is the standard fan construction scale. This allows the modelers to use more parts to create more realistic details.

It is also worth paying attention to how LEGO and non-LEGO elements integrate. Some models spectacularly embrace real buildings; others are full-blown animatronics with motors and special effects mechanisms hidden inside. The use of light and composition often play key roles in the success of some models. Spend some time trying to figuring out the tricks behind the ones that you like best.

Many of the models in the parks are simply too large to contemplate building at home (even inside the computer), but there is no stopping us replicating the smaller ones. Bring a camera and plenty of film!

There are currently three LEGO parks in the world. The original one is located in the village of Billund, Denmark, at the site of the first LEGO headquarters, which is still the firm's headquarters (the Danish village now has an international airport). LEGO initially set it up as a garden for visitors, but its growing popularity moved the company to start considering it as an important part of their business. In 1996, LEGO opened a second park in Windsor, England. The third opened in 1999 in Carlsbad, near San Diego, in Southern California. Over Christmas of 2002–2003, a new LEGO park will open in Günzburg, in Southern Germany. LEGO is obviously expanding this branch of the company, but the location of future parks remains largely unknown.

Your Own World

Throughout the process of writing this book, one of the major concerns for the writers and editors was where to draw the finish line. We could have provided more detail on POV-Ray, how to create parts for the library, or how to model certain objects with LEGO parts. Our problem of where to end came from the fact that the available software offers as many possibilities as real-life LEGO bricks do. Virtual LEGO design borrows directly from classical disciplines such as sculpture, architecture, and painting, and it uses the techniques of newer disciplines such as photography and cinematography.

But this is only the tip of the iceberg. The LEGO-like images you've learned how to create in this book are based on sophisticated equations that must be harnessed in a way accessible to casual users of the programs used to create them. Beyond the basics, it is impossible to try to explore all the possibilities in one book.

Even if you are new to LEGO and/or computer 3D software, you have arrived at the right time. The work done by fans in the past few years is now beginning to produce some very impressive results in both fields, and the resulting technology has greatly empowered the user. Not only is building virtual worlds with (or without) LEGO easier than ever, but we also have access to the tools that used to be exclusively in the hands of professionals.

What is the best way to take advantage of all this wealth? LEGO and 3D illustration are very engaging and rewarding hobbies; don't be afraid to explore and make mistakes, because that is the best way to learn. Build freely on the works of others, but remember and acknowledge their contributions and try to give something back to the community. Always keep in mind that the sky is the limit. In theory, with the software described in this book you can create images or even movies that approach the high level of quality seen in films by such companies as Disney or Lucasfilm.

This is hardly a dream. All you have to do to achieve it is follow the philosophy of the man who started it all, Ole Kirk Chirstiansen. Take the high-quality approach, never be completely satisfied with your work or your knowledge, and don't forget that there is always room for improvement.

Figure 10.12 was created by stretching the software available with this book to its limits—using experimental radiosity settings in POV-Ray to achieve the highest possible photorealistic effects. As you can see, perhaps the best thing about this hobby is that perfection is always right around the corner.

Figure 10.12 Radiosity Rendering Created with POV

Summary

The tools offered in this book offer entryways into two very exciting worlds: virtual LEGO modeling and 3D illustration software. Both worlds have taken advantage of the Internet, allowing fans of both to pool the work of individuals and recycle it as support for all others.

The evolution of the international LEGO community has been spectacular, especially if we consider the high quality of the resources created by amateur fans. The LEGO Users Group Network Web site, known as LUGNET, is the central starting point to explore the LEGO online world via the works of other users, forums, contests, and the like.

LUGNET offers thousands of links to other sites, some of which are themselves central points for LEGO subcategories, such as LDraw for LEGO CAD or Brickfilms for LEGO animation. These subcategories cater to just about every conceivable taste (good taste, that is).

Yet another layer lies below these often interwoven subcategories: the numerous personal LEGO sites fans use to highlight their interests and creations. Do a little Web surfing with your favorite search engine for whatever LEGO-related theme interests you, and we promise you won't be disappointed.

The Internet has also made it possible for local clubs to organize themselves easily, which in turn has created a growth in LEGO conventions. Meeting other LEGO fans via these venues is always an interesting and enjoyable experience.

Finally, LEGO parks constitute a window into the work of professional LEGO modelers. The same parts that we use in MLCad are used in new scales, new themes, and new ways. A good modeler is always alert to these details and interested in seeing what other modelers are up to.

Solutions Fast track

A Thriving Online Community

☑ The International LEGO Users Group Network at www.lugnet.com is the main Internet resource for any LEGO fan, including specific support for local associations.

☑ The site at www.brickshelf.com holds an ever-changing gallery of LEGO-related images and a repository for scans of official LEGO printed material, with instructions booklets for official kits and catalogs.

- ☑ The www.ldraw.org site gives support to the LEGO CAD fan by offering all sorts of technical information and support for the LDraw system.
- ☑ The www.brickfilms.com site offers resources for LEGO animators, whether stop-motion or 3D CGI creators. The emphasis is in cinematography, or at least kinematics. If it moves, send it here.
- ☑ From Bricks to Bothan, www.fbtb.com, is a site that revolves around a theme (LEGO *Star Wars*), not a specific technique.
- ☑ The www.bricklink.com site is a collection of "by fans, for fans" online stores. With the advent of the Internet, many new ways of trading LEGO have appeared.
- ☑ The www.firstlegoleague.org site is an educational contest aimed at children and teenagers from all over the world.
- ☑ The www.lego.com site is the official and extensive Web site of the LEGO Company, oriented to the core (and very demanding!) users of the product: children and teenagers.

Other Resources

- ☑ LEGO clubs and conventions are a great way to meet people who share a common interest in all things LEGO.
- ☑ Through the sum of individual efforts, LEGO clubs can carry out activities that single fans or an online presence simply cannot provide.
- ☑ LEGO parks offer a view into the fabulous work of professional modelers. They are good places to see models built with themes and scale that are not available elsewhere. Make sure you look carefully!

Frequently Asked Questions

The following Frequently Asked Questions, answered by the authors of this book, are designed to both measure your understanding of the concepts presented in this chapter and to assist you with real-life implementation of these concepts. To have your questions about this chapter answered by the author, browse to **www.syngress.com/solutions** and click on the **"Ask the Author"** form.

Q: What is the best Web site to start my journey into the online LEGO community?

A: The International LEGO Users Group Network at www.lugnet.com is the main Internet resource for any LEGO fan. It offers a combination of important resources of its own, forums open to all LEGO users (and fans!), as well as practical links to other sites. Anything LEGO related will ultimately be found here.

Q: Is LEGO a collectible toy?

A: LEGO is *the* collectible toy. LEGO products of all kinds have a clear market value, set in a manner similar to stock exchanges. (Yes, there can be crashes.) Before the Internet, LEGO fans had few options when it came to actually purchasing parts: They could go to a store and buy official LEGO kits, possibly find some LEGOs secondhand at yard sales, or purchase them directly from LEGO. Today, the scene has changed completely. The Internet site eBay holds *hundreds* of auctions of LEGO material daily. This is the stock exchange equivalent and applies to many other collectible items. But the case of LEGO is somewhat unique: The daily trading is massive and greatly varied. For instance, unlike many collectible toys, the value of LEGO does not necessarily go down once we use it. Old sets in sealed boxes can occasionally reach thousands of dollars at auction, but a piece of pre-Technic chain will also fetch between 25 and 50 U.S. cents *per link*, no matter the state it is in. (It's too scarce for buyers to be picky.) Obviously, the "traditional" presence has also grown strongly. There is no better deal for the fan, whether from a collectible point of view or from a building point of view, than buying *new* LEGO at discounted prices. Visit the official LEGO store to see *their* deals!

Q: How do I become involved in LEGO contests?

A: The easiest way is to think up a challenge and tell people about it. You will soon have drawn a very opinionated crowd and often hilarious results! Contests come in many sizes and shapes. Some have an online presence, others are affairs between friends or clubs. A large part of the LEGO crowd loves contests (read: showing their creations), so do not think twice about inquiring in the LUGNET forums about it.

Q: I have a child who is crazy about LEGO. What should I do?

A: Congratulations! Your child has enlisted in a very positive and safe activity. By design, LEGO is a toy aimed at triggering the imagination. Using LEGO, children quickly develop instinctive spatial perception, a structured approach to tasks, and many other *intellectual* tools. LEGO, like listening to Mozart, makes children more intelligent. But this is only the beginning. Since LEGO is an excellent training tool applicable to many fields, it is a way to raise the interest of children and young adults in all sorts of fields. Not only that, you have an intergenerational tool at your disposal: LEGO fans *always* have something to talk about, even if they are interested in different LEGO themes. One last valuable piece of advice: Look at storage solutions before you collect a large amount of LEGO. LEGO modeling is like working in a lab: Things sometimes have to be messy for ideas to flow!

Q: I am a teacher who is thinking of using LEGO with my students—how do I go about it?

A: There are plenty of resources, so in principle the best solution is to set your didactical objectives first and then see what the available options are. If you do not immediately see anything suited to your needs, ask in the LUGNET forums, or ask your students to *design* a LEGO-based course. This strategy tends to bring out the best in kids. If you are completely out of ideas but want to create some sort of extracurricular activity, we suggest setting up a robotics course. Before you balk at the task, we assure you that the available amount of support material is huge. The basis of any and all LEGO courses follows the key guidelines established by Fred G. Martin, a scientist with degrees in computer science and mechanical engineering and a Ph.D. in media arts and sciences, all from the MIT. Martin worked on the R&D for the MINDSTORMS LEGO robotic elements. Part of his plan's attractiveness is that it automatically engages students of all types in team-oriented tasks.

Gifted mathematicians will depend on mechanical engineers, who will listen attentively to poets and philosophers. The MINDSTORMS robotics system introduced by LEGO in 1998 is much more than gear wheels and thermic sensors. It allows the creation of fully independent robots. Students do not "drive" the 'bots, students "educate" them. Designing the strategy for a brain (and for the rest of the robot) requires programming skills and a way to couple microprocessors to axles and levers. But it also requires noticing how nature can provide the functions if we just take a moment to think about it—and this is an area where the minds of poets and philosophers shine brighter. To boot this automatic integration of "human resources" into teams, robotics courses are often structured around team competitions, with invariably interesting results to both the creators and audience (any kid will want to look at *real* robots, any time), adding yet another attractive twist to the activity. To organize a successful course, do a bit of research in LUGNET and, if possible, contact a local LEGO organization. The Ferraris' book, *Building Robots with LEGO MINDSTORMS* (Syngress Publishing, ISBN: 1-928994-67-9), also offers good insight both into the activity (it is a good classroom textbook) and into organizing contests.

Chapter 11

AT-ST Building Instructions

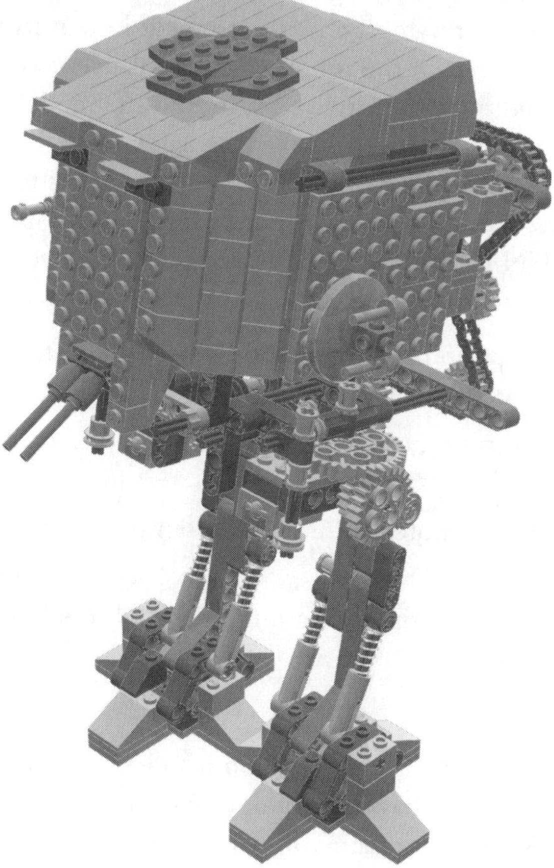

We've included a complete set of building instructions for an AT-ST model built by Kevin Clague in this final chapter for two reasons. First, it's an intricate and entertaining project (especially for fans of the Star Wars movies), and we thought it would be an additional bonus for readers of this book. Second, this set of instructions serves as an excellent example of what you can do with the applications you have been learning about throughout this book. All of the applications we've covered were used to make these building instructions. This AT-ST was modeled using LDraw and MLCad, the flexible elements were synthesized with LSynth, then the entire model was run through L3P and POV-Ray using LPub to produce the final building instruction images you will see on the following pages. Notice how I've used text where appropriate to more clearly explain some of the building concepts involved. With the new additions of buffer exchange, ghosting, and part list meta-commands, the imagery here was all produced with the tools described in this book.

AT-ST is a two motor, weight shifting biped that can walk forward and turn right and left. AT-ST is one of my most ambitious bipeds in that its feet are relatively small compared to the rest of the robot. AT-ST's balance is very finely tuned. If you try to add additional decorative features to the model, they might throw off its delicate balance. One motor makes the legs turn right and left, the other makes AT-ST lean right and left. AT-ST uses touch sensors to detect when its leg stride is complete. It uses a rotation sensor to know how far it is leaning to the left or right. A TECHNIC chain is used to drive the leg movement, as well as the left and right leaning motion. AT-ST walks using this sequence:

1. Leans left, using top motor, until rotation sensor says it has gone far enough in that direction.
2. Right leg strides forward, using bottom motor, until touch sensors say it has gone far enough.
3. Leans right, using top motor, until rotation sensor says it has gone far enough in that direction.
4. Left leg strides forward, using bottom motor, until touch sensors say it has gone far enough.
5. Repeat

AT-ST is programmed in NQC and takes commands via the LEGO Remote Control. It has no autonomous behaviors. Think of AT-ST as a remote control car. Before you can run AT-ST's program you need to turn the RCX brick on, which can be tricky when AT-ST is fully assembled because the RCX On button is buried under AT-ST's face. You can either remove the face, turn the RCX on, and replace the face, or you can use a long axle to reach into the underside of AT-ST's head and press the On button.

You can run AT-ST's program using the LEGO Remote Control. When you first run AT-ST's program, AT-ST just stands there waiting for a command from the remote. You need to have both AT-ST's feet flat on the ground, and side by side as shown in the building instructions. If you do not do this, the motor might rip the chain apart.

The LEGO Remote Control can send three messages, numbered 1, 2, and 3. AT-ST's program uses these messages to know what to do. These messages tell AT-ST when to start walking, stop walking, start turning, and stop turning. Table 11.1 shows the meanings of the message buttons depending on what AT-ST is currently doing:

Table 11.1 Message meanings

Current Activity	Message Number	New Activity
Stopped	1	Turn left
Stopped	2	Walk forward
Stopped	3	Turn right
Walking forward	1	Turn left
Walking forward	2	Stop
Walking forward	3	Turn right
Turning left	1	No change
Turning left	2	Walk forward
Turning left	3	Walk forward
Turning right	1	Walk forward
Turning right	2	Walk Forward
Turning right	3	No change

The LDraw file for AT-ST (AS-ST.mpd), as well as the program source (AT-ST.nqc) and the downloadable program (AT-ST.rcx), are installed as part of the Examples installation package you installed in Chapter 2. The LDraw file and program are installed in the LDraw\Models\PowerTools directory.

Bill of Materials

AT-ST is composed of parts from the LEGO MINDSTORMS Robotic Invention System 1.5, the LEGO MINDSTORMS Dark Side Developer Kit, the LEGO MINDSTORMS Ultimate Accessories Set, some parts from my son Andrew's LEGO collection, LEGO TECHNIC chain, and copious amounts of sloped bricks that were purchased separately. Figure 11.1 is a Bill of Materials picture that lists all the parts needed to make AT-ST.

Figure 11.1 AT-ST Bill of Materials

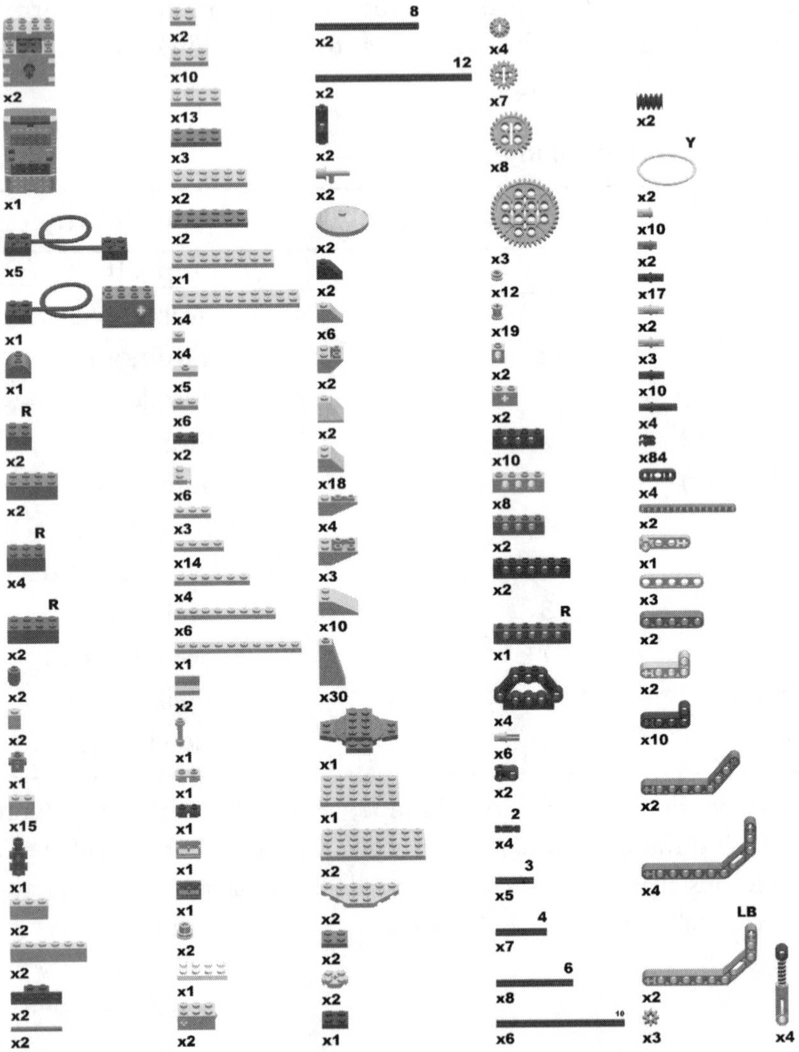

The Feet

This is AT-ST's foot sub-assembly. You will need to build **two** of these.

Foot Step 0

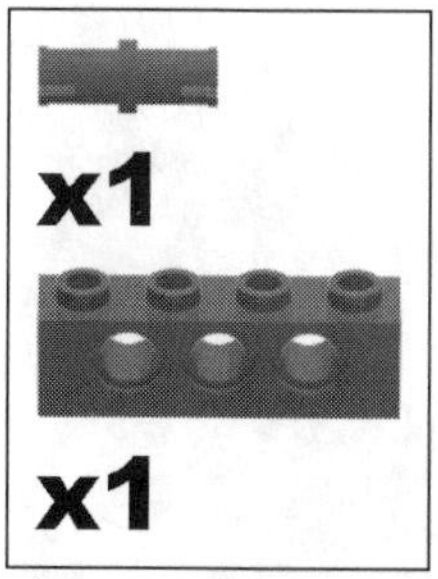

Insert a TECHNIC pin with friction into the middle hole of a black 1x3 TECHNIC brick.

Foot Step 1

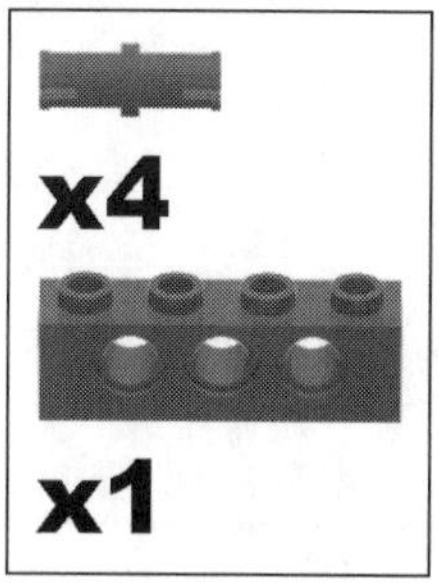

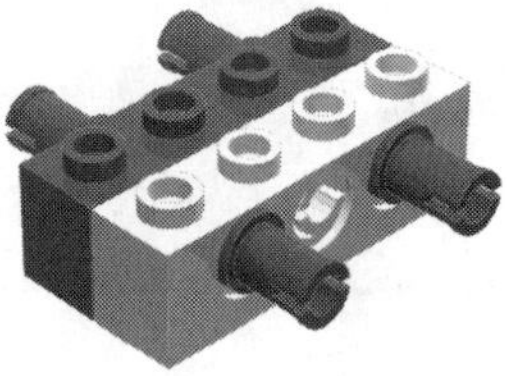

Snap a second 1x3 TECHNIC brick onto the pin, then attach four TECHNIC pins with friction as shown.

Foot Step 2

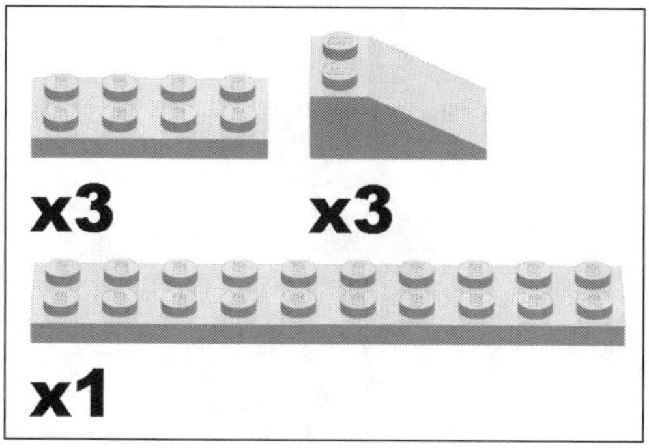

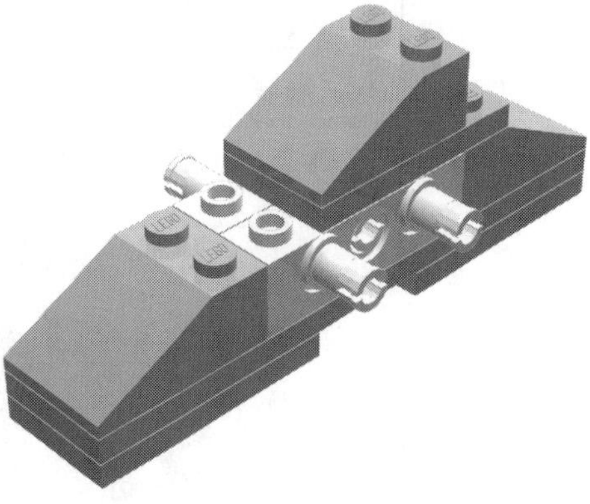

Foot Step 3

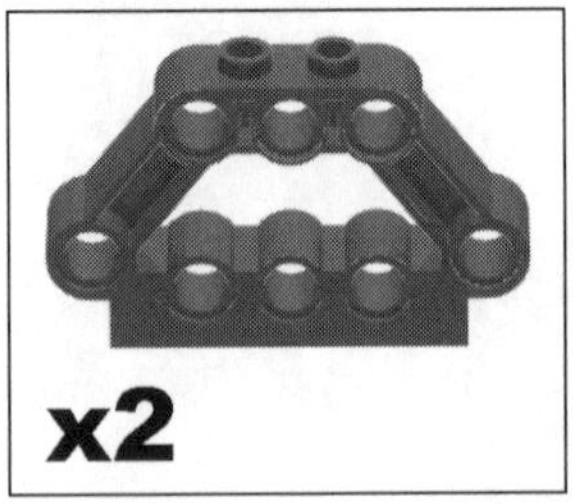

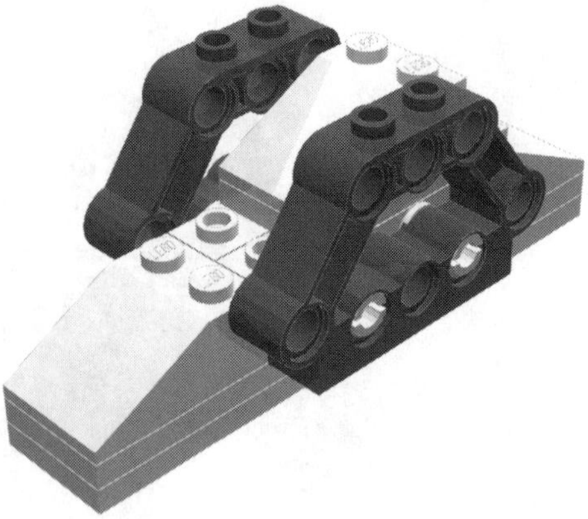

Foot Step 4

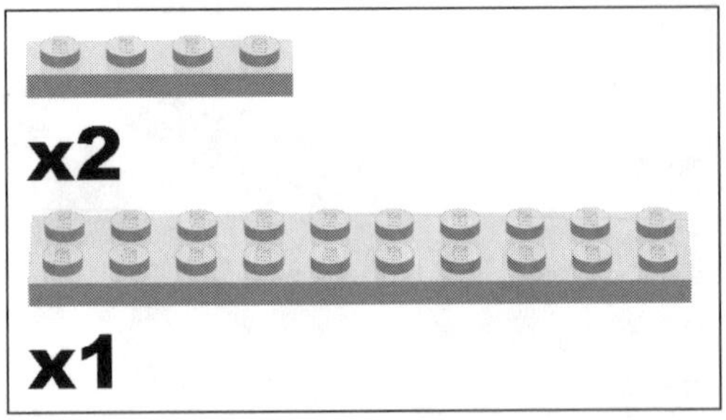

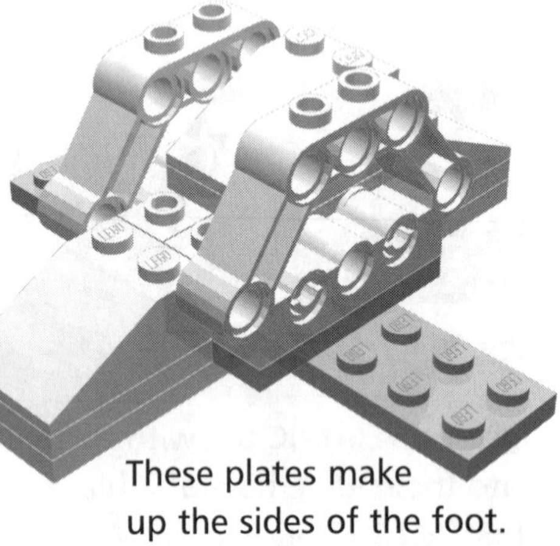

Foot Step 5

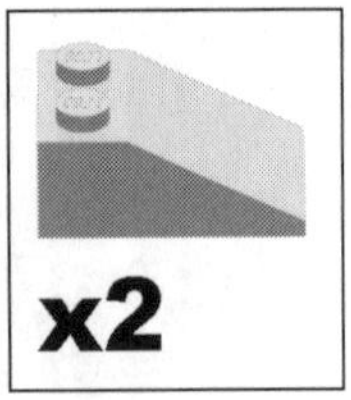

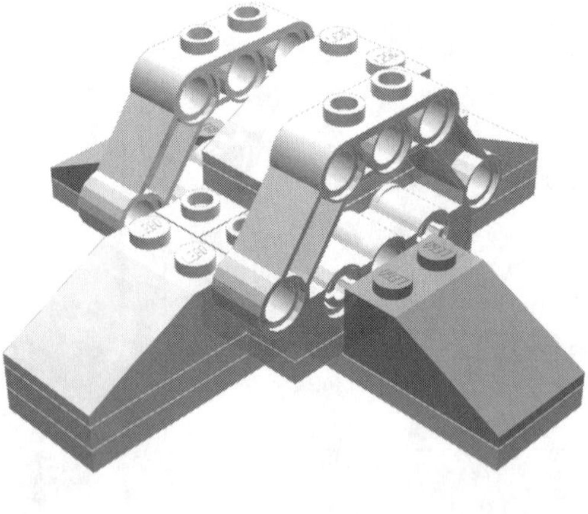

Foot Step 6

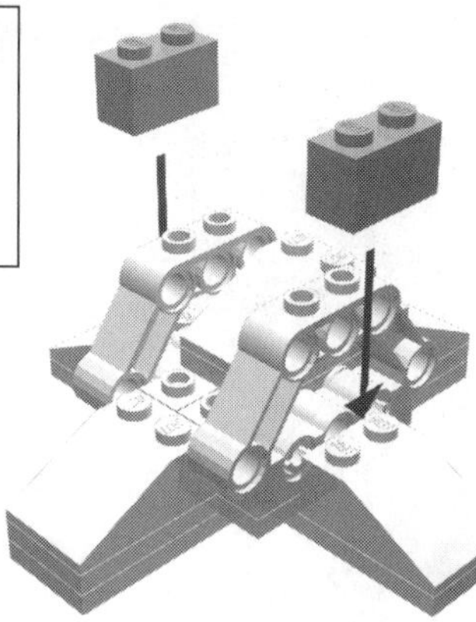

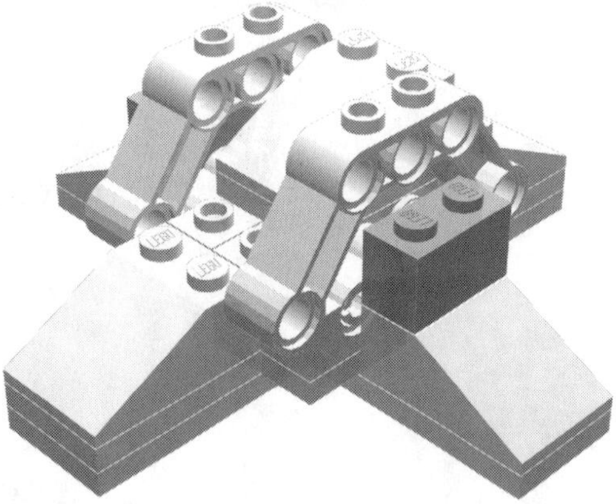

Remember that you will have to build **two** foot sub-assemblies for your AT-ST.

Creating the Instructions…

Annotating with arrows

The arrows in the smaller image for **Foot Step 6** were added by LPub. At the start of this step I used an MLCad BUFEXCHG STORE A to remember the design before the arrows and bricks were added. I used a quadrilateral and a triangle to make up each arrow. The bricks are shown out of position so you can see them before they are added to the design. I had to GHOST these parts so they didn't show up when this part was used to make the legs.

In the larger image I used a BUFEXCHG RETREIVE A to "forget" the arrows and out of place bricks, then added the bricks back in their final position. I surrounded these bricks with PLIST BEGIN IGN and PLIST END so that they didn't show up in the part list image again.

The Right Leg

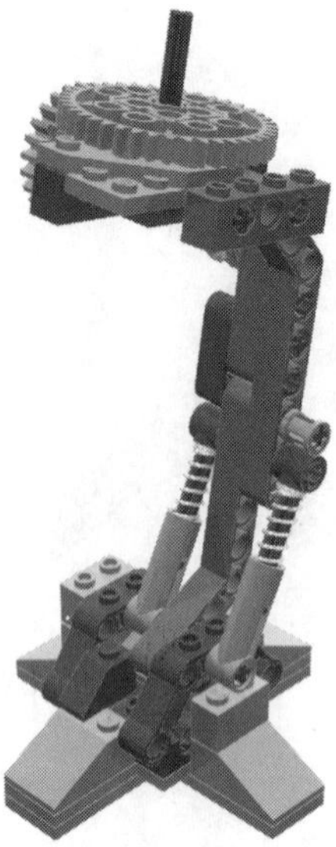

Now begin construction of the right leg sub-assembly.

Right Leg Step 0

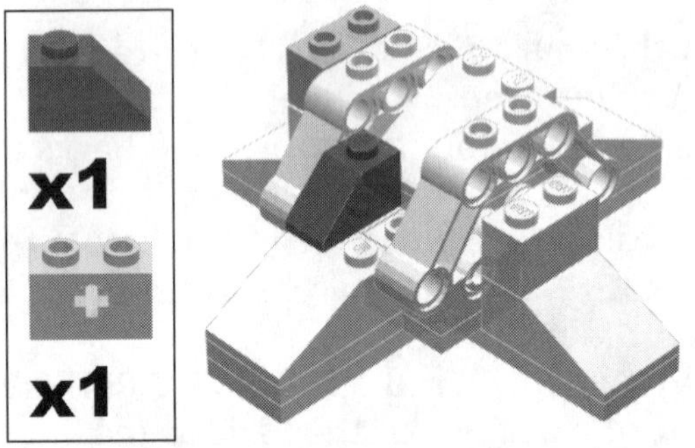

Take one of the feet sub-assemblies you just built and begin building the ankle.

Right Leg Step 1

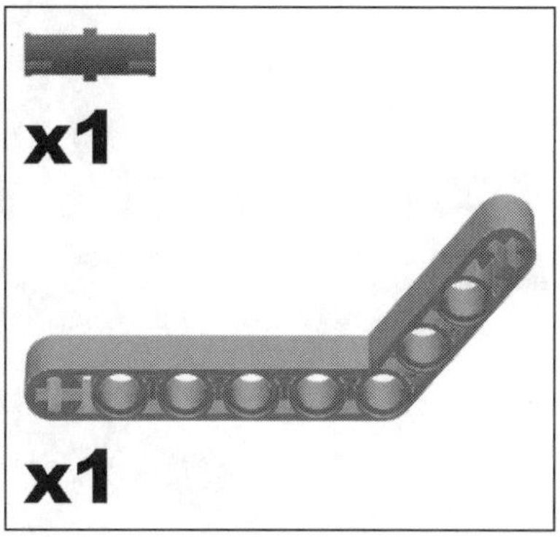

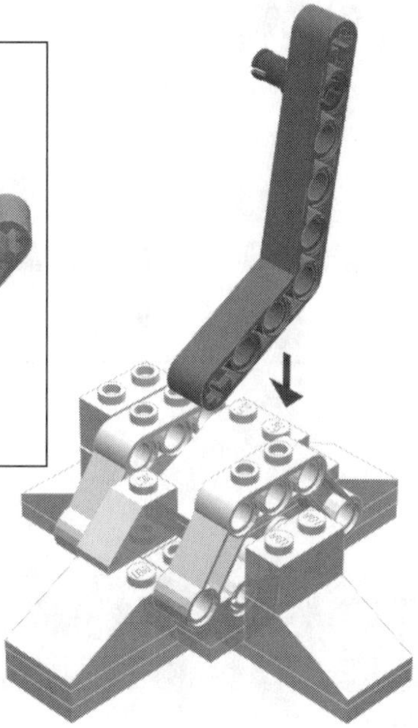

Creating the Instructions…

Hidden Pin

I had originally had steps 1 and 2 for the legs as one step, but doing so would mean adding a pin and covering both ends in the same step, making it hard to identify where the pin went. Instead, I broke this into two steps, using an arrow to show where the liftarm and pin should be placed in **Right Leg Step 1**.

Right Leg Step 2

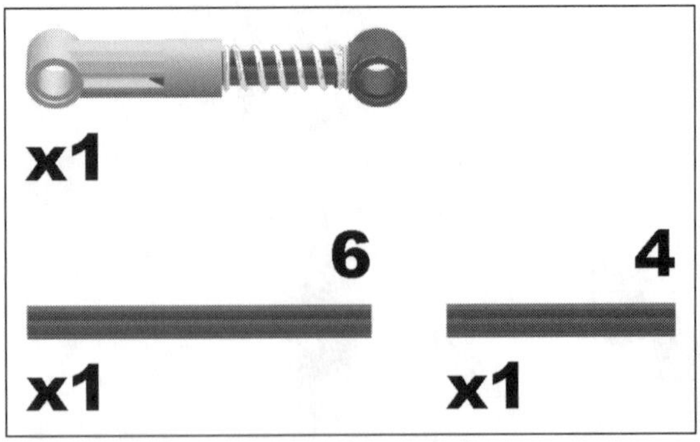

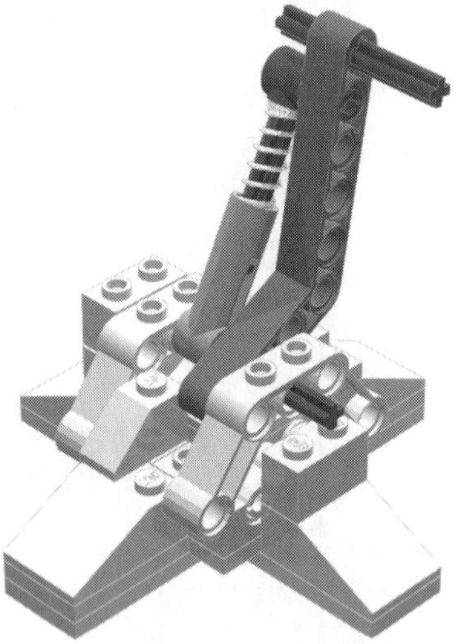

Creating the Instructions…

The Shock Absorbers

Since the shock absorbers used in AT-ST's construction are each made of two LDraw parts, I used PLIST BEGIN SUB and PLIST END to make sure the shock absorber showed up in the part list image as one part.

Right Leg Step 3

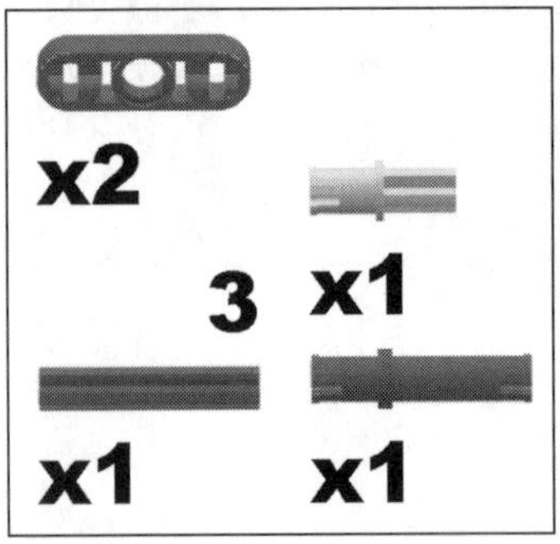

Build the connection between the lower and upper leg.

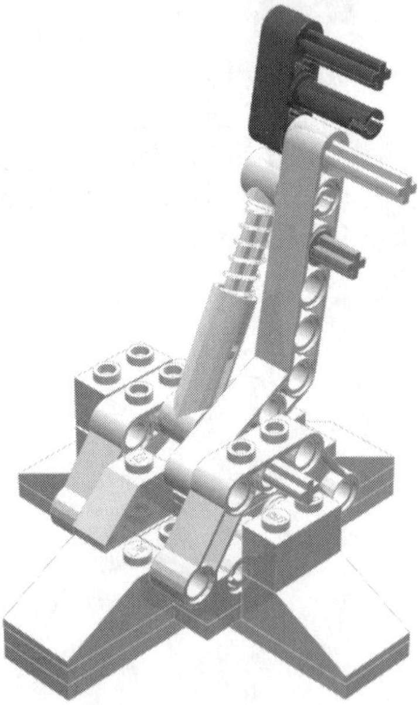

Right Leg Step 4

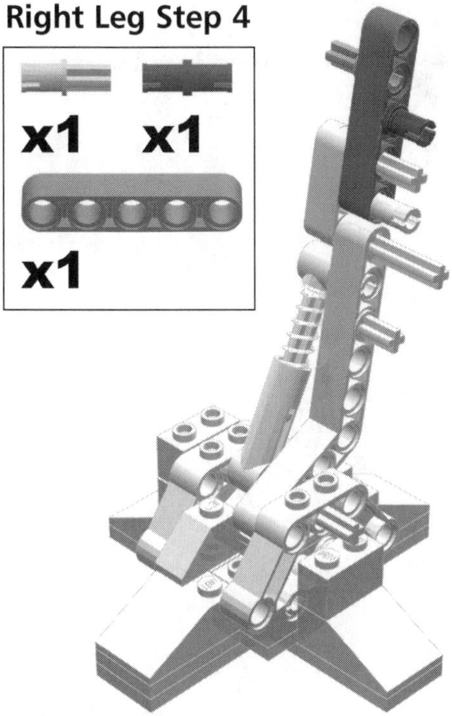

Right Leg Step 5

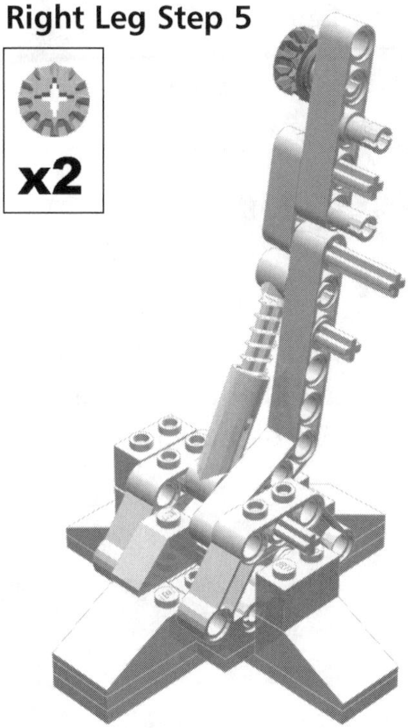

Right Leg Step 6

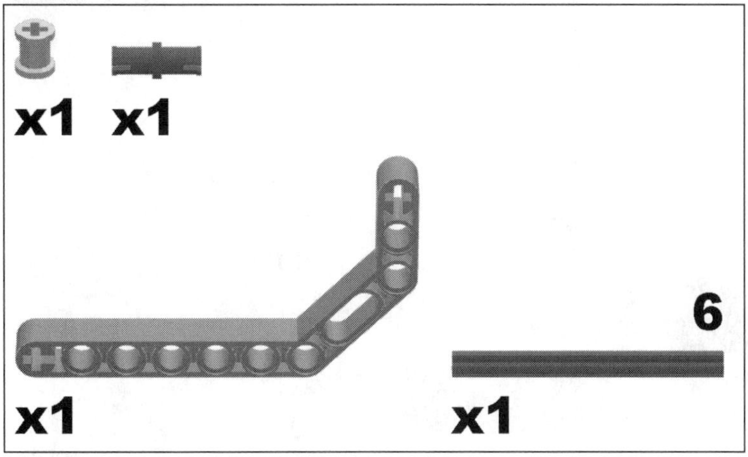

Attach the upper leg.

Right Leg Step 7

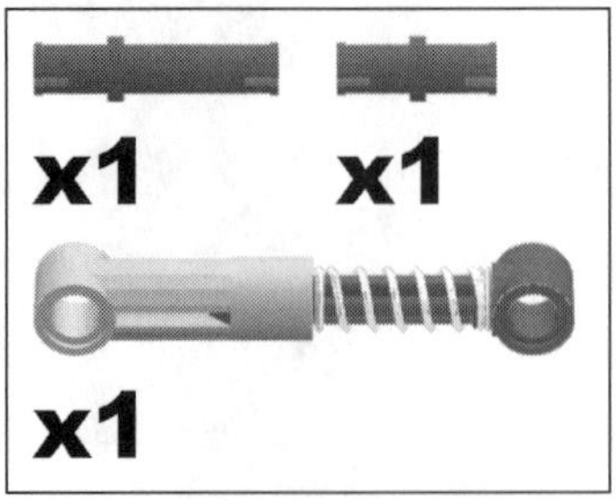

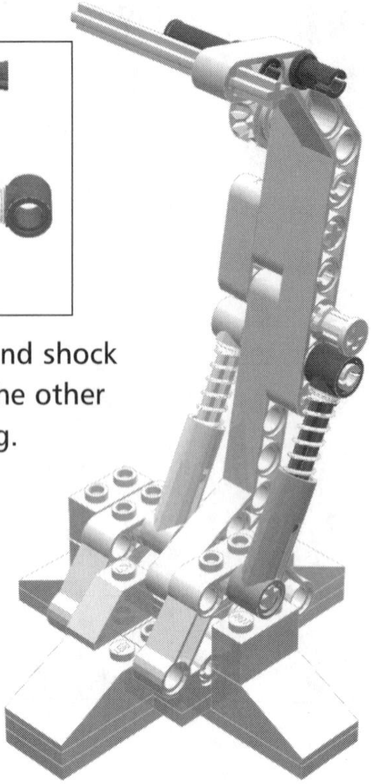

Attach a second shock absorber to the other side of the leg.

Right Leg Step 8

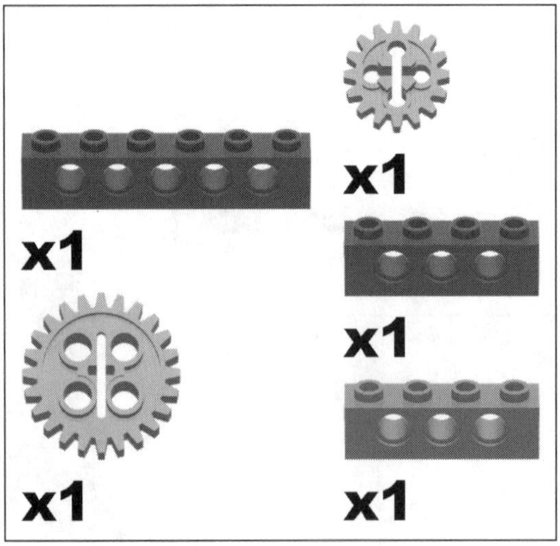

Start building the hip at the top of the leg.

Right Leg Step 9

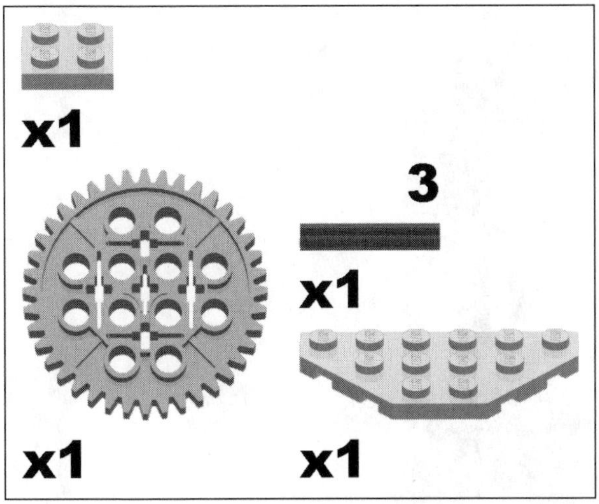

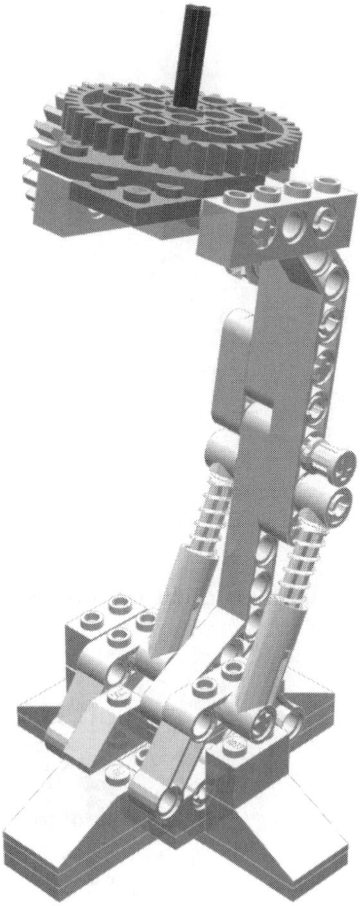

The 40t gear is what makes the leg stride forward and backwards. The #3 axle is used to attach the leg to the chassis.

The Left Leg

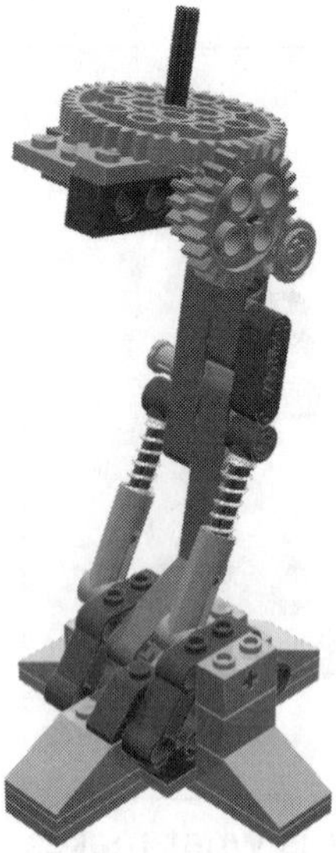

Start building the left leg sub-assembly.

Left Leg Step 0

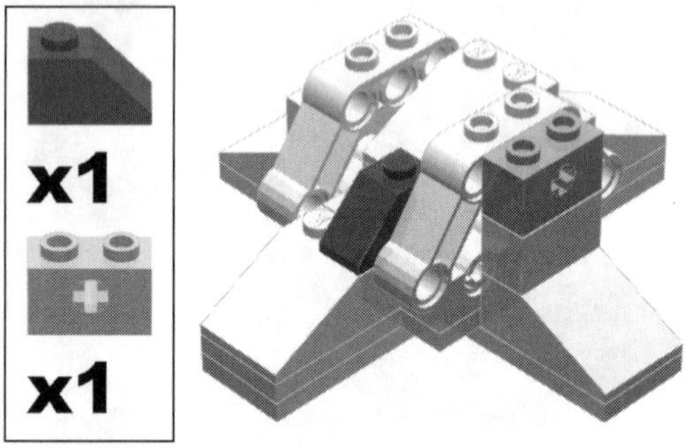

Take the second foot sub-assembly you just built and begin building the ankle.

Left Leg Step 1

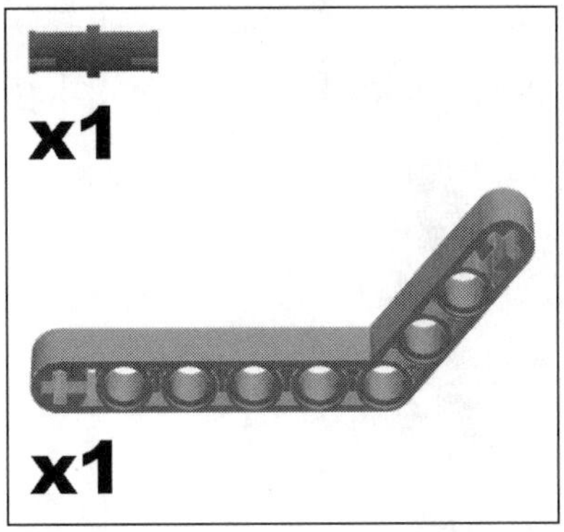

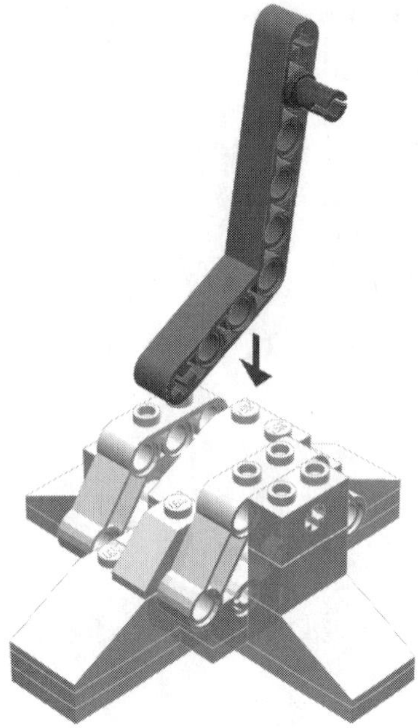

Left Leg Step 2

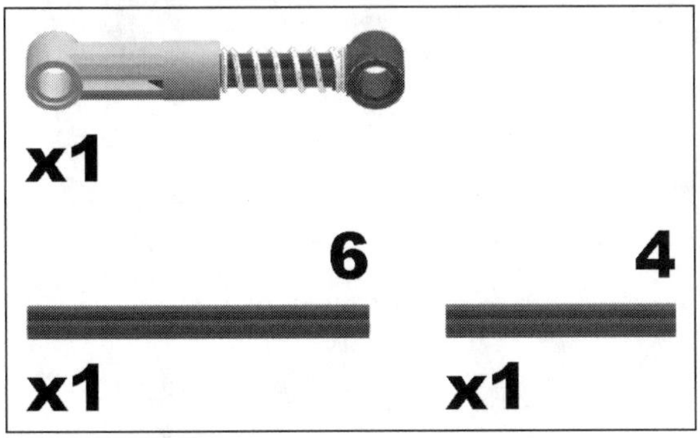

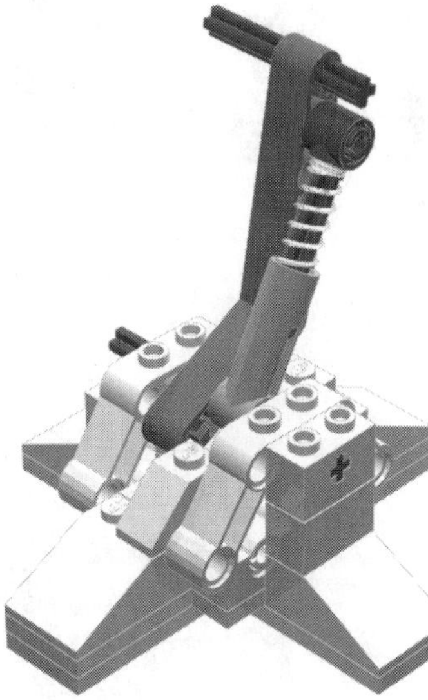

Left Leg Step 3

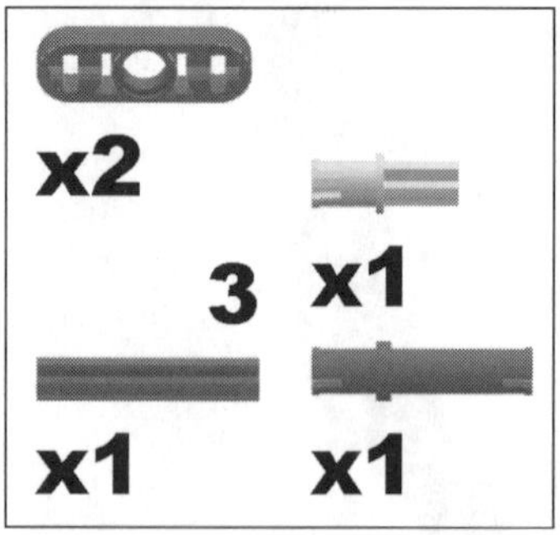

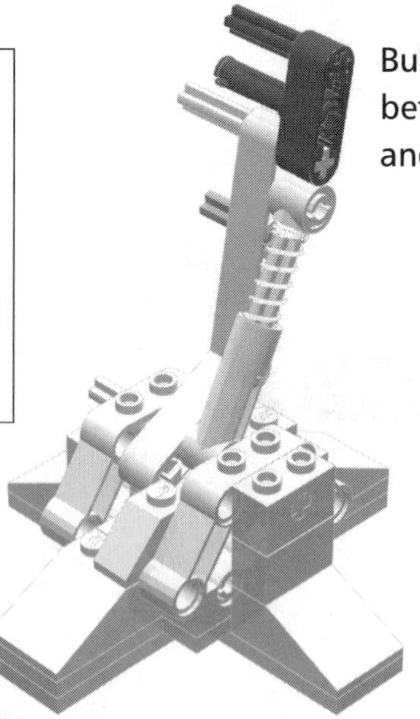

Build the connection between the lower and upper leg.

Left Leg Step 4

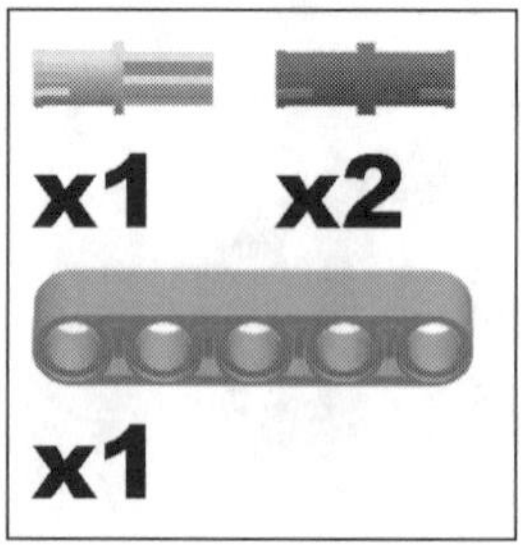

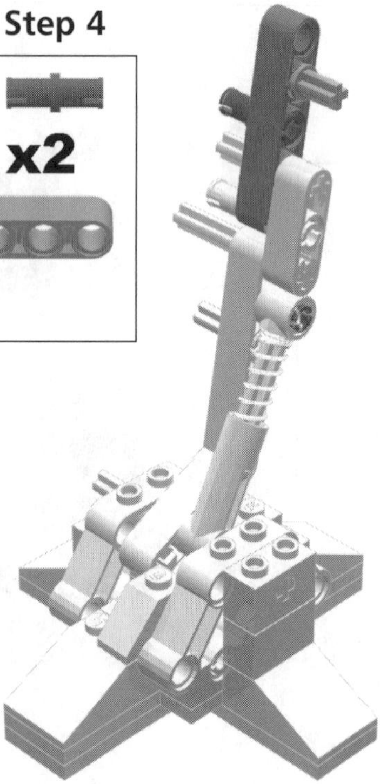

Left Leg Step 5

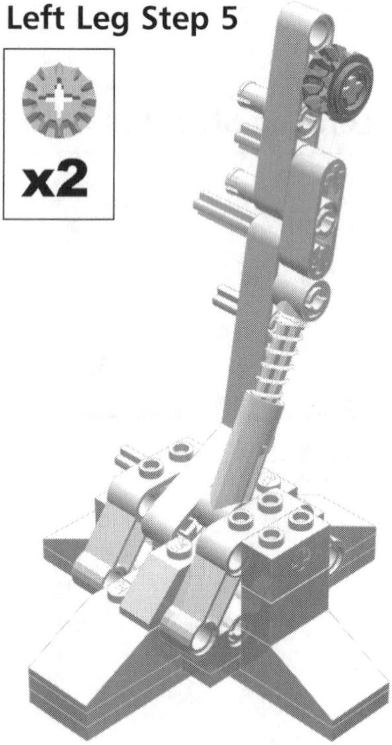

Left Leg Step 6

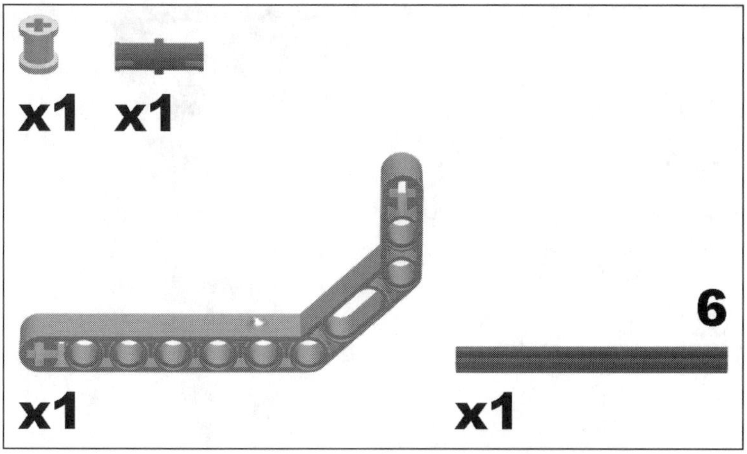

Attach the upper leg.

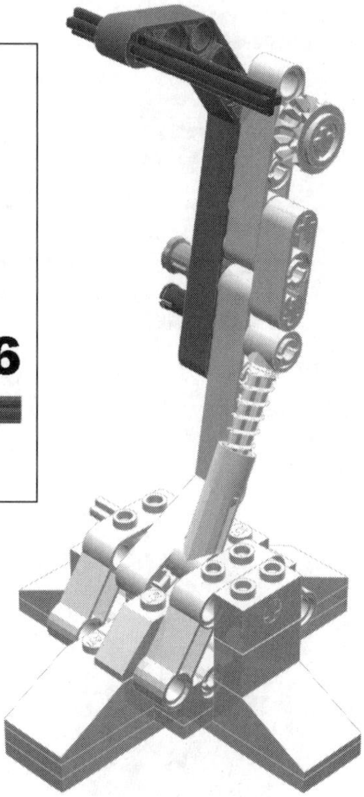

Left Leg Step 7

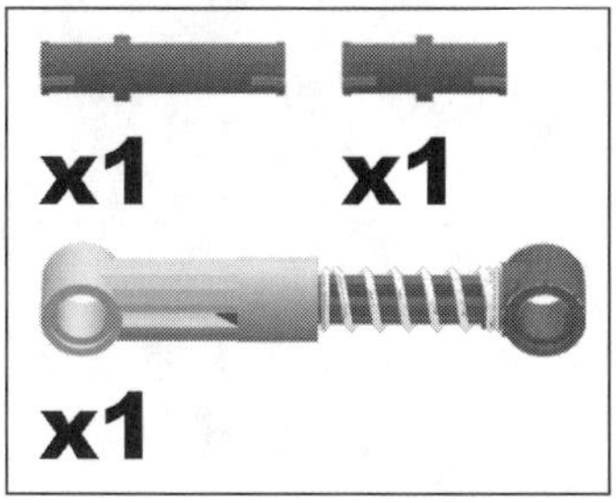

Attach a second shock absorber to the other side of the leg.

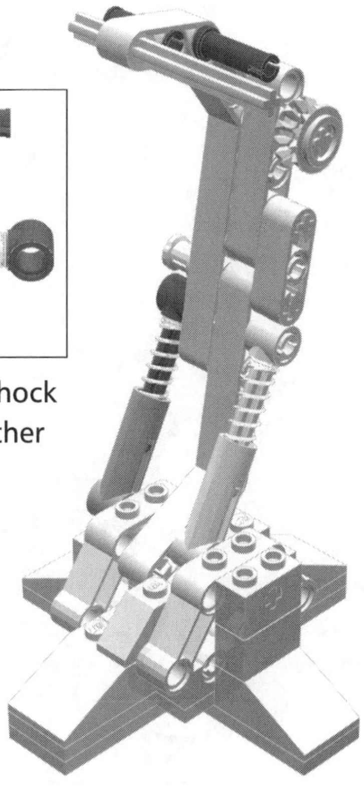

Left Leg Step 8

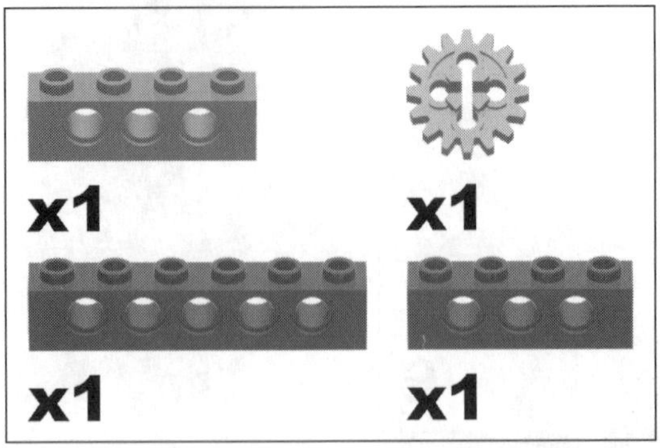

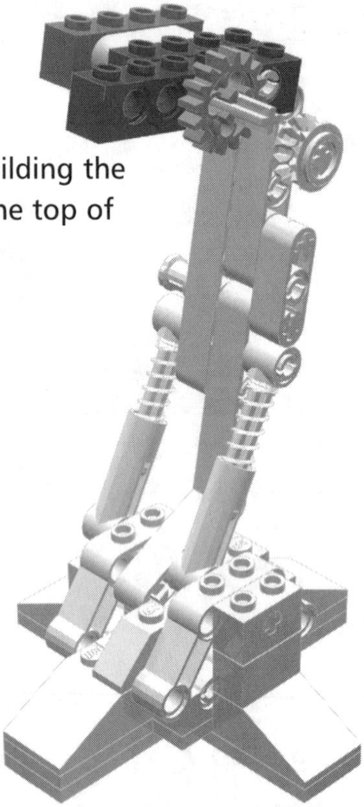

Start building the hip at the top of the leg.

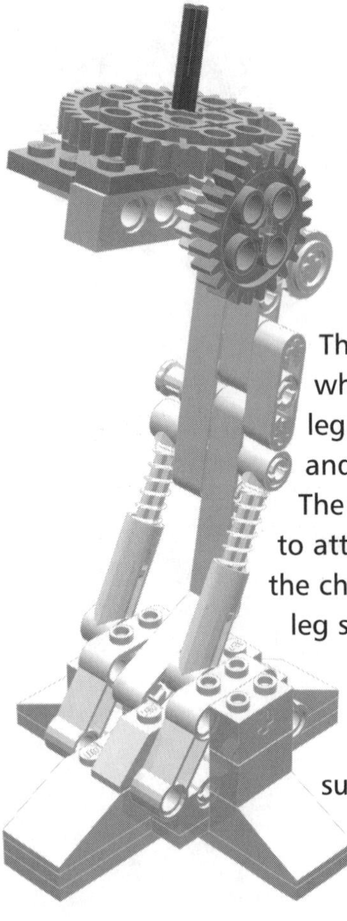

Left Leg Step 9

The 40t gear is what makes the leg stride forward and backwards. The #3 axle is used to attach the leg to the chassis. This left leg sub-assembly should be a mirror image of the right leg sub-assembly.

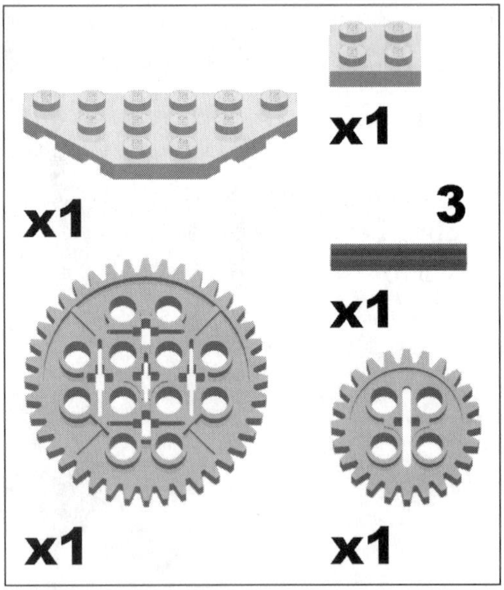

The Chassis

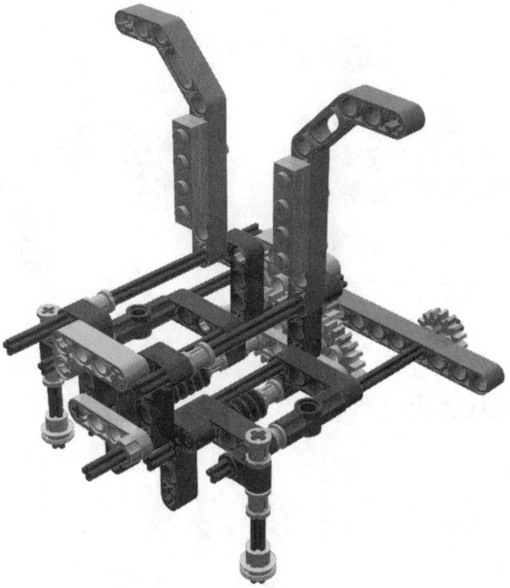

The chassis sub-assembly attaches to the legs and holds the RCX brick.

Chassis Step 0

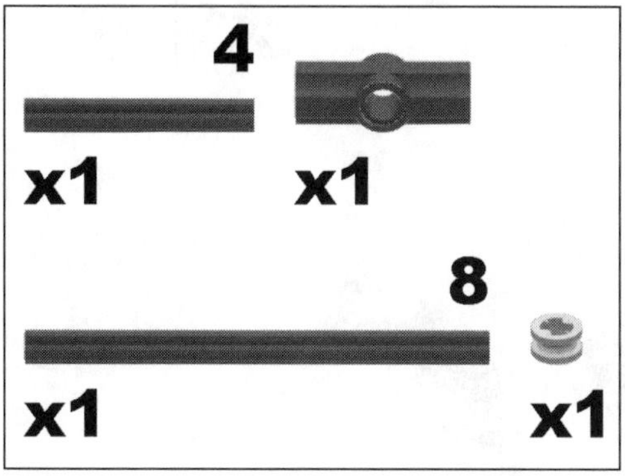

Start by constructing the left hip of the chassis using the axles and connectors as shown.

Chassis Step 1

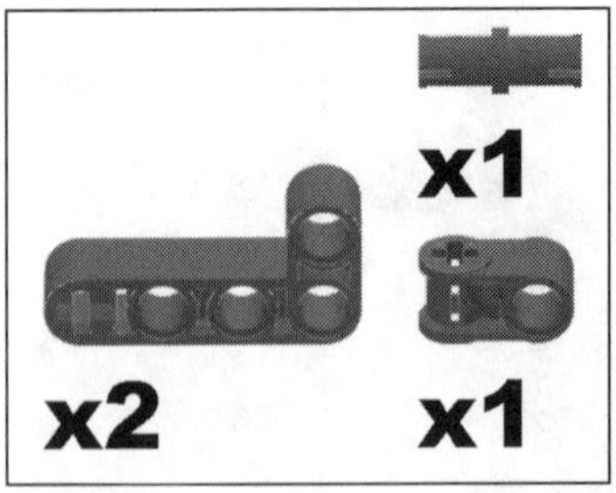

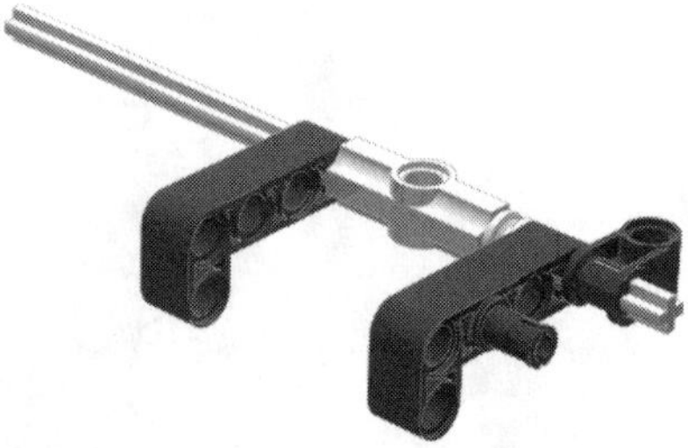

Chassis Step 2

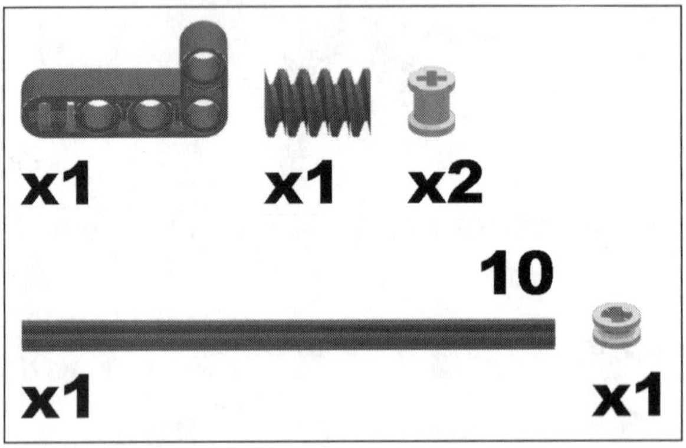

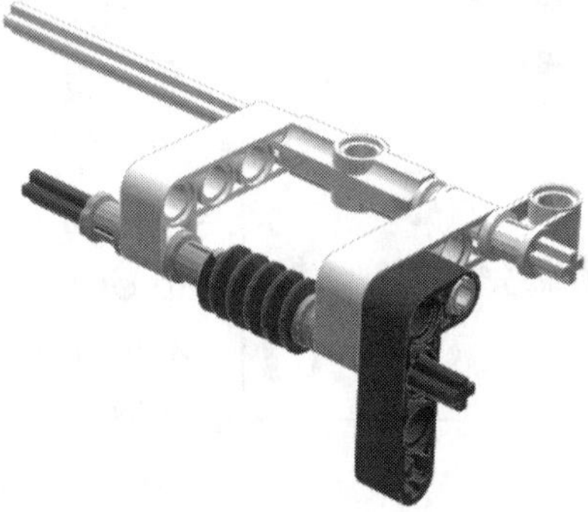

The worm gear you add in this step turns the left leg once it is attached.

Chassis Step 3

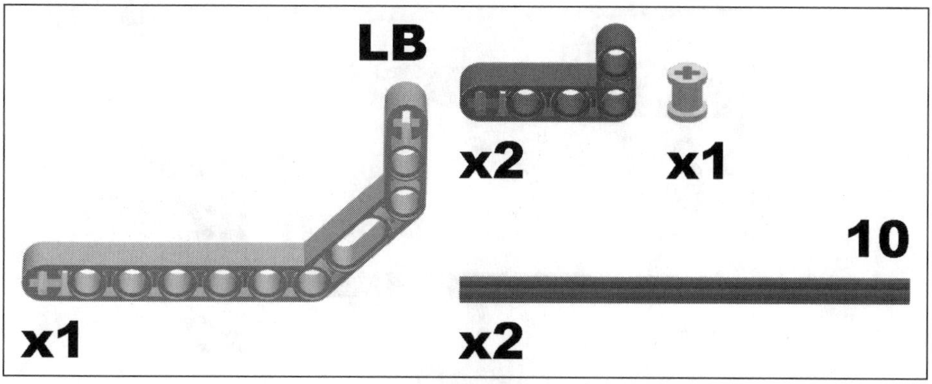

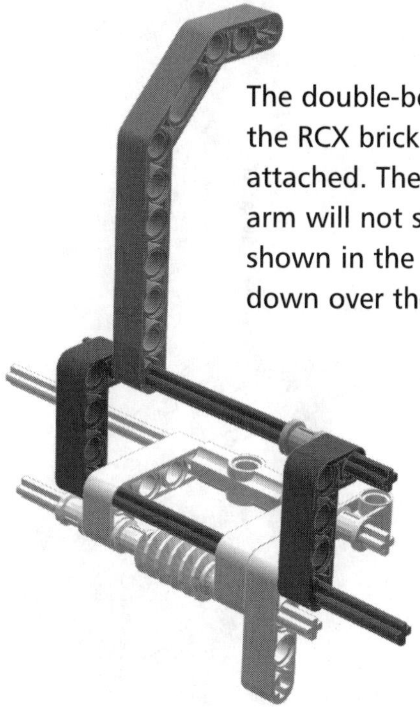

The double-bent lift arm holds the RCX brick once it is attached. The double bent lift arm will not stay standing up as shown in the picture; let if flop down over the hip for now.

Chassis Step 4

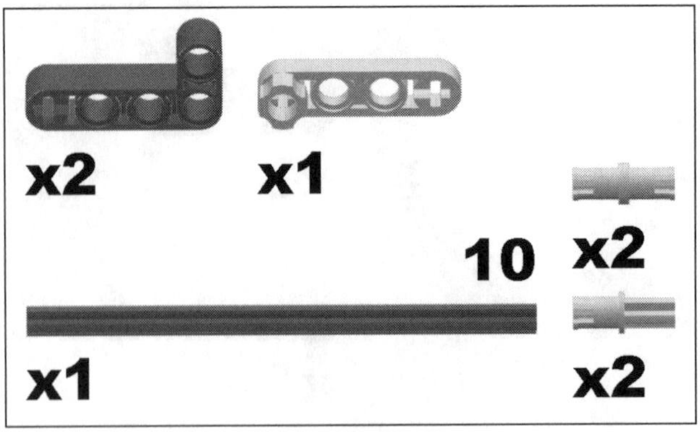

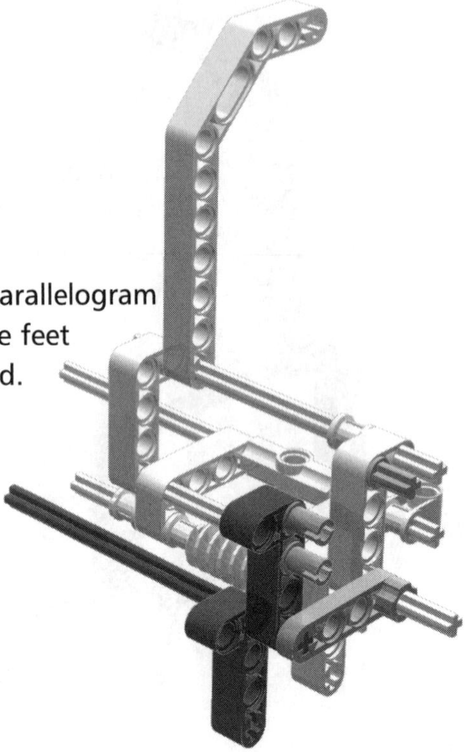

These parts form a parallelogram that always keeps the feet parallel to the ground.

Chassis Step 5

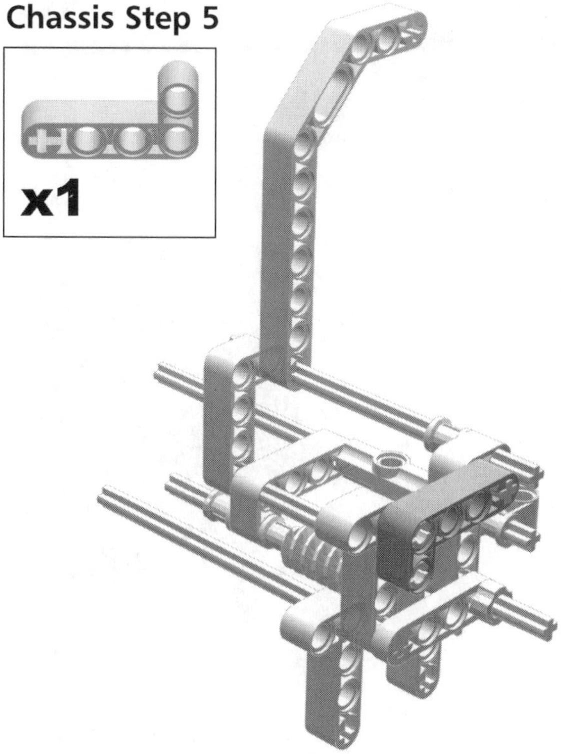

Chassis Step 6

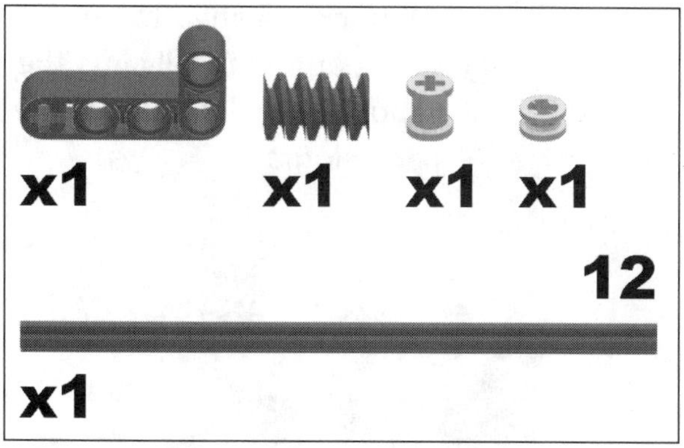

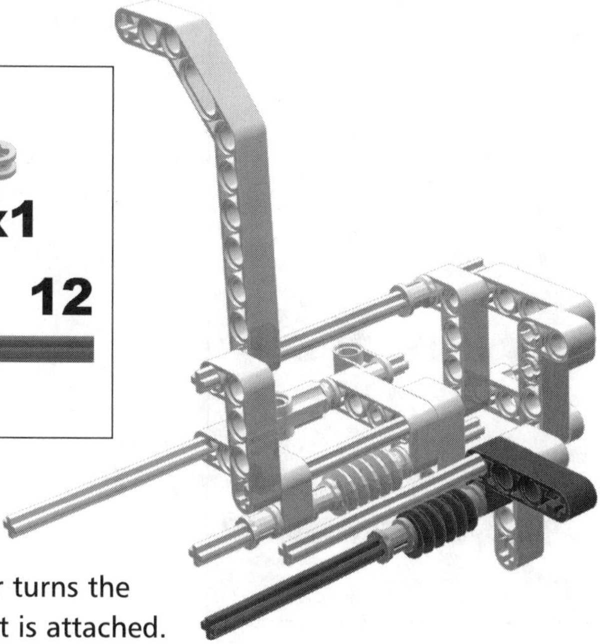

This worm gear turns the right leg once it is attached.

Chassis Step 7

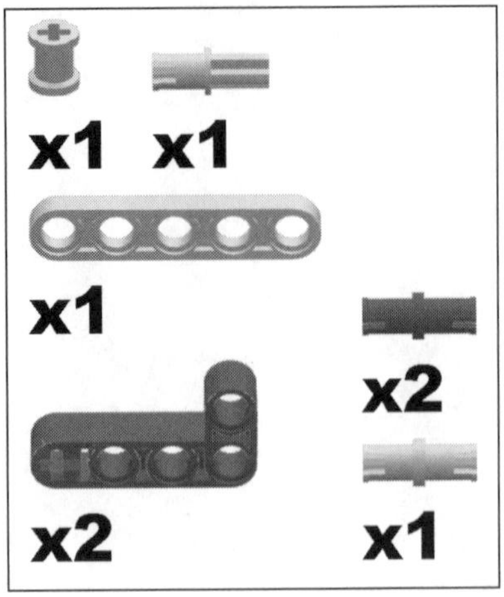

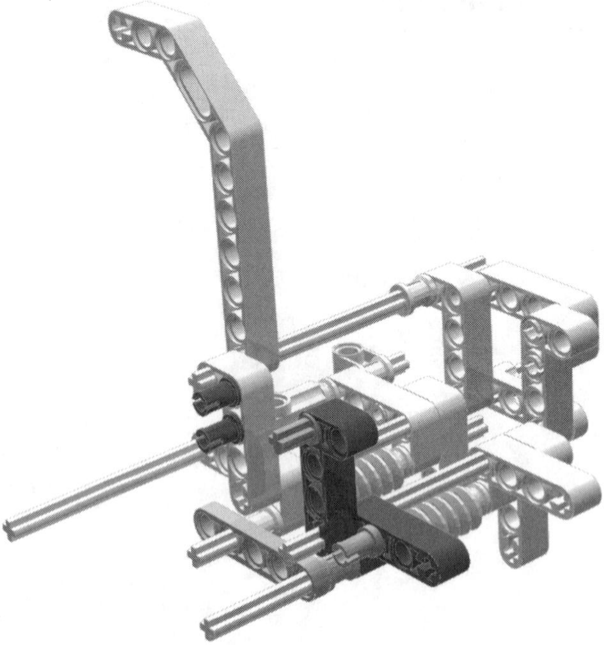

Chassis Step 8

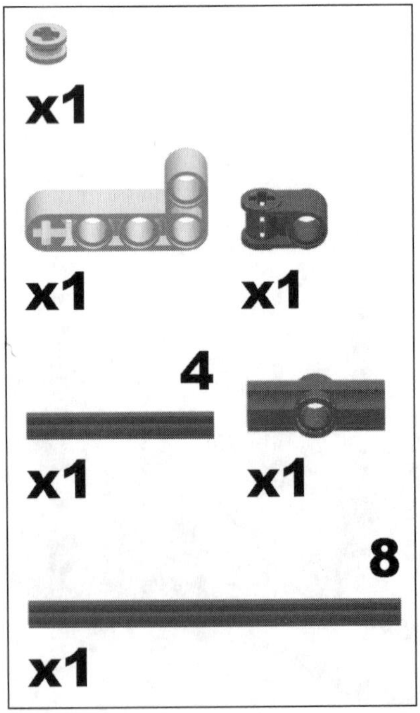

The parallelogram is complete. Notice how you can raise one hip with respect to the other. In the complete assembly, when the RCX brick is leaned to one side, it automatically lifts the opposite side leg due to this parallelogram.

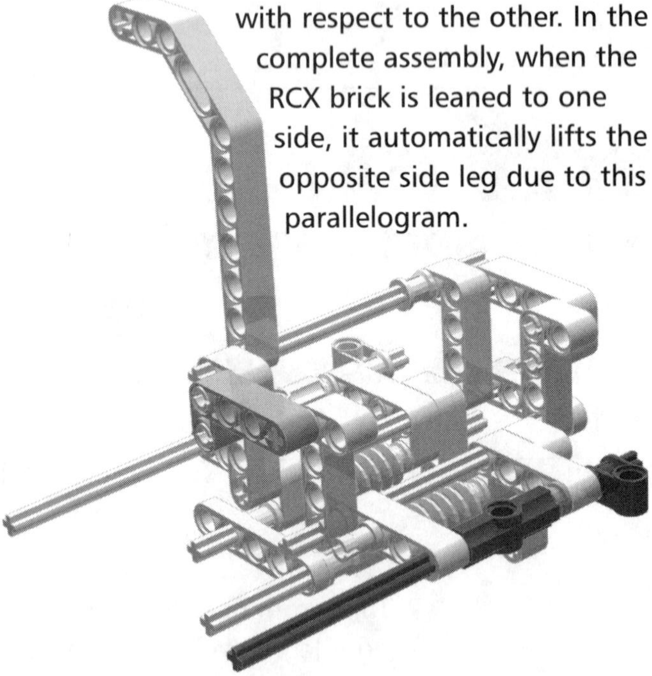

Chassis Step 9

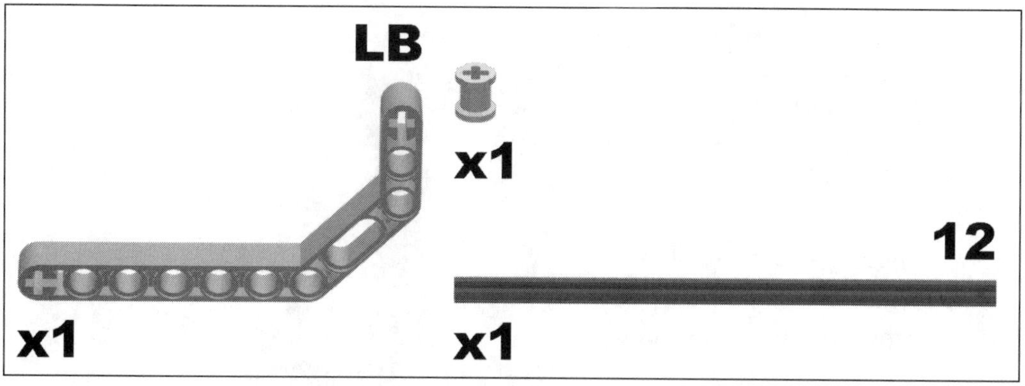

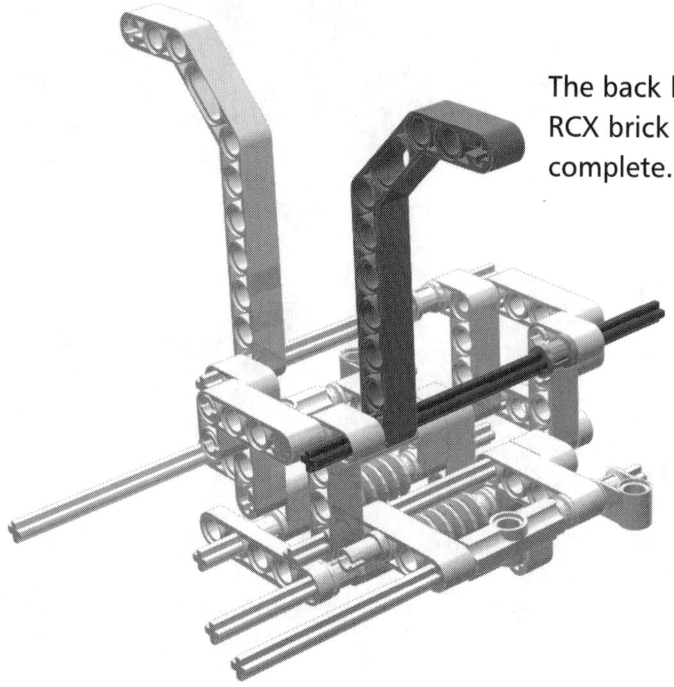

The back half of the RCX brick support is complete.

Chassis Step 10

These gears make sure that when one leg strides, the other leg does the same.

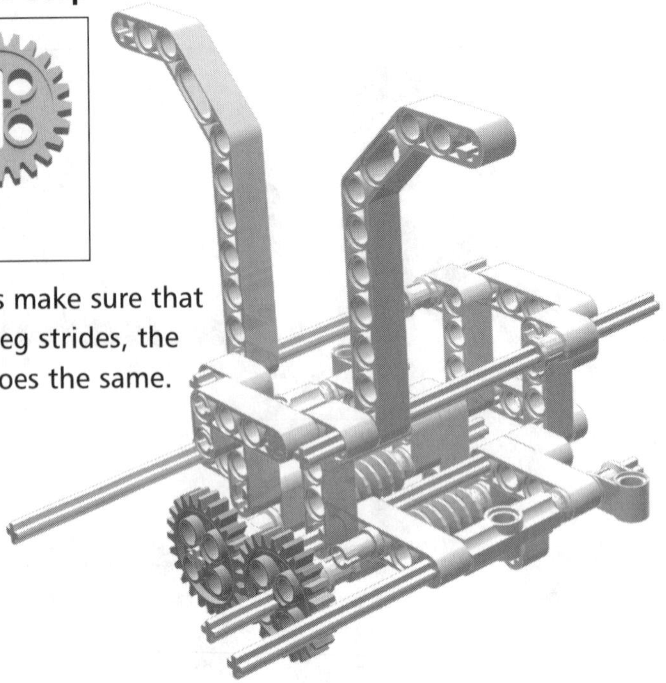

Chassis Step 11

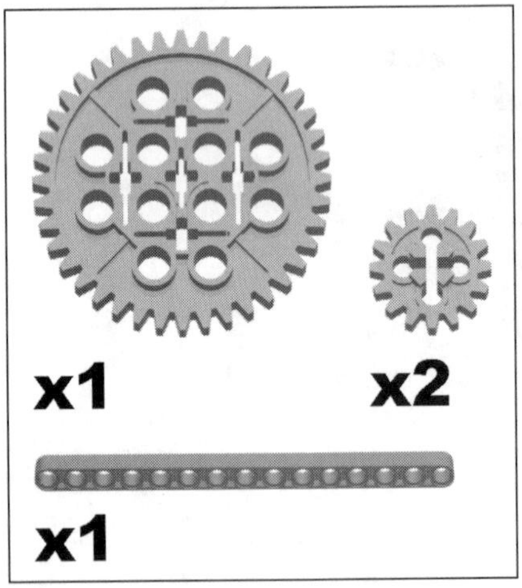

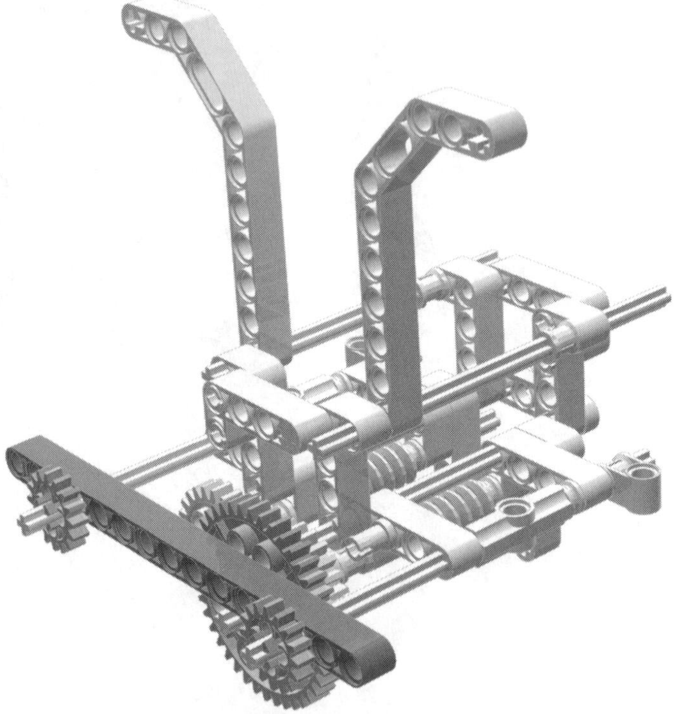

Chassis Step 12

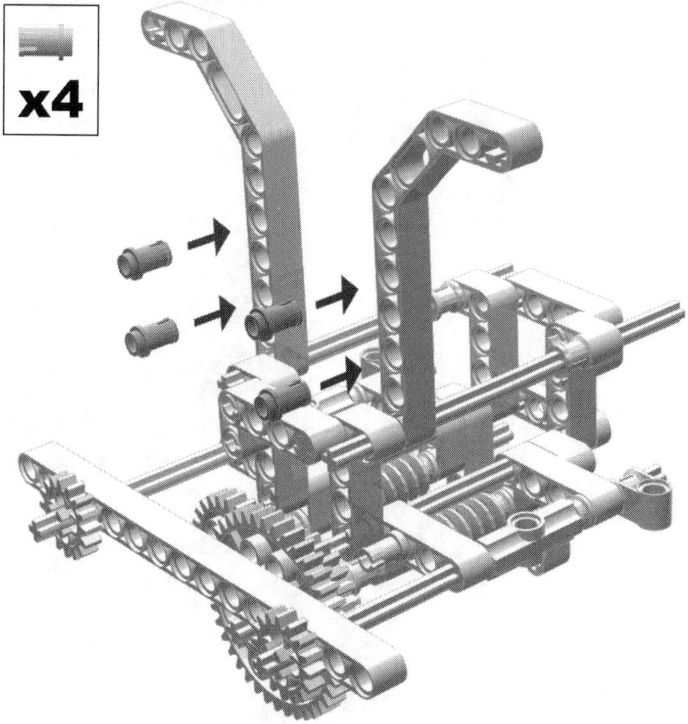

Chassis Step 13

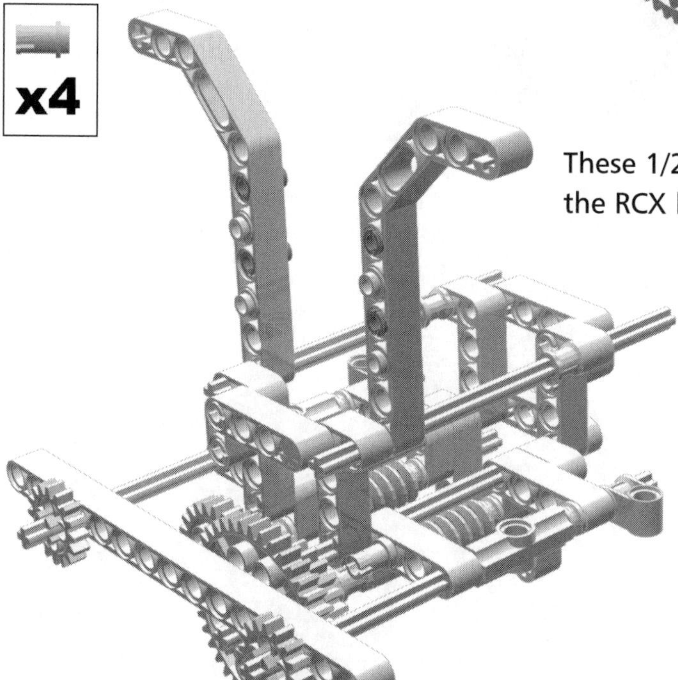

These 1/2 pins will help hold the RCX brick in place.

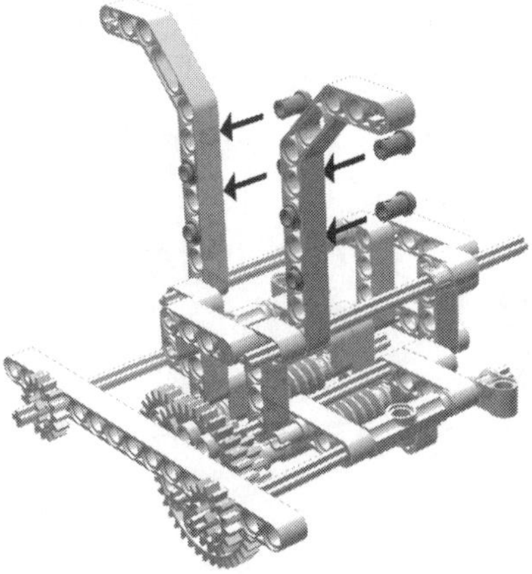

Chassis Step 14

Turn the chassis around and add these plates.

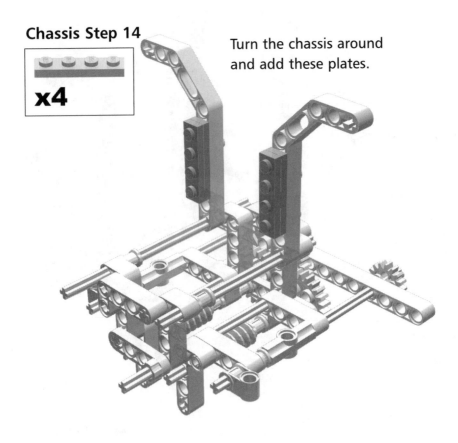

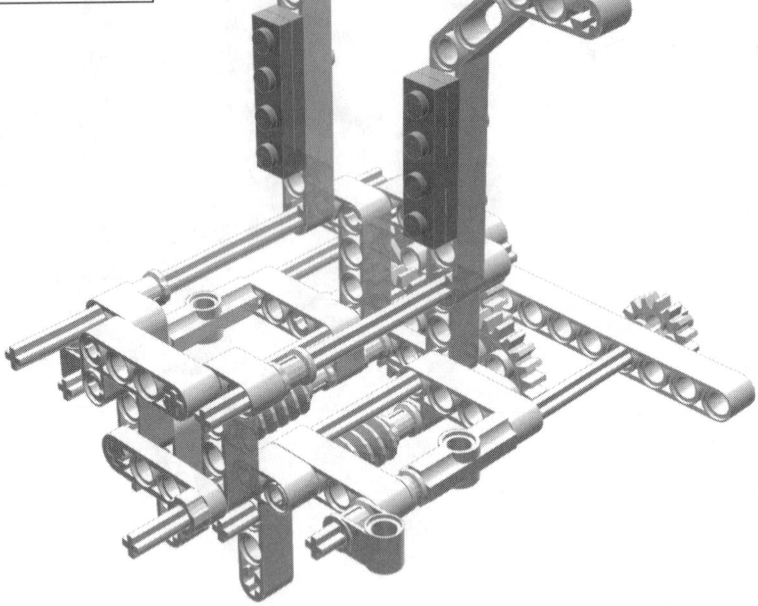

Chassis Step 15

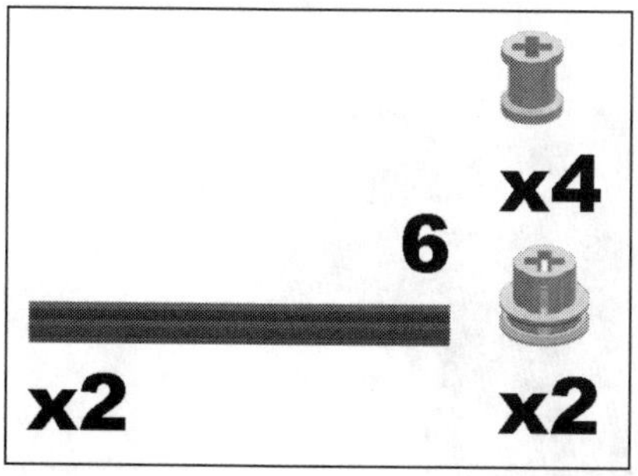

The parts you add in this step will bump into touch sensors on the legs, forming a limit switch so that the RCX brick can tell when it is done striding.

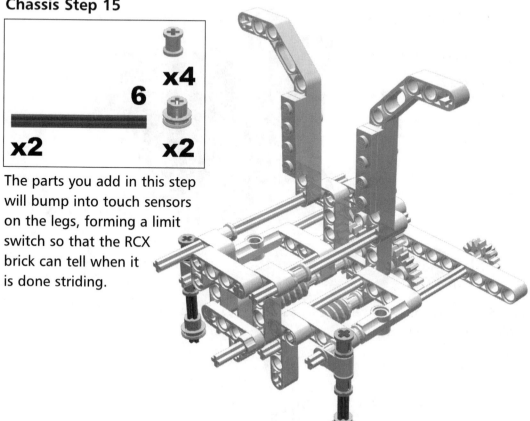

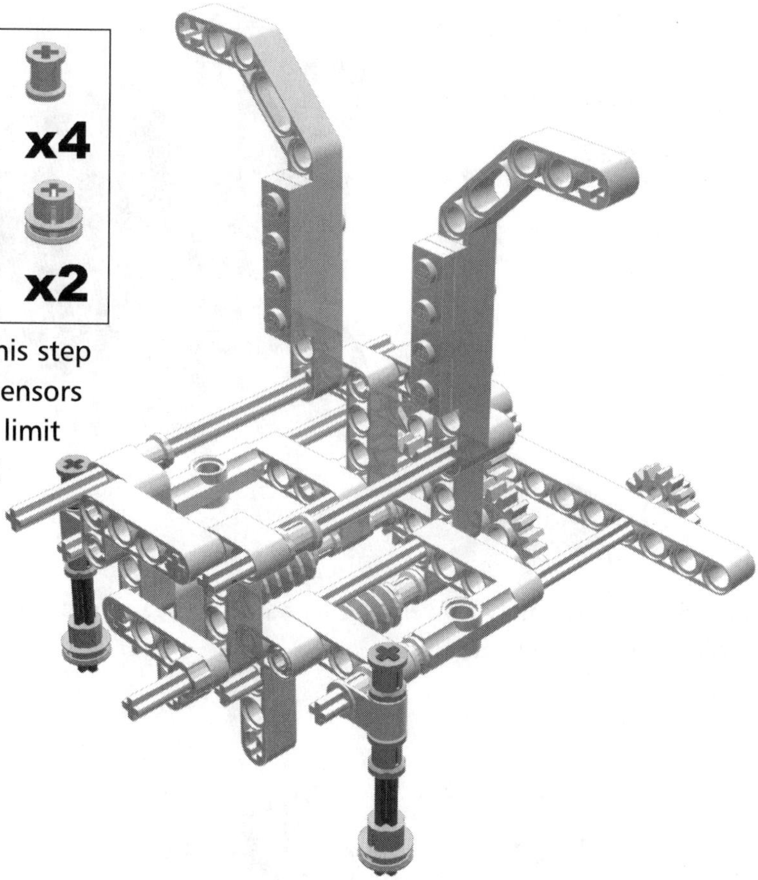

The Drive

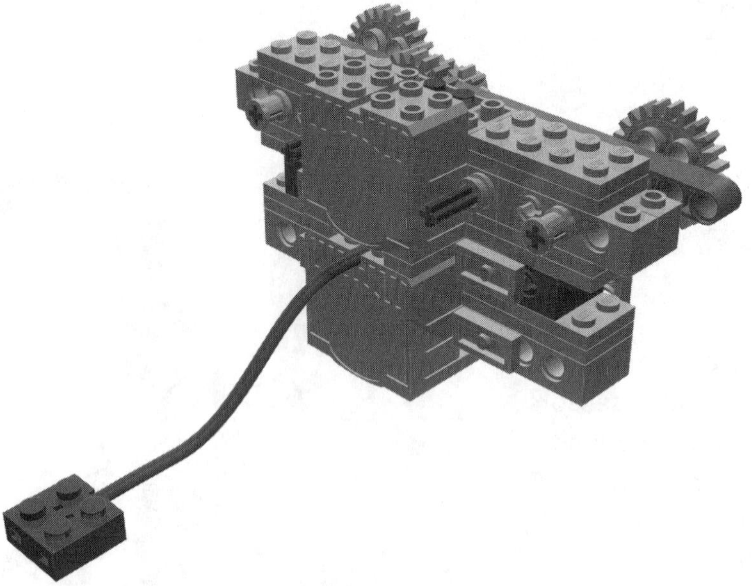

The drive sub-assembly powers the legs and the leaning mechanism that shifts the AT-ST's weight from side to side.

Drive Step 0

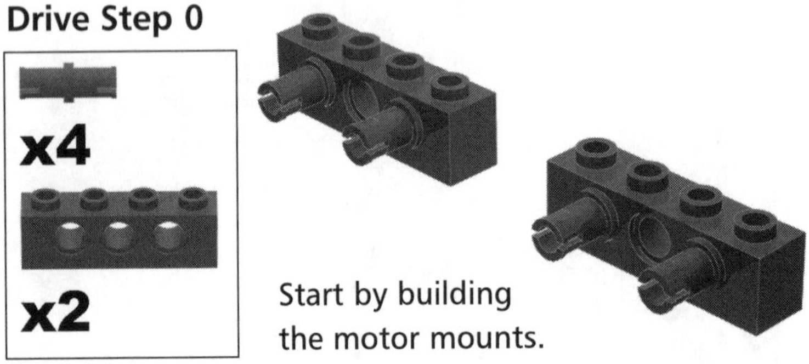

Start by building the motor mounts.

Drive Step 1

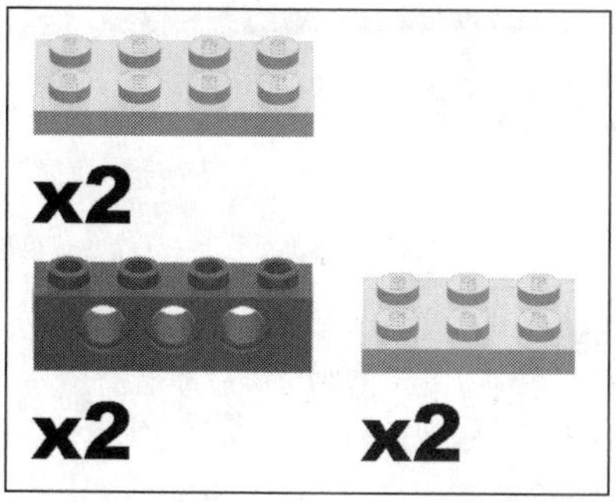

Drive Step 2

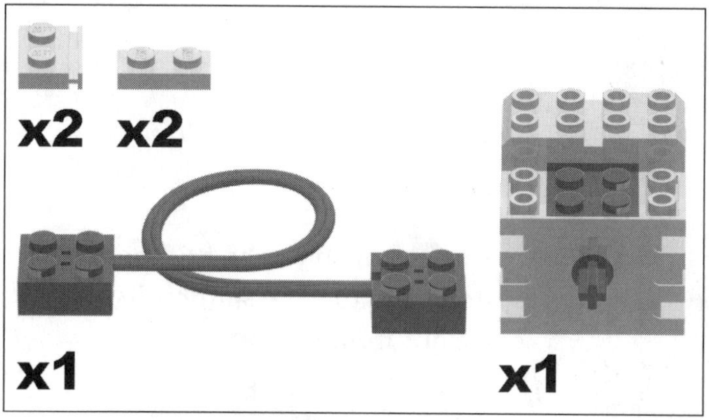

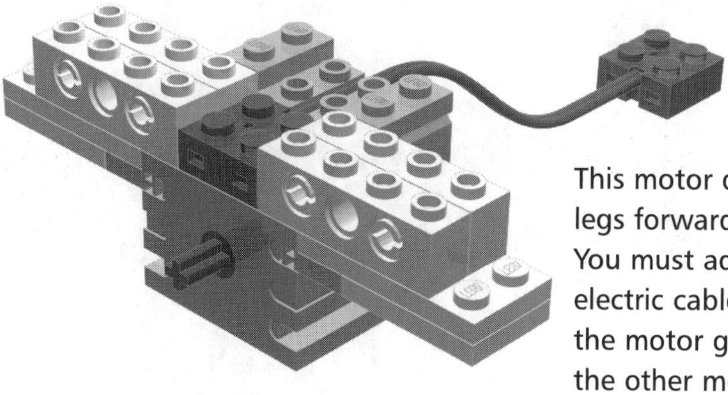

This motor drives the legs forward and back. You must add the electric cable now before the motor gets buried by the other motor.

Creating the Instructions…

Ghosting the Cable

I added the cable in **Drive Step 02** and ghosted it using MLCad. The position of the cable in the construction image for this step is not the final position it will be in when it is hooked to the RCX brick. Ghosting the cable makes it disappear when the drive sub-assembly is added to the complete assembly later in the instructions. I created the electrical cables using LSynth.

Drive Step 3

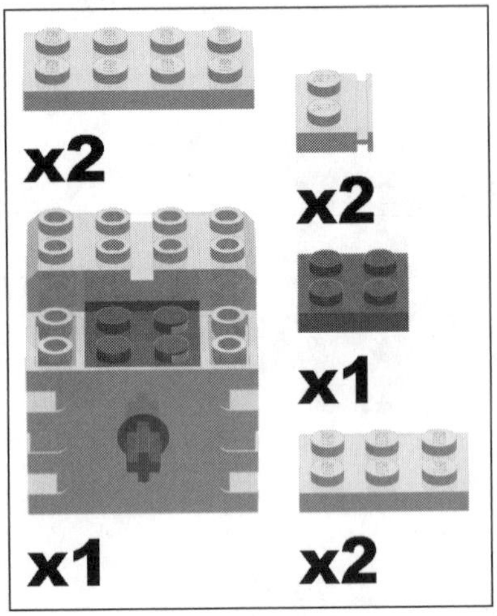

This motor shifts the RCX brick from side to side, shifting its weight from foot to foot.

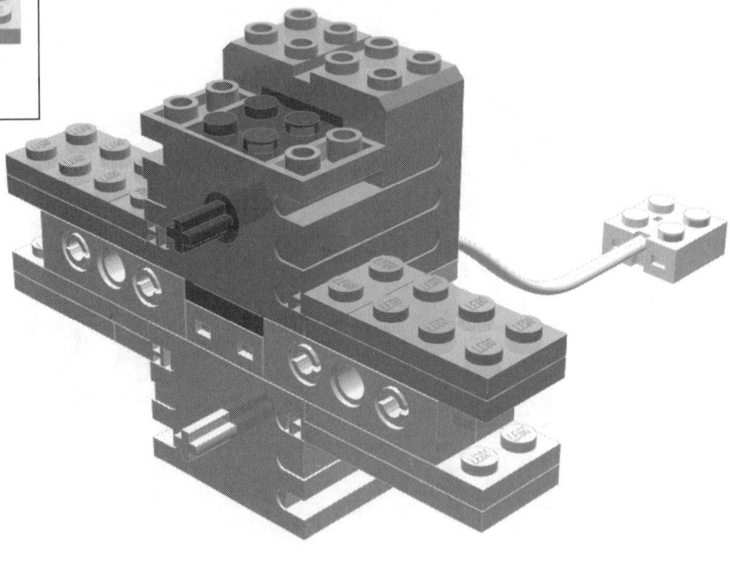

Drive Step 4

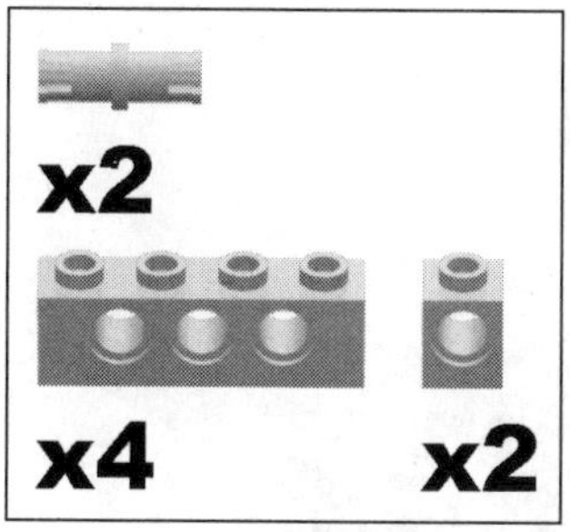

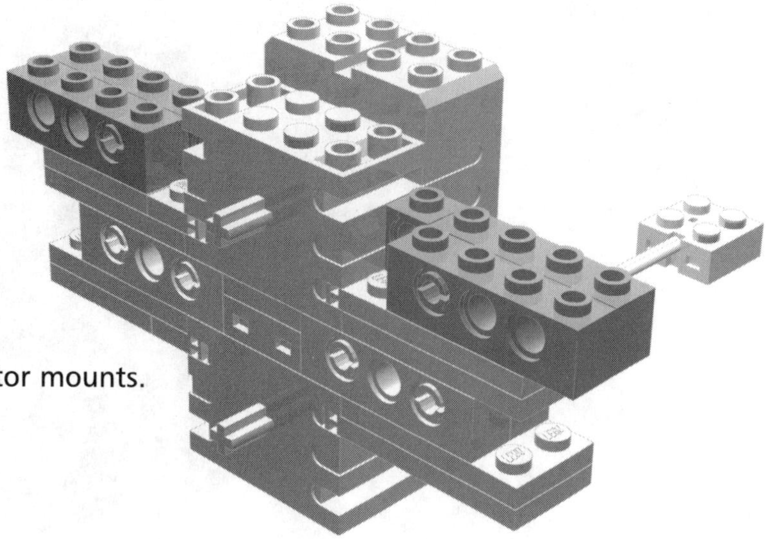

Finish the motor mounts.

Drive Step 5

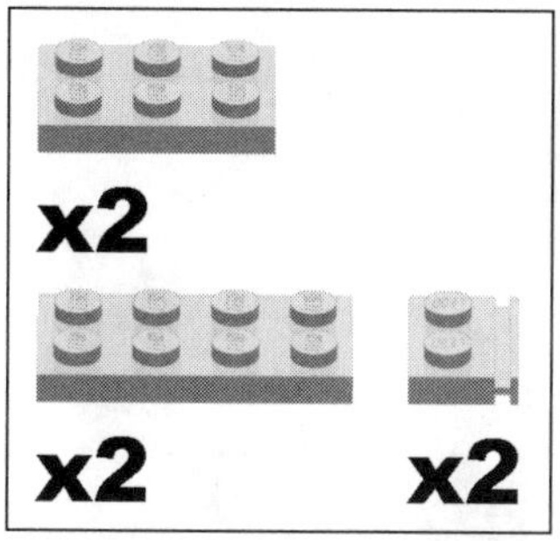

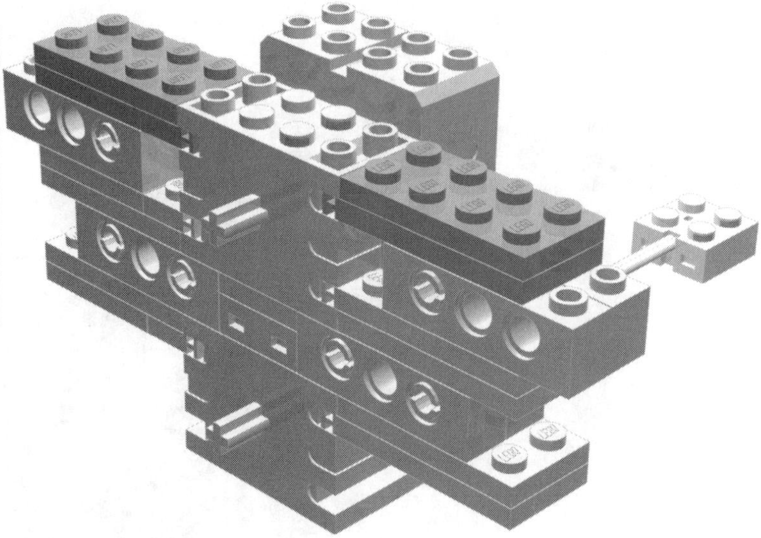

Drive Step 6

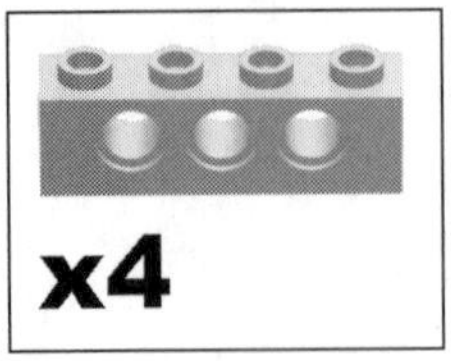

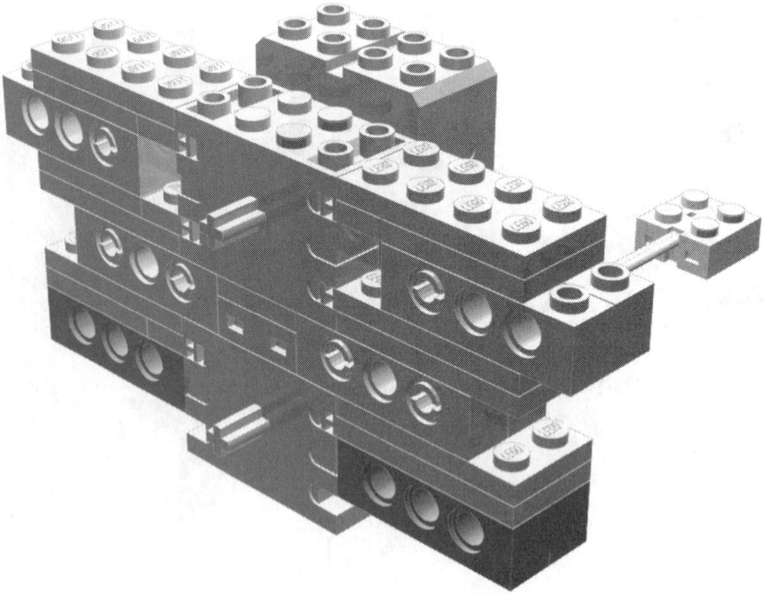

Drive Step 7

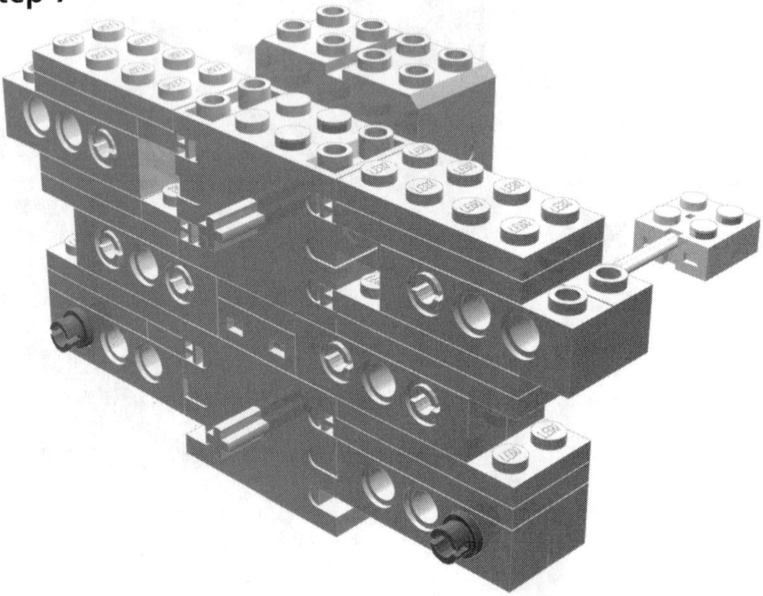

Drive Step 8

x1

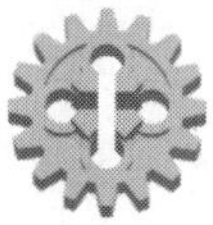

x1

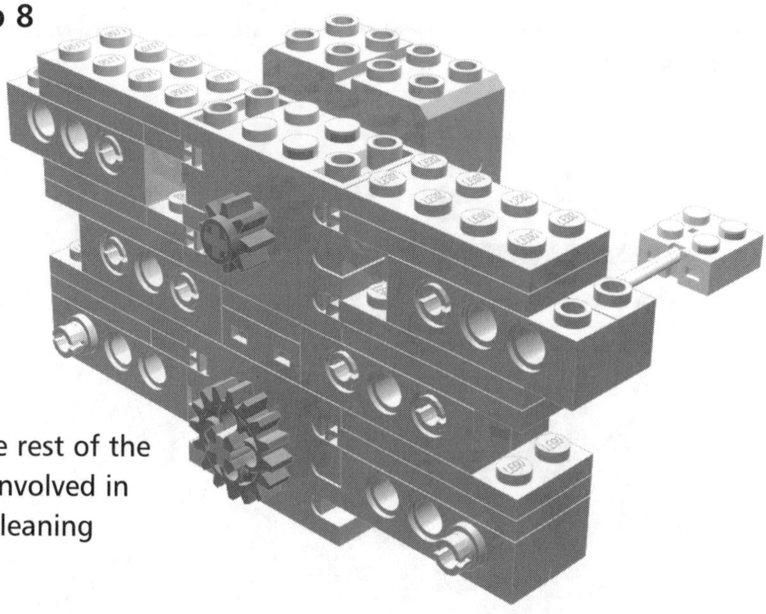

Build up the rest of the drive train involved in the AT-ST's leaning motion.

Drive Step 9

x2

4

x2

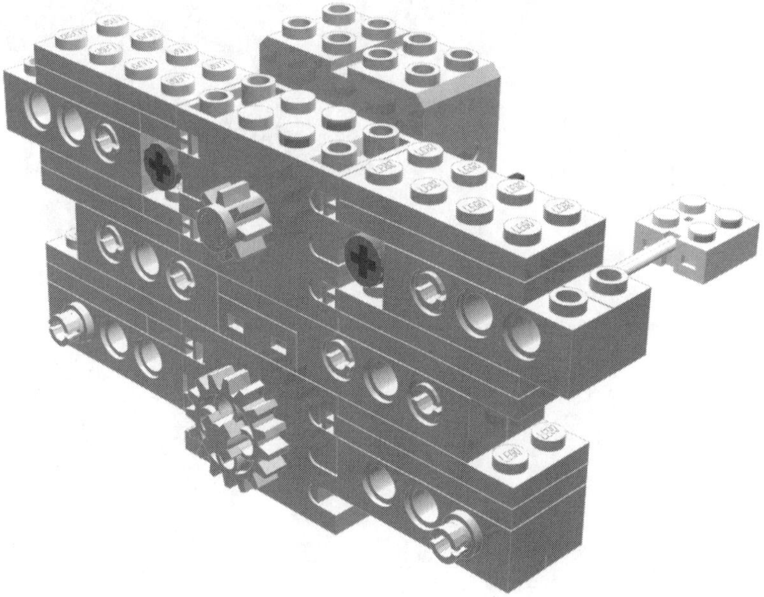

Drive Step 10

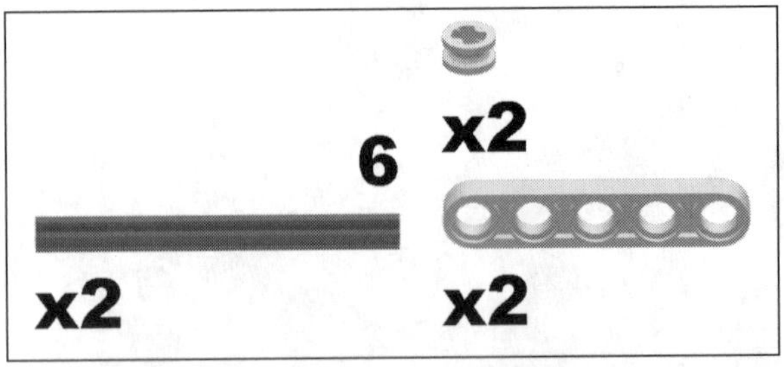

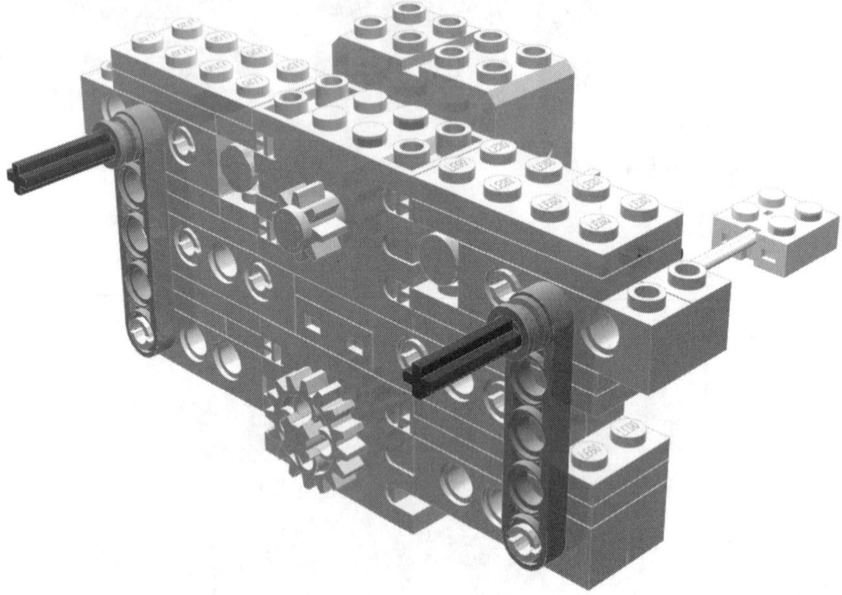

Drive Step 11

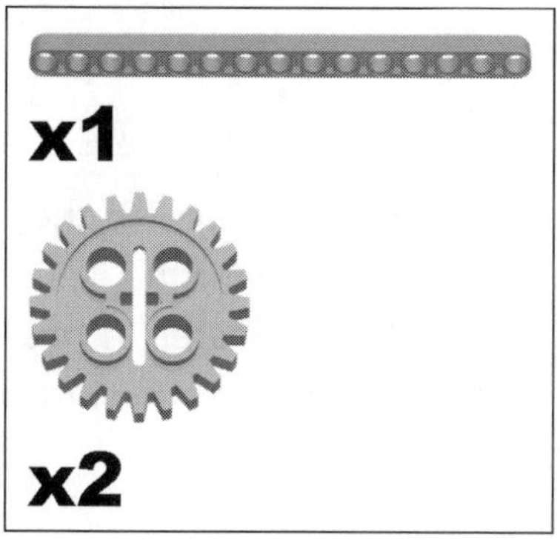

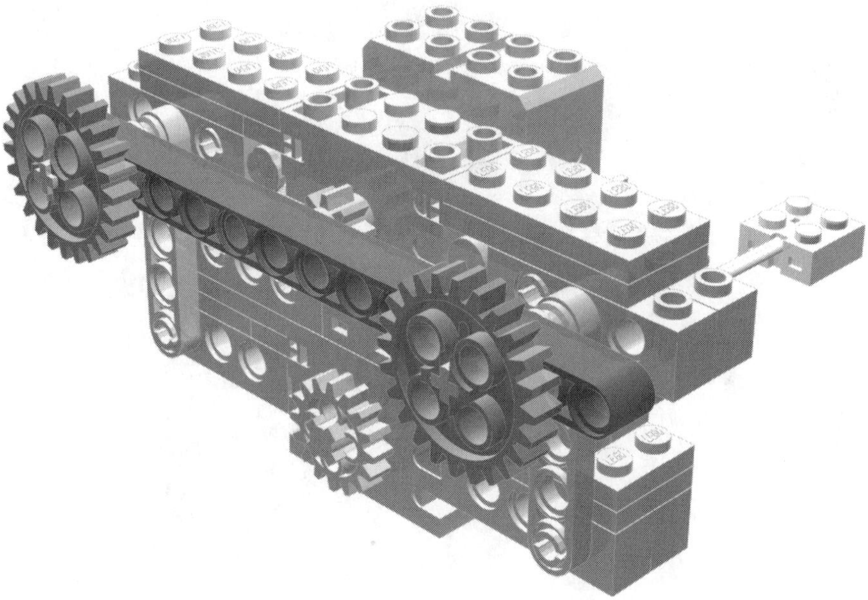

Drive Step 12

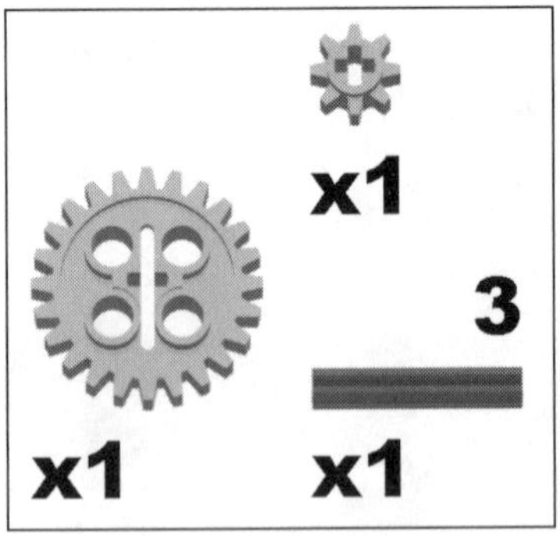

These are the last of the gears for the drive train's leaning mechanism. Turning the 24t gear you install in this step should turn both the motor and the neighboring 24t gear.

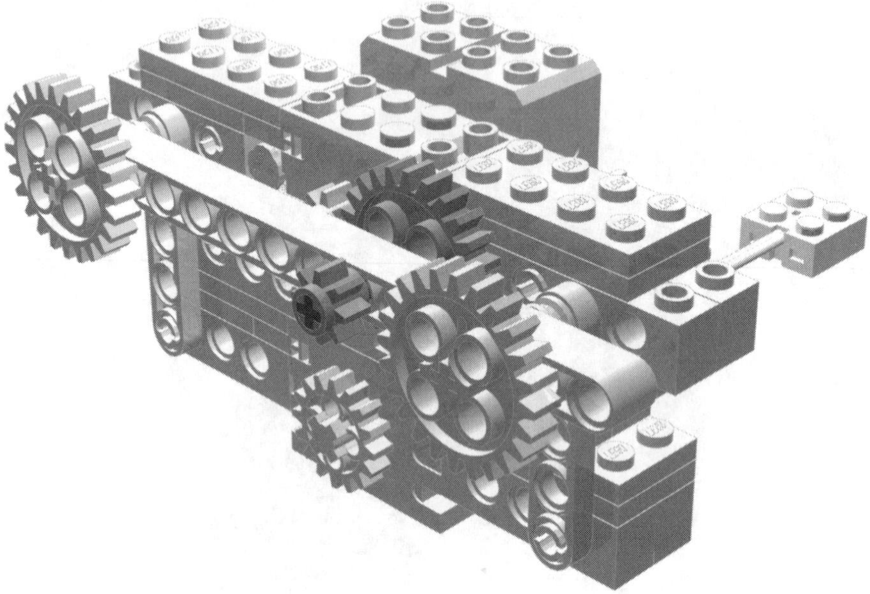

Drive Step 13

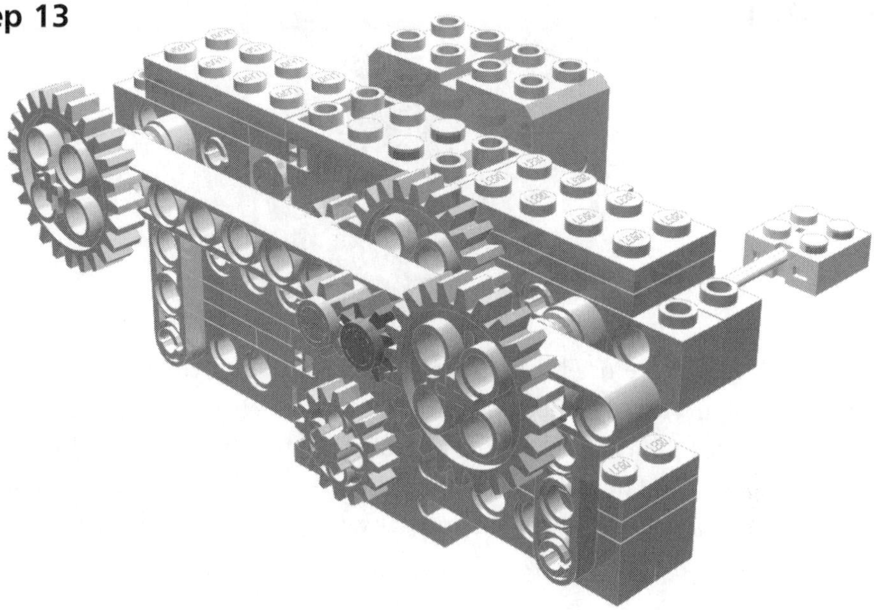

Drive Step 14

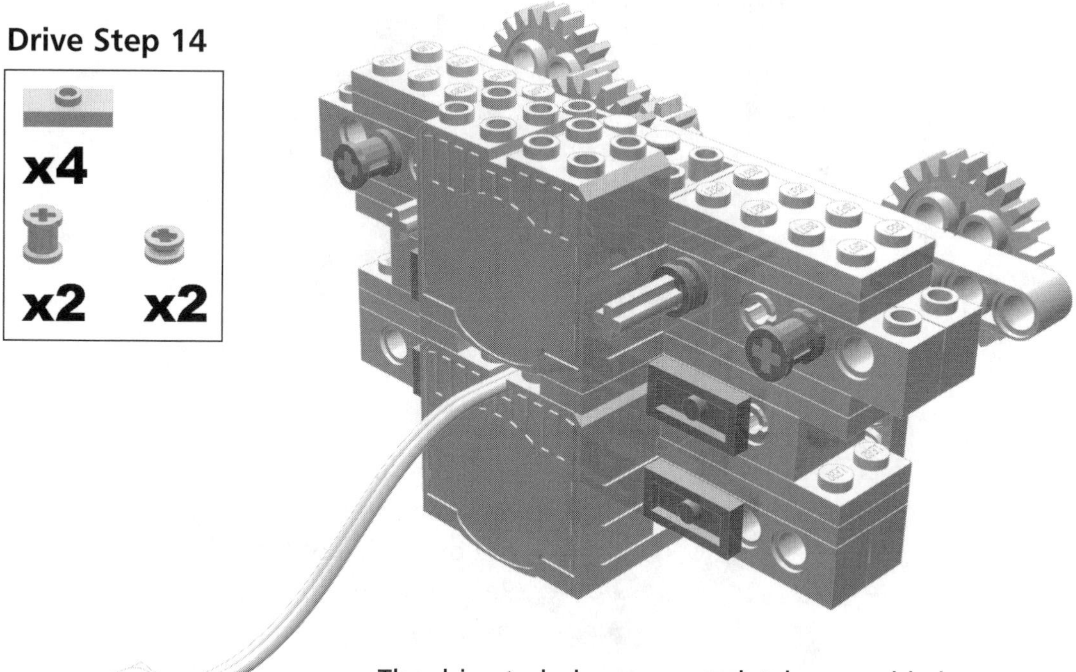

The drive train is now completely assembled.

The Sprocket

TECHNIC chains are used to connect the drive train gears to the chassis gears that drive the legs and leaning mechanism.

Make two of these sprocket sub-assemblies. They will be pulled towards each other using rubber bands. The combination of sprockets and rubber bands make a chain tightener which makes sure the chain is always tight against the gears.

Sprocket Step 0

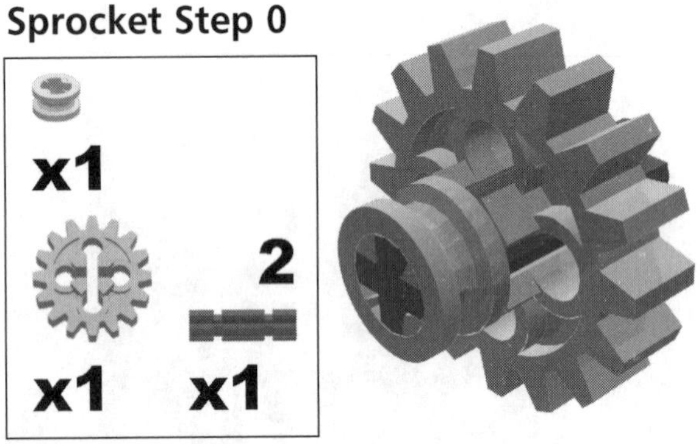

Sprocket Step 1

Turn the sprocket around and add the half bushing to the other side of the axle. Remember to build two sprocket sub-assemblies!

The Face

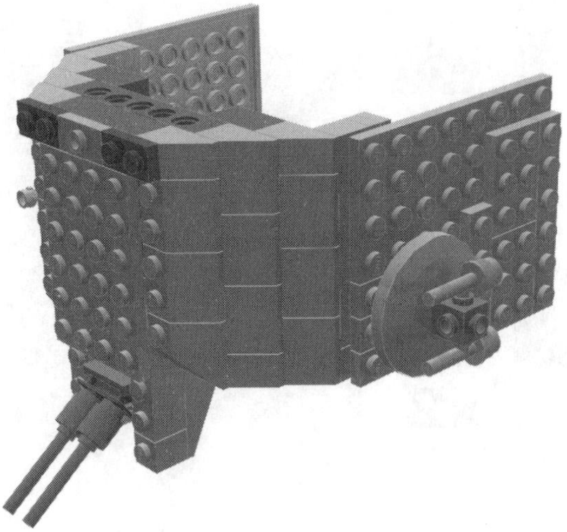

The face sub-assembly plays a critical role in counter-balancing the weight of the motors.

Face Step 0

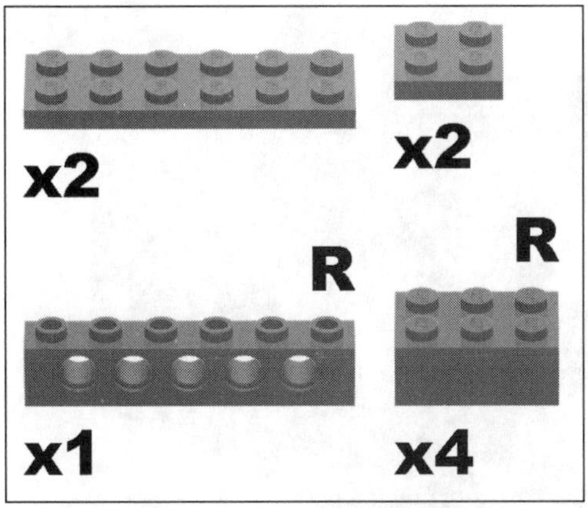

Even though the part list calls for red bricks in this step, you can use whatever color you want because these bricks are all hidden under the rest of the face.

Face Step 1

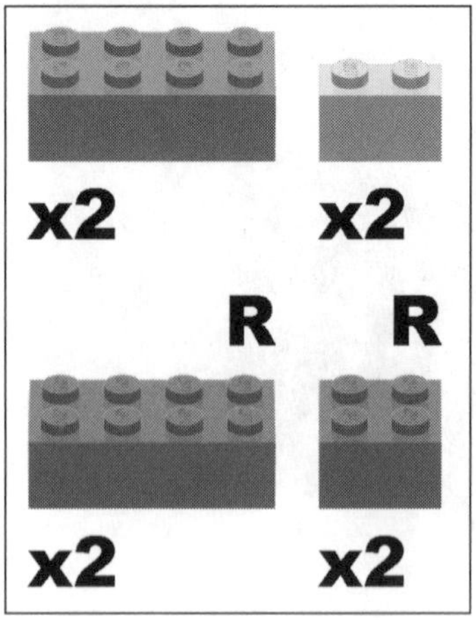

Face Step 2

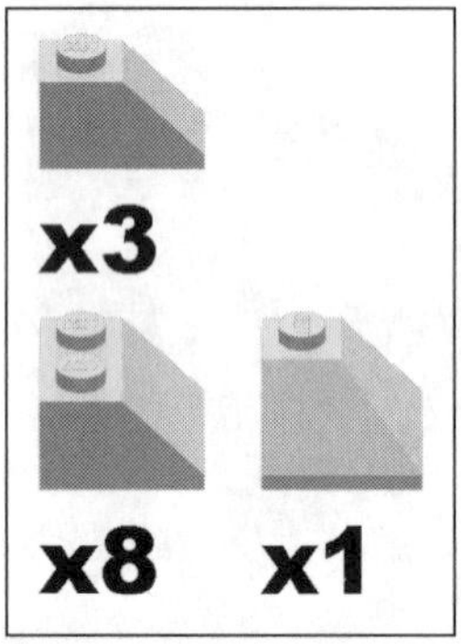

Construction Tip...

Slope Parts

AT-ST requires quite a few slopes to build. I bought many of the slopes I used in AT-ST's construction on BrickLink.com in bulk.

Face Step 3

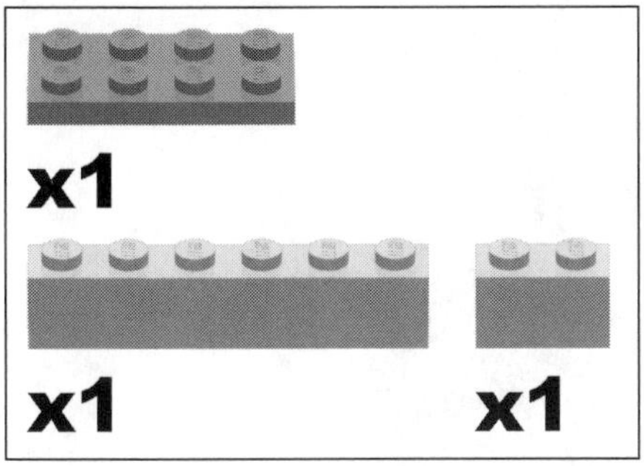

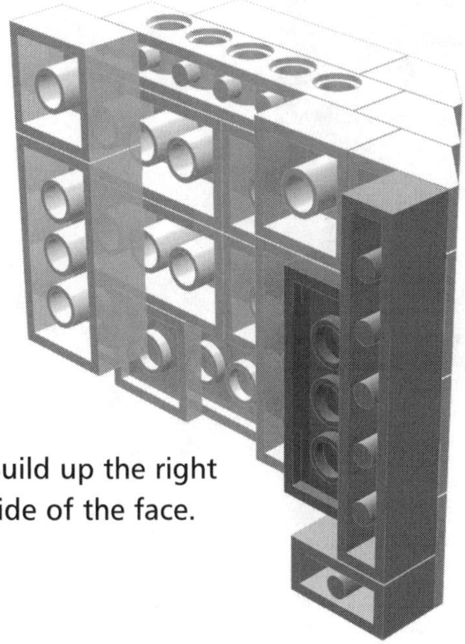

Build up the right side of the face.

Face Step 4

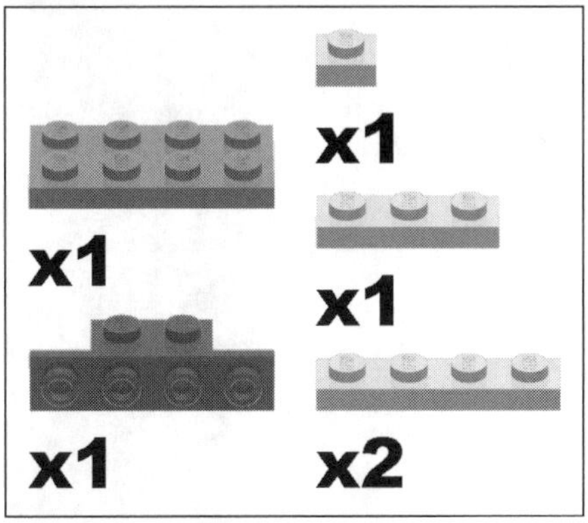

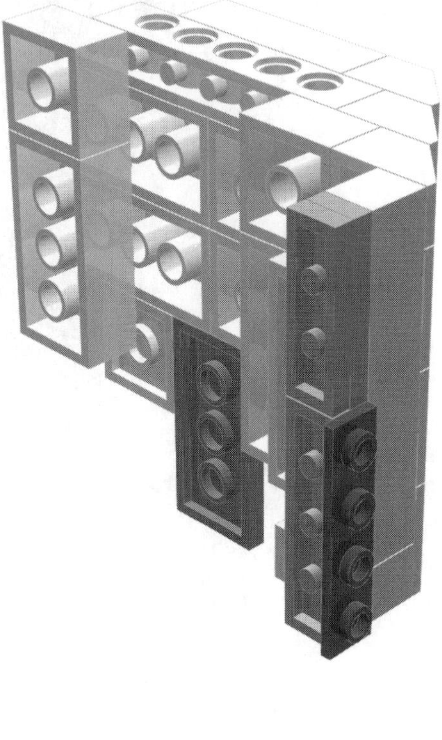

Face Step 5

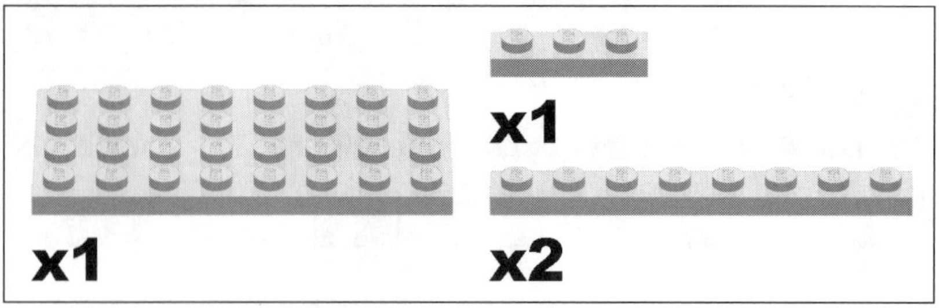

Face Step 6

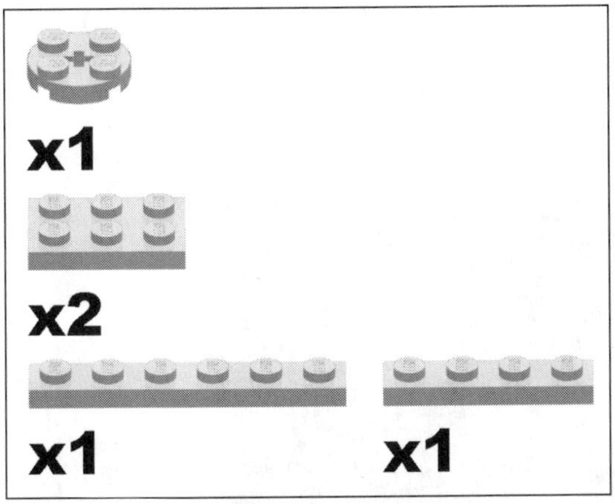

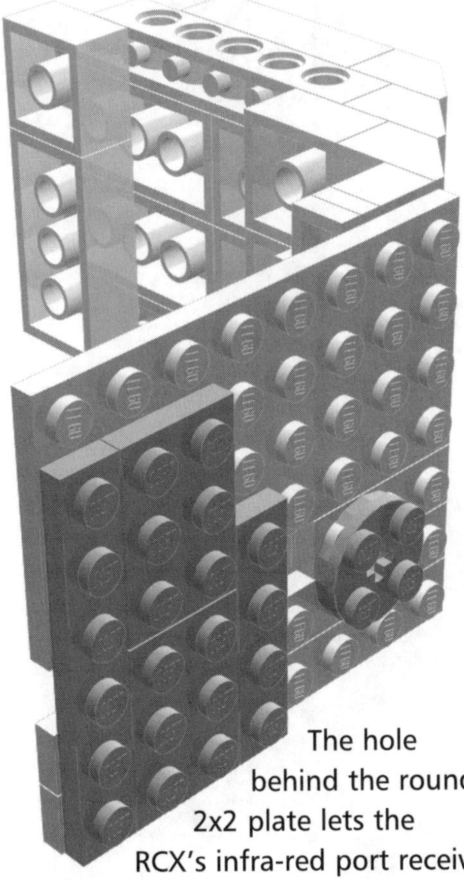

The hole behind the round 2x2 plate lets the RCX's infra-red port receive input from the LEGO Remote Control.

Face Step 7

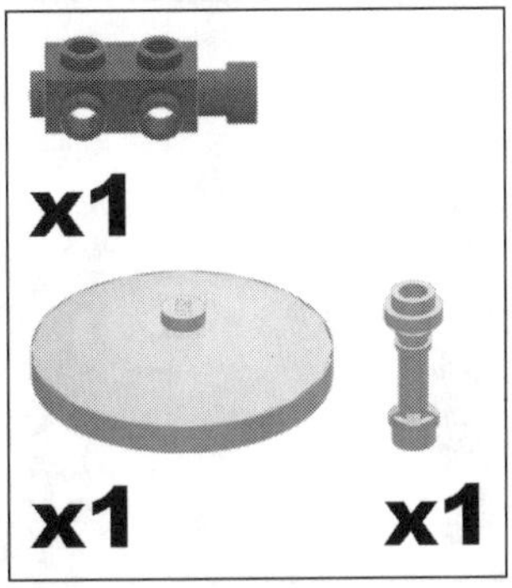

You must have guns on an AT-ST!

Face Step 8

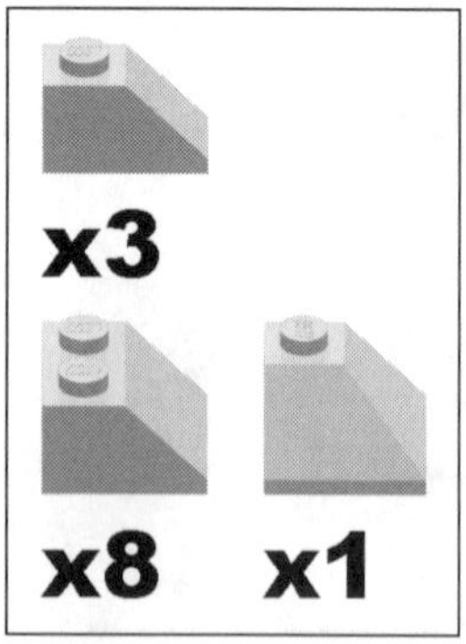

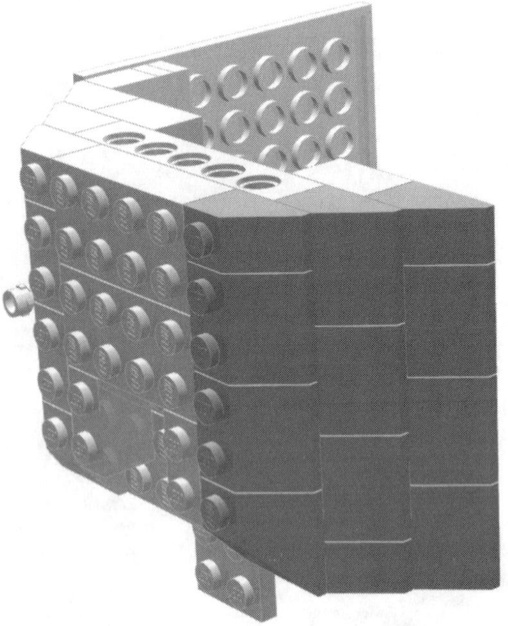

More slopes!

Face Step 9

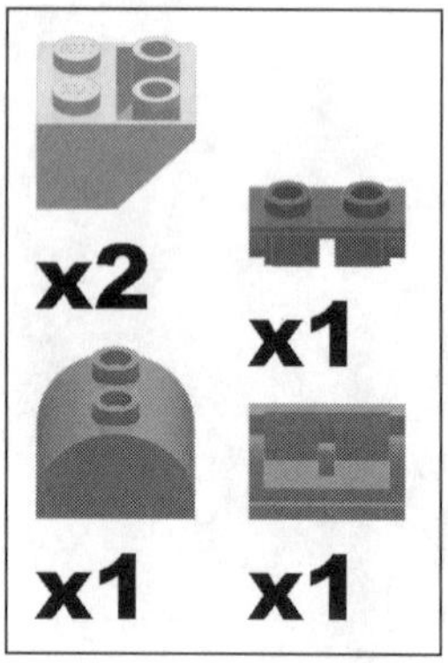

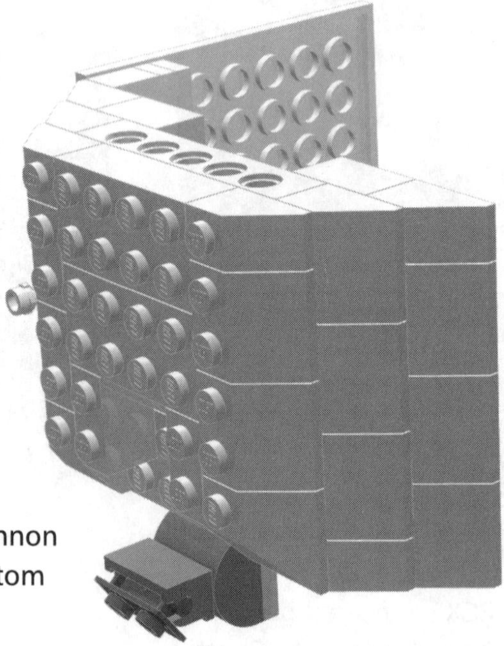

Add the blaster cannon mounts to the bottom of the face.

Face Step 10

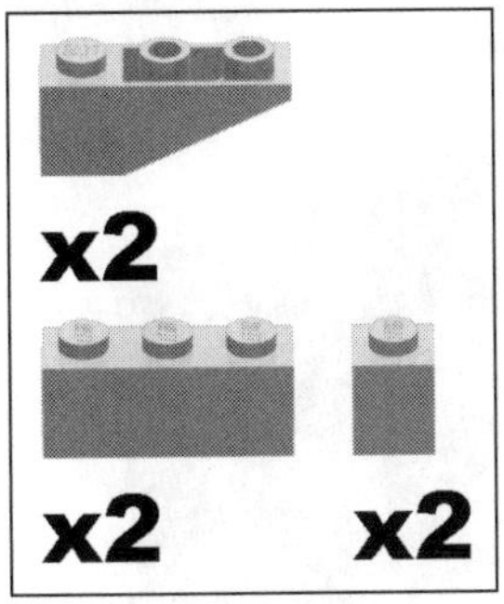

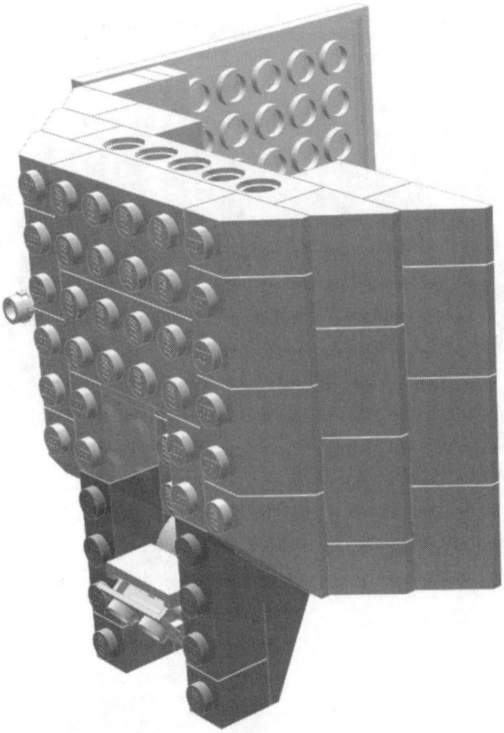

Face Step 11

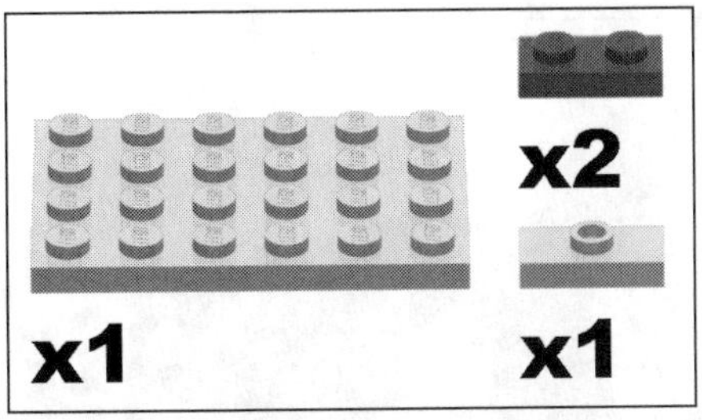

Add the "eyes" of the AT-ST. In the movies, these were actually ports that the drivers could look out of.

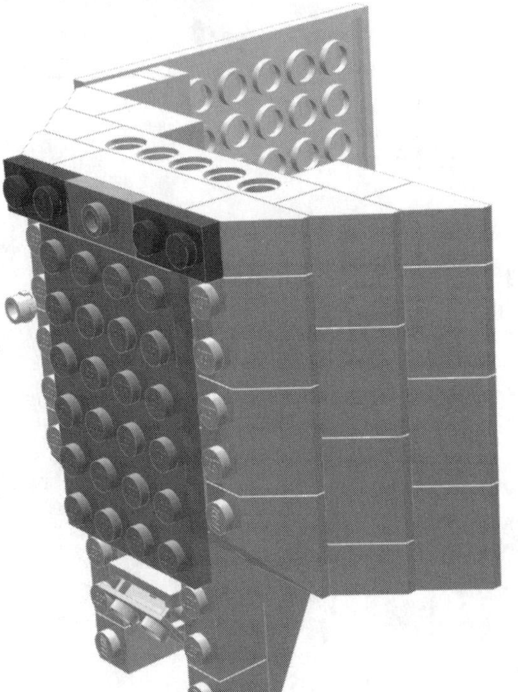

Face Step 12

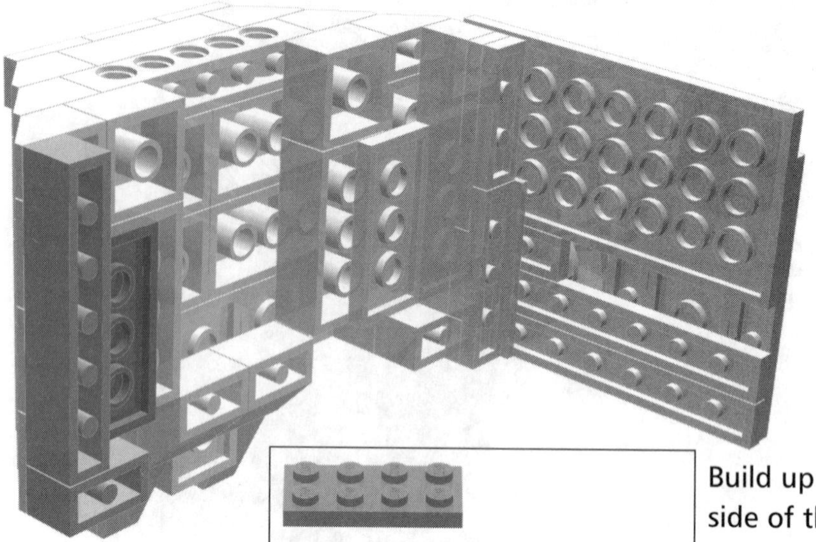

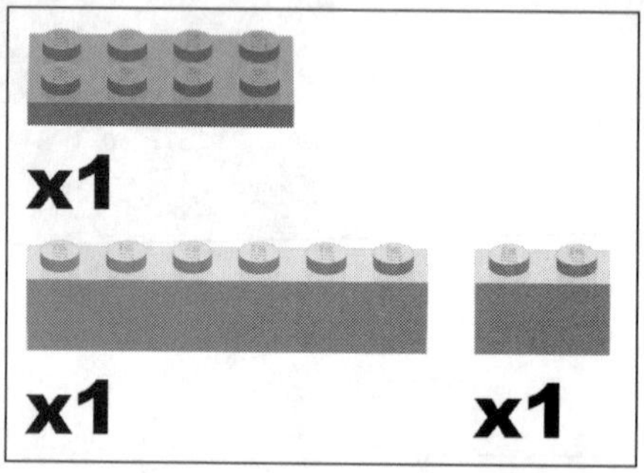

Build up the left side of the face.

Face Step 13

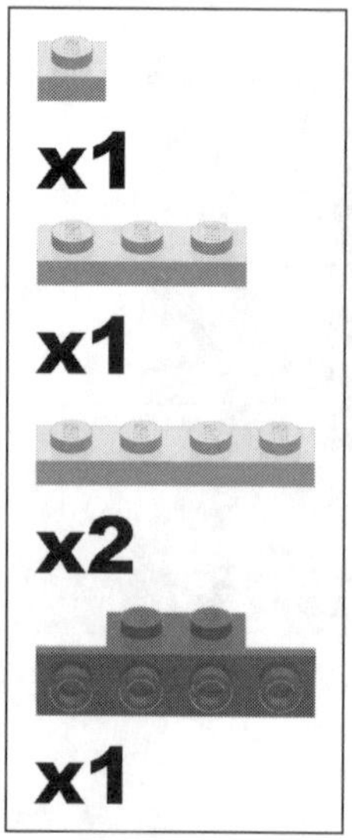

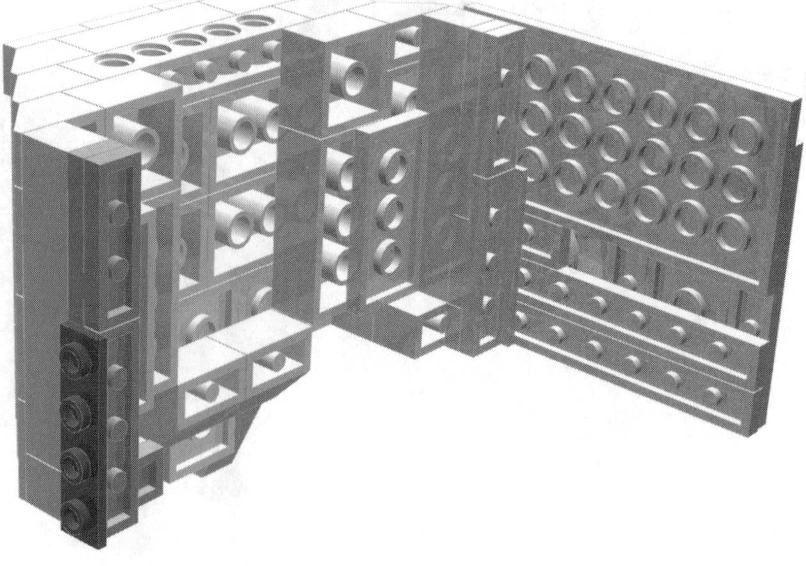

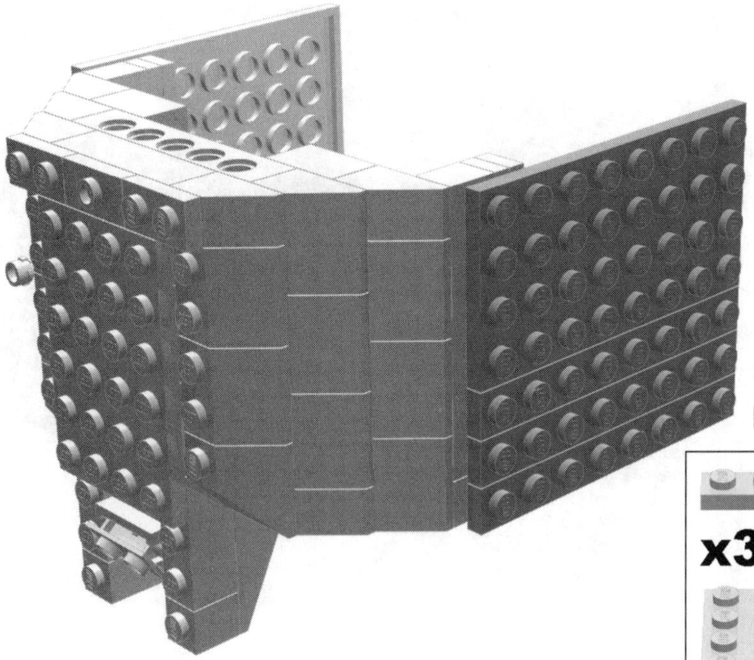

Face Step 14

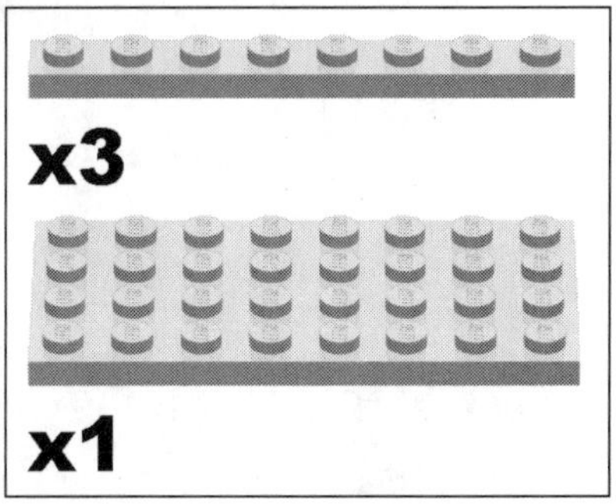

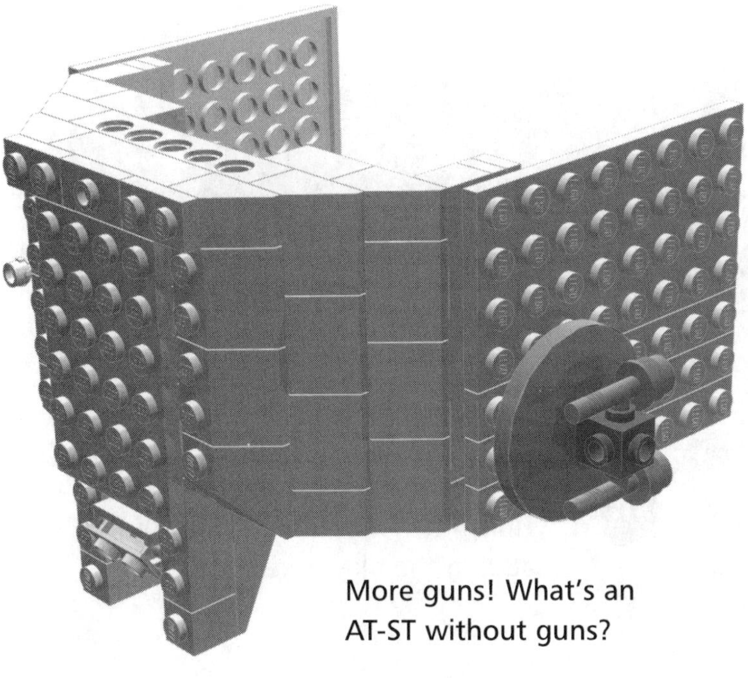

Face Step 15

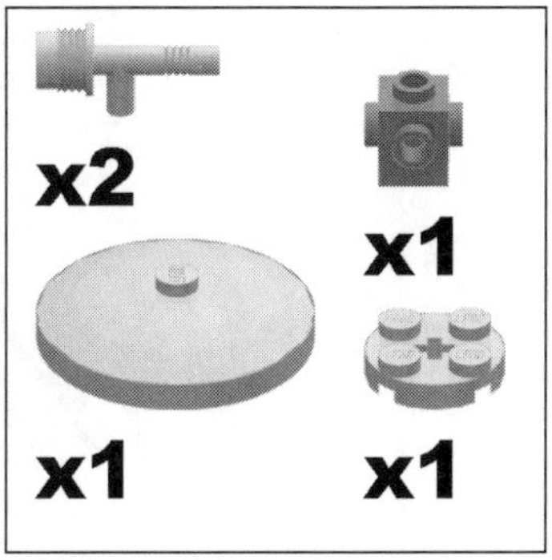

More guns! What's an AT-ST without guns?

Face Step 16

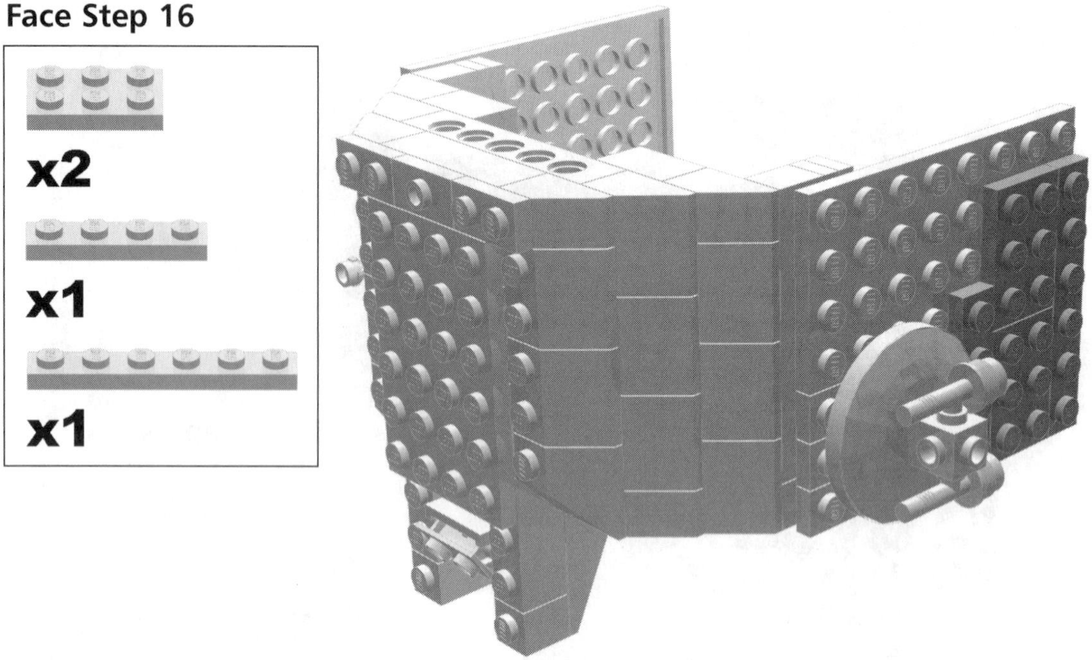

Face Step 17

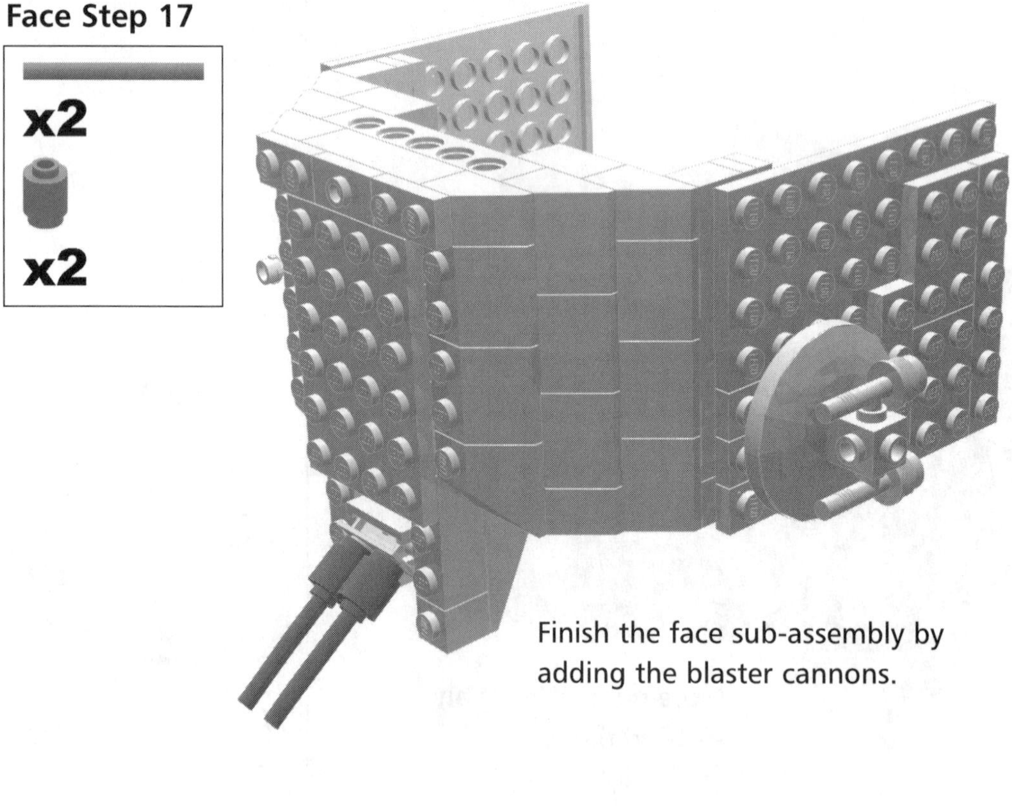

Finish the face sub-assembly by adding the blaster cannons.

The Roof

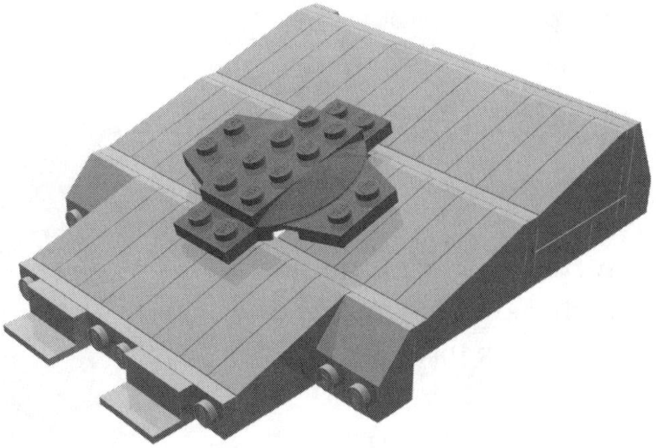

This is the roof sub-assembly. It requires even more slopes.

Roof Step 0

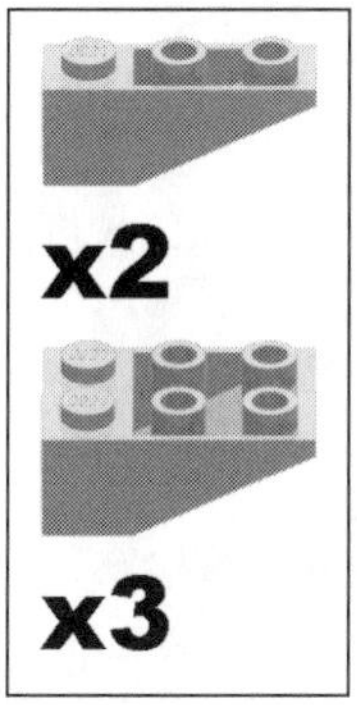

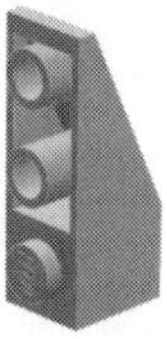

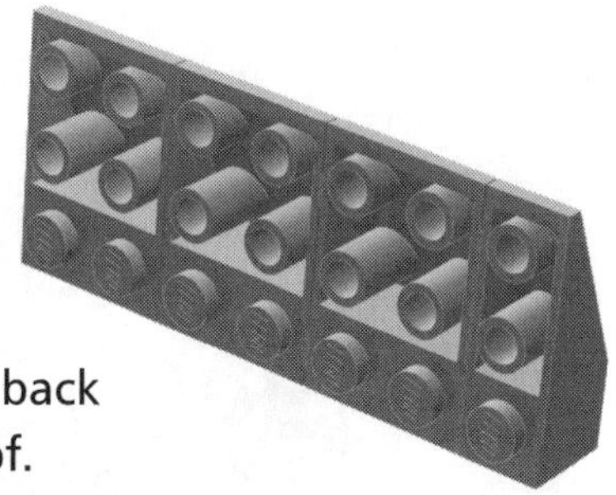

Start with the back side of the roof.

Roof Step 1

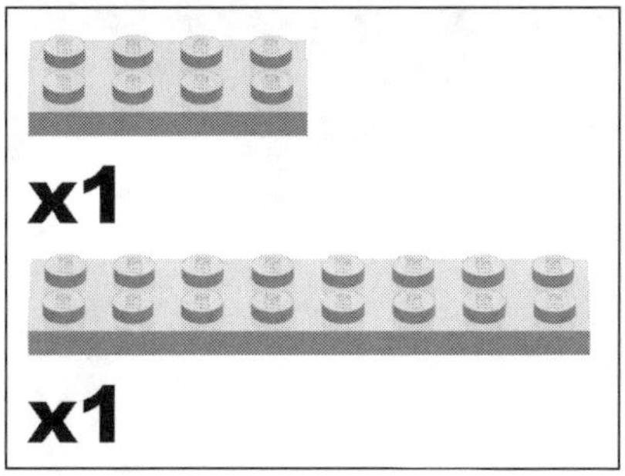

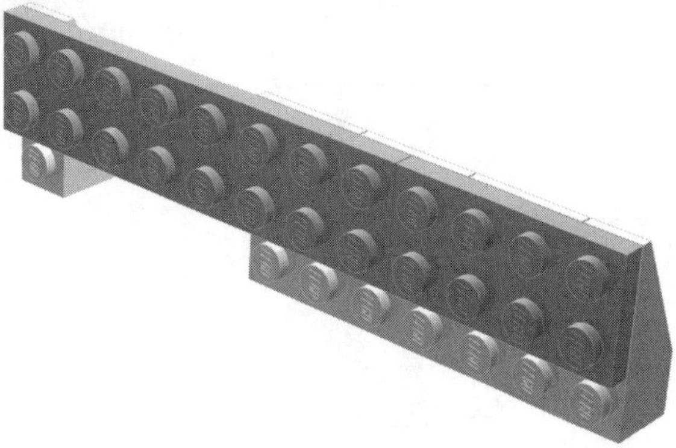

Roof Step 2

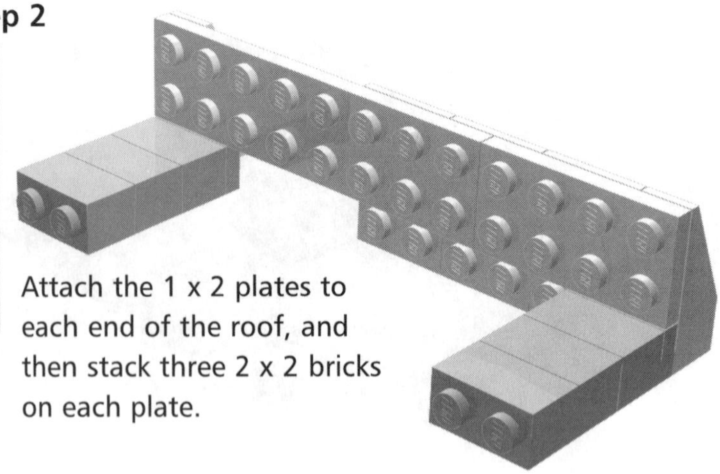

Attach the 1 x 2 plates to each end of the roof, and then stack three 2 x 2 bricks on each plate.

Roof Step 3

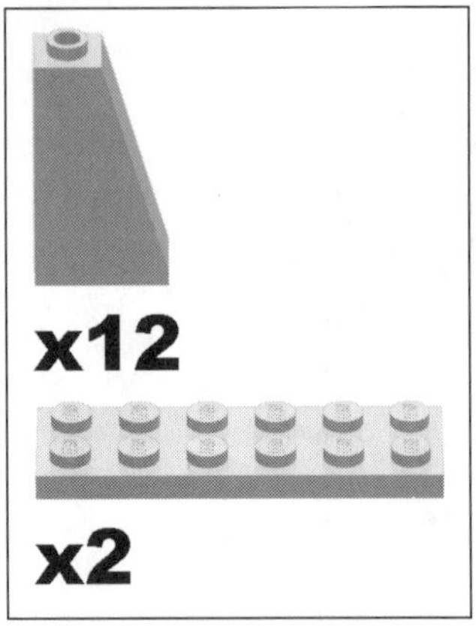

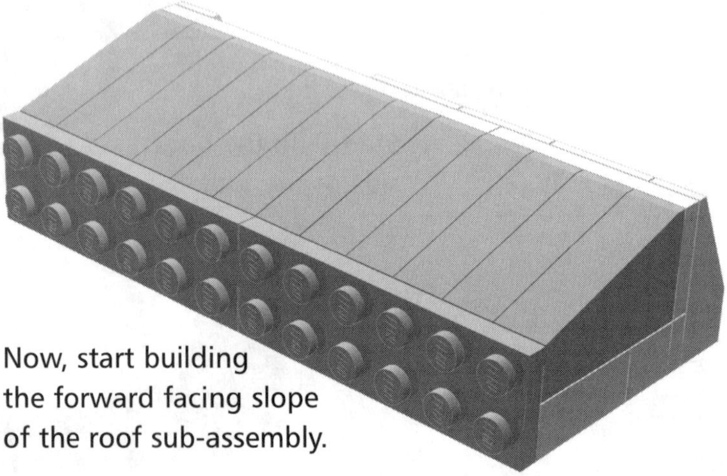

Now, start building the forward facing slope of the roof sub-assembly.

Roof Step 4

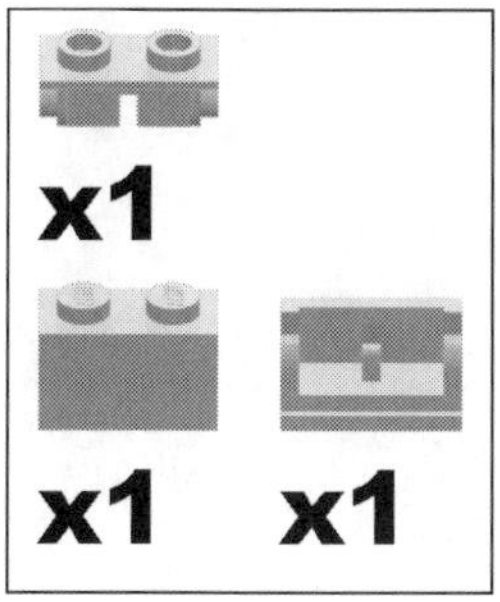

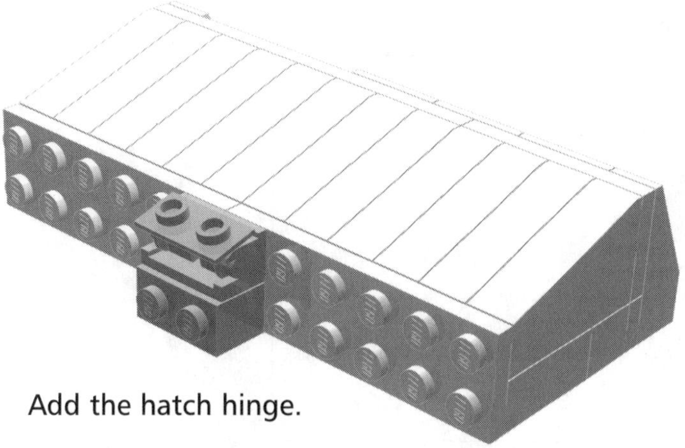

Add the hatch hinge.

Roof Step 5

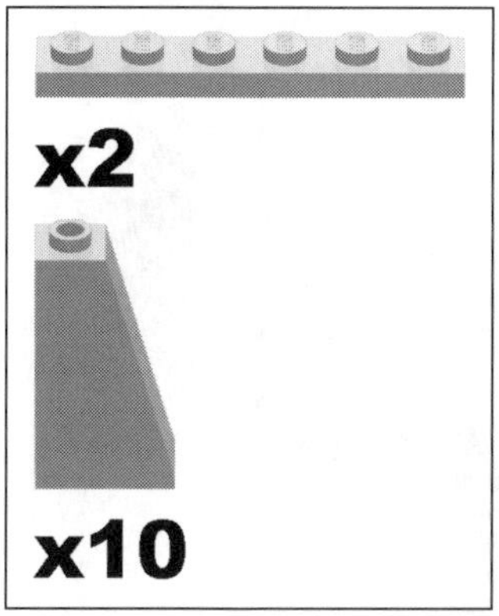

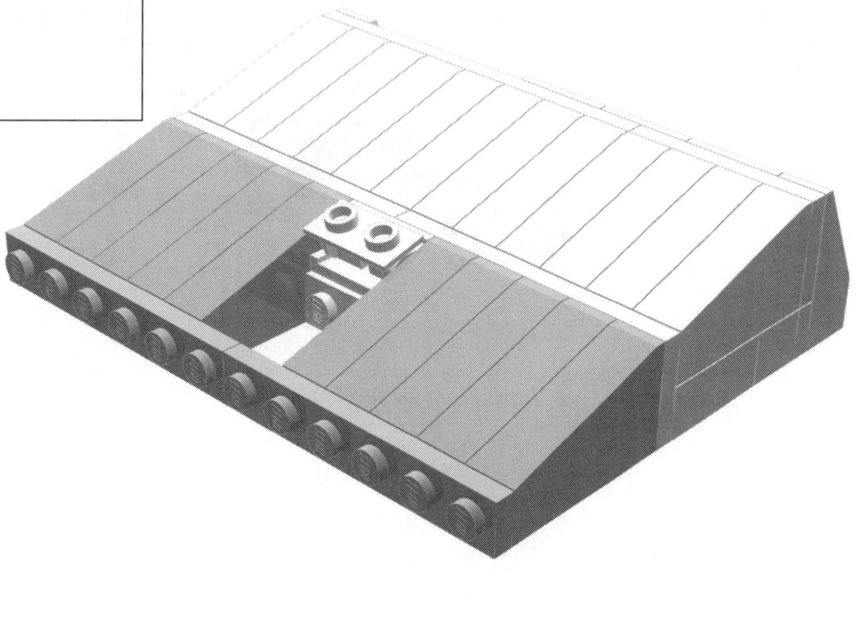

Roof Step 6

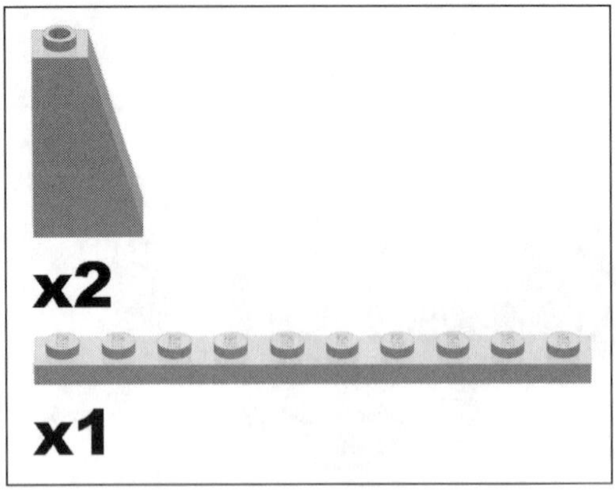

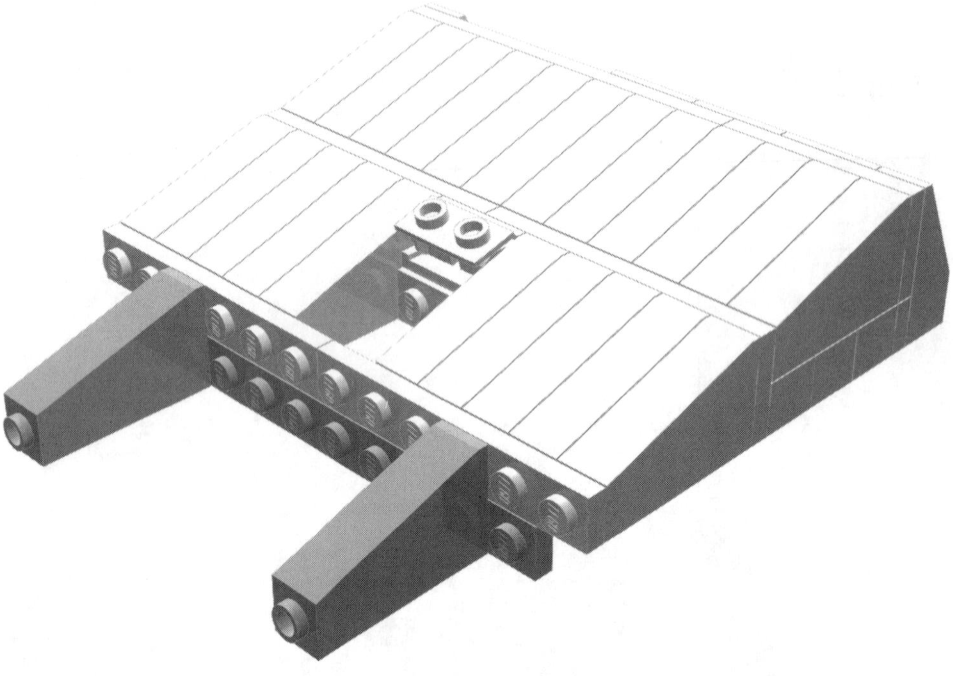

Roof Step 7

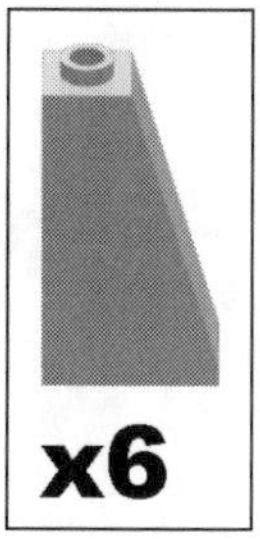

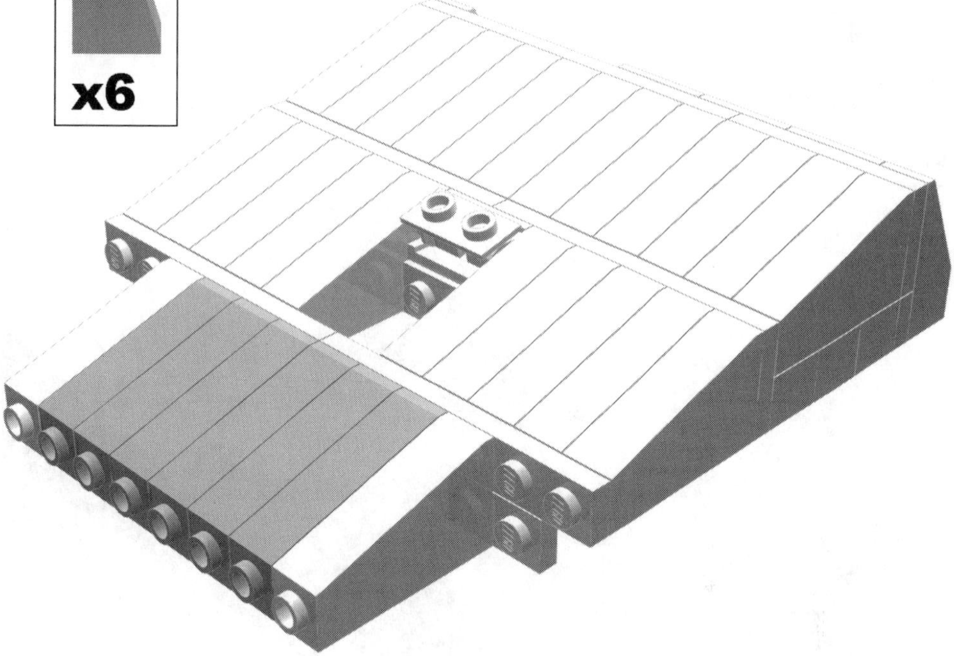

Roof Step 8

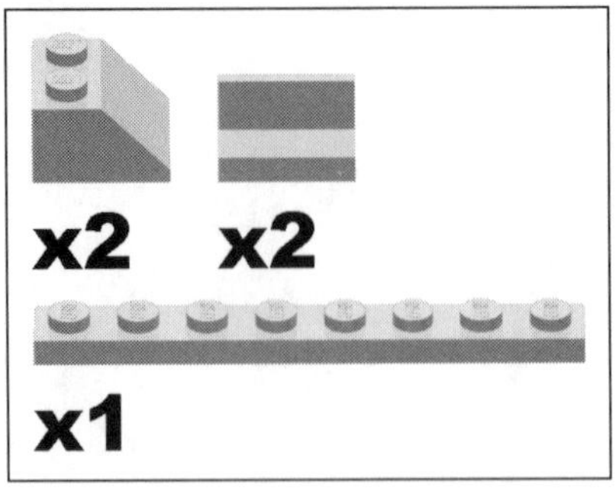

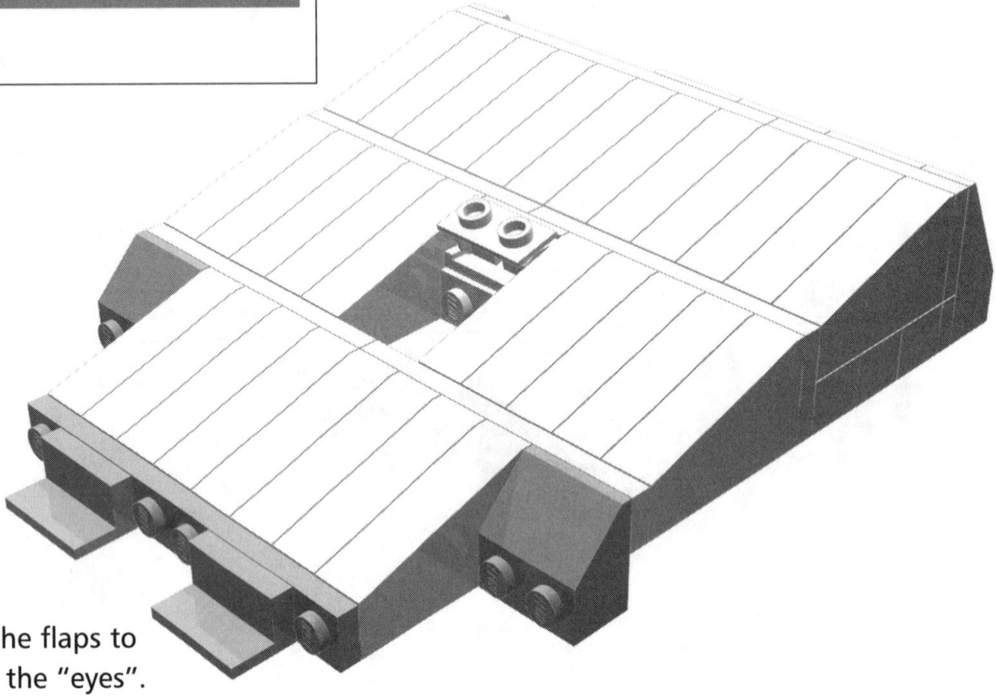

Add the flaps to cover the "eyes".

Roof Step 9

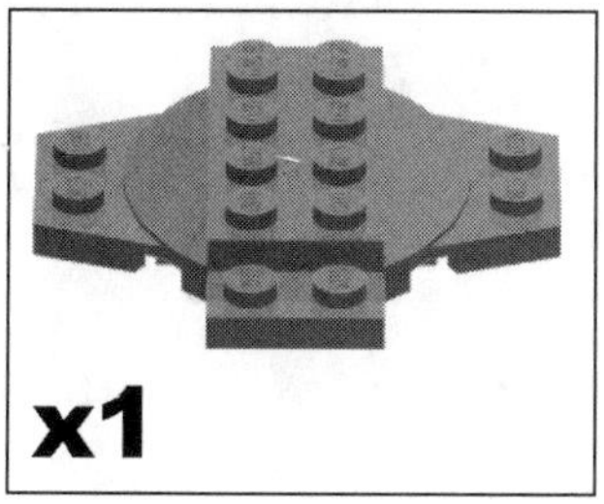

Add the hatch.

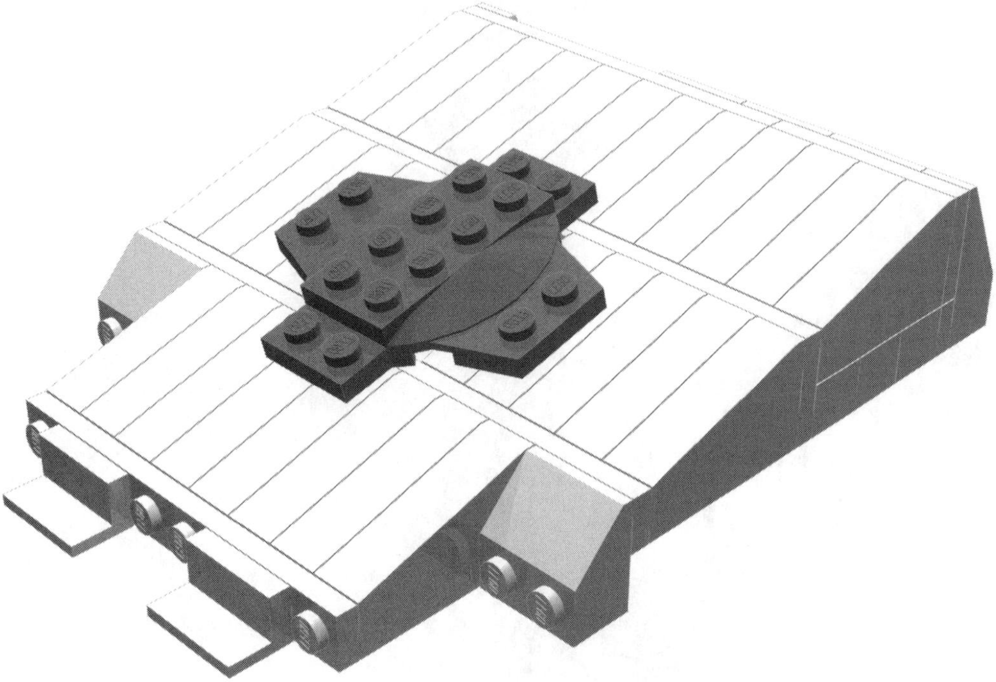

Putting It All Together

Now it is time to assemble the complete AT-ST model using the sub-assemblies you've just built.

Final Step 0

Locate the right and left leg sub-assemblies. Position them as shown.

Creating the Instructions…

Overlapping Parts

The AT-ST's feet overlap one another where they meet in the instruction images. In the real world, this does not happen. When walking, one foot is above the other. When resting, AT-ST takes a half stride that makes the feet even. When it drops the up foot, the inside of the down foot makes the up foot slide to the side.

Final Step 1

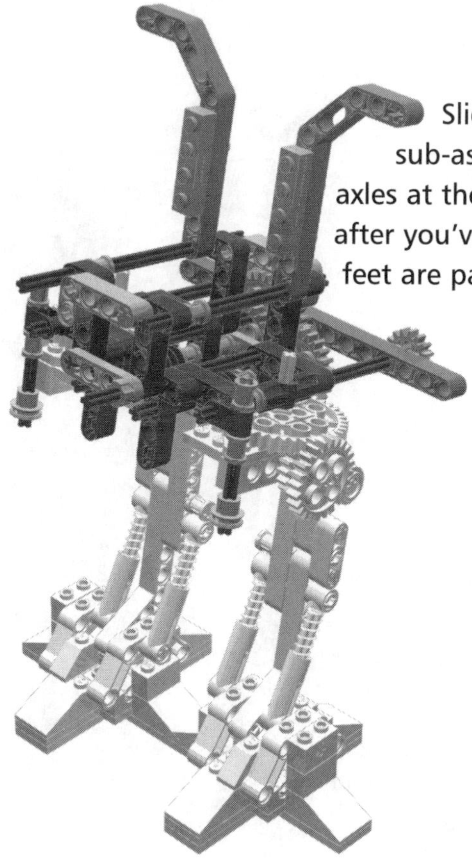

Slide the chassis sub-assembly over the axles at the top of the legs after you've made sure the feet are parallel.

Final Step 02

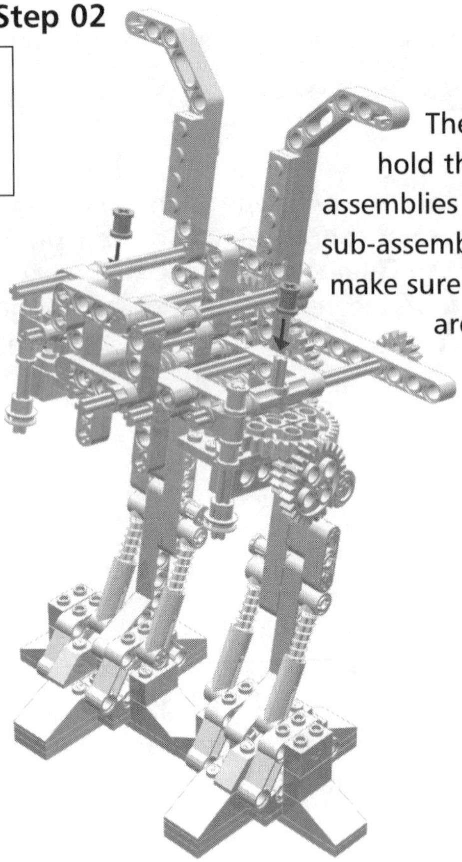

These bushings hold the leg sub-assemblies to the chassis sub-assembly. Again, make sure that the legs are parallel.

Final Step 3

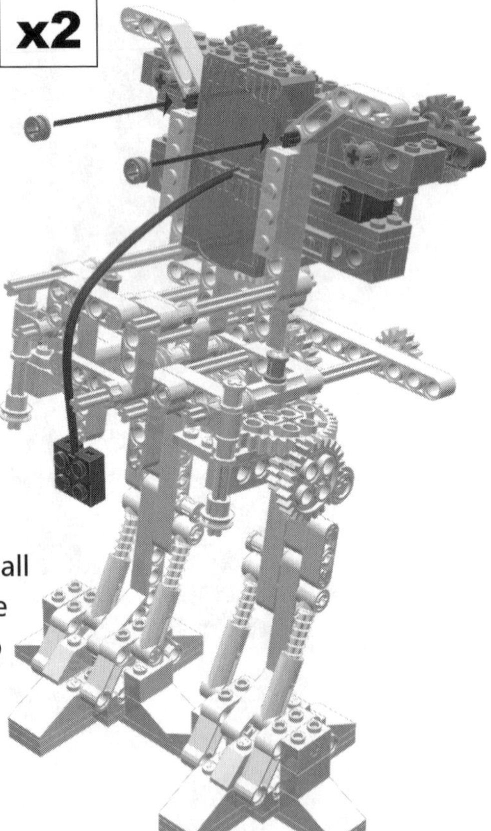

Now, add the drive sub-assembly to the chassis sub-assembly. Use the small pulleys to attach the drive to the chassis. Hold on to the model because it will fall over if you don't.

Creating the Instructions…

Tricks With Cables

The cable pictured here is not the same image as the one pictured in the drive sub-assembly images, because the cable in the drive sub-assembly is *ghosted*. After the drive sub-assembly is added, an MLCad BUFEXCHG STORE is used to remember the design before we add the electrical cable.

Final Step 4

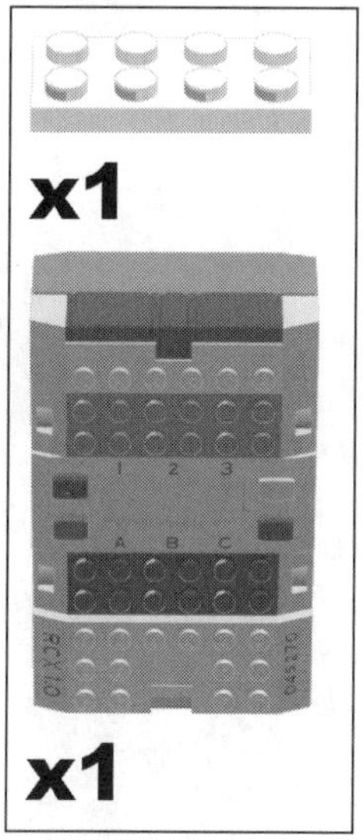

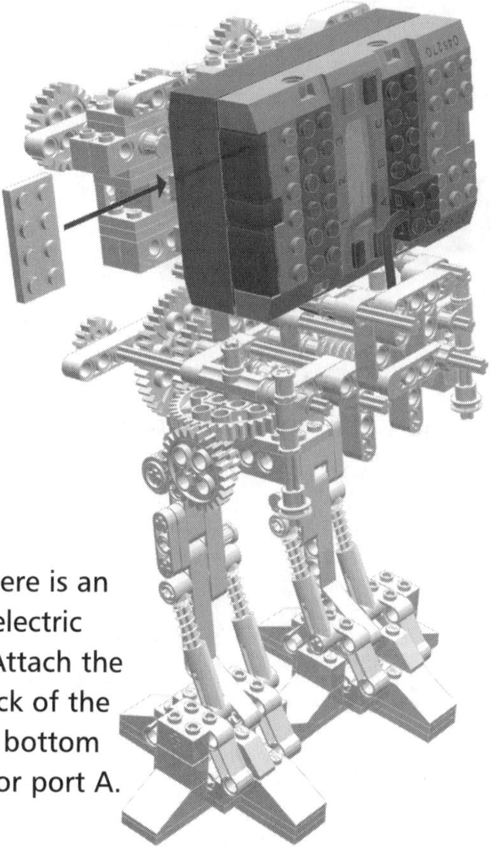

The 2 x 4 plate part added here is an electric plate, used to hook electric cables together electrically. Attach the 2 x 4 electric plate to the back of the RCX, as shown. Connect the bottom motor cable to the RCX motor port A.

Creating the Instructions…

More Tricks With Cables

I used BUFEXCHG RETRIEVE to have LPub "remember" what was in the design before I added the cable in **Final Step 4**. I used LSynth to create the final form of the connected cable, surrounded by PLIST BEGIN IGN and PLIST END meta-commands so the cable did not show up in the part list image.

Final Step 5

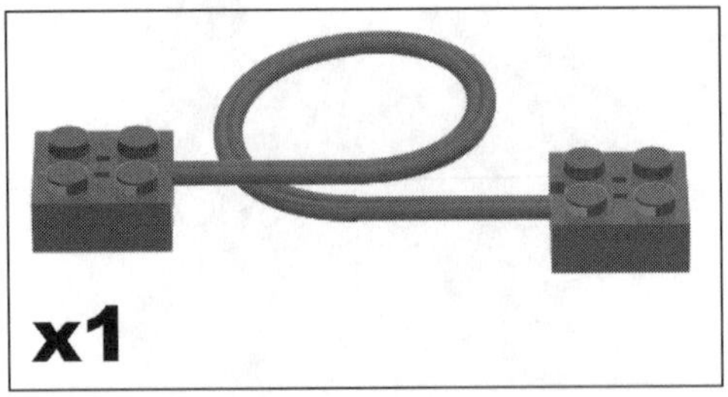

Connect the short electrical cable from sensor port 3 to the electric plate.

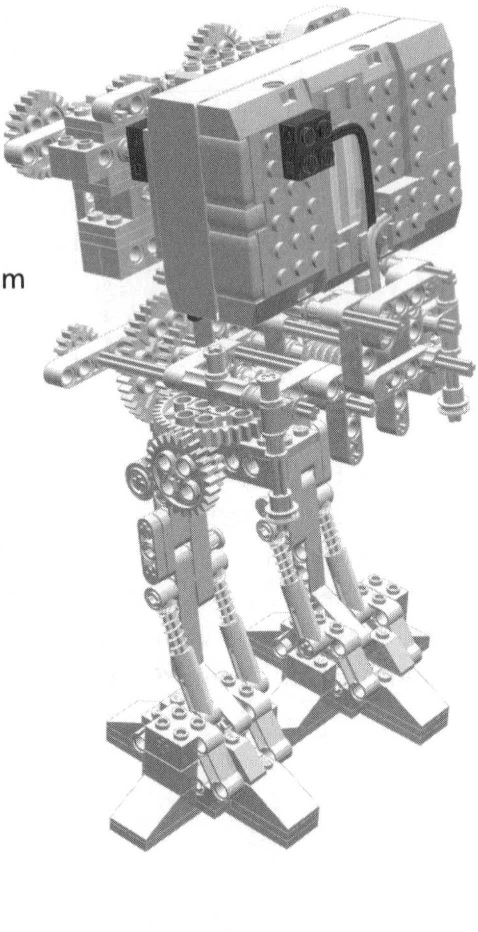

Final Step 6

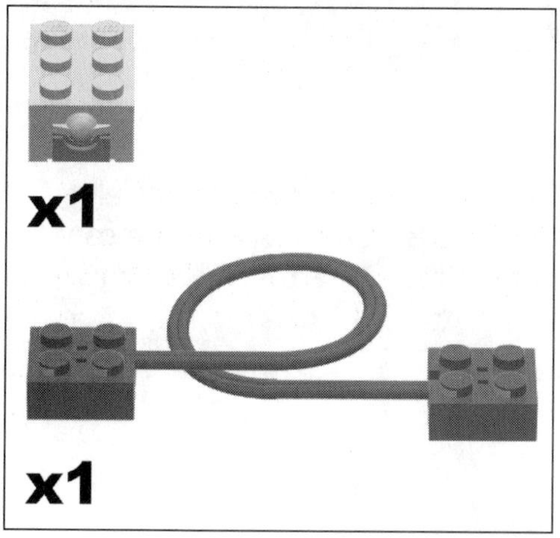

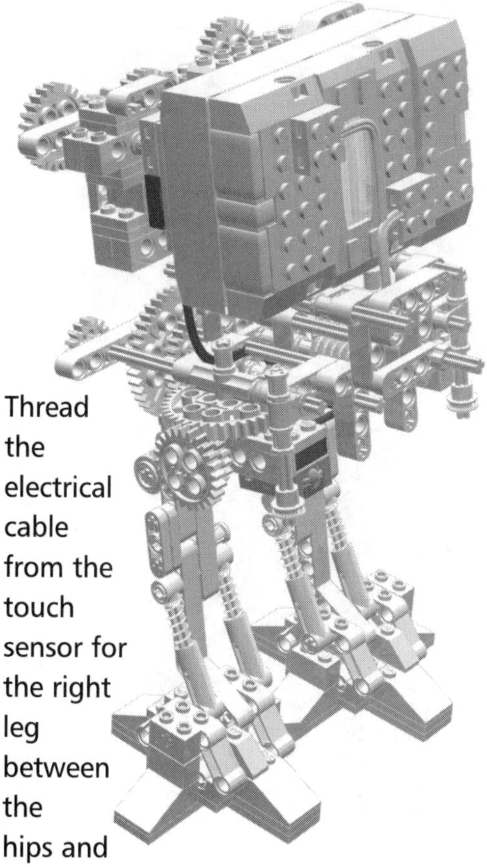

Thread the electrical cable from the touch sensor for the right leg between the hips and up to the electric plate on the back of the RCX brick.

Final Step 7

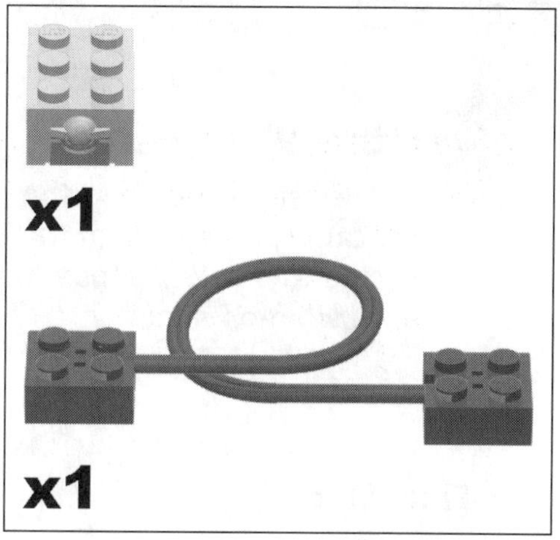

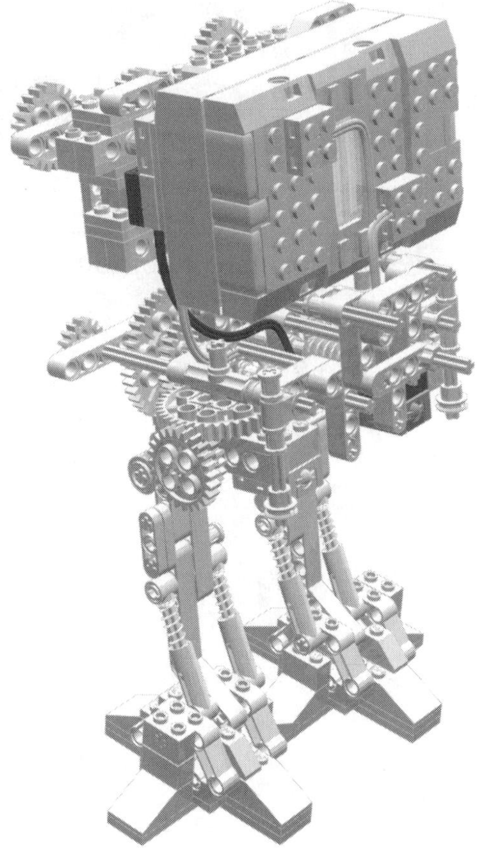

Do the same for the left leg's touch sensor.

Construction Tips…

Cable Placement

Notice that by hooking the left leg's electric cable to the right leg's electric cable at the back of the RCX, both of the legs' touch sensors are connected to one RCX sensor port. In electronic circuitry terms this is known as a *wired-or* circuit.

Final Step 8

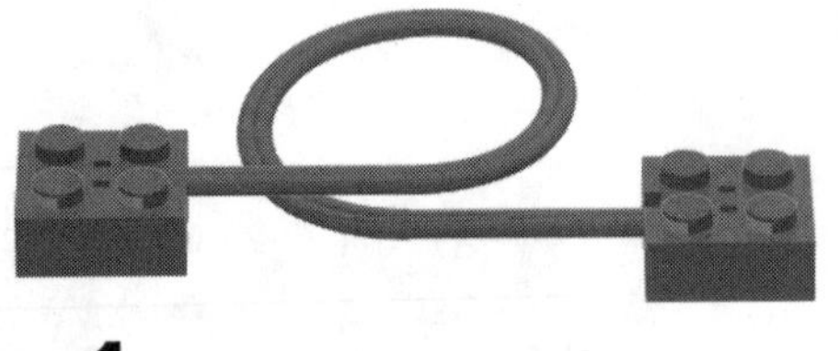

x1

Connect the top motor to the RCX brick.

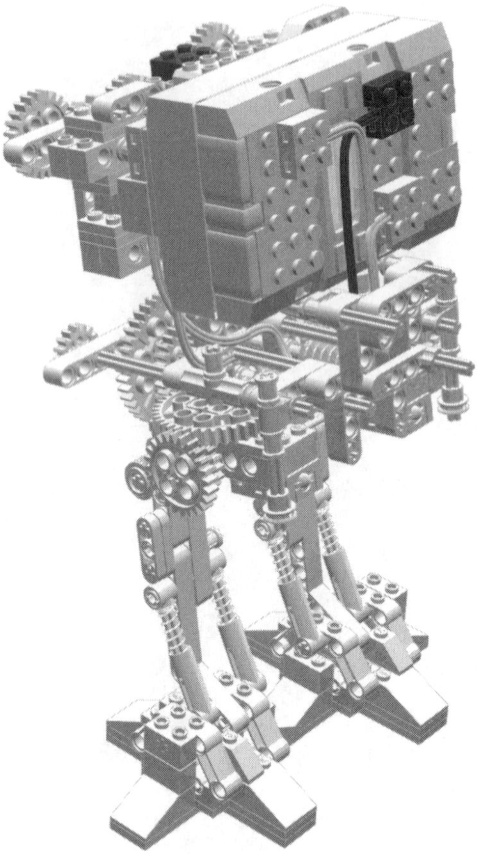

Final Step 9

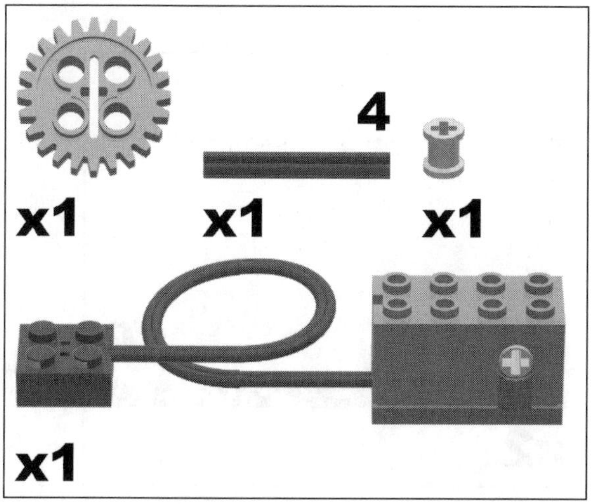

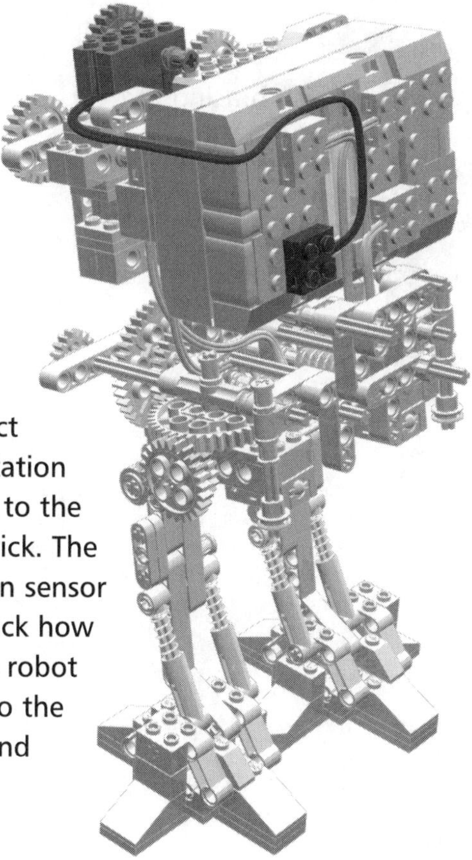

Connect the rotation sensor to the RCX brick. The rotation sensor will track how far the robot leans to the right and left.

Final Step 10

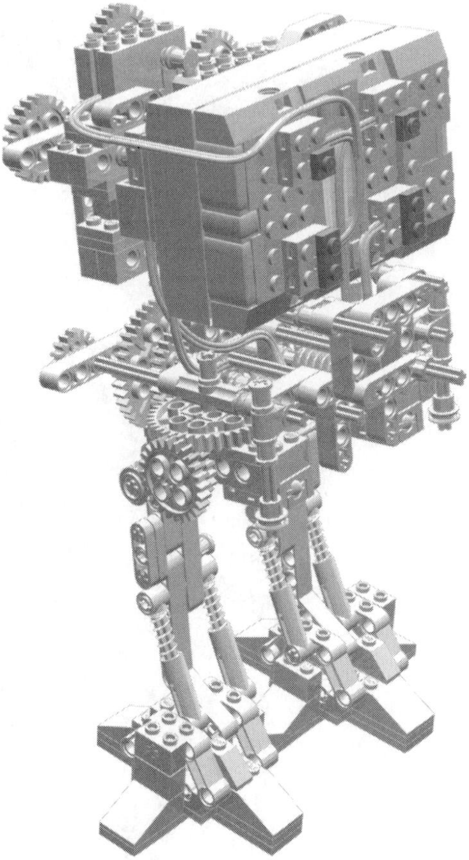

Final Step 11

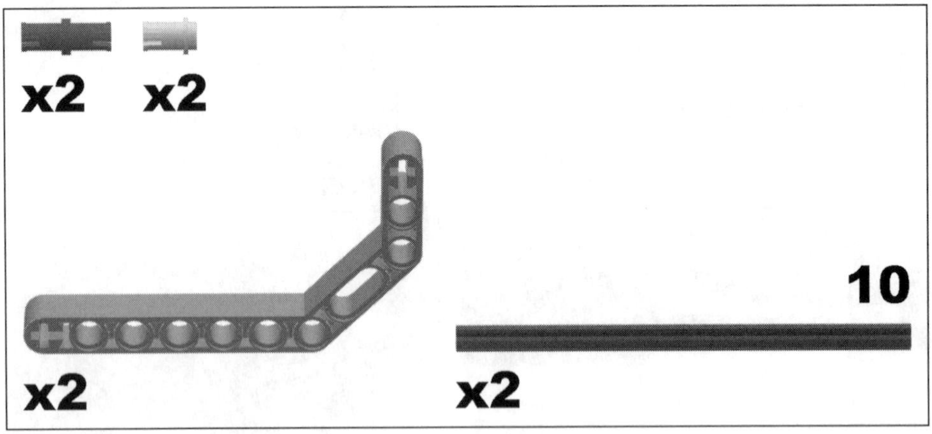

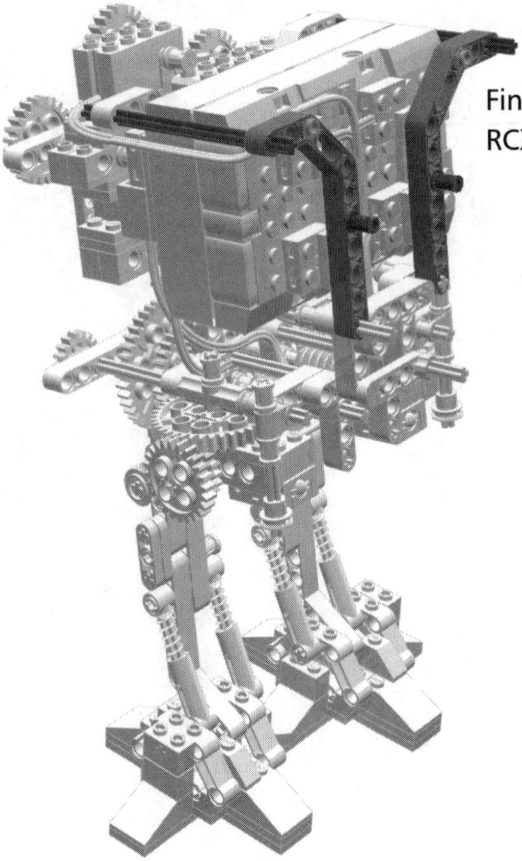

Finish bracing the RCX brick in place.

Final Step 12

Add the chain that makes the legs stride.

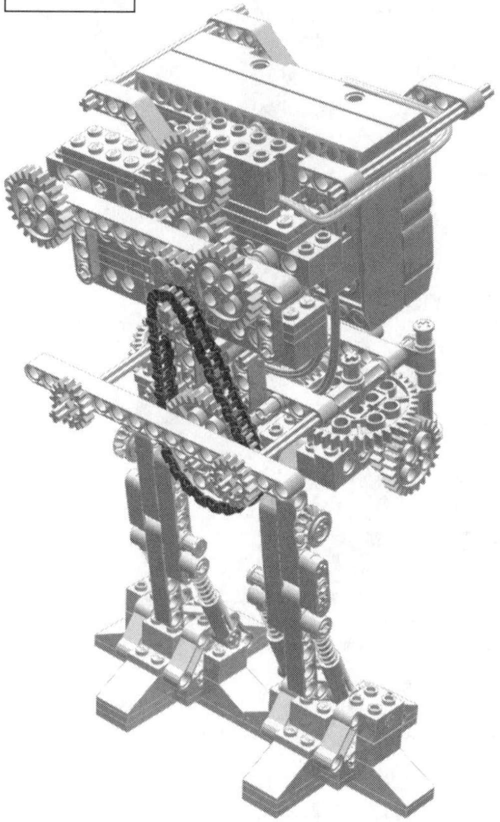

Final Step 13

Add the chain that controls leaning side to side.

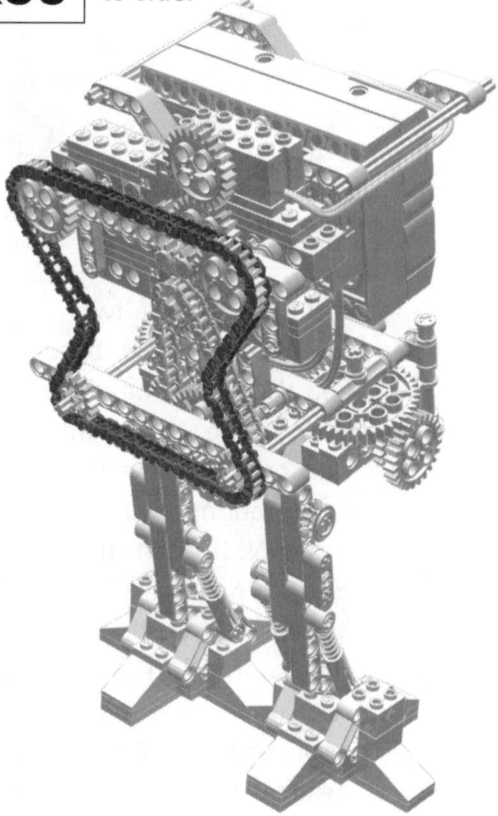

Creating the Instructions…

Using LSynth

The chains used in AT-ST's LDraw design files were synthesized using LDraw chain link parts in LSynth.

Final Step 14

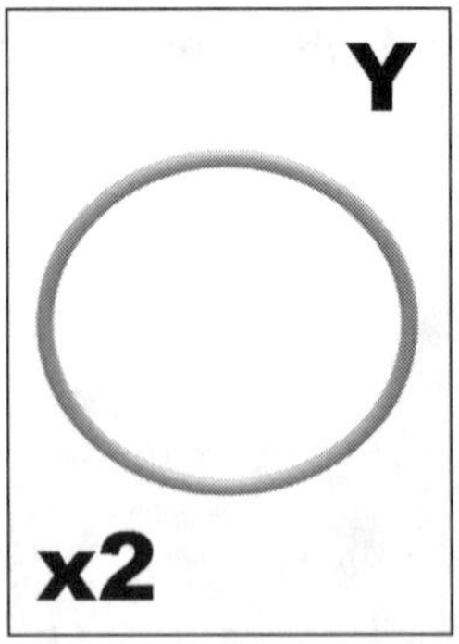

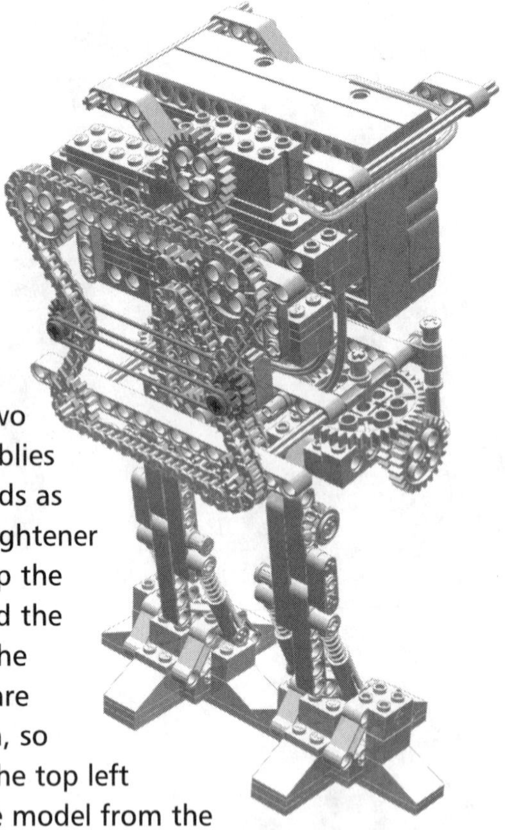

You now add the two sprocket sub-assemblies and the rubber bands as shown. The chain tightener they create will keep the chain taught around the gears. Notice that the bottom two gears are locked into position, so they cannot turn. The top left gear (looking at the model from the angle shown in **Final Step 14**) turns freely. The top right gear pulls the chain right or left, which makes the robot shift weight from one foot to the other. The rotation sensor tracks the amount of chain pulled and stops the motor before it rips the chain apart.

Final Step 15

Add the face sub-assembly.

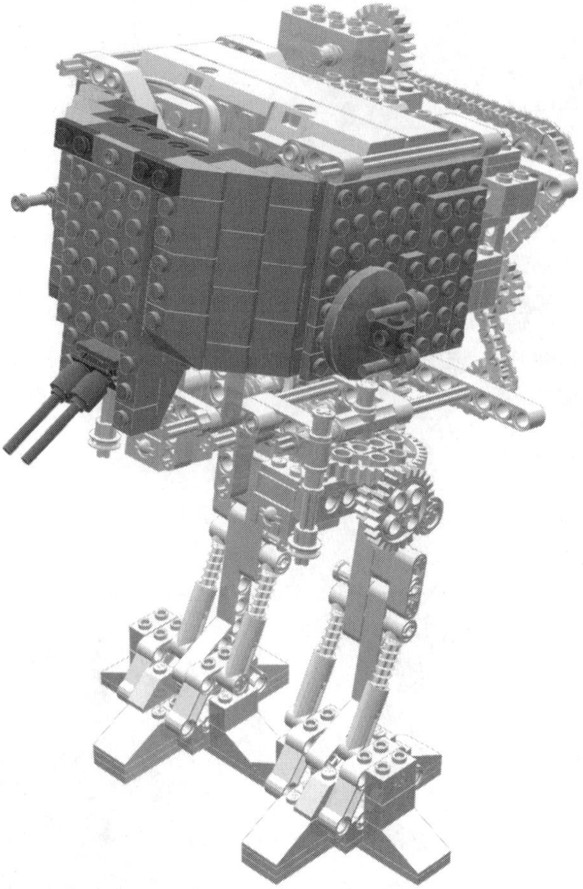

Final Step 16

Finish your construction by adding the roof sub-assembly and your AT-ST is complete!

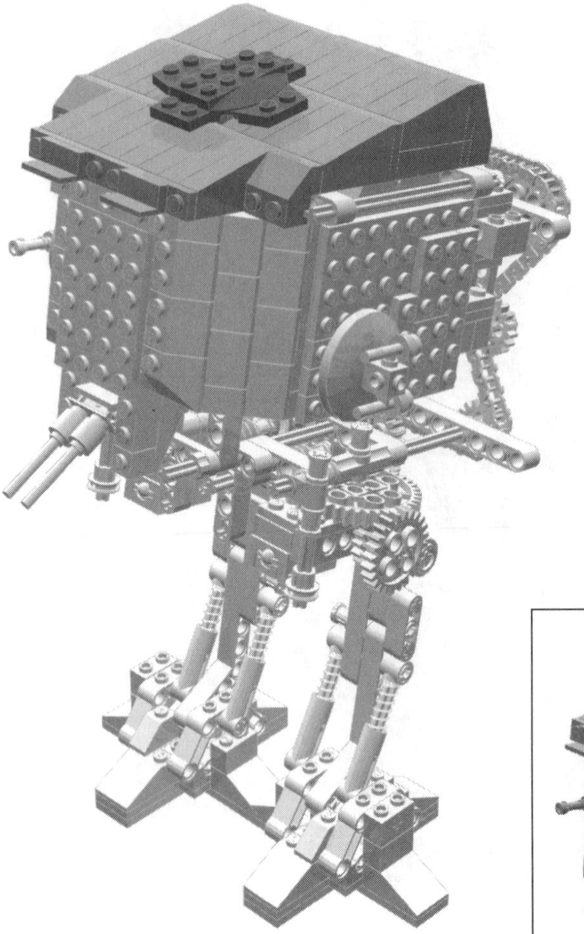

Index

C

D

E

F

G

H

I

M

T

Y

Z

GNU GENERAL PUBLIC LICENSE

Version 2, June 1991

Copyright © 1989, 1991 Free Software Foundation, Inc.
59 Temple Place - Suite 330, Boston, MA 02111-1307, USA

Everyone is permitted to copy and distribute verbatim copies
of this license document, but changing it is not allowed.

Preamble

The licenses for most software are designed to take away your freedom to share and change it. By contrast, the GNU General Public License is intended to guarantee your freedom to share and change free software—to make sure the software is free for all its users. This General Public License applies to most of the Free Software Foundation's software and to any other program whose authors commit to using it. (Some other Free Software Foundation software is covered by the GNU Library General Public License instead.) You can apply it to your programs, too.

When we speak of free software, we are referring to freedom, not price. Our General Public Licenses are designed to make sure that you have the freedom to distribute copies of free software (and charge for this service if you wish), that you receive source code or can get it if you want it, that you can change the software or use pieces of it in new free programs; and that you know you can do these things.

To protect your rights, we need to make restrictions that forbid anyone to deny you these rights or to ask you to surrender the rights. These restrictions translate to certain responsibilities for you if you distribute copies of the software, or if you modify it.

For example, if you distribute copies of such a program, whether gratis or for a fee, you must give the recipients all the rights that you have. You must make sure that they, too, receive or can get the source code. And you must show them these terms so they know their rights.

We protect your rights with two steps: (1) copyright the software, and (2) offer you this license which gives you legal permission to copy, distribute and/or modify the software.

Also, for each author's protection and ours, we want to make certain that everyone understands that there is no warranty for this free software. If the software is modified by someone else and passed on, we want its recipients to know that what they have is not the original, so that any problems introduced by others will not reflect on the original authors' reputations.

Finally, any free program is threatened constantly by software patents. We wish to avoid the danger that redistributors of a free program will individually obtain patent licenses, in effect making the program proprietary. To prevent this, we have made it clear that any patent must be licensed for everyone's free use or not licensed at all.

The precise terms and conditions for copying, distribution and modification follow.

TERMS AND CONDITIONS FOR COPYING, DISTRIBUTION AND MODIFICATION

0. This License applies to any program or other work which contains a notice placed by the copyright holder saying it may be distributed under the terms of this General Public License. The "Program", below, refers to any such program or work, and a "work based on the Program" means either the

Program or any derivative work under copyright law: that is to say, a work containing the Program or a portion of it, either verbatim or with modifications and/or translated into another language. (Hereinafter, translation is included without limitation in the term "modification".) Each licensee is addressed as "you".

Activities other than copying, distribution and modification are not covered by this License; they are outside its scope. The act of running the Program is not restricted, and the output from the Program is covered only if its contents constitute a work based on the Program (independent of having been made by running the Program). Whether that is true depends on what the Program does.

1. You may copy and distribute verbatim copies of the Program's source code as you receive it, in any medium, provided that you conspicuously and appropriately publish on each copy an appropriate copyright notice and disclaimer of warranty; keep intact all the notices that refer to this License and to the absence of any warranty; and give any other recipients of the Program a copy of this License along with the Program.

You may charge a fee for the physical act of transferring a copy, and you may at your option offer warranty protection in exchange for a fee.

2. You may modify your copy or copies of the Program or any portion of it, thus forming a work based on the Program, and copy and distribute such modifications or work under the terms of Section 1 above, provided that you also meet all of these conditions:

a) You must cause the modified files to carry prominent notices stating that you changed the files and the date of any change.

b) You must cause any work that you distribute or publish, that in whole or in part contains or is derived from the Program or any part thereof, to be licensed as a whole at no charge to all third parties under the terms of this License.

c) If the modified program normally reads commands interactively when run, you must cause it, when started running for such interactive use in the most ordinary way, to print or display an announcement including an appropriate copyright notice and a notice that there is no warranty (or else, saying that you provide a warranty) and that users may redistribute the program under these conditions, and telling the user how to view a copy of this License. (Exception: if the Program itself is interactive but does not normally print such an announcement, your work based on the Program is not required to print an announcement.)

These requirements apply to the modified work as a whole. If identifiable sections of that work are not derived from the Program, and can be reasonably considered independent and separate works in themselves, then this License, and its terms, do not apply to those sections when you distribute them as separate works. But when you distribute the same sections as part of a whole which is a work based on the Program, the distribution of the whole must be on the terms of this License, whose permissions for other licensees extend to the entire whole, and thus to each and every part regardless of who wrote it. Thus, it is not the intent of this section to claim rights or contest your rights to work written entirely by you; rather, the intent is to exercise the right to control the distribution of derivative or collective works based on the Program.

In addition, mere aggregation of another work not based on the Program with the Program (or with a work based on the Program) on a volume of a storage or distribution medium does not bring the other work under the scope of this License.

3. You may copy and distribute the Program (or a work based on it, under Section 2) in object code or executable form under the terms of Sections 1 and 2 above provided that you also do one of the following:

a) Accompany it with the complete corresponding machine-readable source code, which must be distributed under the terms of Sections 1 and 2 above on a medium customarily used for software interchange; or,

b) Accompany it with a written offer, valid for at least three years, to give any third party, for a charge no more than your cost of physically performing source distribution, a complete machine-readable copy of the corresponding source code, to be distributed under the terms of Sections 1 and 2 above on a medium customarily used for software interchange; or,

c) Accompany it with the information you received as to the offer to distribute corresponding source code. (This alternative is allowed only for noncommercial distribution and only if you received the program in object code or executable form with such an offer, in accord with Subsection b above.)

The source code for a work means the preferred form of the work for making modifications to it. For an executable work, complete source code means all the source code for all modules it contains, plus any associated interface definition files, plus the scripts used to control compilation and installation of the executable. However, as a special exception, the source code distributed need not include anything that is normally distributed (in either source or binary form) with the major components (compiler, kernel, and so on) of the operating system on which the executable runs, unless that component itself accompanies the executable.

If distribution of executable or object code is made by offering access to copy from a designated place, then offering equivalent access to copy the source code from the same place counts as distribution of the source code, even though third parties are not compelled to copy the source along with the object code.

4. You may not copy, modify, sublicense, or distribute the Program except as expressly provided under this License. Any attempt otherwise to copy, modify, sublicense or distribute the Program is void, and will automatically terminate your rights under this License. However, parties who have received copies, or rights, from you under this License will not have their licenses terminated so long as such parties remain in full compliance.

5. You are not required to accept this License, since you have not signed it. However, nothing else grants you permission to modify or distribute the Program or its derivative works. These actions are prohibited by law if you do not accept this License. Therefore, by modifying or distributing the Program (or any work based on the Program), you indicate your acceptance of this License to do so, and all its terms and conditions for copying, distributing or modifying the Program or works based on it.

6. Each time you redistribute the Program (or any work based on the Program), the recipient automatically receives a license from the original licensor to copy, distribute or modify the Program subject to these terms and conditions. You may not impose any further restrictions on the recipients' exercise of the rights granted herein. You are not responsible for enforcing compliance by third parties to this License.

7. If, as a consequence of a court judgment or allegation of patent infringement or for any other reason (not limited to patent issues), conditions are imposed on you (whether by court order, agreement or otherwise) that contradict the conditions of this License, they do not excuse you from the conditions of this License. If you cannot distribute so as to satisfy simultaneously your obligations under this License and any other pertinent obligations, then as a consequence you may not distribute the Program at all. For example, if a patent license would not permit royalty-free redistribution of the Program by all those who receive copies directly or indirectly through you, then the only way you could satisfy both it and this License would be to refrain entirely from distribution of the Program.

If any portion of this section is held invalid or unenforceable under any particular circumstance, the balance of the section is intended to apply and the section as a whole is intended to apply in other circumstances.

It is not the purpose of this section to induce you to infringe any patents or other property right claims or to contest validity of any such claims; this section has the sole purpose of protecting the integrity of the free software distribution system, which is implemented by public license practices. Many people have made generous contributions to the wide range of software distributed through that system in reliance on consistent application of that system; it is up to the author/donor to decide if he or she is willing to distribute software through any other system and a licensee cannot impose that choice.

This section is intended to make thoroughly clear what is believed to be a consequence of the rest of this License.

8. If the distribution and/or use of the Program is restricted in certain countries either by patents or by copyrighted interfaces, the original copyright holder who places the Program under this License may add an explicit geographical distribution limitation excluding those countries, so that distribution is permitted only in or among countries not thus excluded. In such case, this License incorporates the limitation as if written in the body of this License.

9. The Free Software Foundation may publish revised and/or new versions of the General Public License from time to time. Such new versions will be similar in spirit to the present version, but may differ in detail to address new problems or concerns.

Each version is given a distinguishing version number. If the Program specifies a version number of this License which applies to it and "any later version", you have the option of following the terms and conditions either of that version or of any later version published by the Free Software Foundation. If the Program does not specify a version number of this License, you may choose any version ever published by the Free Software Foundation.

10. If you wish to incorporate parts of the Program into other free programs whose distribution conditions are different, write to the author to ask for permission. For software which is copyrighted by the Free Software Foundation, write to the Free Software Foundation; we sometimes make exceptions for this. Our decision will be guided by the two goals of preserving the free status of all derivatives of our free software and of promoting the sharing and reuse of software generally.

NO WARRANTY

11. BECAUSE THE PROGRAM IS LICENSED FREE OF CHARGE, THERE IS NO WARRANTY FOR THE PROGRAM, TO THE EXTENT PERMITTED BY APPLICABLE LAW. EXCEPT WHEN OTHERWISE STATED IN WRITING THE COPYRIGHT HOLDERS AND/OR OTHER PARTIES PROVIDE THE PROGRAM "AS IS" WITHOUT WARRANTY OF ANY KIND, EITHER EXPRESSED OR IMPLIED, INCLUDING, BUT NOT LIMITED TO, THE IMPLIED WARRANTIES OF MERCHANTABILITY AND FITNESS FOR A PARTICULAR PURPOSE. THE ENTIRE RISK AS TO THE QUALITY AND PERFORMANCE OF THE PROGRAM IS WITH YOU. SHOULD THE PROGRAM PROVE DEFECTIVE, YOU ASSUME THE COST OF ALL NECESSARY SERVICING, REPAIR OR CORRECTION.

12. IN NO EVENT UNLESS REQUIRED BY APPLICABLE LAW OR AGREED TO IN WRITING WILL ANY COPYRIGHT HOLDER, OR ANY OTHER PARTY WHO MAY MODIFY AND/OR REDISTRIBUTE THE PROGRAM AS PERMITTED ABOVE, BE LIABLE TO YOU FOR DAMAGES, INCLUDING ANY GENERAL, SPECIAL, INCIDENTAL OR CONSEQUENTIAL DAMAGES ARISING OUT OF THE USE OR INABILITY TO USE THE PROGRAM (INCLUDING BUT NOT LIMITED TO LOSS OF DATA OR DATA BEING RENDERED INACCURATE OR LOSSES SUSTAINED BY YOU OR THIRD PARTIES OR A FAILURE OF THE PROGRAM TO OPERATE WITH ANY OTHER PROGRAMS), EVEN IF SUCH HOLDER OR OTHER PARTY HAS BEEN ADVISED OF THE POSSIBILITY OF SUCH DAMAGES.

END OF TERMS AND CONDITIONS

<u>How to Apply These Terms to Your New Programs</u>

If you develop a new program, and you want it to be of the greatest possible use to the public, the best way to achieve this is to make it free software which everyone can redistribute and change under these terms.

To do so, attach the following notices to the program. It is safest to attach them to the start of each source file to most effectively convey the exclusion of warranty; and each file should have at least the "copyright" line and a pointer to where the full notice is found.

one line to give the program's name and an idea of what it does.
Copyright (C) *yyyy* *name of author*

This program is free software; you can redistribute it and/or
modify it under the terms of the GNU General Public License
as published by the Free Software Foundation; either version 2
of the License, or (at your option) any later version.

This program is distributed in the hope that it will be useful,
but WITHOUT ANY WARRANTY; without even the implied warranty of
MERCHANTABILITY or FITNESS FOR A PARTICULAR PURPOSE. See the
GNU General Public License for more details.

You should have received a copy of the GNU General Public License
along with this program; if not, write to the Free Software
Foundation, Inc., 59 Temple Place - Suite 330, Boston, MA 02111-1307, USA.
Also add information on how to contact you by electronic and paper mail.

If the program is interactive, make it output a short notice like this when it starts in an interactive mode:

Gnomovision version 69, Copyright (C) *year name of author*
Gnomovision comes with ABSOLUTELY NO WARRANTY; for details
type 'show w'. This is free software, and you are welcome
to redistribute it under certain conditions; type 'show c'
for details.

The hypothetical commands 'show w' and 'show c' should show the appropriate parts of the General Public License. Of course, the commands you use may be called something other than 'show w' and 'show c'; they could even be mouse-clicks or menu items—whatever suits your program.

You should also get your employer (if you work as a programmer) or your school, if any, to sign a "copyright disclaimer" for the program, if necessary. Here is a sample; alter the names:

Yoyodyne, Inc., hereby disclaims all copyright
interest in the program 'Gnomovision'
(which makes passes at compilers) written
by James Hacker.

signature of Ty Coon, 1 April 1989
Ty Coon, President of Vice

This General Public License does not permit incorporating your program into proprietary programs. If your program is a subroutine library, you may consider it more useful to permit linking proprietary applications with the library. If this is what you want to do, use the GNU Library General Public License instead of this License.

SYNGRESS PUBLISHING LICENSE AGREEMENT

THIS PRODUCT (THE "PRODUCT") CONTAINS PROPRIETARY SOFTWARE, DATA AND INFORMATION (INCLUDING DOCUMENTATION) OWNED BY SYNGRESS PUBLISHING, INC. ("SYNGRESS") AND ITS LICENSORS. YOUR RIGHT TO USE THE PRODUCT IS GOVERNED BY THE TERMS AND CONDITIONS OF THIS AGREEMENT.

LICENSE: Throughout this License Agreement, "you" shall mean either the individual or the entity whose agent opens this package. You are granted a limited, non-exclusive and non-transferable license to use the Product subject to the following terms:

(i) If you have licensed a single user version of the Product, the Product may only be used on a single computer (i.e., a single CPU). If you licensed and paid the fee applicable to a local area network or wide area network version of the Product, you are subject to the terms of the following subparagraph (ii).

(ii) If you have licensed a local area network version, you may use the Product on unlimited workstations located in one single building selected by you that is served by such local area network. If you have licensed a wide area network version, you may use the Product on unlimited workstations located in multiple buildings on the same site selected by you that is served by such wide area network; provided, however, that any building will not be considered located in the same site if it is more than five (5) miles away from any building included in such site. In addition, you may only use a local area or wide area network version of the Product on one single server. If you wish to use the Product on more than one server, you must obtain written authorization from Syngress and pay additional fees.

(iii) You may make one copy of the Product for back-up purposes only and you must maintain an accurate record as to the location of the back-up at all times.

PROPRIETARY RIGHTS; RESTRICTIONS ON USE AND TRANSFER: All rights (including patent and copyright) in and to the Product are owned by Syngress and its licensors. You are the owner of the enclosed disc on which the Product is recorded. You may not use, copy, decompile, disassemble, reverse engineer, modify, reproduce, create derivative works, transmit, distribute, sublicense, store in a database or retrieval system of any kind, rent or transfer the Product, or any portion thereof, in any form or by any means (including electronically or otherwise) except as expressly provided for in this License Agreement. You must reproduce the copyright notices, trademark notices, legends and logos of Syngress and its licensors that appear on the Product on the back-up copy of the Product which you are permitted to make hereunder. All rights in the Product not expressly granted herein are reserved by Syngress and its licensors.

TERM: This License Agreement is effective until terminated. It will terminate if you fail to comply with any term or condition of this License Agreement. Upon termination, you are obligated to return to Syngress the Product together with all copies thereof and to purge and destroy all copies of the Product included in any and all systems, servers and facilities.

DISCLAIMER OF WARRANTY: THE PRODUCT AND THE BACK-UP COPY OF THE PRODUCT ARE LICENSED "AS IS". SYNGRESS, ITS LICENSORS AND THE AUTHORS MAKE NO WARRANTIES, EXPRESS OR IMPLIED, AS TO RESULTS TO BE OBTAINED BY ANY PERSON OR ENTITY FROM USE OF THE PRODUCT AND/OR ANY INFORMATION OR DATA INCLUDED THEREIN. SYNGRESS, ITS LICENSORS AND THE AUTHORS MAKE NO EXPRESS OR IMPLIED WARRANTIES OF MERCHANTABILITY OR FITNESS FOR A PARTICULAR PURPOSE OR USE WITH RESPECT TO THE PRODUCT AND/OR ANY

INFORMATION OR DATA INCLUDED THEREIN. IN ADDITION, SYNGRESS, ITS LICENSORS AND THE AUTHORS MAKE NO WARRANTY REGARDING THE ACCURACY, ADEQUACY OR COMPLETENESS OF THE PRODUCT AND/OR ANY INFORMATION OR DATA INCLUDED THEREIN. NEITHER SYNGRESS, ANY OF ITS LICENSORS, NOR THE AUTHORS WARRANT THAT THE FUNCTIONS CONTAINED IN THE PRODUCT WILL MEET YOUR REQUIREMENTS OR THAT THE OPERATION OF THE PRODUCT WILL BE UNINTERRUPTED OR ERROR FREE. YOU ASSUME THE ENTIRE RISK WITH RESPECT TO THE QUALITY AND PERFORMANCE OF THE PRODUCT.

LIMITED WARRANTY FOR DISC: To the original licensee only, Syngress warrants that the enclosed disc on which the Product is recorded is free from defects in materials and workmanship under normal use and service for a period of ninety (90) days from the date of purchase. In the event of a defect in the disc covered by the foregoing warranty, Syngress will replace the disc.

LIMITATION OF LIABILITY: NEITHER SYNGRESS, ITS LICENSORS NOR THE AUTHORS SHALL BE LIABLE FOR ANY INDIRECT, INCIDENTAL, SPECIAL, PUNITIVE, CONSEQUENTIAL OR SIMILAR DAMAGES, SUCH AS BUT NOT LIMITED TO, LOSS OF ANTICIPATED PROFITS OR BENEFITS, RESULTING FROM THE USE OR INABILITY TO USE THE PRODUCT EVEN IF ANY OF THEM HAS BEEN ADVISED OF THE POSSIBILITY OF SUCH DAMAGES. THIS LIMITATION OF LIABILITY SHALL APPLY TO ANY CLAIM OR CAUSE WHATSOEVER WHETHER SUCH CLAIM OR CAUSE ARISES IN CONTRACT, TORT, OR OTHERWISE. Some states do not allow the exclusion or limitation of indirect, special or consequential damages, so the above limitation may not apply to you.

U.S. GOVERNMENT RESTRICTED RIGHTS. If the Product is acquired by or for the U.S. Government then it is provided with Restricted Rights. Use, duplication or disclosure by the U.S. Government is subject to the restrictions set forth in FAR 52.227-19. The contractor/manufacturer is Syngress Publishing, Inc. at 800 Hingham Street, Rockland, MA 02370.

GENERAL: This License Agreement constitutes the entire agreement between the parties relating to the Product. The terms of any Purchase Order shall have no effect on the terms of this License Agreement. Failure of Syngress to insist at any time on strict compliance with this License Agreement shall not constitute a waiver of any rights under this License Agreement. This License Agreement shall be construed and governed in accordance with the laws of the Commonwealth of Massachusetts. If any provision of this License Agreement is held to be contrary to law, that provision will be enforced to the maximum extent permissible and the remaining provisions will remain in full force and effect.

***If you do not agree, please return this product to the place of purchase for a refund.**

Ride the LEGO® Wave with Syngress!

AVAILABLE NOW!
ORDER at
www.syngress.com/bricks

Building & Programming LEGO® Mindstorms™ Robots KIT

Demystifies robotics, programming, and engineering for LEGO Mindstorms. This 2-book box set contains the best selling *Building Robots with LEGO Mindstorms* (ISBN: 1-928994-67-9) and *Programming LEGO Mindstorms with Java* (ISBN: 1-928994-55-5).

ISBN: 1-931836-71-X

Price: $54.95 USA $84.95 CAN

AVAILABLE NOW!
ORDER at
www.syngress.com/bricks

30 Cool LEGO® Mindstorms™ Projects KIT

Provides all levels of Mindstorms enthusiasts with 30 cool robots to build in less than an hour. Written by LEGO experts, this kit is perfect for all Mindstorms fans. This kit includes:

10 Cool LEGO MINDSTORMS Dark Side Robots, Transports, and Creatures: Amazing Projects You Can Build in Under an Hour (ISBN: 1-931836-59-0)

10 Cool LEGO MINDSTORMS Ultimate Builders Projects: Amazing Projects You Can Build in Under an Hour (ISBN: 1-931836-60-4)

10 Cool LEGO MINDSTORMS Robotics Invention System 2.0 Projects: Amazing Projects You Can Build in Under an Hour (ISBN: 1-931836-61-2)

ISBN: 1-931836-62-0

Price: $69.95 USA $108.95 CAN

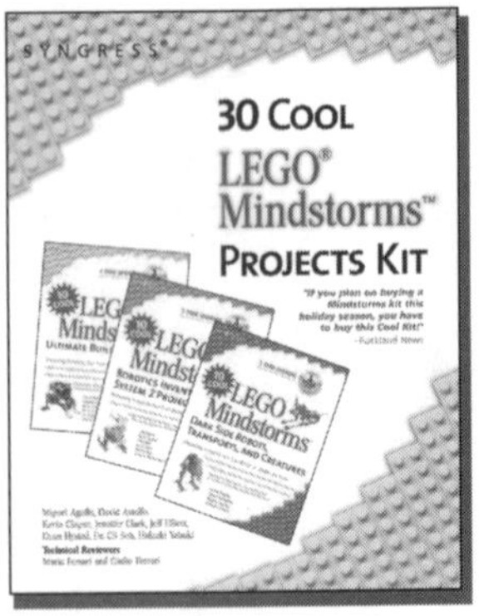

AVAILABLE JANUARY 2003
ORDER at
www.syngress.com/bricks

LEGO® Mindstorms™ Masterpieces: Building and Programming Advanced Robots

The world's leading LEGO Mindstorms inventors share their knowledge and development secrets. Written for owners of LEGO Mindstorms kits who are searching for advanced creations to build and adapt. Includes CD!

ISBN: 1-931836-75-2

Price: $49.95 USA $77.95 CAN

www.syngress.com/bricks

SYNGRESS®